Control in Robotics and Automation

Sensor-Based Integration

ACADEMIC PRESS SERIES IN ENGINEERING

Series Editor
J. David Irwin
Auburn University

Designed to bring together interdependent topics in electrical engineering, mechanical engineering, computer engineering, and manufacturing, the Academic Press Series in Engineering provides state-of-the-art handbooks, textbooks, and professional reference books for researchers, students, and engineers. This series provides readers with a comprehensive group of books essential for success in modern industry. Particular emphasis is given to the applications of cutting-edge research. Engineers, researchers, and students alike will find the Academic Press Series in Engineering to be an indispensable part of their design toolkit.

Published books in the series:
Industrial Controls and Manufacturing, 1999, E. Kamen
DSP Integrated Circuits, 1999, L. Wanhammar
Time Domain Electromagnetics, 1999, S. M. Rao
Single and Multi-Chip Microcontroller Interfacing, 1999, G. J. Lipovski

Control in Robotics and Automation

Sensor-Based Integration

Edited by

B. K. GHOSH
Department of Systems Science and Mathematics
Washington University
St. Louis, Missouri

NING XI
Department of Electrical Engineering
Michigan State University
East Lansing, Michigan

T. J. TARN
Department of Systems Science and Mathematics
Washington University
St. Louis, Missouri

ACADEMIC PRESS
SAN DIEGO / LONDON / BOSTON / NEW YORK / SYDNEY / TOKYO / TORONTO

This book is printed on acid-free paper. ∞

ACADEMIC PRESS
A Division of Harcourt Brace & Company
525 B Street, Suite 1900, San Diego, CA 92101-4495
http://www.apnet.com

Academic Press
24–28 Oval Road, London NW1 7DX
http://www.hbuk.co.uk/ap/

Library of Congress Cataloging-in-Publication Data
Control in robotics and automation : sensor-based integration / B.K. Ghosh, editor, Ning Xi, editor, T.J. Tarn, editor.
p. cm. -- (Academic Press series in engineering)
Includes bibliographical references and index.
ISBN 0-12-281845-8
1. Robots--Control systems. 2. Automatic control. I. Ghosh, B. K. 1956– . II. Xi, Ning. III. Tarn, Tzyh-Jong, 1937– IV. Series.
TJ211.35.C65 1999 98-38975
629.8′92--dc21 CIP

Printed in the United States of America

99 00 01 02 03 DSG 9 8 7 6 5 4 3 2 1

Contents

Preface

In recent years there has been a growing interest in the need for sensor fusion to solve problems in control and planning for robotic systems. The application of such systems would range from assembly tasks in industrial automation to material handling in hazardous environments and servicing tasks in space. Within the framework of an event-driven approach, robotics has found new applications in automation, such as robot-assisted surgery and microfabrication, that pose new challenges to control, automation, and manufacturing communities.

To meet such challenges, it is important to develop planning and control systems that can integrate various types of sensory information and human knowledge in order to carry out tasks efficiently with or without the need for human intervention. The structure of a sensing, planning, and control system and the computer architecture should be designed for a large class of tasks rather than for a specific task. User-friendliness of the interface is essential for human operators who pass their knowledge and expertise to the control system before and during task execution. Finally, robustness and adaptability of the system are essential.

The system we propose should be able to perform in its environment on the basis of prior knowledge and real-time sensory information. We introduce a new task-oriented approach to sensing, planning, and control. As a specific example of this approach, we discuss an event-based method for system design. In order to introduce a specific control objective, we introduce the problem of combining task planning and three-dimensional modeling in the execution of remote operations. Typical remote systems are teleoperated and provide work efficiencies that are on the order of 10 times slower than what is directly achievable by humans. Consequently, the effective integration of automation into teleoperated remote systems offers the potential to improve their work efficiency.

In the realm of autonomous control, we introduce visually guided control systems and study the role of computer vision in autonomously guiding a robot system. As a specific example, we study problems pertaining to a manufacturing work cell. We conclude with a discussion of the role of modularity and sensor integration in a number of problems involving robotic and telerobotic control systems.

Portions of this book are an outgrowth of two workshops in two international conferences organized by the editors of this book. The first one, "Sensor-Referenced Control and Planning: Theory and Applications," was held at the IEEE International Conference on Decision and Control, New Orleans, 1995 and the second one, "Event-Driven Sensing, Planning and Control of a Robotic System: An Integrated Approach," was held at the IEEE/RSJ International Conference on Intelligent Robots and Systems, Osaka, Japan, 1996.

In summary, we believe that the sensor-guided planning and control problems introduced in this book involve state-of-the-art knowledge in the field of sensor-guided automation and robotics.

Contributors

R. Anderson, Intelligent Systems and Robotics Center, Sandia National Laboratories, P.O. Box 5800, MS1125, Albuqerque, New Mexico, 87185

S. Arimoto, Department of Robotics, Ritsumeikan University, 1-1-1 Nojihigashi, Kusatsu, Shiga, 525-877, Japan

J. E. Baker, Oak Ridge National Laboratory, One Bethel Valley Road, Oak Ridge, Tennessee, 37831

B. Ghosh, Department of Systems Science and Mathematics, Campus Box 1040, One Brookings Drive, Washington University, St. Louis, Missouri, 63130

W. R. Hamel, University of Tennessee, 414 Dougherty Engineering Building, Knoxville, Tennessee, 37996

K. Hashimoto, Department of Systems Engineering, Okayama University, 3-1-1 Tsushima-naka, Okayama 700, Japan

A. Isidori, Department of Systems Science and Mathematics, Campus Box 1040, One Brookings Drive, Washington University, St. Louis, Missouri, 63130

P. K. Khosla, Department of Electrical and Computer Engineering, Carnegie Mellon University, Pittsburgh, Pennsylvania, 15213

S. Lee, Department of Computer Science, University of Southern California, Los Angeles, CA 90089

M. Lei, RVSI Acuity CiMatrix, 5 Shawmut Road, Canton, Massachusetts, 02021

B. J. Nelson, Department of Mechanical Engineering, University of Minnesota, 111 Church Street SE, Minneapolis, Minnesota, 55455

N. P. Papanikolopoulos, Department of Computer Science and Engineering, University of Minnesota, 4-192 EE/CS Building, 200 Union Street SE, Minneapolis, Minnesota, 55455

I. Pavlidis, Department of Computer Science and Engineering, University of Minnesota, 4-192 EE/CS Building, 200 Union Street SE, Minneapolis, Minnesota, 55455

F. G. Pin, Oak Ridge National Laboratory, One Bethel Valley Road, Oak Ridge, Tennessee, 37831

S. Ro, Department of Computer Science, University of Southern California, Los Angeles, CA 90089

R. Singh, Department of Computer Science, University of Minnesota, 4-192 EE/CS Building, 200 Union Street SE, Minneapolis, Minnesota, 55455

M. J. Sullivan, Department of Computer Science and Engineering, University of Minnesota, 4-192 EE/CS Building, 200 Union Street SE, Minneapolis, Minnesota, 55455

T. J. Tarn, Department of Systems Science and Mathematics, Campus Box 1040, One Brookings Drive, Washington University, St. Louis, Missouri, 63130

R. Y. Wu, Computer Sciences Corporation, Fairview Heights, Illinois, 62208

N. Xi, Department of Electrical Engineering, 2120 Engineering Building, Michigan State University, East Lansing, Michigan, 48824

D. Xiao, Department of Systems Science and Mathematics, Campus Box 1040, One Brookings Drive, Washington University, St. Louis, Missouri, 63130

Z. Yu, Department of Systems Science and Mathematics, Campus Box 1040, One Brookings Drive, Washington University, St. Louis, Missouri, 63130

SECTION I

Introduction

CHAPTER 1

Sensor-Based Planning and Control for Robotic Systems: An Event-Based Approach

NING XI
Department of Electrical and Computer Engineering, Michigan State University, East Lansing, Michigan

TZYH-JONG TARN
Department of Systems Science and Mathematics, Washington University, St. Louis, Missouri

1 INTRODUCTION

1.1 Motivation

There is growing interest in the development of intelligent robotic systems. The applications of such systems range from assembly tasks in industrial automation to material handling in hazardous environments and servicing tasks in space.

The intelligence of a robotic systems can be characterized by three functional abilities. First, the robotic system should be controlled directly at the task level; that is, it should take task-level commands directly, without any planning type decomposition to joint-level commands. Second, the control systems of robots should be designed for a large class of tasks rather than for a specific task. In this respect, the design of the control system can be called task independent. Finally, the robotic system should be able to handle some unexpected or uncertain events.

Traditionally, robots were designed in such a way that action planning and the controller were treated as separate issues. Robotic system designers concentrated on the controller design, and the robotic action planning was largely left as a task for the robot users. To some extent, this is understandable, because action planning is heavily dependent on the task and task environment.

The split between robot controller design and robot action planning, however, becomes a real issue, because the action planner and a given control system usually have two different reference bases. Normally, the action planner, a human operator or an automatic planner, thinks and plans in terms of events. That is, the planner's normal reference base is a set of *events*. On the other hand, when it comes to the execution of planned actions, the usual reference frame for existing robot control systems is a time-based or *clocked* trajectory, typically a polynomial representation or decomposition of joint space or task space motions with time as a *driver* or independent variable. Eventually, this *clocked* trajectory representation can be combined with some expected or desired sensed events at the end of the trajectory. However, the main motion or action reference base of existing industrial robot control systems is *time*.

The two different reference bases for robot action planning and robot action execution or control (events versus time) cause unwanted complications and represent a bottleneck for creating *intelligent* robot control and *intelligent* robotic workstations. *Intelligent* robot control depends to a large extent on the capability of the robotic system to acquire, process, and utilize sensory information in order to plan and execute actions in the presence of various changing or uncertain events in the robot's work environment. Note that sensed events in a robotic work environment do not appear on a precise time scale. Hence, in reality, motion trajectories from start to destination cannot be planned on the basis of time alone. Instead, the executable representation of robot motion or action plans should be referenced to other variables to which sensed events are normally related. This would make the plan representation for control execution compatible with the normal reference base of the applied sensors.

The main motivation of this thesis work is to take a step toward intelligent robotic systems through the combination of event-based motion planning and nonlinear feedback control.

1.2 Review of Previous Work

There exists voluminous literature on the subject of motion planning. Motion planning consists of two basic problems, path planning and trajectory planning. Latombe [1] and Hwang and Ahuja [2] give excellent surveys and pertinent references in this area. Basically, there are two major approaches. One is based on the configuration space ideas proposed by Lozano-Perez and Wesley [3]. In order to use the configuration space approach, complete knowledge of environment is required, so the most useful results with this approach are for off-line path planning. The other approach uses a potential field method pioneered by Khatib [4]. It can be applied to real-time motion planning. However, to get the potential field of an environment again requires complete knowledge of the robot work space. Therefore, it is very difficult to apply this approach to a changing environment. The issues of motion planning in a dynamic environment are discussed by Fujimura [5]. However, most of the results were obtained under very strict assumptions, such as "the robot velocity is greater than all obstacle velocities," and they are valid only for a two-dimensional work space.

The common limitations of the existing motion planning schemes are twofold:

1. The planned motions are described as a function of time.
2. Complete knowledge of the work environment is assumed.

These limitations make it impossible to modify or adjust a motion plan during execution on the basis of sensory or other on-line information. Therefore, these schemes cannot accommodate a dynamic environment consisting of not sharply defined or unexpected events, such as the appearance of an obstacle. Of course, if some kind of logic function is incorporated in the time-based plan, it may be able to respond to some unexpected events. However, because of the very nature of time-based plans, complete replanning of the motion after a change in the environment or occurrence of an unexpected obstacle is needed in order to reach the final goal.

Some effort has been made to develop a path planning scheme based on sensory information [6]. This method is, however, purely geometric and is not integrated with the control execution.

The results of pioneering research on non–time-based robot motion analysis, planning, representation, and control execution have appeared in the robotic literature. In [7] and [66], the velocity-versus-position phase space technique is introduced, using harmonic functions to relate velocity to position along a given geometric path. Phase space concepts are applied in [8], [9], and [10] to find the optimal joint space trajectory of an arbitrary robot manipulator that has to follow a prescribed path. In [11], a phase space variable is used to obtain a dynamic model of a tricycle-type mobile robot, which can then easily be linearized by feedback. In [12], a phase space approach is applied to the path following control of a flexible joint robot. In these methods, the phase space technique is used as a analytical tool to find an optimal time-based trajectory. In fact, phase space (velocity versus position) has been widely used in physics and in early control theories to describe motion trajectories.

The real challenge in motion planning is to develop a planning scheme integrated with a control system that is able to detect and recognize unexpected events on the basis of sensory information and adjust and modify the base plan at a high rate (same as the dynamic control loop) to cope with time and location variations in the occurrence of events without replanning. The first technical difficulty is the development of a mathematical model to describe the plan so that it is inherently flexible relative to the final task goal and can be easily adjusted in real time according to task measurements. The second difficulty is the development of an efficient representation of a sensory information updating scheme that can be used to transmit the task measurement to the planner at a high rate (same as the control feedback rate). The third difficulty is the integration of the planner and controller to achieve a coordinated action and avoid deadlocks or infinite loops.

2 EVENT-BASED PLANNING AND CONTROL

2.1 Introduction

A traditional planning and control system can be described as in Figure 1.1. The core of the system is the feedback control loop, which ensures the system's stability, robustness, and performance. The feedback turns the controller into an investigation–decision component. The planning process, however, is done off line, which is understandable because the task is usually predefined. The plan is described as a function of time, and the planner gives the desired input to the system according to the original plan. Therefore it could be considered as a memory component for storing the predefined plan. All uncertainty and unexpected events that were not considered in planning are left to the feedback control loop to handle. If a system works in a complicated environment, the controller alone is not able to ensure that the system achieves satisfactory performance.

In the past 5 years, considerable effort has been made to improve the planner and controller in order to handle unexpected or uncertain events, in other words, to achieve intelligent planning and control. The concept of intelligent control was introduced as an interdisciplinary name for artificial intelligence and automatic control systems [13]. Saridis [14] and Saridis and Valavanis [15] proposed a three-layer hierarchy for the controller and planner. Since then, based on a similar idea, various "intelligent" planning and control schemes have been developed [16–18]. The basic idea of existing schemes is to add to the

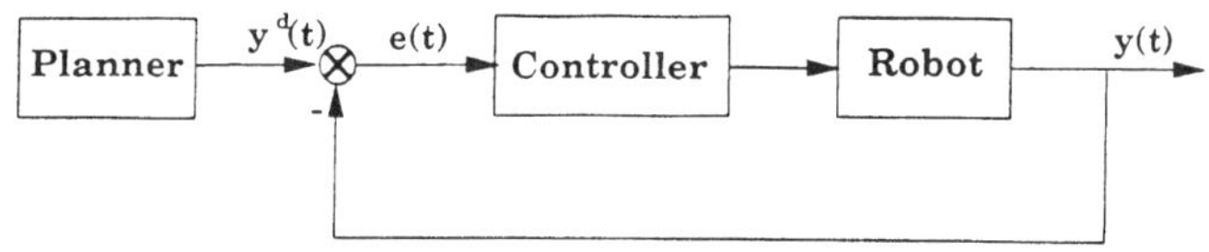

FIGURE 1.1
Traditional planning and control system.

basic system in Figure 1.1 high-level monitoring layers that monitor the performance of the system. When some unexpected discrete event, such as a system component failure or outside disturbance, happens, the high-level layer either replans the desired input or switches it to some predefined contingency plan.

However, for some high-speed systems, which may also work in very complicated environments, it is almost impossible to replan the motion in real time and it is extremely difficult to predefine the contingency plans without knowing the nature of unexpected events. Furthermore, besides discrete events, there are also continuous unexpected events. For example, the error of a system is a cumulation with respect to time. The high-level layer is not able to detect it and take any action until it exceeds a certain threshold. This significantly reduces the precision of the system. In addition, the high-level layers in existing schemes are implemented by different heuristic techniques. The computation is usually time consuming. As a result, the sampling rate of the high-level layer is much lower than that of a real-time control loop. Therefore, it is not able to deal efficiently with continuous unexpected events.

The real challenge is to develop a planning and control scheme that is able to detect and recognize both discrete and continuous events and adjust and modify the original plan at a high rate (same as the feedback control loop) to recover from errors or unwanted situations and eventually to achieve superior performance.

The first technical difficulty is the development of a mathematical model to describe the plan so that it can be easily adjusted and modified in real time according to system output measurements. The second is the development of an efficient representation for sensory information updating that can be used to transmit the system output measurements to the planner at the same high rate as the control feedback loop. The last is the integration of the planner and controller to achieve stable and robust system performance.

2.2 New Motion Reference and Integration of Planning and Control

The event-based planning and control scheme will be able to overcome the preceding difficulties and to meet the challenge. The basic idea of the theory is to introduce a new motion reference variable different from time and related directly to the measurement of system output. Instead of time, the plan–desired system input is parameterized by the new motion reference variable. The motion reference variable is designed to carry efficiently the sensory information needed for the planner to adjust or modify the original plan to form a desired input. As a result, for any given time instant, the desired input is a function of the system output. This creates a mechanism for adjusting and modifying the plan on the basis of the output measurement. More important, it makes the planning a closed-loop, real-time process. The event-based planning and control scheme can be shown as in Figure 1.2.

In Figure 1.2, the function of Motion Reference is to compute the motion reference variable on the basis of the system output measurement. The planner then gives a desired input according to the motion reference. It can be seen that the planning becomes an

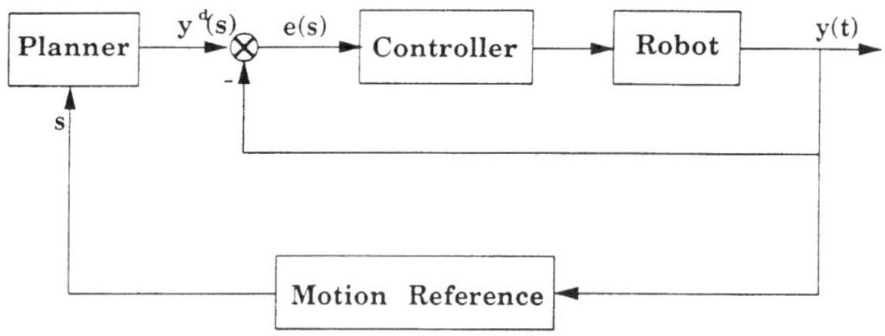

FIGURE 1.2
Event-based planning and control scheme.

investigation–decision component in the sense of feedback. Therefore, the event-based planning and control scheme has an ability to deal with unexpected or uncertain events.

In addition, the motion reference variable is calculated at the same rate as feedback control. In other words, the original plan is adjusted and modified at a very high rate. As a result, it is able to deal not only with discrete unexpected or uncertain events but also with continuous unexpected and uncertain events, such as cumulation of error and system parameter drifting.

Furthermore, the high-level heuristic layer could still be added, which would be compatible with the event-based planning and control scheme.

In considering Figure 1.2, some theoretical questions arise. First, after a motion reference loop is introduced, how does it affect the stability of the system? Second, how does it affect the dynamic performance of the system, and how can such a system be designed to achieve a desired performance?

2.3 Stability in the Event-Based Reference Frame

If a system is asymptotically stable with time t as its motion reference base, and if the new motion reference s is a (monotone increasing) nondecreasing function of time t, then the system is (asymptotically) stable with respect to the new motion reference base s.

If the system is asymptotically stable with respect to t, by the converse theorem [19], we can find a Liapunov function $L(X(t))$ such that

1. $L(X(t))$ is positive definite.
2. $\dfrac{dL(X(t))}{dt}$ is negative definite.

If the motion of the system, is referenced to s, then $L(X(s))$ is still positive definite.

In addition,

$$\frac{dL(X(t))}{dt} = \frac{dL(X(s))}{dt} = \frac{dL(X(s))}{ds}\frac{ds}{dt}$$

If s is a (monotone increasing) nondecreasing function of t, then

$$\left(\frac{ds}{dt} > 0\right), \quad \frac{ds}{dt} \geqslant 0$$

Thus,

$$\frac{dL(X(s))}{ds}$$

is (negative definite) negative semidefinite. Therefore, the system is (asymptotically) stable with respect to s.

2.4 Equivalence of Time-Based and Event-Based Controllers

Two methods could be used for designing a controller. The first one is based on a time-based dynamic model

$$\begin{cases} \dfrac{dx}{dt} = f(x) + g(x)u & x, u \in R^m \\ y = h(x) & y \in R^m \end{cases}$$

Second, the event-based motion plan could be introduced into a dynamic model. A control system could then be designed on the basis of the event-based dynamic model

$$\begin{cases} \dfrac{dx}{ds} = \dfrac{1}{V(s)} f(x) + \dfrac{1}{V(s)} g(x)u & x, u \in R^m \\ y = h(x) & y \in R^m \end{cases}$$

The event-based dynamic model has the same motion reference as the planner. It can be linearized by introducing a proper trajectory plan [11]. Since the event-based dynamic model depends on the trajectory plan, the control law becomes trajectory dependent. For designing a task-independent controller, the time-based dynamic model is adequate because it is independent of the trajectory plan. The most important issue is to synchronize the two references for the planner and controller.

If the nonlinear feedback control algorithm is applied to both time-based and event-based dynamic models and the linearized systems have same pole placements, then no matter what dynamics model is used, the system receives an identical control command.

Time-based nonlinear feedback is given as

$$u_t = \alpha_t(x) + \beta_t(x)\omega_t$$

The corresponding linear model is

$$\begin{cases} \dfrac{d\xi_{1t}}{dt} = \xi_{2t} & \xi_{1t}, \xi_{2t} \in R^m \\ \dfrac{d\xi_{2t}}{dt} = \omega_t & \omega_t \in R^m \end{cases}$$

and

$$\alpha_t(x) = -\begin{bmatrix} L_g L_f h_1 \\ \vdots \\ L_g L_f h_m \end{bmatrix}^{-1} \begin{bmatrix} L_f^2 h_1 \\ \vdots \\ L_f^2 h_m \end{bmatrix}, \quad \beta_t(x) = \begin{bmatrix} L_g L_f h_1 \\ \vdots \\ L_g L_f h_1 \end{bmatrix}^{-1}$$

Event-based nonlinear feedback can also be described as

$$u_s = \alpha_s(x) + \beta_s(x)\omega_s$$

The corresponding linear model is

$$\begin{cases} \dfrac{d\xi_{1s}}{ds} = \xi_{2s} & \xi_{1s}, \xi_{2s} \in R^m \\ \dfrac{d\xi_{2s}}{ds} = \omega_s & \omega_s \in R^m \end{cases}$$

and

$$\alpha_s(x) = -\begin{bmatrix} L_{g/v}L_{f/v}h_1 \\ \vdots \\ L_{g/v}L_{f/v}h_m \end{bmatrix}^{-1} \begin{bmatrix} L_{f/v}^2 h_1 \\ \vdots \\ L_{f/v}^2 h_m \end{bmatrix}, \quad \beta_s(x) = \begin{bmatrix} L_{g/v}L_{f/v}h_1 \\ \vdots \\ L_{g/v}L_{f/v}h_1 \end{bmatrix}^{-1}$$

Therefore,

$$\alpha_s(x) = -v^2 I_{m\times m} \begin{bmatrix} L_g L_f h_1 \\ \vdots \\ L_g L_f h_m \end{bmatrix}^{-1} \frac{1}{v^2} I_{m\times m} \begin{bmatrix} L_f^2 h_1 \\ \vdots \\ L_f^2 h_m \end{bmatrix} = -\begin{bmatrix} L_g L_f h_1 \\ \vdots \\ L_g L_f h_m \end{bmatrix}^{-1} \begin{bmatrix} L_f^2 h_1 \\ \vdots \\ L_f^2 h_m \end{bmatrix} = \alpha_t(x)$$

and

$$\beta_s(x) = v^2 I_{m\times m} \begin{bmatrix} L_g L_f h_1 \\ \vdots \\ L_g L_f h_m \end{bmatrix} = \beta_t(x) v^2 I_{m\times m}$$

Then

$$\begin{aligned} u_s &= \alpha_t(x) + \beta_t(x) v^2 I_{m\times m} \omega_s \\ &= \alpha_t(x) + \beta_t(x)(v^2 \omega_s) \end{aligned}$$

So the corresponding linear model can be written as

$$\begin{cases} \dfrac{d\xi_{1t}}{dt} = \xi_{2t} & \xi_{1t}, \xi_{2t} \in R^m \\ \dfrac{d\xi_{2t}}{dt} = v^2\omega_s & \omega_s \in R^m \end{cases}$$

If the poles of linear models are placed at the same locations, we have

$$\omega_t = v^2 \omega_s$$

Thus,

$$u_s = \alpha_t(x) + \beta_t(x)\omega_t = u_t$$

Therefore, no matter what motion reference is used for a dynamic model, the robot sees that u_t and u_s are formally the same.

The preceding results lay down a foundation for applying the event-based planning and control scheme to practical systems, especially robotic systems. Obviously, different motion reference variables could be chosen based on the nature of the systems and the control objectives. Designing the motion reference becomes the first and the most important task in developing an integrated event-based planning and control scheme.

3 EVENT-BASED MOTION PLANNING AND CONTROL FOR A ROBOT ARM

3.1 Event-Based Robot Motion Description

In general, the motion reference should be closely related to the objective of the system, should properly reflect the performance, and should efficiently carry the sensory information. In robot planning and control, one of the most important problems is to control the robot to track a given path. In a robot tracking problem the major system event is the path tracking itself. Therefore, the most natural reference to this event is the distance traveled, s, along the given path, S. If s is chosen as the reference, then the motion along the given path can be written as

$$\begin{cases} \dfrac{ds}{dt} = v \\ \dfrac{dv}{dt} = a \end{cases} \tag{1.1}$$

where v and a are velocity and acceleration, respectively, along the given path S.

Based on the results of kinematic and dynamic work space analysis [20, 21, 64], the trajectory constraints could be stated as

$$\begin{aligned} |v| &\leqslant v_m && \text{velocity constraint} \\ |a| &\leqslant a_m && \text{acceleration constraint} \\ \left|\frac{da}{dt}\right| &\leqslant k && \text{constraint for jerk-free motion} \end{aligned} \tag{1.2}$$

Obviously, during a motion the arc length s is a function of t. Thus, v and a can also be described as a function of s, instead of t, that is, $v = V(s)$, $a = A(s)$.

In order to get a event-based trajectory plan, we will convert (1.1) and (1.2) to the event-based dynamics model.

Let us define $w = v^2$, that is, $w = W(s)$, and $u = \dfrac{da}{ds}$. From (1.1), we then have

$$\begin{cases} \dfrac{dw}{ds} = 2a \\ \dfrac{da}{ds} = u \end{cases} \tag{1.3}$$

The corresponding constraints are

$$\begin{aligned} |w| &\leqslant w_m \quad \text{velocity constraint} \\ |a| &\leqslant a_m \quad \text{acceleration constraint} \\ |u| &\leqslant u_m \quad \text{jerk-free constraint} \end{aligned} \tag{1.4}$$

It is seen in (1.3) that after s is introduced as the motion reference, the model becomes a second-order linear dynamic model with states and input constraints.

Basically, the event-based trajectory planning is to find the velocity profile as a function of path or position, that is, $v = V(s)$, subject to the kinematic and dynamic constraints.

Obviously, for any given initial and terminal conditions s_0, s_f, v_0, and v_f, the trajectory plan is not unique. Using various criteria, different event-based optimal plans could be obtained.

3.2 Event-Based Time-Optimal Plan

It is well known that the time, T, to complete a motion is

$$T = \int_{t_0}^{t_f} dt = \int_{s_0}^{s_f} \frac{1}{v} ds = \int_{s_0}^{s_f} \frac{1}{\sqrt{w}} ds$$

Let us define $x_1 = w$, $x_2 = a$, $c_1 = x_1 - w_m$, $c_2 = -x_1 - w_m$, $c_3 = x_2 - a_m$, $c_4 = -x_2 - a_m$, and

$$X = \begin{bmatrix} x_1 \\ x_2 \end{bmatrix}, \quad F = \begin{bmatrix} 0 & 2 \\ 0 & 0 \end{bmatrix}, \quad B = \begin{bmatrix} 0 \\ 1 \end{bmatrix}, \quad \frac{dX}{ds} = X'$$

Then

$$X' = FX + Bu$$

with constraints $C \leqslant 0$, where $C = [c_1 \ c_2 \ c_3 \ c_4]^T$.

Now the preceding motion planning problem becomes an optimal control problem. It can be stated as follows:

$$\begin{aligned} &\underset{u}{Min}\, J, && J = \int_{s_0}^{s_f} x_2^{-\frac{1}{2}} ds \\ &\text{Subject to} && X' = FX + Bu \\ &C \leqslant 0, && |u| \leqslant u_m \end{aligned} \tag{1.5}$$

with $X(0) = 0$, $X(s_f) = 0$.

The Pontryagin maximum principle [22] can be applied to solve this problem. The Hamiltonian of (1.5) is

$$H = x_1^{-\frac{1}{2}} + \lambda^T(FX + Bu) + \mu_1 c_1'' + \mu_2 c_2'' + \mu_3 c_3' + \mu_4 c_4' \tag{1.6}$$

where $\lambda = [\lambda_1 \ \lambda_2]^T$ satisfies

$$\lambda' = -\frac{\partial H}{\partial H} \tag{1.7}$$

In addition,

$$
\begin{aligned}
c_1'' &= \frac{d^2 c_1}{ds^2} = 2u, \quad & c_2'' &= \frac{d^2 c_2}{ds^2} = -2u \\
c_3' &= \frac{dc_3}{ds} = u, \quad & c_4' &= \frac{dc_4}{ds} = -u
\end{aligned}
\tag{1.8}
$$

and

$$
\begin{aligned}
\mu_1 &= \begin{cases} 0 & \text{if} \quad x_1 < w_m \\ >0 & \text{if} \quad x_1 = w_m \end{cases}, \quad & \mu_2 &= \begin{cases} 0 & \text{if} \quad x_1 > -w_m \\ >0 & \text{if} \quad x_1 = -a_m \end{cases} \\
\mu_3 &= \begin{cases} 0 & \text{if} \quad x_2 < a_m \\ >0 & \text{if} \quad x_2 = a_m \end{cases}, \quad & \mu_4 &= \begin{cases} 0 & \text{if} \quad x_2 > -a_m \\ >0 & \text{if} \quad x_2 = -a_m \end{cases}
\end{aligned}
\tag{1.9}
$$

From (1.6)–(1.9), the time-optimal solution is obtained as

$$
u = \begin{cases}
u_m & s_0 \leqslant s \leqslant s_1 \\
0 & s_1 < s \leqslant s_2 \\
-u_m & s_2 < s \leqslant s_3 \\
0 & s_3 < s \leqslant s_4 \\
-u_m & s_4 < s \leqslant s_5 \\
0 & s_5 < s \leqslant s_6 \\
u_m & s_6 < s \leqslant s_f
\end{cases}
\tag{1.10}
$$

$$
a = \begin{cases}
u_m s - u_m s_0 & s_0 \leqslant s \leqslant s_1 \\
a_m & s_1 < s \leqslant s_2 \\
-u_m s + u_m s_3 & s_2 < s \leqslant s_3 \\
0 & s_3 < s \leqslant s_4 \\
-u_m s + u_m s_4 & s_4 < s \leqslant s_5 \\
-a_m & s_5 < s \leqslant s_6 \\
u_m s - u_m s_f & s_6 < s \leqslant s_f
\end{cases}
\tag{1.11}
$$

$$
w = \begin{cases}
u_m s^2 - 2u_m s_0 s + u_m s_0^2 & s_0 \leqslant s \leqslant s_1 \\
2a_m s + u_m s_1^2 - 2u_m s_0 s_1 + u_m s_0^2 - 2a_m s_1 & s_1 < s \leqslant s_2 \\
-u_m s^2 + 2u_m s_3 s + w_m - u_m s_3^2 & s_2 < s \leqslant s_3 \\
w_m & s_3 < s \leqslant s_4 \\
-u_m s^2 + 2u_m s_4 s + w_m - u_m s_4^2 & s_4 < s \leqslant s_5 \\
-2a_m s + 2a_m s_5 - u_m s_5^2 + 2u_m s_4 s_5^2 + w_m - w_m s_4^2 & s_5 < s \leqslant s_6 \\
u_m s^2 - 2u_m s_f s + u_m s_f^2 & s_6 < s \leqslant s_f
\end{cases}
\tag{1.12}
$$

where

$$
\begin{aligned}
s_1 &= s_0 + \frac{a_m}{u_m}, & s_2 &= s_0 + \frac{w_m}{a_m} \\
s_3 &= s_0 + \frac{w_m}{a_m} + \frac{a_m}{u_m}, & s_4 &= s_f - \frac{w_m}{a_m} - \frac{a_m}{u_m} \\
s_5 &= s_f - \frac{w_m}{a_m}, & s_6 &= s_f - \frac{a_m}{u_m}
\end{aligned}
\tag{1.13}
$$

Since we have $w = v^2$ and $w_m = v_m^2$, the time-optimal velocity profile is

$$
v = \begin{cases}
(u_m s^2 - 2u_m s_0 s + u_m s_0^2)^{\frac{1}{2}} & s_0 \leqslant s \leqslant s_1 \\
(2a_m s + u_m s_1^2 - 2u_m s_0 s_1 + u_m s_0^2 - 2a_m s_1)^{\frac{1}{2}} & s_1 < s \leqslant s_2 \\
(-u_m s^2 + 2u_m s_3 s + v_m^2 - u_m s_3^2)^{\frac{1}{2}} & s_2 < s \leqslant s_3 \\
v_m & s_3 < s \leqslant s_4 \\
(-u_m s^2 + 2u_m s_4 s + v_m^2 - u_m s_4^2)^{\frac{1}{2}} & s_4 < s \leqslant s_5 \\
(-2a_m s + 2a_m s_5 - u_m s_5^2 + 2u_m s_4 s_5^2 + w_m^2 - v_m^2 s_4^2)^{\frac{1}{2}} & s_5 < s \leqslant s_6 \\
(u_m s^2 - 2u_m s_f s + u_m s_f^2)^{\frac{1}{2}} & s_6 < s \leqslant s_f
\end{cases}
\tag{1.14}
$$

It can be seen that this time-optimal trajectory is a closed-form solution that is essential for real-time implementation. The velocity and acceleration profiles are shown in Figure 1.3.

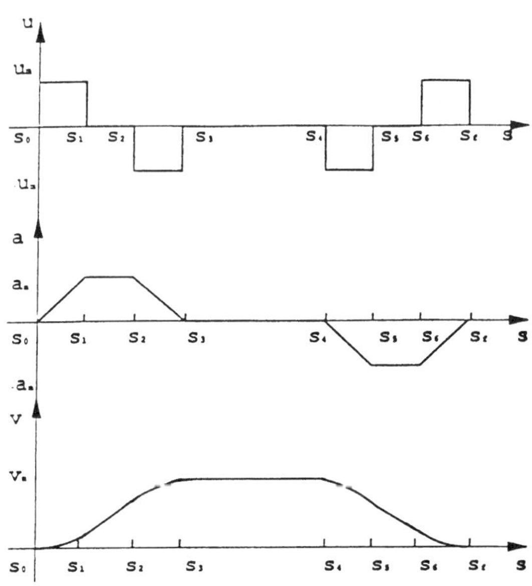

FIGURE 1.3
Velocity and acceleration profiles of the time-optimal motion plan.

3.3 Event-Based Minimum-Energy Plan

In some trajectory planning problems, instead of giving the maximum velocity along the given path, v_m, the total desired time to complete the path, t_f, is given. In this case, the minimum-energy plan can be found. Let

$$J = \int_{s_0}^{s_f} |u| \, ds$$

The minimum-energy problem can be stated as

$$\begin{gathered} \underset{u}{Min}\, J \\ \text{Subject to} \quad X' = FX + Bu \\ |a| \leqslant a_m, \; |u| \leqslant u_m \end{gathered} \tag{1.15}$$

with $X(s_0) = 0$, $X(s_f) = 0$, and given t_f.

As with the time-optimal planning problem, the Pontryagin maximum principle could be applied to find a solution for (1.15). The solution has the same form as (1.10)–(1.14). The only thing left is to determine the v_m, as it is not given here.

As the final time t_f is given,

$$t_f = \int_0^{t_f} dt = \int_{s_0}^{s_f} \frac{1}{v} \, ds \tag{1.16}$$

Based on (1.13) and (1.14), Eq. (1.16) can be solved and

$$v_m = \frac{a_m u_m t_f - a_m^2 - \sqrt{(a_m^2 - a_m u_m t_f)^2 - 4u_m^2 a_m s_f}}{2u_m} \tag{1.17}$$

Therefore, the minimum-energy trajectory is same as (1.14) and has velocity and acceleration profiles similar to those shown in Figure 1.3, except that v_m is given by (1.17).

Several Remarks

- The initial and final conditions, $X(s_0)$ and $X(s_f)$, are not necessarily to be zero. Since (1.5) and (1.15) are linear dynamic models, all the preceding results can easily be extended to nonzero cases.
- The bounds w_m, a_m are not necessarily to be constant. If they are functions of s, the preceding methods could still be used to find the solutions.
- The solutions do not necessarily have profiles such as (1.14). If $a_m^2 \geqslant w_m u_m$, then s_2 and s_1 will become a single point. There will not be a period with constant acceleration a_m. Using the same argument, if

$$\frac{w_m}{a_m} + \frac{a_m}{u_m} \geqslant \frac{s_f - s_0}{2}$$

  then s_3 and s_4 will become a single point. The period with the constant velocity in the velocity profile will vanish.

3.4 Cartesian Space Decomposition of Event-Based Plans

In the preceding two parts, the event-based motion plan $v = V(s)$ and $a = A(s)$ have been obtained. In order to get the following task space plan:

$$\begin{cases} \dot{x} = V_x(s) \\ \dot{y} = V_y(s) \\ \dot{z} = V_z(s) \end{cases} \tag{1.18}$$

and

$$\begin{cases} \ddot{x} = A_x(s) \\ \ddot{y} = A_y(s) \\ \ddot{z} = A_z(s) \end{cases} \tag{1.19}$$

the $v = V(s)$ and $a = A(s)$ will be decomposed in the Cartesian space according to the given path.

Any geometric path given in the task space can be approximated by a combination of several straight lines and circular arcs. Hence, it is necessary only to find the decompositions for the straight line and the circular path segments.

Straight Line Path

Suppose that the straight line path in task space is given and has a direction cosine (m, n, p), with initial point (x_0, y_0, z_0) and final point (x_f, y_f, z_f). It is easy to find a decomposition of the event-based plan for the given straight line path,

$$\begin{cases} \dot{x} = mV(s) \\ \dot{y} = nV(s) \\ \dot{z} = pV(s) \end{cases} \tag{1.20}$$

and

$$\begin{cases} \ddot{x} = mA(s) \\ \ddot{y} = nA(s) \\ \ddot{z} = pA(s) \end{cases} \tag{1.21}$$

Circular Path

First, it is assumed that the circle is in the xy plane of the task space and the center of the circle is at the origin. The radius of the circle is r, and the equation of the circle in the task space is given by

$$\begin{cases} x = r\cos(s/r) \\ y = r\sin(s/r) \\ z = 0 \end{cases}$$

Therefore, the Cartesian space decomposition of the event-based plan is

$$\begin{cases} \dot{x} = -\dfrac{y}{r} V(s) \\ \dot{y} = \dfrac{x}{r} V(s) \\ \dot{z} = 0 \end{cases} \tag{1.22}$$

and

$$\begin{cases} \ddot{x} = -\dfrac{x}{r^2}(V(s))^2 - \dfrac{y}{r} A(s) \\ \ddot{y} = -\dfrac{y}{r^2}(V(s))^2 + \dfrac{x}{r} A(s) \\ \ddot{z} = 0 \end{cases} \tag{1.23}$$

In a general situation, the circular path is centered at (x_0, y_0, z_0) and tilted in the task space. It is, however, always possible to build a new coordinate (x_r, y_r, z_r) such that the given circle is in the $x_r y_r$ plane and centered at $x_r = 0$, $y_r = 0$. It is very easy to find a constant transformation matrix T to satisfy

$$T\left(\begin{bmatrix} x \\ y \\ z \end{bmatrix} - \begin{bmatrix} x_0 \\ y_0 \\ z_0 \end{bmatrix}\right) = \begin{bmatrix} x_r \\ y_r \\ z_r \end{bmatrix}$$

Then

$$\begin{bmatrix} x \\ y \\ z \end{bmatrix} = T^{-1} \begin{bmatrix} x_r \\ y_r \\ z_r \end{bmatrix} + \begin{bmatrix} x_0 \\ y_0 \\ z_0 \end{bmatrix}$$

Therefore, the Cartesian space decomposition for a general circular path in task space is

$$\begin{bmatrix} \dot{x} \\ \dot{y} \\ \dot{z} \end{bmatrix} = T^{-1} \begin{bmatrix} \dot{x}_r \\ \dot{y}_r \\ \dot{z}_r \end{bmatrix}, \quad \begin{bmatrix} \ddot{x} \\ \ddot{y} \\ \ddot{z} \end{bmatrix} = T^{-1} \begin{bmatrix} \ddot{x}_r \\ \ddot{y}_r \\ \ddot{z}_r \end{bmatrix} \tag{1.24}$$

where $[\dot{x}_r \ \dot{y}_r \ \dot{z}_r]^T$ and $[\ddot{x}_r \ \ddot{y}_r \ \ddot{z}_r]^T$ are given in (1.22) and (1.23).

3.5 Event-Based Control

The dynamic model of a robot arm with six degrees of freedom (DOF) is given by

$$\tau = D(q)\ddot{q} + C(q, \dot{q}) + G(q) \tag{1.25}$$

The position and orientation output is given by

$$Y = h(q) = [h_1(q) \quad h_2(q) \quad h_3(q) \quad h_4(q) \quad h_5(q) \quad h_6(q)]^T \tag{1.26}$$

where $Y = [x \ y \ z \ O \ A \ T]^T$

q = joint angle vector

$D(q)$ = inertia matrix

$C(q, \dot{q})$ = centripetal and Coriolis terms

$G(q, \dot{q})$ = gravity loading

τ = joint torque vector

and $q, \tau \in \mathbb{R}^6$.

Equations (1.25) and (1.26), the robot dynamic model, are nonlinear equations. It is very difficult to design a control law directly. Instead, the nonlinear feedback technique [32] will be used to linearize and decouple the dynamic model and convert the nonlinear control problem to a linear control problem.

Let $x_1 = q$, $x_2 = \dot{q}$, and $E(x_1, x_2) = C(x_1, x_2) + G(x_1)$. Equations (1.25) and (1.26) can be rewritten in a standard nonlinear state space form

$$\begin{bmatrix} \dot{x}_1 \\ \dot{x}_2 \end{bmatrix} = \begin{bmatrix} x_2 \\ -D^{-1}(x_1)E(x) \end{bmatrix} + \begin{bmatrix} 0 \\ D^{-1}(x_1) \end{bmatrix} \tau$$

Therefore, the robot dynamics model could be stated as

$$\begin{cases} \dot{x} = f(x) + g(x)\tau \\ y = h(x_1) \end{cases} \tag{1.27}$$

where $x = [x_1, x_2]^T$.

Using results of differential geometric control theory [61], there exist a diffeomorphic state transformation $T(x)$ and a nonlinear feedback law $\tau = \alpha(x) + \beta(x)v$ that linearizes and decouples the robot dynamics. The diffeomorphic state transformation $T(x)$ is given by

$$z = T(x) = [h_1(x_1), L_f h_1(x_1), \ldots, h_6(x_1), L_f h_6(x_1)]^T$$

and the nonlinear feedback law is

$$\tau = \alpha(x) + \beta(x)v$$

with

$$\alpha(x) = -D(x_1)J_h^{-1} \begin{bmatrix} L_f^2 h_1(x_1) \\ \vdots \\ L_f^2 h_6(x_1) \end{bmatrix} = -D(x_1)J_h^{-1}[\dot{J}_h \dot{q} - J_h D^{-1}(x_1)E(x)] \tag{1.28}$$

$$\beta(x) = D(x_1)J_h^{-1} \tag{1.29}$$

where h_i is the ith component of $h(q)$, L_f^k denotes the kth Lie derivative of $h(x)$ along the vector field $f(x)$, and J_h is the output Jacobian matrix of $h(x_1)$.

In the transformed state z with the auxiliary input v, Eq. [1.27] appear in the Brunowsky canonical form as follows:

$$\dot{z} = Az + Bv \tag{1.30}$$

$$y = Cz \tag{1.31}$$

Here A, B, C are block diagonal matrices. To see the structure of these equations, we write them in a more detailed fashion. Equations (1.30) and (1.31) represent six linear and decoupled subsystems in the form

$$\dot{Z}_i = \begin{bmatrix} 0 & 1 \\ 0 & 0 \end{bmatrix} z_i + \begin{bmatrix} 0 \\ 1 \end{bmatrix} v_i$$

$$y_i = [1 \quad 0] z_i$$

where $z_i = [h_i \;\; L_f h_i]^T$. Each identical subsystem has double poles at the origin; therefore the system is not asymptotically stable.

Introducing the feedback law

$$v_i^* = v_i - F_i z_i, \quad i = 1, \ldots, 6$$

where $F_i = [f_{i_2} \; f_{i_2}]$, the final form of the closed loop is as follows:

$$\dot{z}_i = \begin{bmatrix} 0 & 1 \\ -f_{i_1} & -f_{i_2} \end{bmatrix} z_i + \begin{bmatrix} 0 \\ 1 \end{bmatrix} v_i^*$$

$$y_i = [1 \quad 0] z_i$$

Note that F_i represents a linear Proportional-plus-Derivative (PD) controller.

Therefore, the nonlinear feedback control law is given by

$$\tau = D(q) J_h^{-1} [\ddot{Y}^d(t) + K_v \dot{e}(t) + K_p e(t) - \dot{J}_h \dot{q}] + C(q, \dot{q}) + G(q) \tag{1.32}$$

where

$$e(t) = Y^d(t) - Y(t)$$

$$\dot{e}(t) = \dot{Y}^d(t) - \dot{Y}(t)$$

From this, it can be seen that for a time-based plan, the reference base of input and measurement is time t. For any time instance t, a measurement $Y(t)$, $\dot{Y}(t)$; a desired input $Y^d(t)$, $\dot{Y}^d(t)$; and errors $e(t)$, $\dot{e}(t)$ can be obtained. However, for an event-based plan, the time is no longer a reference base. The input of the system is parameterized by the event-based motion reference s. According to the new motion reference s, the error e and $\dot{e}$ must be redefined in order to get a event-based control law.

In essence, for a digital sampled data control system we could determine the corresponding $Y^d(s)$, $\dot{Y}^d(s)$ for each sampling time $n_i \Delta t$ by first computing the desired velocity and then

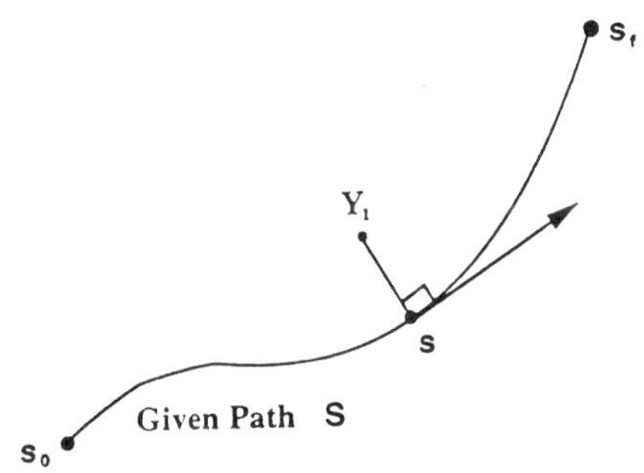

FIGURE 1.4
The event-based error definition.

integrating the velocity to determine the corresponding desired position as done in [7] and [66]. Instead of this technique, we choose a new procedure illustrated in Figure 1.4.

In this figure, $Y_1 = [x, y, z]^T$ is a measurement, and the point s corresponds to a point in the given path that, in our technique, has the minimum distance from Y_1 to the given path, that is, the orthogonal projection of Y_1. The Cartesian space coordinate of point s is considered as a desired position $Y^d(s)$.

Based on s, a desired velocity $\dot{Y}^d(s)$ and desired acceleration $\ddot{Y}^d(s)$ can be obtained from the event-based plan. Therefore, the new error definitions are

$$\begin{aligned} e(s) &= Y^d(s) - Y(s) \\ \dot{e}(s) &= \dot{Y}^d(s) - \dot{Y}(s) \end{aligned} \tag{1.33}$$

It can be seen that the new error definitions minimize the position error and make all errors independent of time. If a robot arm is stopped unexpectedly during a motion, since the motion reference base s depends only on the position of the robot instead of the time increment, it stops increasing as well. Therefore, the errors will remain unchanged, which makes it possible for the planner to modify the original plan to deal with the unexpected events. It should be noticed that in this situation, the error would keep increasing if the scheme as described in [7] and [66] was implemented. This is because, in spite of the fact that the robot arm has stopped, the desired inputs of the system are still updated along with the increase in the time. As a result, errors will keep increasing. Eventually, the system will become unstable. Therefore the time is still a "driving force" for the system.

Finally, Eq. (1.33) and $\ddot{Y}^d(s)$ can be put into Eq. (1.32) to obtain an event-based control law.

The event-based planning and control scheme is shown in Figure 1.5. The most important part of Figure 1.5 is the motion reference block. For every measurement point Y, the motion reference block calculates the orthogonal projection point on the given path in order to get the corresponding motion reference variable.

3.6 Experimental Results

Trajectory tracking of both minimum-time and minimum-energy motion plans have been tested on a PUMA 560 arm. The details of the experimental setup will be described in Section 5.

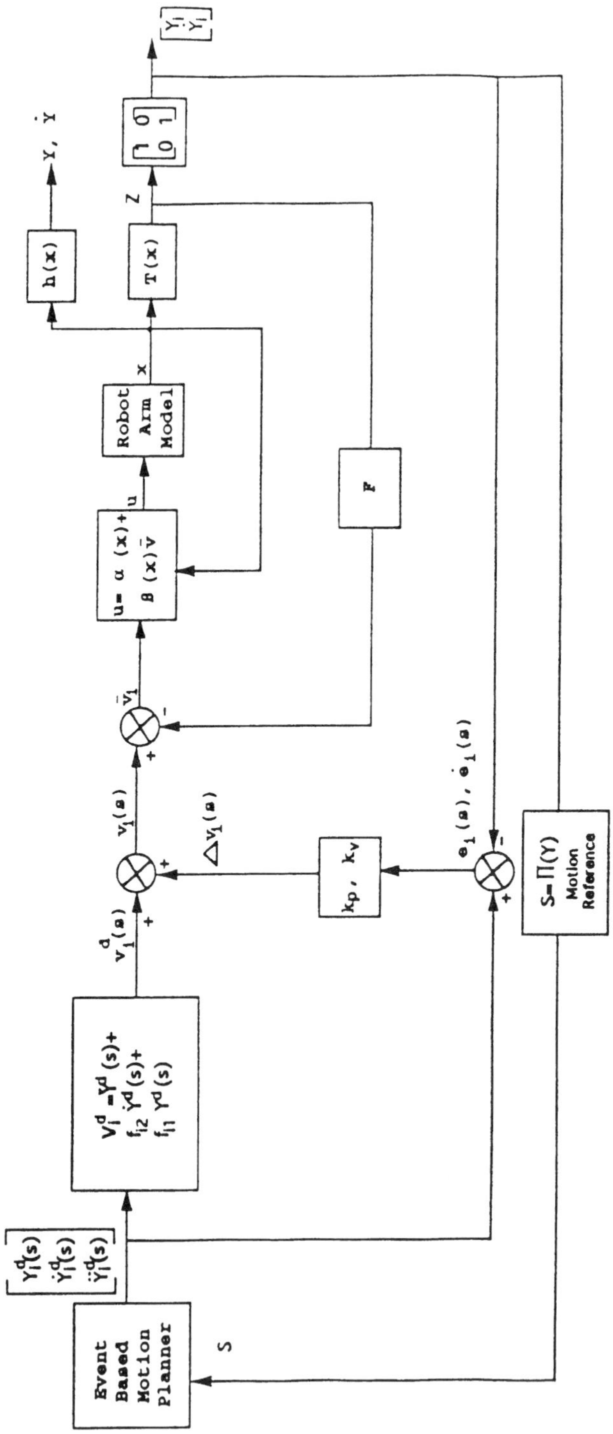

FIGURE 1.5
The event-based planning and control scheme for a single robot arm.

The sampling rate and feedback rate were 1000 hertz (1 millisecond) and the plots were made with sample points taken every 100 milliseconds. All plots correspond to the best possible gain values experimentally obtained for the task.

In the following plots, the absolute position error is defined as

$$e_{\text{pos}} = \sqrt{(x^d(s) - x(s))^2 + (y^d(s) - y(s))^2 + (z^d(s) - z(s))^2}$$

and the absolute orientation error is defined as

$$e_{\text{orin}} = \arccos(\tfrac{1}{2}(t_r R - 1))$$

where R is a rotation matrix between the actual orientation and the desired orientation.

Figure 1.6 shows the performance plots for four-circle tracking using the time-optimal event-based plan. The radius of the circle is 0.1 m. It is tilted at 45°. In addition, $v_m = 0.2$ m/s, $a_m = 0.3$ m/s^2. It is seen from the performance plots that the peak absolute error is less than 1 millimeter. In particular, the velocity error has been reduced comparing with a time-based planning and control scheme [35]. In addition, the steady-state error has been significantly reduced to less than 0.5 millimeter. The basic reason for obtaining a smaller steady-state error is that the time t is no longer a motion reference base, and the new reference base, arc

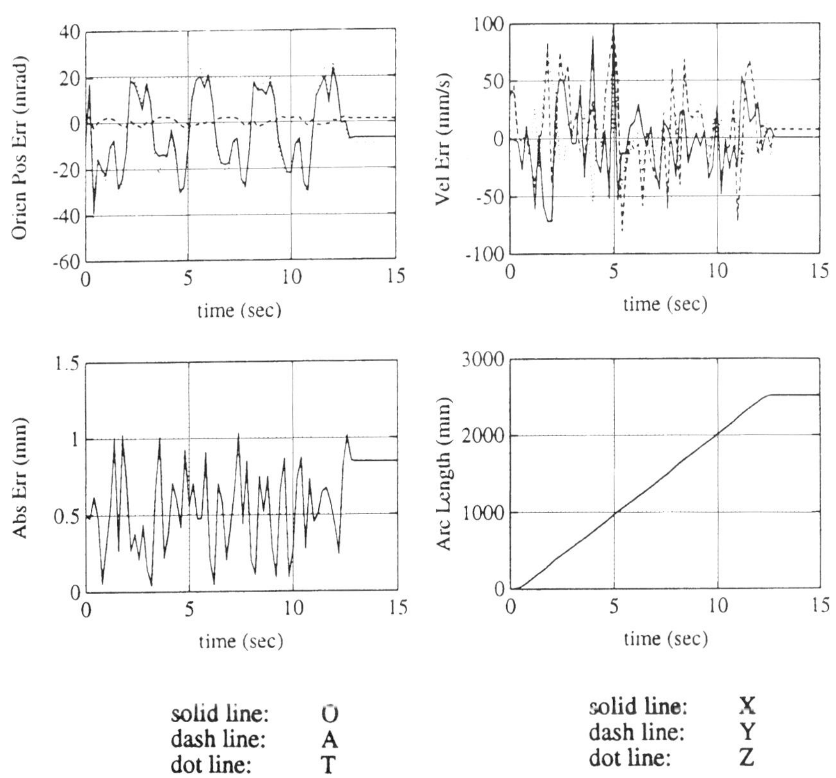

FIGURE 1.6
Four-circle tracking based on the time-optimal plan.

length s, is directly related to the position. The event-based error definition ensures minimization of the position error.

The arc length plots in the figure give the profiles of s versus time. It is seen that s is a monotone increasing function of t.

In Figure 1.7, the trajectory constraints are increased to $v_m = 0.4\,\text{m/s}$, $a_m = 0.4\,\text{m/s}^2$. Because the errors have been reduced through the implementation of event-based planning and control, the robot arm was able to track the four circles within 10 seconds. This cannot be achieved by a time-based fifth-order polynomial motion plan [35].

Figure 1.8 is the result of using a time-optimal trajectory along a straight line path from (0.6 m, 0.0, −0.4 m) to (0.0, 0.6 m, −0.2 m) and $v_m = 0.2\,\text{m/s}$ and $a_m = 0.3\,\text{m/s}^2$.

The minimum-energy event-based plan for two-circle tracking was also tested. The circles are tilted at 45° and $A_m = 0.3\,\text{m/s}^2$. The results for different terminal times t_f are given in Figures 1.9–1.11. Because the motion reference base is not time, the desired final times t_f are not precisely achieved.

Figure 1.12 presents the results of an interesting experiment. During a straight line motion, an unexpected obstacle stopped the robot motion. If the time-based plan were implemented, the errors would keep increasing and eventually result in instability. However, it is shown that the errors remained constant when the motion stopped, and once the obstacle was removed, the robot completed the rest of the planned motion without replanning. This demonstrates that the event-based planning and control scheme provides the robot with the

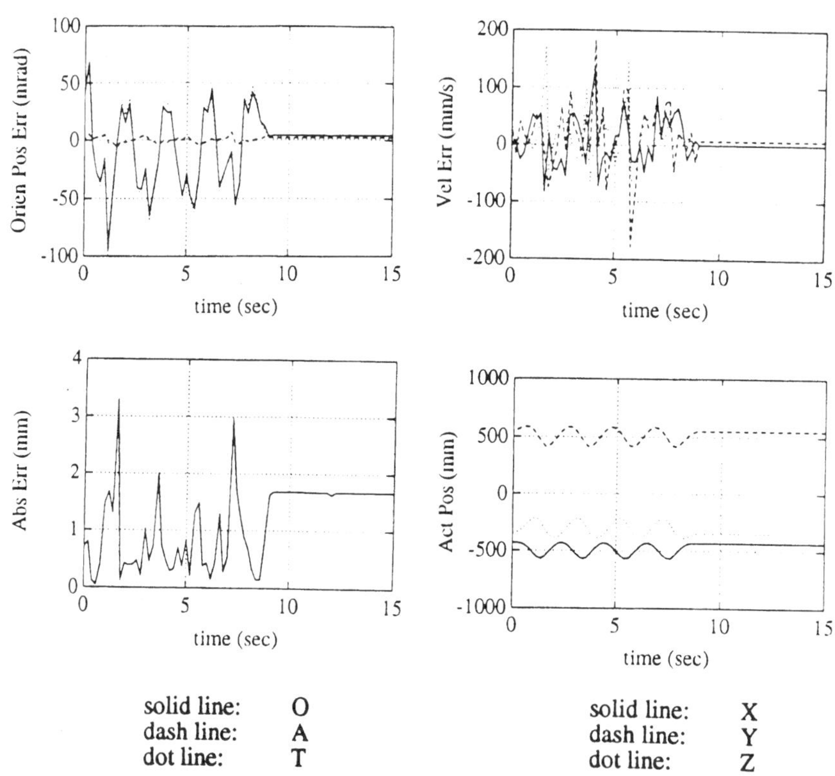

FIGURE 1.7

Four-circle high-velocity tracking based on a time-optimal plan.

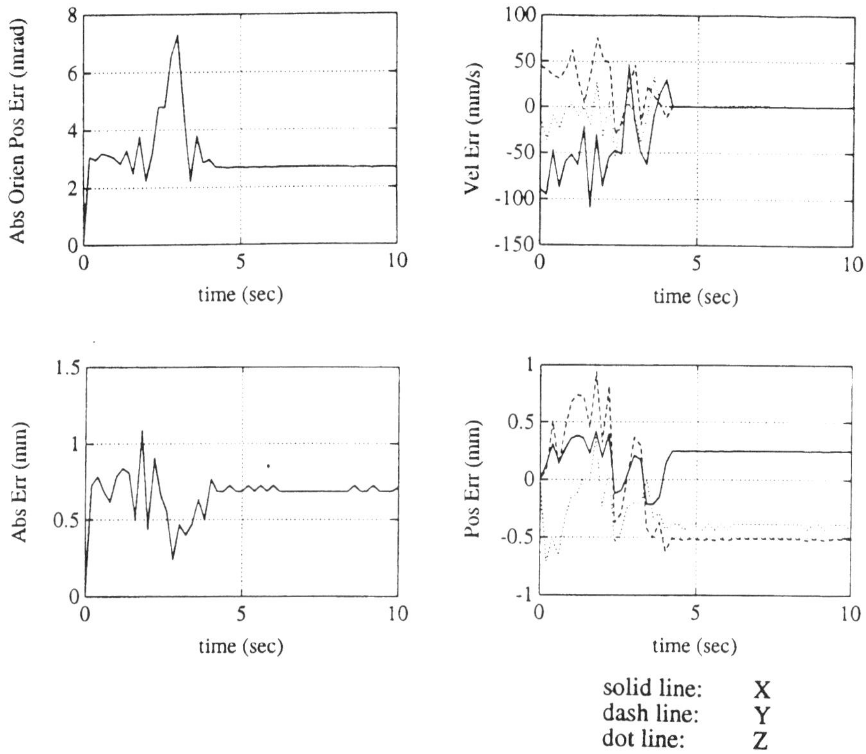

FIGURE 1.8
Straight line path tracking based on a time-optimal plan.

ability to handle an unexpected event. It significantly improves the safety and reliability of the robotic system.

Figure 1.13 presents the results of a similar experiment for a circular path.

The preceding experimental results indicate that the performance of the event-based planning and control scheme is comparable to that of the time-based motion planning and control scheme. *It is even better.* The important point, however, is that it provides a natural reference base for sensor-based planning and control.

4 EVENT-BASED PLANNING AND CONTROL FOR MULTIROBOT COORDINATION

4.1 Introduction

An important issue in multirobot systems is coordinated control. To achieve intelligence of multirobot systems, it is essential to develop a proper planning and control scheme for coordination.

Multirobot coordinated control has been a research subject for several years. Various coordination schemes have been proposed. In [23] and [49], the master–slave coordination scheme was proposed. The hybrid position–force control theory was extended to multiarm coordinated control [36–38]. Control algorithms for multiarm object handling that take into

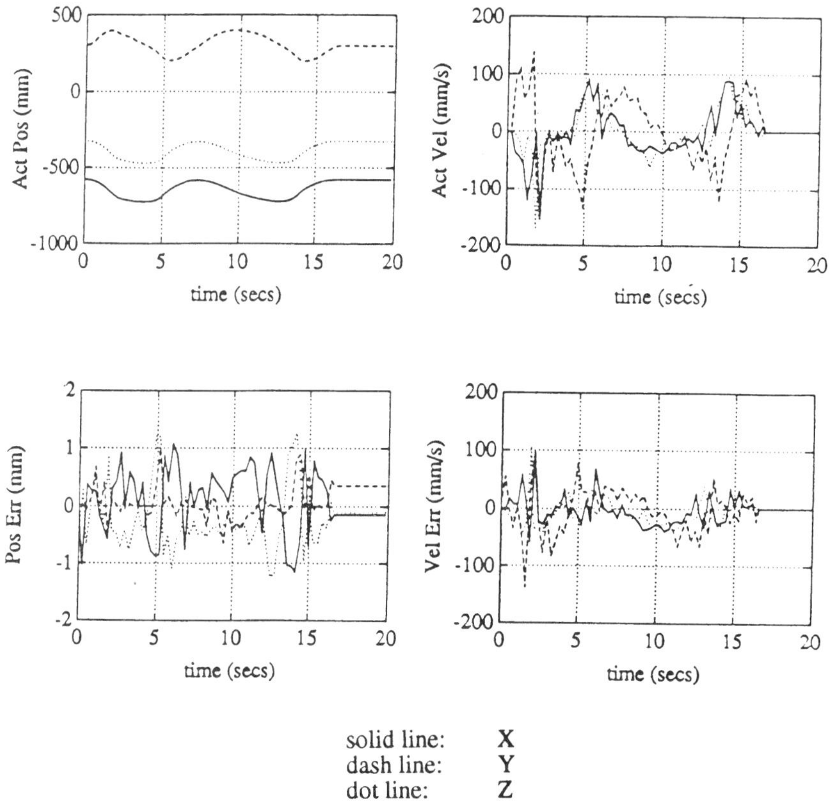

FIGURE 1.9
Two-circle tracking based on a minimum-energy plan, $t_f = 16$ s.

account the object dynamics and achieve simultaneous position and internal force control appear in [24], [33], [40], [41], and [42]. The coordination of a multifingered robot has also been widely discussed in [44], [45], [46], and [47]. In a multirobot system, redundancy becomes even more important. The related results can be found in [53], [54], [55], [56], [57], and [58]. Dual-arm situations have been intensively investigated in [43], [48], [50], [52], and [63]. An experimental evaluation of master–slave and hybrid position–force control schemes was presented in [51].

In this section, issues in multirobot rigid-object handling are discussed. First, a new event-based motion reference for a multirobot system is introduced. Then time- and energy-optimal motion plans are obtained on the basis of this new motion reference. A general task space is defined. Based on the nonlinear feedback technique, the multirobot system including the robots' joint motor dynamics is linearized and decoupled with respect to the general output defined in the general task space. Then a task projection operator is introduced. It projects the general output to a controllable subspace, that is, to the actual task space for each individual robot. Finally, experimental results for a dual-arm coordination task are presented.

The ultimate goal is to develop an intelligent planning and control scheme for multiarm coordination that can be conveniently implemented in a distributed computing architecture.

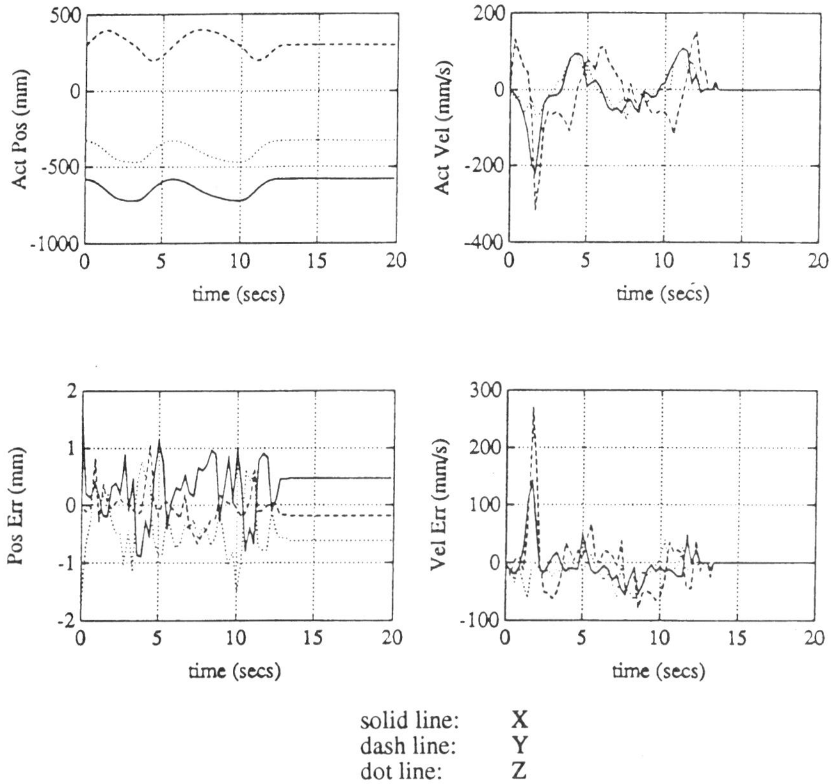

FIGURE 1.10
Two-circle tracking based on a minimum-energy plan, $t_f = 12$ s.

4.2 Event-Based Coordination

An event-based motion planning and control scheme was successfully applied to a single robot arm system in the preceding section. It is extended here for coordination planning and control of multirobot systems. The most important step is to introduce a proper motion reference variable to carry the coordination information efficiently to the planner such that the best coordination can be achieved.

We consider a rigid object b handled by k robots that transport it in free space along a given path S, which is the path of the center of gravity of the object.

In Figure 1.14,

K_w = world reference frame

K_b = body-attached frame at center of gravity of object

K_i = frame fixed at contact point of the ith robot that coincides with the hand coordinate frame of the ith robot

$r_b \in \mathbb{R}^6$ = generalized object coordinate with respect to K_w

$r_i \in \mathbb{R}^6$ = generalized coordinate for the ith robot with respect to K_w

In addition, $r_i = h_i(r_b)$ is the coordinate transformation from the body-attached frame to the

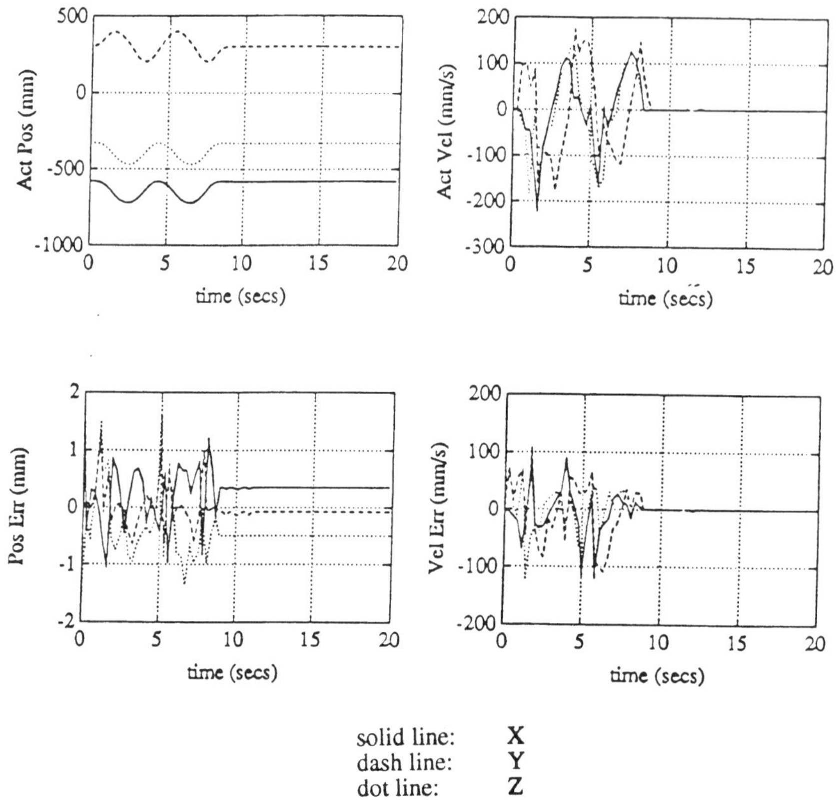

FIGURE 1.11
Two-circle tracking based on a minimum-energy plan, $t_f = 8$ s.

ith contact frame. We can assume that all robots can apply enough wrenches to control the object in $\mathbb{R}^6$.

The event-based motion reference s is defined as the distance that the center of gravity of the object travels along the given path S. The techniques used in the last section can be applied here to find a time- or energy-optimal motion plan for the object as a function of s. Therefore, the desired velocity and acceleration of the object are

$$\begin{aligned} \dot{r}_b^d &= [\dot{x}_b^d(s)\dot{y}_b^d(s)\dot{z}_b^d(s)\dot{O}_b^d(s)\dot{A}_b^d(s)\dot{T}_b^d(s)]^T \\ \ddot{r}_b^d &= [\ddot{x}_b^d(s)\ddot{y}_b^d(s)\ddot{z}_b^d(s)\ddot{O}_b^d(s)\ddot{A}_b^d(s)\ddot{T}_b^d(s)]^T \end{aligned} \tag{1.34}$$

Hence, based on the given coordinate transformations, the event-based motion plan for the ith robot can be computed as

$$\dot{r}_i^d(s) = J_{h_i}(r_b^d)\dot{r}_b^d(s) \tag{1.35}$$

$$\ddot{r}_i^d(s) = J_{h_i}(r_b^d)\ddot{r}_b^d(s) + \dot{J}_{h_i}(r_b^d)\dot{r}_b^d(s) \tag{1.36}$$

where $J_{h_i}(r_b^d) = \dfrac{\partial h_i}{\partial r_b}$ is the Jacobian matrix for the coordinate transformation.

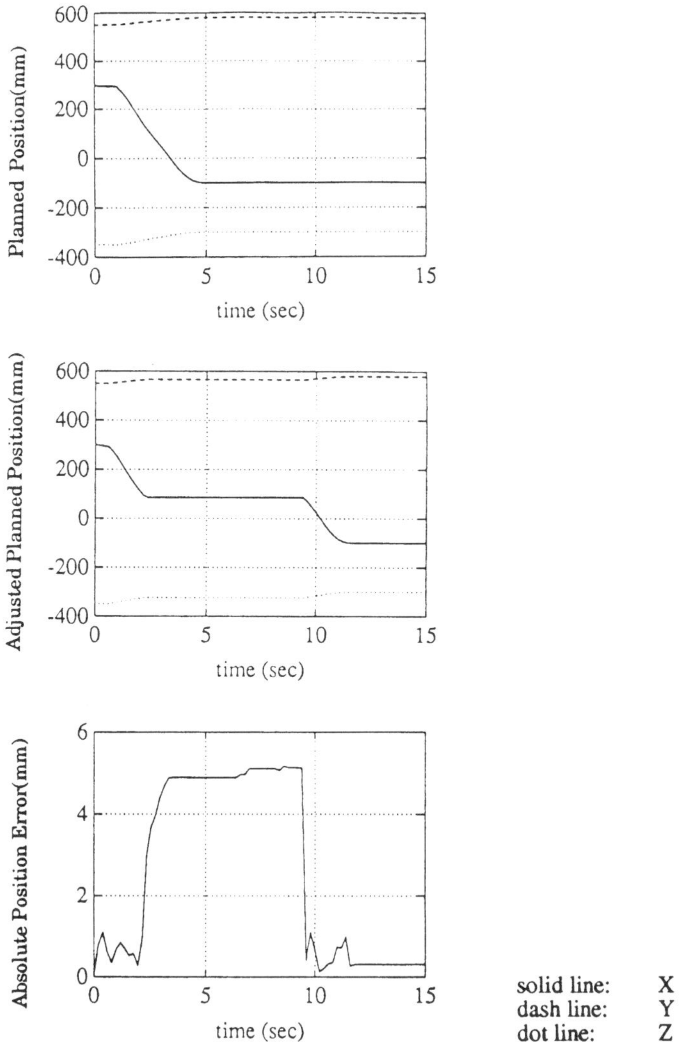

FIGURE 1.12
Straight line motion with an unexpected obstacle.

In addition, the internal force exerted on the object must be controlled in order to keep the contact between the robots and the object or to optimize the load distribution.

Let $f_i = [f_{1i}\, f_{2i}\, f_{3i}\, f_{4i}\, f_{5i}\, f_{6i}]^T$, $i = 1, 2, \ldots k$, be the general force with respect to K_w exerted on the object by the ith robot, and

$$I_j^+ = \{i | f_{ji} \geq 0\}, \quad I_j^- = \{i | f_{ji} < 0\}, \quad j = 1, 2, \ldots 6$$

$$f_{j+} = \sum_{i \in I_i^+} f_{ji}, \quad f_{j-} = \sum_{i \in I_i^-} f_{ji}, \quad j = 1, 2, \ldots 6$$

where f_{j+} and f_{j-} are the summations of all positive and negative forces along the jth

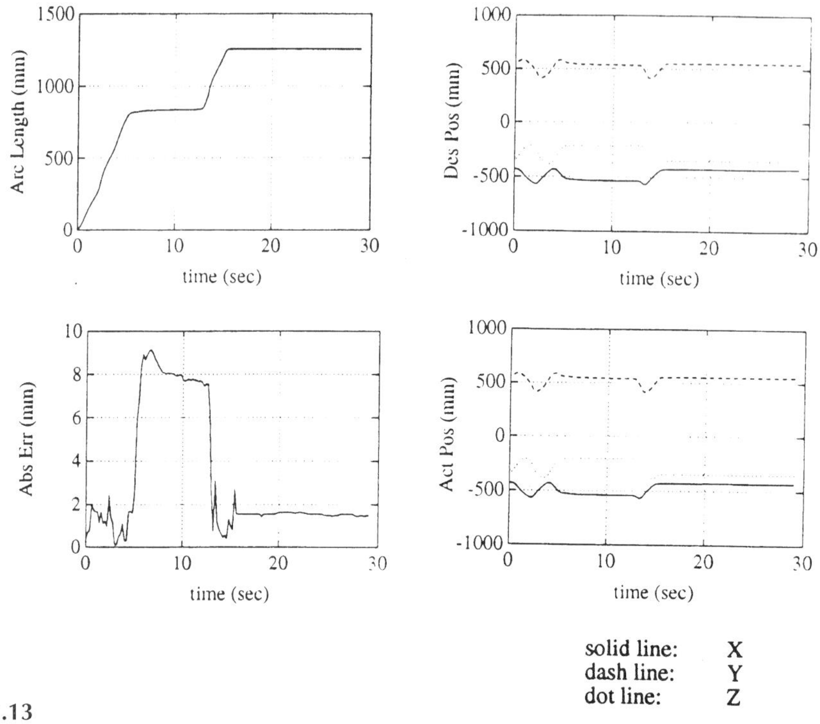

FIGURE 1.13
Circular motion with an unexpected obstacle.

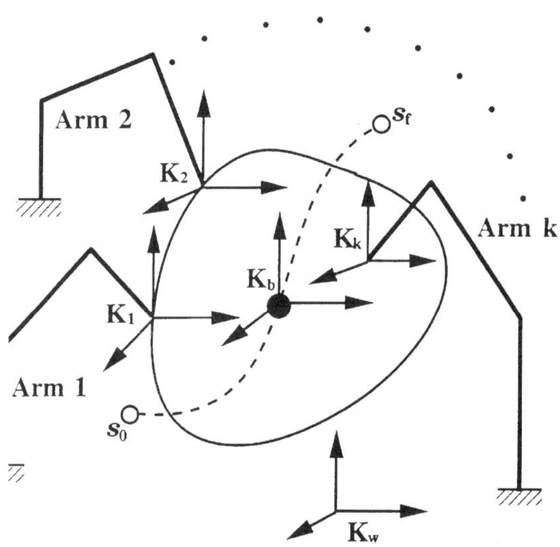

FIGURE 1.14
A rigid object handled by multiple robots.

direction on the object. The internal force is then

$$f_{int} = [f_{1\,int}, f_{2\,int}, f_{3\,int}, f_{4\,int}, f_{5\,int}, f_{6\,int}]^T$$

where

$$f_{j\,int} = \begin{cases} f_{j+} & f_{j+} + f_{j-} \leqslant 0 \\ f_{j-} & f_{j+} + f_{j-} > 0 \end{cases} \quad j = 1, 2, \ldots 6$$

Based on the contact condition and the physical properties of the object, $f_{j\,int}$ can be planned as a function of event-based motion reference s,

$$f^d_{j\,int} = F_{j\,int}(s), \quad j = 1, 2, \ldots 6 \tag{1.37}$$

However, in most tasks, the $f_{j\,int}$ can simply be planned as a constant value in order to keep the contact between the robots and the object.

The event-based description of task plans does not simply involve replacing the commonly used motion reference—time. Because the event-based motion reference is directly related to the states of the system, the planner is driven by the states of the system. As a result, planning becomes part of a real-time, closed-loop process.

Equations (1.35), (1.36), and (1.37) give the task plans for each individual robot in the multiarm system. Instead of time, they are driven by the event-based motion reference s, which is directly related to the coordination of the system. The coordination requires that all the robots in the system follow the motion of the object in a planned manner. To achieve coordinated control, it is necessary to obtain information about the motions of the object and robots. This information can be passed to the planner of each robot through the motion reference. Based on the current state of the system, the motion reference s is designed to evolve in such a way that the system can achieve the best possible coordinated control.

The coordinated control can be expressed by the requirement that for any point along the path of the object, the robots should be in the states determined by the task planner, which is driven by the event-based motion reference. For given measurements r_i, f_i, $i = 1, 2, \ldots k$, define a coordination criterion

$$J = \sum_{i=1}^{k} [(r^d_i(s) - r_i)^T W_{r_i}(r^d_i(s) - r_i) + (f^d_i(s) - f_i)^T W_{f_i}(f^d_i(s) - f_i)] \tag{1.38}$$

where W_{r_i} and W_{f_i}, $i = 1, 2, \ldots k$, are weight matrices. They can weight the coordination errors in different directions to ensure efficient coordinated control in some specific directions determined by a given task.

The optimal motion reference s^*, which can achieve the best coordination, is the solution of

$$\min_{s \in S} J \tag{1.39}$$

The closed-form solution can be easily obtained for most paths, such as straight lines and circles, by solving

$$\frac{\partial J}{\partial s} = 0$$

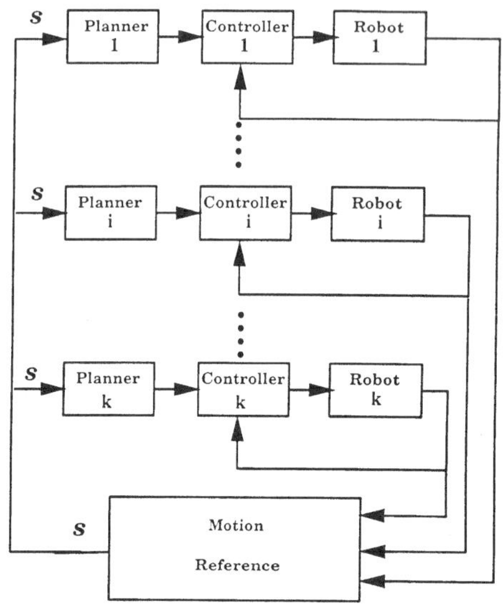

FIGURE 1.15
Event-based planning and control scheme for multiarm coordination.

Once the motion reference s is determined, the planner of each robot can calculate the desired inputs based on the given s. It can be seen that the planners driven by the event-based motion reference always give the optimal plan to minimize the coordination error. The event-based planning and control scheme for multiarm coordination is shown in Figure 1.15.

Several Remarks

1. The motion planner driven by the event-based motion reference is no longer a memory component driven by time. Based on the information about the current state of the system, the planner gives the best possible desired input for the system to achieve coordination. The planning becomes a closed-loop, real-time process, and the planner becomes an investigation–decision component.
2. In Figure 1.15, similarly to the master–slave scheme in [23], each robot has its own independent controller and planner. This gives flexibility to the structure of the multiarm system and makes it convenient to implement it in a distributed computing architecture.
3. In addition, similarly to the hybrid position–force control scheme in [24], each individual robot has information about the current states of the other robots. This makes it possible to achieve better coordination and internal force control [51].
4. Because the task planner is driven by s instead of time, if the motion is stopped by an unexpected event, such as an obstacle, the motion reference s stops increasing. Therefore, the errors remain constant and the coordination can still be maintained according to the original plan. Once the obstacle is removed, s starts to evolve again. The motion can be completed without replanning. Hence, this event-based coordination scheme has the ability to handle some unexpected events.
5. The criterion (1.37) can easily be extended to consider other factors, such as minimum internal force and optimal load distribution.

4.3 Effect of Motor Dynamics on Force Control

It is essential to control the motion of a robot and the output force of the robot simultaneously in multirobot coordination. It was shown that motor dynamics played a very important role in the motion control of a robot manipulator [25].

In order to achieve both a good transient response and a small steady-state tracking error, it is necessary to include the motor dynamics in the hybrid position–force control model.

Without considering the joint motor dynamics, the nonredundant robot dynamics can be modeled as

$$D(q)\ddot{q} + C(q, \dot{q}) + G(q) + J^T(q) f = \tau \tag{1.40}$$

where

$$\begin{aligned} q^{\#} &= \text{joint angle vector} \\ D(q) &= \text{inertia matrix} \\ C(q, \dot{q}) &= \text{centripetal and Coriolis terms} \\ G(q) &= \text{gravity loading} \\ \tau &= \text{joint torque vector} \\ J(q) &= \text{Jacobian matrix, such that } \dot{y} = J(q)\dot{q} \\ y &= \text{position and orientation in task space} \\ f &= \text{force output in task space} \end{aligned}$$

Applying the well-known nonlinear feedback control law,

$$\tau = D(q)J^{-1}(q)[V_1 - \dot{J}(q)\dot{q}] + C(q, \dot{q}) + G(q) + J^T(q)V_2 \tag{1.41}$$

if $y \perp f$, which implies that the force and position are not commanded in the same axis of task space, the linearized model is

$$\begin{cases} \ddot{y} = V_1 \\ f = V_2 \end{cases} \tag{1.42}$$

where

$$\begin{aligned} V_1 &= \text{position and orientation command} \\ V_2 &= \text{force command} \end{aligned}$$

It is evident that there are no dynamics between the input command and the output force in (1.42). The command affects the output instantaneously.

The commonly used force control law [24, 34] is

$$V_2 = f^d + K_{fp}(f^d - f) + K_{fi}\int_0^t (f^d - f)\,dt \tag{1.43}$$

where f^d is the desired force. K_{fp} and K_{fi} are the proportional and integral feedback gains.

It can easily be shown that based on the linearized model $f = V_2$, the feedback law

$$V_2 = f^d + K_{fp}(f^d - f) + K_{fi} \int_0^t (f^d - f)\, dt$$

is equivalent to

$$V_2 = f^d + \frac{K_{fi}}{1 + K_{fp}} \int_0^t (f^d - f)\, dt \tag{1.44}$$

Therefore, the proportional control actually plays no role in the force feedback control law. As a result, it would be very difficult to achieve both a small steady-state tracking error and a quick transient response. Furthermore, if there is no integral feedback, that is, $K_{fi} = 0$, the force control turns into an open-loop system. Because there are no dynamics between the input and output of the force control loop, the motor dynamics become the dominant part. Hence they must be included in the dynamic model in order to achieve stable and robust performance.

4.4 Hybrid Position–Force Control for Coordinated Robots

The controllers in Figure 1.15 perform hybrid position–force control or position control alone, depending on the nature of the tasks. However, it is desirable to have as small a change as possible in the controllers for different tasks. This is one of the important characteristics of an intelligent robotic system.

It was shown in the last section that the dynamics of the joint motors play an important role in hybrid position–force control. It is important to include the dynamics in the system's dynamic model.

Let the ith robot in the system have n_i joints. Its dynamic model, including the joint motor dynamics, can be written as

$$\begin{cases} D_i(q_i)\ddot{q}_i + C_i(q_i, \dot{q}_i) + G_i(q_i) + \tau_{fi} = \tau_i \\ \dfrac{d\tau_i}{dt} = -T_i^{-1}\tau_i + K_{ei}\dot{q}_i + u_i \end{cases} \tag{1.45}$$
$$i = 1, 2, \ldots k$$

where

τ_f = joint torque that produces output force f_i

T_i = $n \times n$ diagonal matrix whose entries are the time constant of the joint motors

K_{ei} = voltage constant of the motor (back electromotive force)

u_i = motor armature voltage

and $q_i, \tau_i \in \mathbb{R}^{n_i}$.

The general task space for the ith robot is defined as

$$\mathcal{T}_i = \mathcal{Y}_i \oplus \mathcal{F}_i \oplus \mathcal{Z}_i$$

where

$$\mathscr{Y}_i \in \mathbb{R}^6 \quad \text{is the general position space.}$$
$$\mathscr{F}_i \in \mathbb{R}^6 \quad \text{is the general force space.}$$
$$\mathscr{Z}_i \in \mathbb{R}^{n_i} \quad \text{is the general redundant joint space.}$$

$$\mathscr{T}_i = \left\{ \begin{bmatrix} y_i \\ f_i \\ q_i \end{bmatrix} \in \mathbb{R}^{12+n_i} \right\}$$

and $\oplus$ denotes the orthogonal direct sum. For any given task, it will be a subspace of the general task space.

Because only n_i joints are available, the robot can be controlled only in a subspace of $\mathscr{T}_i$, $\mathscr{T}_{ic} \in \mathbb{R}^{n_i}$. Hence, the actual output is in $\mathscr{T}_{ic}$.

$\mathscr{T}_{ic}$ is defined as

$$\mathscr{T}_{ic} = \mathscr{Y}_{ic} \oplus \mathscr{F}_{ic} \oplus \mathscr{Z}_{ic}$$

where

$$\mathscr{Y}_{ic} \in \mathbb{R}^{l_{i1}}$$
$$\mathscr{F}_{ic} \in \mathbb{R}^{l_{i2}}$$
$$\mathscr{Z}_{ic} \in \mathbb{R}^{n_i - (l_{i1} + l_{i2})}$$

The task projection operator $B_{ic}: \mathscr{T}_i \mapsto \mathscr{T}_{ic}$ is the basis of $\mathscr{T}_{ic}$ and is described as

$$B_{ic} = [Y_{ic} \quad F_{ic} \quad Z_{ic}]$$

The joint space torque that produces the output force f can be written as [39]

$$f = (J^T(q))^{\#} \tau_f$$

where "#" denotes a pseudoinverse.

Because of redundancy, the joint torque is not unique for a given f_i and can be formulated as [39]

$$\tau_{f_i} = J_i^T(q_i) f_i + [I_{n_i} - J_i^T(q_i)(J_i^T(q_i))^{\#}]\Gamma_{f_i}$$

where I_{n_i} is an $n_i \times n_i$ identity matrix; $J_i(q_i)$ is a Jacobian matrix, $\dot{r}_i = J_i(q_i)\dot{q}_i$; "#" denotes a pseudoinverse; Γ_{f_i} is any $n_i \times 1$ vector; and $[I_{n_i} - J_i^T(q_i)(J_i^T(q_i))^{\#}]\Gamma_{f_i}$ is a vector in the null space of $(J_i^T(q_i))^{\#}$, which describes the redundancy of the robot.

By a similar argument,

$$\ddot{q}_i = J_i^{\#}(q_i)(\ddot{r}_i - \dot{J}_i(q)\dot{q}_i) + [I_{n_i} - J_i^{\#}(q_i)J_i(q_i)]\Gamma_{r_i}$$

where Γ_{r_i} is any $n_i \times 1$ vector and $[I_{n_i} - J_i^{\#}(q_i)J_i(q_i)]\Gamma_{r_i}$ is a vector in the null space of $J_i(q_i)$, which also describes the redundancy of the robot.

By choosing $\Gamma_{r_i} = 0$ and $\Gamma_{f_i} = \ddot{q}_i$, (1.45) can be written as

$$\begin{cases} D_i(q_i)J_i^{\#}(q_i)(\ddot{r}_i - \dot{J}_i(q_i)\dot{q}_i) + C_i(q_i, \dot{q}_i) + G_i(q_i) + J_i^T(q_i)f_i + [I_{n_i} - J_i^T(q_i)(J_i^T(q_i))^{\#}]\ddot{q}_i = \tau_i \\ \dfrac{d\tau_i}{dt} = -T_i^{-1}\tau_i + K_{e_i}\dot{q} + u_i \end{cases} \tag{1.46}$$

Let

$$\tau_{i1} = D_i(q_i)J_i^{\#}(q_i)(\ddot{r}_i - \dot{J}_i(q_i)\dot{q}_i) + C_i(q_i, \dot{q}_i) + G_i(q_i)$$
$$\tau_{i2} = J_i^T(q_i)f_i$$
$$\tau_{i3} = [I_{n_i} - J_i^T(q_i)(J_i^T(q_i))^{\#}]\ddot{q}_i$$

Then $\tau_i = \tau_{i1} + \tau_{i2} + \tau_{i3}$.

In addition, let

$$u_{i1} = \frac{d\tau_{i1}}{dt} + T_i^{-1}\tau_{i1} - K_{ei}\dot{q}_i$$
$$u_{i2} = \frac{d\tau_{i2}}{dt} + T_i^{-1}\tau_{i2}$$
$$u_{i3} = \frac{d\tau_{i3}}{dt} + T_i^{-1}\tau_{i3}$$

Then $u_i = u_{i1} + u_{i2} + u_{i3}$. As a result of these definitions, the dynamic model (1.46) has been separated in the general task space $\mathscr{T}_i$.

In the general position space $\mathscr{Y}_i$:

$$\begin{cases} D_i(q_i)J_i^{\#}(\ddot{r}_i - \dot{J}_i(q_i)\dot{q}_i) + C_i(q_i, \dot{q}_i) + G_i(q_i) = \tau_{i1} \\ \dfrac{d\tau_{i1}}{dt} = -T_i^{-1}\tau_{i1} + K_{ei}\dot{q}_i + u_{i1} \end{cases} \tag{1.47}$$

In the general force space $\mathscr{F}_i$:

$$\begin{cases} J_i^T(q_i)f_i = \tau_{i2} \\ \dfrac{d\tau_{i2}}{dt} = -T_i^{-1}\tau_{i2} + u_{i2} \end{cases} \tag{1.48}$$

In the general redundant joint space $\mathscr{L}_i$:

$$\begin{cases} [I_{n_i} - J_i^T(q_i)(J_i^T(q_i))^{\#}]\ddot{q}_i = \tau_{i3} \\ \dfrac{d\tau_{i3}}{dt} = -T_i^{-1}\tau_{i3} + u_{i3} \end{cases} \tag{1.49}$$

When the motor dynamics is considered, acceleration measurements are necessary for linearization and decoupling in the general position and redundant joint spaces [25]. In practice, it is very difficult to get accurate acceleration information. Therefore, the motor dynamics is considered only in the general force space and can be ignored in (1.47) and (1.49).

After ignoring the motor dynamics in the general position and redundant joint spaces, (1.47) and (1.49) can be rewritten as

$$D_i(q_i)J_i^{\#}(q_i)(\ddot{r}_i - \dot{J}_i(q_i)\dot{q}_i) + C_i(q_i, \dot{q}_i) + G_i(q_i) = T_i K_{ei}\dot{q}_i + T_i u_{i1} \tag{1.50}$$

and

$$[I_{n_i} - J_i^T(q_i)(J_i^T(q_i))^{\#}]\ddot{q}_i = T_i u_{i3} \tag{1.51}$$

Based on the well-known nonlinear feedback technique, the nonlinear controls

$$\begin{aligned}
u_{i1} &= T_i^{-1}(D_i(q_i)J_i^{\#}(q_i)V_{i1} - D_i(q_i)J_i^{\#}(q_i)\dot{J}_i(q_i)\dot{q}_i + C_i(q_i, \dot{q}_i) + G_i(q_i) - T_i K_{ei}\dot{q}_i)\\
u_{i2} &= (T_i^{-1}J_i^T(q_i) + \dot{J}_i^T(q_i))f_i + J_i^T(q_i)V_{i2}\\
u_{i3} &= T_i^{-1}[I_{n_i} - J_i^T(q_i)(J_i^T(q_i))^{\#}]V_{i3}
\end{aligned}$$

can linearize and decouple the systems (1.47), (1.48), and (1.49) in $\mathscr{Y}_i$, $\mathscr{F}_i$, and $\mathscr{L}_i$. The linearized models are

$$\begin{aligned}
\ddot{r} &= V_{i1}\\
\dot{f}_i &= V_{i2}\\
\ddot{q}_i &= V_{i3}
\end{aligned}$$

where $V_{i1}, V_{i2} \in \mathbb{R}^6$ and $V_{i3} \in \mathbb{R}^{n_i}$ are auxiliary inputs.

Since only n_i controls are available, this linearization and decoupling are not actually feasible in $\mathscr{T}_i$. But if we consider only the controllable subspace $\mathscr{T}_{ic}$, the linearization and decoupling can be achieved.

It can be proved that, if the position and force are not controlled in the same task space axis for a robot, then the nonlinear feedback control,

$$\begin{aligned}
u_i = T_i^{-1}(D_i(q_i)J_i^{\#}(q_i)\bar{V}_{i1} &- D_i(q_i)J_i^{\#}(q_i)\dot{J}_i(q_i)\dot{q}_i + C_i(q_i, \dot{q}_i)\\
&+ G_i(q_i) - T_i K_{e_i}\dot{q}_i) + T_i^{-1}J_i^T(q_i) + \dot{J}_i^T(q_i))f_i\\
&+ J_i^T(q_i)\bar{V}_{i2} + T_i^{-1}[I_{n_i} - J_i^T(q_i)(J_i^T(q_i))^{\#}]\bar{V}_{i3}
\end{aligned} \tag{1.52}$$

linearizes and decouples the dynamic system (1.46) in $\mathscr{T}_c$. The linearized and decoupled system is

$$\begin{aligned}
\ddot{\bar{r}}_i &= \bar{V}_{i1}\\
\dot{\bar{f}}_i &= \bar{V}_{i2}\\
\ddot{\bar{q}}_i &= \bar{V}_{i3}
\end{aligned}$$

where

$$\begin{bmatrix} \bar{r}_i \\ \bar{f}_i \\ \bar{q}_i \end{bmatrix} = B_{ic}^T \begin{bmatrix} r_i \\ f_i \\ q_i \end{bmatrix}$$

and

$$\begin{bmatrix} \bar{V}_{i1} \\ \bar{V}_{i2} \\ \bar{V}_{i3} \end{bmatrix} = B_{ic}^T \begin{bmatrix} V_{i1} \\ V_{i2} \\ V_{i3} \end{bmatrix}$$

Then the linear controllers

$$\begin{aligned} \bar{V}_{i1} &= \ddot{\bar{r}}_i^d(s) + k_{iv}(\dot{\bar{r}}_i^d(s) - \dot{\bar{r}}_i) + k_{ip}(\bar{r}_i^d(s) - \bar{r}_i) \\ \bar{V}_{i2} &= \dot{\bar{f}}_i^d(s) + k_{if}(\bar{f}_i^d(s) - \bar{f}_i) \\ \bar{V}_{i3} &= \ddot{\bar{q}}_i^d(s) + k_{i\dot{q}}(\dot{\bar{q}}_i^d(s) - \dot{\bar{q}}_i) + k_{iq}(\bar{q}_i^d(s) - \bar{q}) \end{aligned} \tag{1.53}$$

can be used to stabilize and control position, force, and redundant joints. All desired values in (1.53) are given by event-based task planners.

Several Remarks

1. Because the preceding control law can directly take task space commands, it is a task-level controller.
2. For a different task or a different system configuration, the structure of controler (1.52) is same. Changes must be made only in the task projection operator B_{ic}. In this respect, the controller structure is task independent and is suitable for multirobot systems working on complex tasks.
3. The management of redundancy consists of designing B_{ic} to achieve certain goals. In addition, once the redundant joints are determined, their motion can be planned and controlled based on some secondary optimization criteria, such as obstacle avoidance and load sharing [24, 59].

4.5 Experimental Results

Hybrid Position–Force Control

The preceding hybrid position–force control algorithm was implemented and tested on a 6-DOF PUMA 560 robot arm equipped with a FSA-3254 six-axis force–torque sensor. The force–torque was measured at a rate of 1000 Hz, and the filtered force–torque was computed at a rate of 500 Hz. The sampling rate for control and joint position–velocity measurement was 1000 Hz. In the experiments the positions in the x and y directions and orientation (O, A, T) were commanded. In the z direction, only force was commanded. Two different controllers, one with consideration of the joint motor dynamics (third-order model) and the other without consideration of the joint motor dynamics (second-order model), were used to perform various tasks.

In Figure 1.16, the constant desired force was tracked. It can be seen that the third-order controller gave better force tracking. In addition, both controllers maintained good x and y position tracking. Since the z position was not commanded, the error in the z direction was much bigger than that in x and y.

In Figure 1.17 and Figure 1.18, results for tracking variant desired forces are given. The advantage of the third-order controller was more obviously shown in these cases. The control law (1.43) was used in the second-order controller. Obviously, it took some time to establish the feedback from the integration term. As a result, it performed poorly when tracking a variant input.

The time constant of the motors for the first three joints of the PUMA 560 robot is 3 ms,

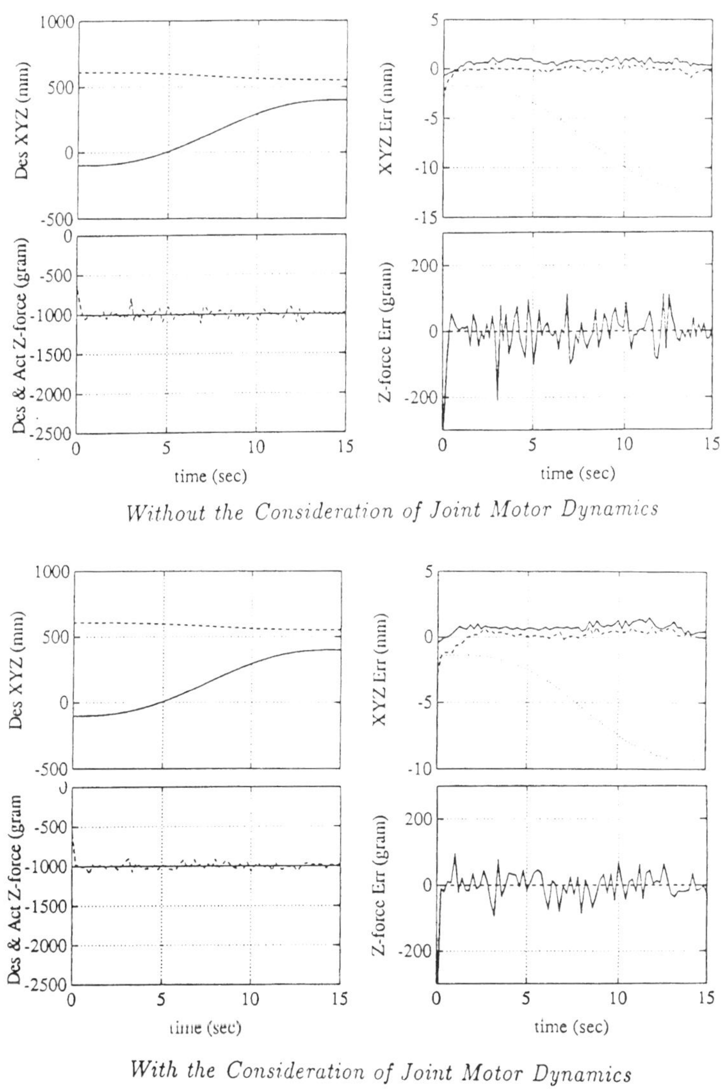

FIGURE 1.16
Constant desired force tracking.

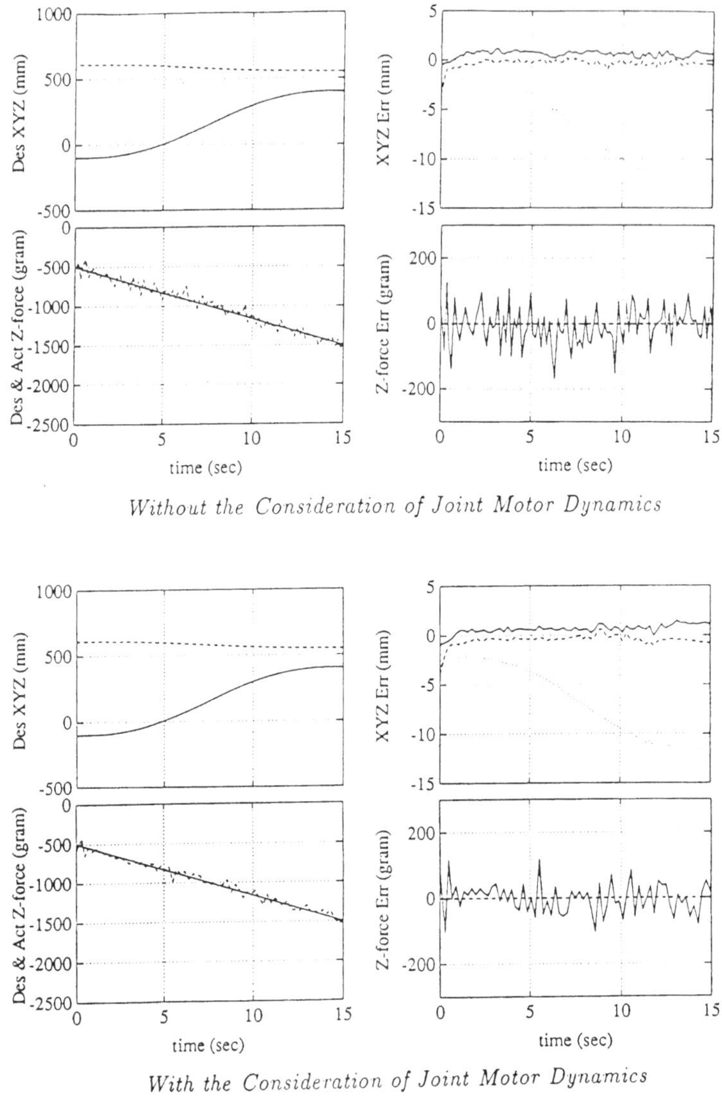

FIGURE 1.17
Ramp desired force tracking.

and that for the last three joints is 1 ms. Therefore, if the higher sampling rate were applied in force measurement, better results would be expected. However, the most important point is that the experimental results clearly demonstrate the significance of considering the joint motor dynamics in hybrid position–force control.

Dual-Robot Coordination

The given task, as shown in Figure 1.19, was to transport a carton, weighing 0.45 kg, by squeezing it. The sampling rate for position and velocity measurements was 1000 Hz and for force–torque was 1000 Hz. The feedback is computed at a rate of 1000 Hz.

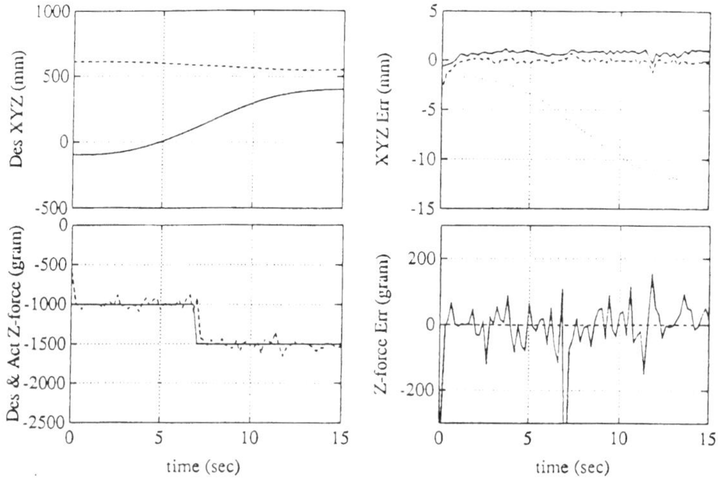

Without the Consideration of Joint Motor Dynamics

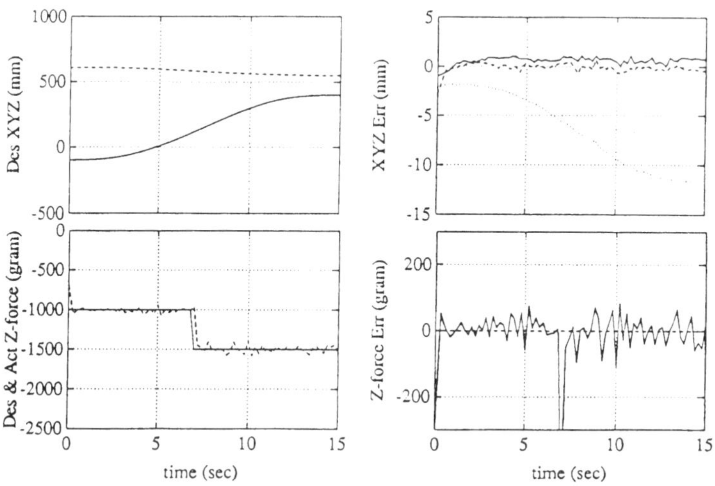

With the Consideration of Joint Motor Dynamics

FIGURE 1.18
Step desired force tracking.

In the following figures, the coordination error is defined as

$$e_{coord\ i} = ((x_1(s) - x_2(s))^2 + (y_1(s) - y_2(s))^2 + (z_1(s) - z_2(s))^2)^{\frac{1}{2}} - ((x_1^d(s) - x_2^d(s))^2 + (y_1^d(s) - y_2^d(s))^2 + (z_1^d(s) - z_2^d(s))^2)^{\frac{1}{2}}$$

which is the difference between the actual and planned distances of the two contact surfaces while squeezing the object.

In Figure 1.20, the motion is in the z direction and internal force is controlled in the y

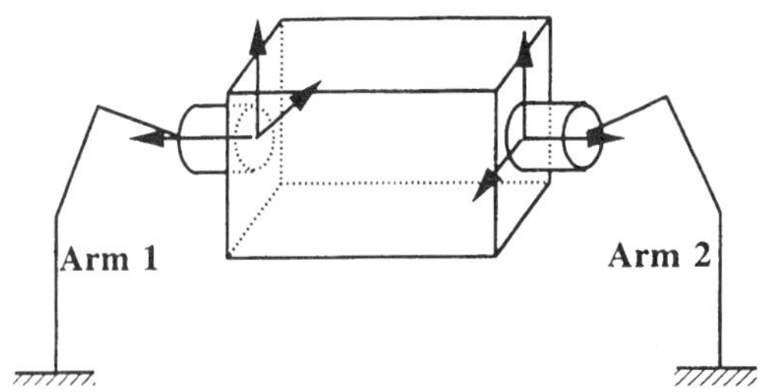

FIGURE 1.19
Transporting a carton by dual arms.

direction. The position and orientation of the left arm were controlled. For the right arm, the position in the x and z directions, the force in the y direction, and the orientation were controlled. It can be seen that the internal force was well maintained, and good coordination was achieved.

In Figure 1.21, both motion and internal force were in the y direction. Very good coordination control and internal force control were achieved.

Figure 1.22 compares the experimental results of using the event-based coordination scheme and using the undistinguished scheme with a time-based motion plan [24]. The motion direction is the same as the internal force direction. Therefore, the coordination error and internal force error are strongly coupled. It can be seen that the time-based scheme is unable to complete the task because the coordination and internal force errors increase without bounds. However, in the event-based scheme, the coordination and internal force are well maintained.

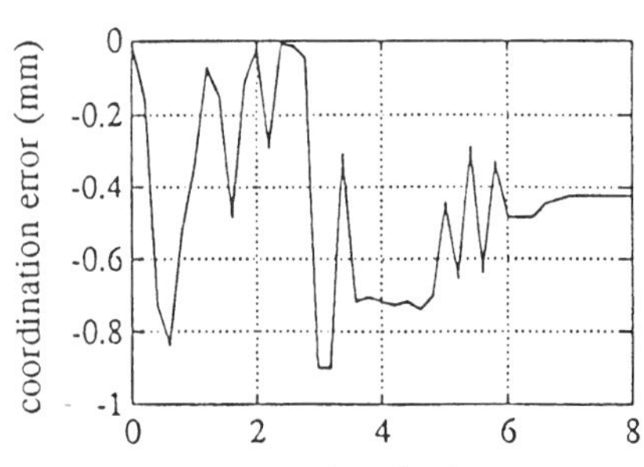

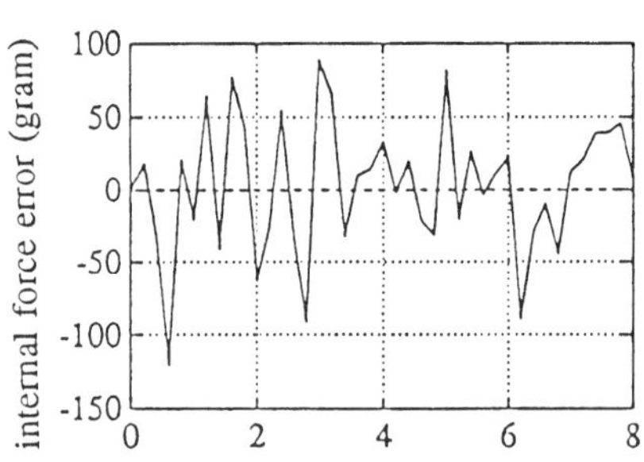

FIGURE 1.20
The motion direction orthogonal to the internal force direction.

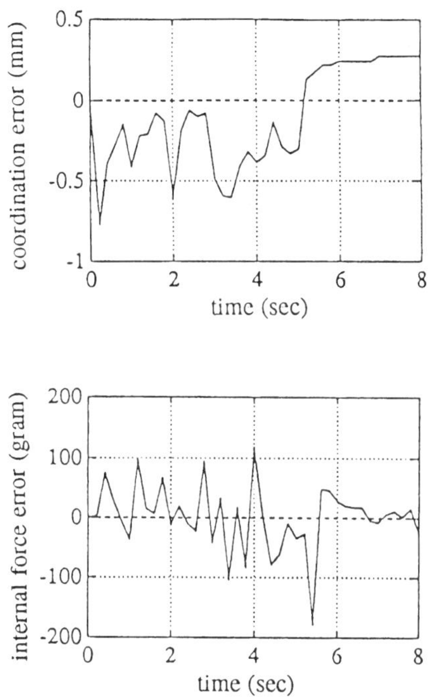

FIGURE 1.21
Motion and internal force in the same direction.

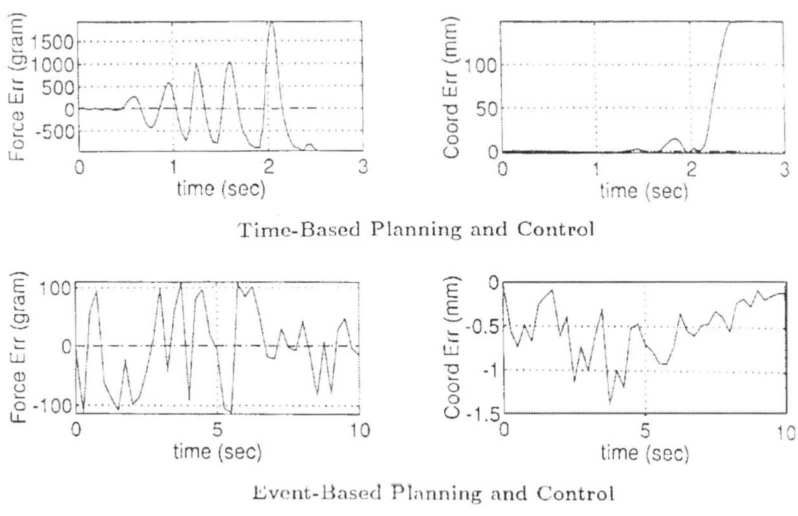

FIGURE 1.22
Comparison of time-based and event-based coordinated control.

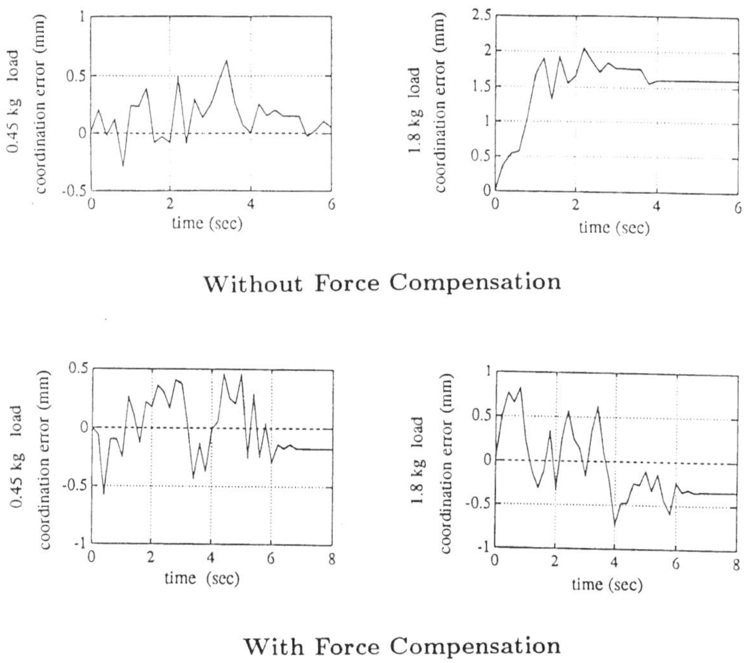

FIGURE 1.23
Compensation of unknown load by force–torque feedback.

In Figure 1.23, two different boxes, weighing 0.45 kg and 1.8 kg, were lifted by the dual-arm system without force feedback in their controllers. It is shown that an increase of the load significantly reduces the accuracy of the coordination. However, if force feedback is used, as shown in Figure 1.23, the effect of increased load on the coordination is very small. That is, without knowing the change of load, the use of the force feedback in the nonlinear feedback control law (1.52) can automatically compensate for the unknown load. Therefore, high-accuracy force measurement and feedback play important roles in a multirobot system coordinated control.

These experimental results demonstrate the advantages of the event-based coordination scheme. The significance of the event-based coordination scheme is that it can handle some unexpected events, and the control is carried out on the task level. Furthermore, the structure of the control system is task independent. It makes it possible for the multirobot system to work on complex tasks. Therefore, the event-based coordination scheme can be an important step toward the development of intelligent multirobot systems.

5 IMPLEMENTATION OF EVENT-BASED PLANNING AND CONTROL

5.1 Introduction

The practical implementation of the planning and control scheme is an important step in the development of robotic systems. It consists of two issues. First, one is developing a planning and control scheme that can be easily and efficiently implemented. The second issue is developing an efficient and user-friendly computing architecture for the implementation.

Recently, several distributed computing architectures have been proposed. A hierarchical multimicroprocessor system was designed to control the coordination of two PUMA robots in a master–slave mode [26]. The distributed operating system REKCOR (REal-time Kernel for COordinating Robots) was developed for a hierarchical computing structure [27]. The real-time communication issues in a computer-controlled robotic system are discussed in [28] and [29]. The local area network (LAN) was applied to coordinated control of a multirobot system [30].

Basically, there exist two communication and synchronization schemes, tight and loose coupling for distributed computing systems [31]. A tightly coupled computing system is characterized by its reliance on shared memory as a communication scheme and a single common operating system coordinating and synchronizing the interactions between processors. In contrast, the loosely coupled systems use message-based schemes in accordance with network communication protocols. These systems are often controlled by distributed operating environments. Of course, depending on the amount of coupling, these schemes could also be combined.

In addition, based on the distributed computing architectures, various parallel algorithms have been proposed to compute the robot dynamics [71–75]. In [65], an efficient Jacobian inversion algorithm was proposed. It is an important step to implementing the task-level control in real time. The parallelization of the nonlinear feedback control method was presented in [69].

It can be seen that the event-based planning and control scheme is developed with consideration of its implementability. As a result, the scheme lends itself naturally to a distributed computing architecture. The practical implementation of the event-based planning and control scheme will be discussed in the following sections.

5.2 Description of Experimental System

A dual-arm experimental system, Figure 1.24, is currently operational in the Center for Robotics and Automation at Washington University.

Two 6-DOF PUMA 560 robot arms [77] are equipped with FSA3254 force–torque sensors [70]. Each sensor can measure force–torque in six directions at a 1000-Hz rate.

Each robot arm is controlled by a universal motor controller (UMC) [76], which is capable of up to a 1000 Hz servo rate (without interpolation) on every axis. The UMC controller consists of five parts. The first part is the power supply module, which provides both logic and servo power. The second part is the joint interface module. It contains the pulse width modulation amplifier, which converts the logic-level signals to pulse width–modulated signals at a level appropriate for the particular motor being driven. The third part is the joint processor, which consists of a 10-MHz 32-bit NS 32016 microprocessor and a floating-point coprocessor. It is used to perform the robot arm calibration and configuration setup. The fourth part is the user processor, which is also based on the NS 32016 processor. It has 128 K RAM memory, which is used as a shared memory for communication between the UMC and other high-level computing devices. The last part is the expansion module, which contains the interface board to connect the user processor multibus with the VME bus of high-level computers.

The high-level computing device is a Silicon Graphics SGI 4D/340 VGX computer [78]. It has four symmetric R 3000/3010 RISC processors with a 33-MHz clock rate. It is capable of delivering up to 117 MIPS or 36 MFLOPS. An independent geometry engine provides a real-time graphics capability. The SGI computer is interfaced with both UMC controllers through a shared memory scheme. This allows extremely fast communication

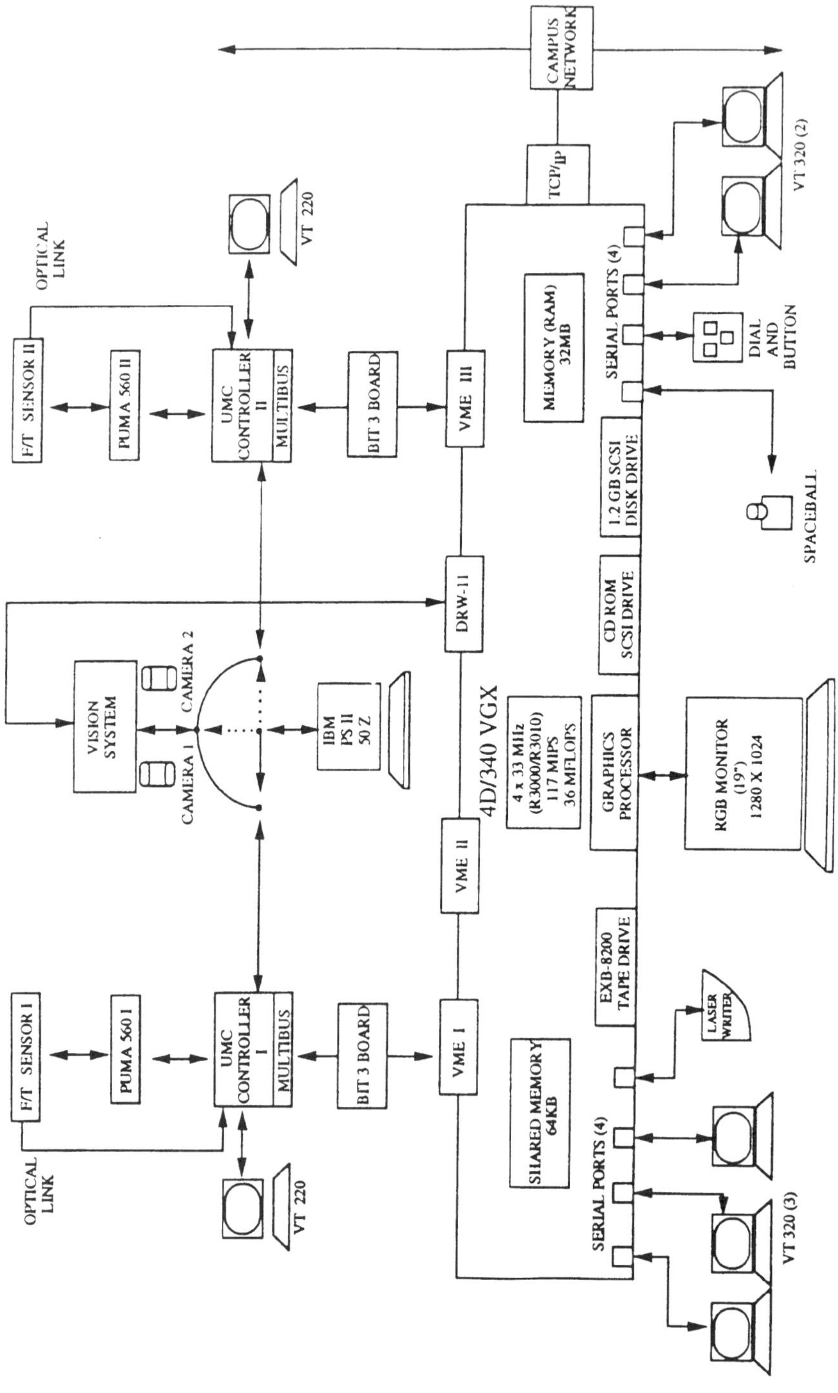

FIGURE 1.24
Dual-arm experimental system at Washington University.

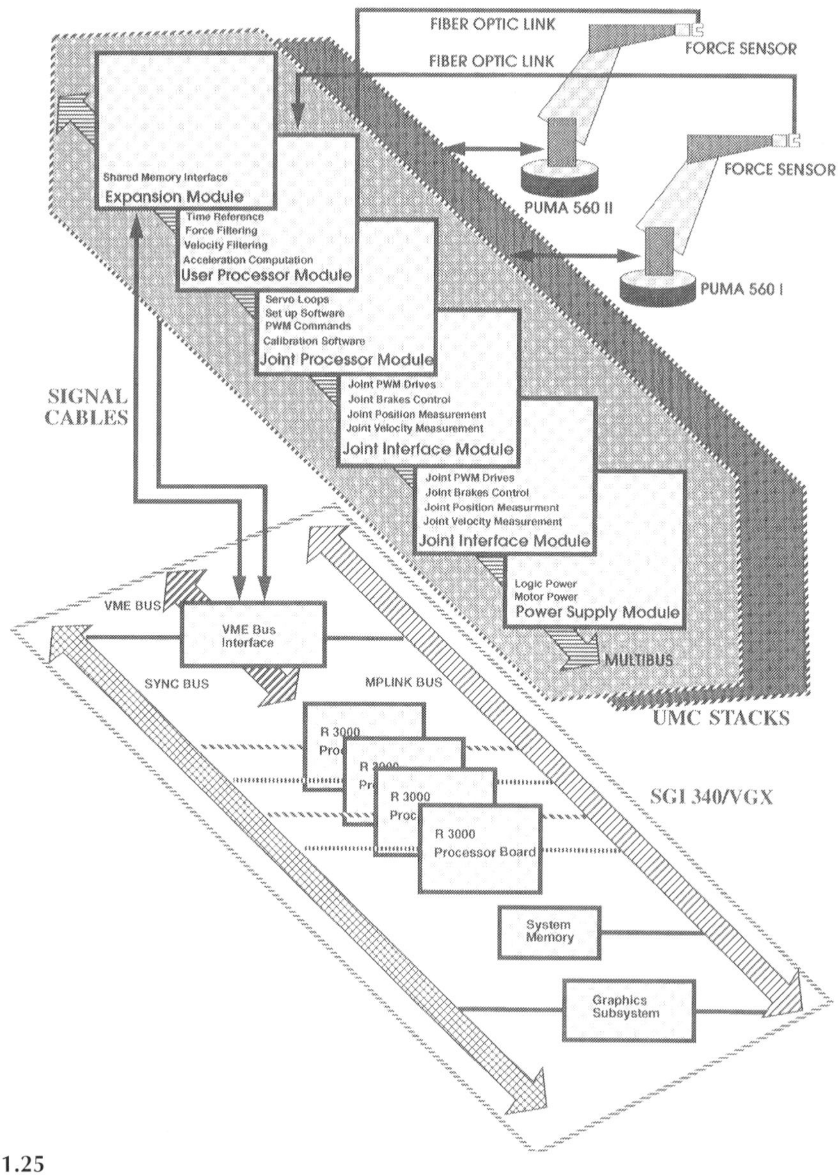

FIGURE 1.25
Bus structure of the control architecture.

between the SGI computer and the UMC controllers. The bus structure of the system is shown in Figure 1.25.

Thus, existing computing and control systems can easily form the basis of development of a distributed computing and control architecture for a multiarm robotic system.

5.3 Computation Load Analysis

In order to implement efficiently the planning and control schemes for multiarm systems, it is important to study the real-time computation load of single-arm planning and control. The

SGI computer provides a profiling tool that can produce detailed information about program execution. Using profiling tools, the areas of code where most of the execution time is spent can be found. In a typical program, a large part of the execution time is spent in relatively few sections of code. It is most profitable to concentrate on improving the implementation in those sections.

The nonlinear control with event-based planning for a single robot basically consists of four parts. First is the data acquisition. It includes getting the measurement data from the sensors and processing them, such as filtering and transforming joint space data to task space through forward kinematics. The second part is the event-based motion planning. It calculates the desired input for the robot based on the curent system outputs. The third part is the nonlinear feedback computation. The last part is for computing the control commands, sending them to the controller, and synchronizing the sampling and control according to the given sampling rate.

The results of computing load analysis for 15 seconds of real-time execution of event-based motion planning and nonlinear control of a single arm are shown in Figure 1.26. The computation was performed by a single processor. The technique used for analysis is the basic lock counting. A basic block is a sequence of instructions that is entered only at the beginning and exits only at the end. Measuring the execution of basic blocks provides statistics on the load of computation.

It can be seen in Figure 1.26 that

- The total number of program cycles in 15 seconds is 225235635.
- The subroutine for joint velocity filtering, *vel filter*, used 118095744 cycles, which is 52.43% of the total number of program cycles. Total execution time for *vel filter* is 7.8730 seconds. It is more than half of the total computing time.
- The *vel filter* used an average of 656 cycles per call and consisted of 1073 bytes of generated code per line of source text. It is in the source file *test_phase.c*.
- The cumulative total of all cycles used by *vel filter* and *nlf* (nonlinear feedback) is 70.61%. They are the major part of the computation load.
- The subroutine *read_4bu* is for reading the clock. The idle time of the CPU was spent in executing *read_4bu*. The total executing time reflects the idle time of the CPU. It can be seen that the idle time is less than 0.1644 second, which is less than 1.09% of the total computing time. Therefore, the CPU was almost saturated.

5.5 Distributed Computing Architecture

Based on the preceding analysis, a single processor will not be able to perform the real-time computation for a multiarm system, which requires additional computing resources to calculate the multiarm motion reference and event-based planning and nonlinear control for additional robots. The total computing time required will be more than the computing time for a single arm multiplied by the number of arms. Therefore, a distributed computing architecture is needed to implement the event-based planning and control for a multiarm system.

The event-based coordination scheme can be conveniently implemented in a distributed computing arhitecture. As shown in Figure 1.15, all controllers and planners are independent entities. Only information exchanges are required for the "motion reference" block to receive the outputs of robots and to send out s. Therefore, it can easily be implemented in a memory sharing bus network, Figure 1.27. The planning and control algorithms for different robots

```
Profile listing generated Fri Jun 25 16:31:46 1993 with:
   prof -pixie a.out a.out.Addrs a.out.Counts

----------------------------------------------------------------------------
*  -p[rocedures] using basic-block counts;                                 *
*  sorted in descending order by the number of cycles executed in each     *
*  procedure; unexecuted procedures are excluded                           *
----------------------------------------------------------------------------

225235635 cycles

    cycles %cycles  cum %      cycles  bytes procedure (file)
                                /call  /line

 118095744   52.43  52.43         656   1072 velfilter (test_phase.c)
  40940458   18.18  70.61        2729  10924 nlf (test_phase.c)
  19007544    8.44  79.05        1268   4744 getstate (test_phase.c)
   8388550    3.72  82.77          63      6 sinf (fsincos.s)
   7000779    3.11  85.88         467   1276 stiction_comp (test_phase.c)
   5593388    2.48  88.36         373   2052 fforward (test_phase.c)
   4195470    1.86  90.23          11     22 fabs (fabs.c)
   3241728    1.44  91.67          18     72 read_4b (test_phase.c)
   2836176    1.26  92.92          21      5 cosf (fsincos.s)
   2740274    1.22  94.14         274   1572 goto_init (test_phase.c)
   2465504    1.09  95.24          16     64 read_4bu (test_phase.c)
   1984268    0.88  96.12          22     88 write_4b (test_phase.c)
   1590424    0.71  96.82           7      6 atanf (atanf.s)
   1326160    0.59  97.41          89    708 gravity_comp (test_phase.c)
   1205977    0.54  97.95          14      6 rint (rint.s)
   1170312    0.52  98.47          26      7 atan2f (atanf.s)
    896161    0.40  98.86         180   1280 stline_phase_planner (test_phase.c)
    545509    0.24  99.11      545509     28 main (test_phase.c)
    455091    0.20  99.31          91    364 save_record (test_phase.c)
    425085    0.19  99.50          85    340 save_record6 (test_phase.c)
    415083    0.18  99.68          83   1028 orin_phase_planner (test_phase.c)
    330573    0.15  99.83      330573   1876 init_st (test_phase.c)
    210084    0.09  99.92          42      6 sqrtf (sqrtf.s)
     63865    0.03  99.95          48     31 _flsbuf (flsbuf.c)
     45175    0.02  99.97         886     18 _doprnt (doprnt.c)
     28622    0.01  99.98        1061     19 _filbuf (filbuf.c)
      6182    0.00  99.99         229     34 number (doscan.c)
      5133    0.00  99.99         191     31 _doscan (doscan.c)
      4518    0.00  99.99         226      5 _dtoa (gen/dtoa.s)
      2814    0.00  99.99         469   1876 umc_grav (test_phase.c)
      2565    0.00  99.99          45     32 _xflsbuf (flsbuf.c)
      2443    0.00  99.99          25     12 fclose (flsbuf.c)
      1998    0.00  99.99          37     17 ungetc (ungetc.c)
      1544    0.00 100.00          78     11 atof (atof.c)
      1530    0.00 100.00          30     20 printf (printf.c)
      1371    0.00 100.00          69     23 cvt (doprnt.c)
      1215    0.00 100.00        1215     19 _cleanup (flsbuf.c)
      1102    0.00 100.00         551   2204 getinit (test_phase.c)
      1084    0.00 100.00          39     26 fflush (flsbuf.c)
       645    0.00 100.00         108      5 _atod (gen/atod.s)
       603    0.00 100.00          31      5 _dwmultu (gen/dwmultu.s)
       486    0.00 100.00          18     36 scanf (scanf.c)
       450    0.00 100.00          30      6 _tenscale (gen/tenscale.s)
       416    0.00 100.00         104    416 umc_com (test_phase.c)
       384    0.00 100.00          20      6 _fp_class_d (gen/_fp_class.s)
       342    0.00 100.00           6      7 _write (sys/_write.s)
       162    0.00 100.00           6      7 _read (sys/_read.s)
       117    0.00 100.00          13     22 nvmatch (getenv.c)
       112    0.00 100.00         112     18 getenv (getenv.c)
        89    0.00 100.00          45     23 _findbuf (flsbuf.c)
        60    0.00 100.00           5     20 write_2bu (test_phase.c)
        36    0.00 100.00          36      4 strcpy (gen/strcpy.s)
        36    0.00 100.00          18     16 _isatty (_isatty.c)
        33    0.00 100.00          33    156 sis_open (test_phase.c)
        32    0.00 100.00          32     23 _wrtchk (flsbuf.c)
        26    0.00 100.00          26     17 _setchrclass (ctype.c)
        26    0.00 100.00          26      5 __start (crt1text.s)
        18    0.00 100.00           6      7 _close (sys/_close.s)
        13    0.00 100.00          13     28 __call_exitfns (atexit.c)
        12    0.00 100.00           6      7 _ioctl (sys/_ioctl.s)
        10    0.00 100.00          10     11 exit (gen/cuexit.c)
         8    0.00 100.00           4     16 _oserror (_oserror.c)
         6    0.00 100.00           6     16 open (sys/open.s)
         6    0.00 100.00           6     16 mmap (sys/mmap.s)
         2    0.00 100.00           2      8 __readenv_sigfpe (gen/stubfpestart.c)
         2    0.00 100.00           2      8 _exit (sys/exit.s)
```

FIGURE 1.26
Computing load analysis of event-based planning and control for a single arm.

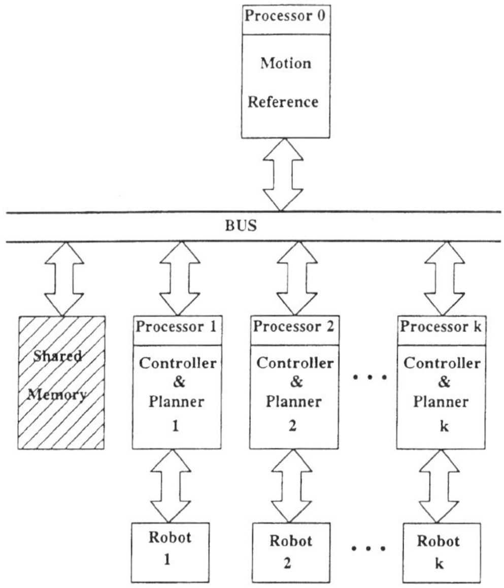

FIGURE 1.27
Memory-sharing bus network.

can be computed in separate processors in SGI, which run in parallel. The motion reference is computed in another processor. Based on the computation load profile in the last part, the bottleneck of the computation is the joint velocity filtering. Part of the velocity filtering will be passed to the user processor in the UMC controller. Because of the limitation of the computation speed of the UMC user processor, the SGI still has to perform part of the joint velocity filtering. Through the high-speed bus, all processors in UMC and SGI use write–read shared memory to exchange information, as shown in Figure 1.27. It combines the tightly and loosely coupled schemes. All processors communicate through a shared memory, but the UMC controller and SGI have their own operating systems and run in an asynchronous manner. This scheme ensures high communication speed and at the same time also simplifies the programming.

It can be seen that changing the number of robots in the system does not affect the overall data processing and computation structure. This gives great flexibility to the multiarm system. The control and planning algorithms are executed in parallel. As a result, the overall computation time will not be significantly changed by increasing the number of robots in the system. Since the event-based coordination scheme naturally lends itself to a distributed computing architecture, it is very efficient for real-time computation.

5.5 Several Issues in Practical Implementation

Robot Dynamic Model

In the dynamic model of the PUMA 560 robot (1.25),

$$D(i, j) = D(j, i)$$

and

$$C(i, j, k) = C(i, k, j)$$
$$C(i, k, k) = -C(k, j, i), \quad i, k \geqslant j$$
$$C(i, j, j) = 0, \quad i \geqslant j$$

After neglecting the less significant terms [67], the nonzero terms are given as follows.

Inertial terms ($Kg - m^2$):

$$D(1, 1) = 2.57 + 1.38C_2C_2 + 0.3S_{23}S_{23} + 0.74C_2S_{23}$$
$$D(1, 2) = S_2 - 0.0057S_{23} - 0.1367C_{23}$$
$$D(1, 3) = -0.0057S_{23} - 0.1367C_{23}$$
$$D(2, 2) = 6.79 + 0.74S_3$$
$$D(2, 3) = 0.3679 + 0.3922S_3 - 0.0134C_3$$
$$D(3, 3) = 1.16$$
$$D(4, 4) = 0.2$$
$$D(5, 5) = 0.18$$
$$D(6, 6) = 0.19$$

Coriolis terms ($Kg - m^2$):

$$C(1, 1, 2) = 0.0174 - 1.362C_2S_2 + 0.3562C_3S_3 - 0.7124S_2S_3S_{23} + 0.0268C_2S_{23} + 0.3922C_2C_{23} - 0.3922S_2S_{23} - 0.046C_{22} - 0.0347C_{23}C_{23} - 0.0112S_3$$
$$C(1, 1, 3) = 0.0174 + 0.3562C_2S_2 + 0.3562C_3S_3 - 0.7124S_2S_3S_{23} + 0.0134C_2S_{23} + 0.3922C_2C_{23} - 0.0347C_{23}C_{23}$$
$$C(1, 2, 2) = 1.8181C_2 + 0.1367S_{23} - 0.0057C_{23}$$
$$C(1, 2, 3) = 0.1367S_{23} - 0.0057C_{23}$$
$$C(2, 2, 3) = 0.3922C_3 + 0.0134S_3$$

Gravity terms ($N - m$):

$$G(1) = 0$$
$$G(2) = -99.8C_2 - 3.73S_2 + 1.08C_{23} - 26.64S_{23}$$
$$G(3) = 1.08C_{23} - 26.64S_{23}$$
$$G(4) = 0.085S_{23}S_4S_5$$
$$G(5) = -0.085(C_{23}S_5 + S_{23}C_4C_5)$$
$$G(6) = 0$$

Most of the experiments were carried out by using the preceding dynamic model. Our experimental results, however, suggest that when the velocity of the robot is confined within the limit set by the manufacturer, neglecting the Coriolis and centripetal forces does not cause performance to deteriorate [35].

Sticktion and Friction Compensation

The joint sticktion and friction are not incorporated in the robot dynamic model. They are compensated by augmenting the voltage commands to the controllers.

PUMA 560 is driven by brush-type joint motors. They typically have a breakaway friction that is about 4 to 6% of the full rated torque of the motor that sets a lower bound on the friction values. The sticktion and friction parameter values were experimentally determined [68].

The compensation of sticktion plays an especially important role in event-based planning and control. It can be seen in Figure 1.3 that the initial feedforward and feedback are zero, since the desired acceleration is zero with no position and velocity errors according to the event-based plan (Figure 1.3). Therefore, the sticktion must be well compensated, and some initial error has to be given in order to start a motion. The modified motion plan creates a velocity error about 10% of the maximum planned velocity at the beginning of the motion. Experimental results show that this scheme can start the motion smoothly and also can overcome the incomplete compensation of the sticktion.

Joint Velocity Measurement Filtering

The joint velocity measurements are derived from the joint angle measurements by differentiation. As a result, the joint velocity measurements are very noisy and cannot be directly used for feedback control.

A multistep velocity filter is designed to estimate the joint velocity measurement. Each step calculates the estimate of joint velocity as

$$\hat{v}(k) = \left(\sum_{i=k-n+1}^{k-1} \hat{v}(i) + v(k) - \max V_k - \min V_k \right) \Big/ (n-2)$$

where

$$V_k = \{v(k), \hat{v}(k-1), \ldots \hat{v}(k-n+1)\}$$

The estimate of the velocity is a running average of prior estimates and new measurements after removing the maximum and minimum values. The window of filtering is n.

Each filter can run independently without synchronizing with the others. It is convenient to implement filters in the distributed computing architecture as shown in Figure 1.25.

In the experiments presented in Sections 3 and 4, three filter units were used for each joint velocity. One unit was implemented in the UMC user processor. The other two units were implemented in SGI.

The main focus of this section has been the implementation of an event-based planning and control scheme in a distributed computing architecture. The system that is described in this section has proved to be an extremely successful research tool for dual-arm coordinated control.

Apart from being versatile, robust, and open ended in its architecture, it can provide sensory information at various levels, according to the requirements of the control strategy.

In particular, the experiments reported in previous sections conclusively demonstrate the effectiveness of the system. Another important feature of the system is that it can also accommodate a variety of control strategies besides the event-based scheme. All control strategies may be implemented easily by coding in high-level languages such as the C programming language.

The computing and control architecture demonstrated here can clearly pave the way for a new generation of commercial robot controllers that are more responsive, more flexible, more efficient, and more robust.

6 CONCLUSIONS

In this research an integrated event-based planning and control method has been developed using a motion reference variable other than time. It has been successfully applied to single-robot arm motion planning and control, as well as multirobot coordination planning and control. The important contributions of this research are as follows:

1. A new planning and control scheme—event-based planning and control—has been developed. Instead of time, the events of a system are used as a motion reference to describe the motion plan and to drive the system. The time is implicitly included inside the motion plan and control process. Therefore, the planning along with the feedback control becomes a real-time dynamic process. The planner becomes an investigation–decision component of the system. As a result, the system has the ability to deal with unexpected and uncertain events. The event-based planning and control method can be an important step toward the development of intelligent planning and control theory.
2. The event-based planning and control theory has been applied to single-robot motion planning and control. An event-based representation of robot arm motion in the task space is proposed. The time and energy optimization techniques are used to determine event-based trajectories. A new event-based error definition and computation scheme has been introduced and combined with a nonlinear feedback control law, which linearizes and decouples the control in the task space.

   The significance of the event-based motion planning and control scheme is its compatibility with sensor-based planning and control, because sensed events in robotic manipulation rarely, if ever, occur on a precise time scale. Obstacle avoidance is an important example. Therefore, it can significantly improve the safety and reliability of robotic systems.

   The event-based planning and control schemes were experimentally implemented and tested on the 6-DOF position and orientation control of a PUMA 560 robot arm with very good results.
3. Based on the event-based planning and control theory, a new coordination scheme for a multirobot system has been developed. The event-based motion reference for the planning and control of coordinated robots has been introduced. It drives the system to achieve optimal coordination. After introducing the general task space, the task-independent controllers for each robot have been designed. The significance of the event-based coordination scheme is that the controls of robots are carried out on the task level and the structure of the control system is task independent. This makes it possible for the multirobot system to work on complex tasks. In addition, the scheme

can be implemented in a distributed computing architecture, which increases the efficiency and flexibility of multirobot system operation.

The event-based coordination scheme has been experimentally implemented and tested for the coordinated control of two 6-DOF PUMA 560 robots with very good results.

4. A new hybrid position–force controller has been developed. It incorporates the robot dynamics as well as the dynamics of robot joint motors. A new stability conclusion has been obtained. It requires the consideration of joint motor dynamics for the design of a stable and improved hybrid position–force controller. The inclusion of the joint motor dynamics improves the force control performance, in particular in the case of tracking the desired variable force.
5. The event-based multirobot coordination scheme is developed with consideration of the implementation. It naturally lends itself to a distributed computing architecture. Based on the existing computing facility at the Center for Robotics and Automation in Washington University, a distributed computing and control architecture has been developed to implement the event-based coordination scheme. It combines the tightly and loosely coupled schemes. High-speed information exchange (1000 Hz) has been achieved. In addition, all processors can be programmed separately. This provides a convenient user interfacing method.

REFERENCES

[1] J. C. Latombe, *Robot Motion Planning*, Kluwer Academic Publishers, Boston, 1991.

[2] Y. K. Hwang and N. Ahuja, Gross motion planning — a survey, *ACM Computing Survey*, Sept. 1992.

[3] T. Lozano-Perez and M A. Wesley, An algorithm for planning collision-free paths among polyhedral obstacles, *ACM Communications*, Vol. 22, No. 10, pp. 560–570, Oct. 1979.

[4] O. Khatib, Real-time obstacle avoidance for manipulators and mobile robots, *International Journal of Robotics Research*, Vol. 5, No. 1, pp. 90–98, 1986.

[5] K. Fujimura, *Motion Planning in Dynamic Environments*, Springer-Verlag, Tokyo, 1991.

[6] B. Donald and J. Jennings, Sensor interpretation and task-directed planning using perceptual equivalence classes, *Proceedings of the 1991 IEEE International Conference on Robotics and Automation*, Sacramento, CA, April 1991.

[7] A. K. Bejczy and Z. F. Szakaly, A harmonic motion generator for telerobotic applications, *Proc. of IEEE Int. Conf. on Robotics and Automation*, Sacramento, CA, April 1991, pp. 2032–2039.

[8] F. Pfeiffer and R. Johanni, A concept for manipulator trajectory planning, *IEEE Journal of Robotics and Automation*, Vol. RA-3, No. 2, April 1987, pp. 115–123.

[9] K. G. Shin and N. D. MeKay, Minimum-time control of robotic manipulators with geometric path constraints, *IEEE Trans. Automatic Contr.*, Vol. AC-30, pp. 531–541, June 1985.

[10] J. E. Bobrow, S. Dubowsky, and J. S. Gibson, Time-optimal control of robotic manipulators along specified paths, *Int. J. Robot Res.*, Vol. 4, pp. 3–17, Fall 1985.

[11] M. Sampei, T. Tamura, T. Itoh, and M. Nakamichi, Path tracking control of Tralier-like mobile robot, IROS '91.

[12] O. Dahl, Path following for a flexible joint robot, Preprints of the IFAC/IFIP/IMACS Symposium on Robot Control, Vienna, Sept. 16–18, 1991, pp. 261–266.

[13] K. S. Fu, Learning control systems and intelligent control systems: An intersection of artificial intelligence and automatic control, *IEEE Trans. Automatic Control*, Vol. AC-16, 1971.

[14] G. N. Saridis, Knowledge implementation: Structure of intelligent control system, *Proceedings of IEEE International Symposium on Intelligent Control*, pp. 9–19, 1987.

[15] G. N. Saridis and K. P. Valavanis, Analytic design of intelligent machines, *IFAC Journal Automatica*, Vol. 24, No. 2, 1982, pp. 123–133.

[16] J. S. Albus, Hierarchical interaction between sensory processing and world modeling in intelligent systems, *Proceedings of 5th IEEE Symposium on Intelligent Control*, Philadelphia, 1990, pp. 53–59.

[17] B. P. Zeigler, *Object-Oriented Simulation with Hierarchical, Modular Models: Intelligent Agents and Endomorphic Systems*, Academic Press, San Diego, 1990.
[18] P. J. Antsaklis, K. M. Passino, and S. J. Wang, Towards intelligent autonomous control systems: architecture and fundamental issues, *Journal of Intelligent and Robotics Systems*, Vol. 1, No. 4, 1989, pp. 315–342.
[19] W. Hahn, *Stability of Motion*, Springer-Verlag, New York, 1967.
[20] Z. Li, T. J. Tarn, and A. K. Bejczy, Dynamic workspace analysis of multiple cooperating robot arms, *IEEE Trans. on Robotics and Automation*, Vol. 7, No. 5, Oct. 1991.
[21] Z. Li, T. J. Tarn, A. K. Bejczy, and B. K. Ghosh, Motion space analysis of an object handled by two robot arms, *Proc. of the 28th IEEE Conference on Decision and Control*, Tampa, FL, Dec. 13–15, 1989.
[22] A. E. Bryson, Jr. and Y. C. Ho, *Applied Optimal Control: Optimization, Estimation and Control*, John Wiley & Sons, New York, 1975.
[23] J. Y. S. Luh and Y. F. Zheng, Constrained relations between two coordinated industrial robots for motion control, *International Journal of Robotics Research*, Vol. 6, No. 3, Fall 1987.
[24] A. K. Ramadorai, T. J. Tarn, and A. K. Bejczy, Task definition, decoupling and redundancy resolution by nonlinear feedback in multi-robot object handling, *Proc. of IEEE Conf. on R & A*, Nice, France, May 1992.
[25] T. J. Tarn, A. K. Bejczy, X. Yun, and Z. Li, Effect of motor dynamics on nonlinear feedback robot arm control, *IEEE Transactions on Robotics and Automation*, Vol. 7, No. 1, Feb. 1991.
[26] C. O. Alford and S. M. Belyey, Coordinated control of two robot arms, Proceedings of 1984 *IEEE International Conference on Robotics and Automation*, March 13–15, 1984, Atlanta, GA, pp. 468–473.
[27] Y. F. Zheng, J. Y. S. Luh, and P. F. Jia, A real-time distributed computer system for coordinated-motion control of two industrial robots, *Proceedings of 1987 IEEE International Conference on Robotics and Automation*, March 31–April 3, 1987, Raleigh, NC, pp. 1236–1241.
[28] K. G. Shin and M. E. Epstein, Intertask communication in an integrated multirobot system, *IEEE Journal of Robotics and Automation*, Vol. RA-3, pp. 90–100, 1991.
[29] K. G. Shin, Real-time communications in a computer-controlled workcell, *IEEE Transactions on Robotics and Automation*, Vol. 7, pp. 105–113, 1991.
[30] Q. Yin and Y. F. Zheng, Performance analysis of token bus LAN in coordinating multiple robots, *Proc. of IEEE Conference on Robotics & Automation*, Nice, France, 1992.
[31] D. Gauthier, P. Freedman, G. Carayannis, and A. Malowany, Interprocess communication for distributed robotics, *IEEE J. of Robotics & Automation*, Vol. RA-3, No. 6, Dec. 1987.
[32] T. J. Tarn, A. K. Bejczy, A. K. Isidori, and Y. L. Chen, Nonlinear feedback in robot arm control, *Proc. of 23rd IEEE Conference on Decision and Control*, Las Vegas, 1984.
[33] T. J. Tarn, A. K. Bejczy, and X. Yun, Coordinated control of two arm robots. *Proc. of IEEE International Conference on Robotics and Automation*, San Francisco, 1986, pp. 1193–1202.
[34] T. J. Tarn, A. K. Bejczy, and X. Yun, Robot arm force control through system linearization by nonlinear feedback, *Proc. of IEEE Conf. on R & A*, Philadelphia, 1988, pp. 1618–1625.
[35] T. J. Tarn, A. K. Bejczy, S. Ganguly, A. K. Ramadorai, and G. T. Marth, Experimental evaluation of the nonlinear feedback robot controller, *Proceedings of the 1991 IEEE Conference of R & A*, Sacramento, April 1991, pp. 1638–1644.
[36] M. H. Railbert and J. J. Craig, Hybrid position/force control of manipulators, *Trans of the ASME*, Vol. 102, June 1981, pp. 126–133.
[37] M. T. Mason, Compliance and force control for computer controlled manipulators, *IEEE Trans. on Systems, Man and Cybernetics*, Vol. SMC-11, No. 6, June 1981, pp. 418–432.
[38] S. Hayati, Hybrid position/force control of multi-arm cooperating robots, *Proc. of IEEE Conf. on R & A*, San Francisco, 1986, pp. 82–89.
[39] O. Khatib, A unified approach for motion and force control of robot manipulators: The operational space formulation, *IEEE Journal of R & A*, Vol. RA-3, No. 1, Feb. 1987, pp. 43–53.
[40] Y. Nakamura, K. Nagai, and T. Yoshikawa, Mechanics of coordinate manipulation by multiple robotic systems, *Proc. of IEEE Conf. on R & A*, Raleigh, 1987, pp. 991–998.
[41] T. Yoshikawa, Dynamic hybrid position/force control of robot manipulators — Description of hand constraints and calculation of joint driving force, *IEEE Trans. on R & A*, Vol. RA-3, No. 5, October 1987, pp. 386–392.
[42] T. Yoshikawa, T. Sugie, and M. Tanaka, Dynamic hybrid position/force control of robot manipulators — Controller design and experiment, *IEEE Journal of Robotics and Automation*, Vol. 4, No. 6, December 1988, pp. 699–705.
[43] M. Uchiyama and P. Dauchez, A symmetric hybrid position/force control scheme for the coordination of two robots, *Proc. of IEEE Conf. on R & A*, Philadelphia, 1988, pp. 350–356.
[44] D. J. Montana, The kinematics of contact and grasp, *Int. Journal of Robotics Res.*, Vol. 7, No. 3, June 1988, pp. 17–32.

[45] H. Liu, T. Iberall, and G. A. Bekey, The multidimensional quality of task requirements for dextrous robot hand control, *Proc. of Int. Conf. on R & A*, Scottsdale, 1989, pp. 452–457.
[46] Z. Li, P. Hsu, and S. Sastry, Grasping and coordinated manipulation by a multifingered robot hand, *Int. Journal of Robotics Res.*, Vol. 8, No. 4, Aug. 1989, pp. 33–49.
[47] Z. Li and S. Sastry, Task-oriented optimal grasping by multifingered robot hands, *IEEE Journal of Robotics and Automation*, Vol. 4, No. 1, Feb. 1988, pp. 32–44.
[48] X. Yun, Nonlinear feedback control of two manipulators in presence of environmental constraints, *Proc. of IEEE Int. Conf. on R & A*, Scottsdale, 1989, pp. 1252–1257.
[49] Y. F. Zheng and J. Y. S. Luh, Control of two coordinated robots in motion, *Proceedings of 1985 IEEE International R & A Conference*, St. Louis, pp. 1761–1766.
[50] X. Yun, Coordination of two-arm pushing, *Proc. of IEEE Int. Conf. on R & A*, Sacramento, 1991, pp. 182–187.
[51] C. D. Kopf and T. Yabuta, Experimental comparison of master/slave and hybrid two-arm position/force control, *Proceedings of the 1988 IEEE R & A Conference*, Philadelphia, pp. 1633–1637.
[52] A. J. Koivo and M. A. Unseren, Reduced order model and decoupled control architecture for two manipulators holding a rigid object, *Transactions of the ASME*, Vol. 113, Dec. 1991, pp. 646–654.
[53] P. Hsu, J. Hauser, and S. Sastry, Dynamic control of redundant manipulators, *Journal of Robotic Systems*, 6(2), 1989, pp. 133–148.
[54] A. De Luca, *Zero Dynamics in Robotics Systems, Nonlinear Synthesis*, Progress in Systems and Control Series, Birkhauser, Boston, 1991, pp. 68–87.
[55] Y. Nakamura, H. Hanafusa, and T. Yoshikawa, Task priority based redundancy control of robot manipulators, *Int. Journal of Robotics Res.*, Vol. 6, No. 2, 1987, pp. 3–15.
[56] A. A. Maciejewski and C. A. Klein, Obstacle avoidance for kinematically redundant manipulators in dynamically varying environments, *Int. Journal of Robotics Res.*, Vol. 4, No. 3, 1985, pp. 109–117.
[57] A. De Luca and G. Oriolo, Efficient dynamic resolution of robot redundancy, *Proc. of American Control Conference*, Boston, 1991, pp. 221–227.
[58] F. A. Mussa-Ivaldi and N. Hogan, Integrable solutions of kinematic redundancy via impedance control, *Int. Journal of Robotics Research*, Vol. 10, No. 5, Oct. 1991, pp. 481–491.
[59] Y. F. Zheng and J. Y. S. Luh, Optimal load distribution for two industrial robots handling a single object, *Proc. of IEEE Int. Conf. on R & A*, Philadelphia, 1988, pp. 344–349.
[60] W. B. Gao, M. Cheng, and D. Xiao, Force control for tracking a set of tasks in presence of constraints, *Proc. of IFAC Symposium on Robot Control*, SYROCO '91, 1991, pp. 389–394.
[61] A. Isidori, *Nonlinear Control Systems*, 2nd edition, Springer-Verlag, New York, 1989.
[62] T. J. Tarn, A. K. Bejczy, G. T. Marth, and A. K. Ramadorai, Kinematic Characterization of the PUMA 560 manipulator, Lab Report, SSM-RL-91-15, Dept. of SSM, Washington University, St. Louis, Dec. 1991.
[63] V. Kumar, X. Yun, E. Paljung, and N. Sarkar, Control of contact conditions for manipulation with multiple robotic systems, *Proceedings of 1991 IEEE Int. Conference of R & A*, Sacramento, April 1991, pp. 170–175.
[64] Z. Li, T. J. Tarn, and A. K. Bejczy, Dynamic workspace analysis of multiple cooperating robot arms, *IEEE Transactions on R & A*, Vol. 7, No. 5, Oct. 1991, pp. 589–596.
[65] A. Fijany and A. K. Bejczy, Efficient Jacobian inversion for the control of simple robot manipulators, *Proceedings of IEEE Int. R & A Conference*, Philadelphia, 1988, pp. 999–1007.
[66] Z. F. Szakaly and A. K. Bejczy, Performance capabilities of a JPL dual arm advanced teleoperation system, *Proceedings of SOAR 1990 Workshop*, Albuquerque, NM.
[67] B. Armstrong, O. Khatib, and J. Burdick, The explicit dynamic model and inertial parameters of the PUMA 560 arm, *Proc. of Int. Conference on R & A*, 1986, pp. 510–518.
[68] S. Ganguly, Discrete time nonlinear feedback method of robot arm control, D.Sc. Dissertation, Washington University, St. Louis, May 1991.
[69] A. K. Ramadorai, T. J. Tarn, and A. K. Bejczy, Feasibility of parallelization of nonlinear feedback method of robot arm control, Preprints of IFAC Symposium on Robot Control, SYROCO '91, Vienna, 1981, pp. 177–182.
[70] User's Manual or FSA-3254-F and FSA 3254-P Force/Torque Sensors, California Cybernetics, 1992.
[71] R. H. Lathrop, Parallelism in manipulator dynamics, *Int. Journal of Robotics Research*, Summer 1985, Vol. 4, No. 2, pp. 80–102.
[72] R. Nigam and C. S. G. Lee, A multiprocessor-based controller for the control of mechanical manipulators, *IEEE Journal of R & A*, Dec. 1985, Vol. RA-1, No. 4, pp. 173–182.
[73] C. S. G. Lee and P. R. Chang, Efficient parallel algorithms for robot forward dynamics computation, *IEEE Trans. on Systems, Man and Cybernetics*, Vol. 18, No. 2, April 1988, pp. 238–251.

[74] A. Fijany, Parallel algorithms and architectures in robotics, Ph.D. dissertation, University of Orsay (Paris XI), Sept. 1988.
[75] S. Geffin and B. Furht, A dataflow multiprocessr system for robot arm control, *Int. Journal of Robotics Res.*, Vol. 9, No. 3, June 1990, pp. 93–103.
[76] Universal Motor Controller Manual, California Cybernetics, 1992.
[77] Unimation 500 Series Equipment Manual, Unimation, a Westinghouse Co., 1984.
[78] Technical report of Silicon Graphics Power Series, Silicon Graphic, Inc., 1991.

SECTION II

Visually Guided Sensing and Control

CHAPTER 2

Observer-Based Visual Servoing

KOICHI HASHIMOTO
Department of Systems Engineering, Okayama University, Okayama, Japan

Visual feedback is a prevalent approach in autonomous manipulation. Conventional visual feedback schemes use the vision sensor to generate the hand trajectory at the stage of environment inspection. Whole manipulation is carried out on the basis of the planned trajectory without the help of the vision sensor. Although this off-line scheme may be useful for structured environments, it is not applicable for dynamic environments in which the objects are moving with time.

In this chapter we deal with a closed-loop control scheme for visually guided robot manipulators. Feedback control systems that incorporate vision sensors in the feedback loop are called *visual servo systems*. Since the robot is guided by the vision system in on-line fashion, a visually servoed robot does not need to know *a priori* the position and orientation of the workpieces in the environment. In a manufacturing environment, for example, visual servoing can eliminate robot teaching and allow tasks that are not strictly repetitive.

Visual servoing schemes are classified into two groups, namely, *position based* and *feature based*. Position-based approaches estimate the object position in real time and use the information to generate the trajectory. Thus approach is a natural extension of the conventional off-line scheme, but the estimation problem tends to be quite sensitive to the image distortion and noise. On the other hand, the feature-based approach uses the object features directly in the visual sensory output without computing the object position and orientation. This approach does not need to reconstruct the three-dimensional information of the object and so thus robust against noise and calibrate error.

One of the biggest problems in feature-based visual servoing is the slow sampling rate of the camera. A typical off-the-shelf CCD camera has a sampling rate of 60 or 50 Hz. On the other hand, to ensure sufficient stability as well as moderate accuracy for the joint angle servo system it is necessary to keep the sampling rate no lower than 500 Hz. Moreover, the CCD camera has sampling delay in principle and a typical vision processing board takes several sampling periods for feature extraction. This delay and slow sampling make the closed-loop system oscillatory or even unstable.

To compensate for the delay and to provide the intersample information to the joint servo, an observer is introduce. The observer estimates the object velocity and updates the visual information with the sampling rate of the joint servo. Also, two observer-based controllers are presented. One is a nonlinear model-based controller and the other is a linearized version of the nonlinear controller. The nonlinear controller is applied to a direct drive robot for which nonlinear dynamics cannot be neglected. The linearized controller is used for a general six-degree-of-freedom geared robot.

The stability of the observer-based control system is presented and the effectiveness of the observer is verified by experiments on a two-link direct drive robot and a PUMA 560.

1 INTRODUCTION

Visual information on tasks and environments is essential for robots to execute flexible and autonomous tasks. A typical task of autonomous manipulation is to track a moving object with a robot hand based on the information from a vision sensor. To carry out this task, the vision sensor must be incorporated in the feedback loop. Figure 2.1 shows an example of a visual feedback task. A camera is mounted on the robot hand and it captures images of the object. An image is considered as a two-dimensional array of gray-level signals whose size is typically 512 × 512 pixels. If the gray levels from all pixels are considered as the measurement signal, the system will not be manageable because the size of the measurement vector is larger than 200,000 and each element has a nonlinear correlation with its neighbors. Therefore, preprocessing of the raw image is necessary. Usually, image features of the object are extracted by preprocessing. A few examples of research based on the stochastic models of two-dimensional observations are found (e.g. [1]), but most visual servoing schemes use the features of the image as an observation. An image feature is any structural feature that can be extracted from an image (e.g., an edge or a corner). Typically, an image feature will correspond to the projection of a physical feature of the object [27]. The robot is controlled on the basis of the image features, and further image processing (e.g., image understanding or recognition) is omitted.

There are two approaches in visual feedback control: *position based* and *feature based* [46]. With position-based schemes, the object position and orientation relative to the camera are computed by using photogrammetric [12, 47, 49], stereo [1, 38, 39], or "depth from motion" techniques [28, 35]. Because the position of the object is available as the output of the image processing part, a conventional position controller can be used to control the manipulator. However, geometric model of the object is required and the camera–robot

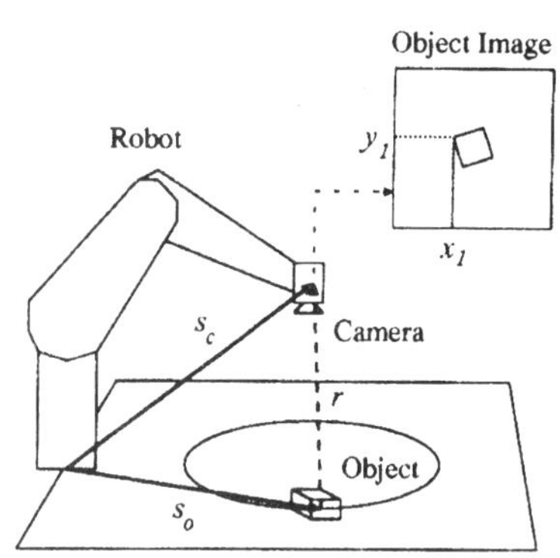

FIGURE 2.1
Visual tracking.

system must be calibrated. Also, the precision of the model and calibration is critical to the system performance. On the other hand, feature-based schemes use the features directly for feedback control; that is, features are controlled in the image plane [10, 11, 18, 29, 36, 46]. Thus, the controller must be modified to close the feedback loop in the image plane or feature space. However, the computational burden is reduced. Also errors in the geometrical model and camera calibration may be eliminated. A comparative study may be found in [19]. There are some good tutorials for visual feedback control [4, 27]. In this chapter, the problems of feature-based visual feedback are treated; however the results can be applied to the position-based schemes with slight modifications.

A vision sensor provides rich information about the object and the environment, but the sampling rate is usually very slow (e.g., 30 Hz) compared with that of mechanical systems (e.g., 1000 Hz). Note that according to Weiss's definition [46], a visual servo system does not have a joint servo loop. Thus assuming the use of a normal CCD camera for a visual servo system as per Weiss, the sampling rate of the joint servo must be reduced to 30 Hz and the controlled accuracy will be degraded considerably. Therefore, it is natural to compose the control system with two feedback loops having different sampling rates. This structure is classified as "dynamic look and move" by Weiss, but in this chapter it is called "visual servo." A block diagram of a feature-based visual servo system is shown in Figure 2.2

A visual sensor is incorporated in the feedback loop and a joint servo loop lies inside the vision loop. Since the inner loop is 30 times faster than the outer loop, the reference command or the joint servo system should be interpolated. A large step change of the reference will saturate the inner loop driver and will degrade the performance due to overshoot and oscillation. To generate the robot trajectory that connects the given set points with continuous velocity or acceleration, interpolation algorithms using polynomials are well known. Polynomial interpolation required boundary condition, such as velocities and accelerations at both edges. However, the boundary condition is not trivial in visual servoing because the target object may move during the sampling interval of vision (Feddema and Mitchell [11] and Koivo and Houshangi [30]).

Dynamic effects in real-time visual feedback were studied by Corke and Good [3, 6–8]. In [5], the inner loop dynamics is modeled by a linear system with delay and a detailed analysis of the visual servo system as a multirate control system. Also it is concluded in [8], that "to achieve high performance visual servoing it is necessary to minimize the open loop latency, have an accurate dynamical model of the system and to employ a feedforward type control strategy."

There are many publications that try to incorporate feedforward structure into the visual servo controller. If the object velocity is constant, a constant-gain velocity estimator is appropriate. For example, Allen *et al.* [1] use an $\alpha - \beta - \gamma$ filter. Another approach is a Kalman filter. For many applications, the target acceleration is assumed to be zero-mean

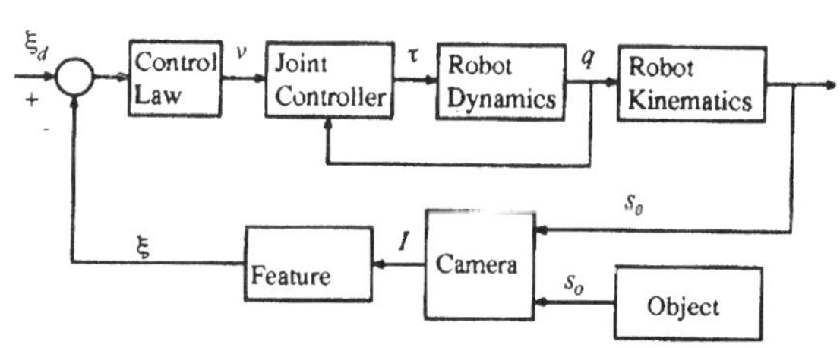

FIGURE 2.2
Feature-based visual tracking.

Gaussian. Then the Kalman filter estimates the target position and velocity with updating the filter gain [2, 47–49]. Similarly, an AR model can be used [30]. Note that these estimates should be executed with the sampling rate of the joint servo, and the input to the filter must be generated appropriately during the vision sample interval.

On the other hand, if a model for object motion is available, a Luenberger-type observer can be adopted. Ghosh *et al.* [14–16] propose observers to estimate the target velocity. Rizzi and Koditchek [37, 40] study a window position predictor for object tracking. The authors use an observer for estimating the object velocity and propose a nonlinear and a linearized controller with an observer [17, 24, 25]. Observers can be used for a large class of motions, including constant-velocity, constant-acceleration, and cyclic motions. Based on the model of the object motion, unknown parameters such as position, direction, velocity, center of circle, and so on are estimated. Most of the observers are formulated as a nonlinear adaptive identification problem.

This chapter discusses the compensation of the delay by estimating the target velocity. A model describing the object motion is introduced and a nonlinear observer is presented. The observer estimates the velocity parameters of the object motion model. The effectiveness of the proposed method is evaluated by simulations and experiments on a two-link planar direct drive robot. The results exhibit the fast convergence of the estimator and the accurate tracking performance of the controller. Since the nonlinear dynamics can be neglected for some manipulators (e.g., PUMA 560), a linearized version of the observer and controller is given. Experimental results with PUMA 560 are also provided.

2 MATHEMATICAL FORMULATION

The system model is considered as a map from the joint angle to the object image, which is composed of the kinematic model of the robot and the imaging model of the camera as shown in Figure 2.3. The camera is assumed to be mounted on the robot hand. Then the kinematic model becomes a map from the joint angle to the camera position and orientation. The camera model is a map from the position and orientation of the camera to the image of the object. The object motion is assumed to be an autonomous system that is independent of the robot motion.

From now on, the position and orientation are called simply the position in this chapter unless otherwise specified. The representation of orientation is not significant, and one can use any representations by using three parameters. However, the rotational velocity must not be considered as their time derivative. It must be the rotational velocity around three axes of the coordinate system in which the position is represented.

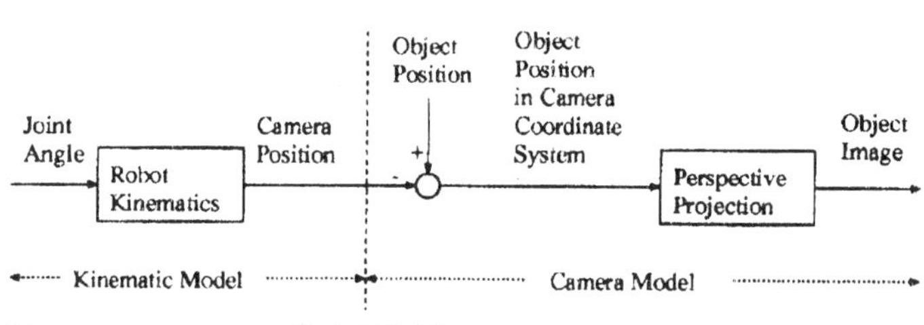

FIGURE 2.3
System model.

2.1 Robot Model

Assume that the robot has m ($\leqslant 6$) joints and the camera is mounted on the robot hand. Let s_{cam} be the (6×1) vector of camera position, then the *kinematic model of the robot* is given by

$$s_{cam} = \phi(q) \tag{2.1}$$

where $q \in \mathbf{R}^m$ is the joint angle. The *dynamic model of the robot is*

$$\frac{d}{dt}\begin{bmatrix} q \\ \dot{q} \end{bmatrix} = \begin{bmatrix} \dot{q} \\ -M^{-1}h \end{bmatrix} + \begin{bmatrix} 0 \\ M^{-1} \end{bmatrix}\tau \tag{2.2}$$

where τ is the actuator torque vector, M is the inertia matrix, and h is the vector representing the Coriolis, centrifugal, and gravity forces [9, 41, 42].

2.2 Object Motion Model

Assume that the object has m_{obj} ($\leqslant 6$) degrees of freedom. Let s_{obj} be the (6×1) vector of object position and p be the $(m_{obj} \times 1)$ vector of generalized coordinates representing the object position. Also assume that the object velocity is generated by an l ($\leqslant m_{obj}$) dimensional parameter vector θ^* such that

$$\dot{p} = W(p)\theta^* \tag{2.3}$$

is satisfied, where $W(p)$ is an $m_{obj} \times l$ matrix function of p. The vector θ^* and the equation (2.3) are called the *velocity parameter* and the *object motion model*. This motion model is simple, but it can model a fairly large class of autonomous motions including straight, circular, oval, and "figure 8" motions. A similar model was studied in [13] and extended to the two-stage estimator in [16].

2.3 Camera Model

The object image is generated by the perspective projection of the relative position between the camera and the object. The perspective projection is a map between two different representations of the position of the object, that is, the representations in the camera coordinate system $r = [X\ Y\ Z\ \alpha\ \beta\ \gamma]^T$ and in the image plane $[x\ y]^T$. Let ${}^cR_w$ be the (6×6) coordinate transformation matrix from the world coordinates to the camera coordinates. Note that ${}^cR_w$ includes the transformation of the orientation parameters. Then r is defined by

$$r = {}^cR_w(s_{obj} - s_{cam}) \tag{2.4}$$

The vector $[x\ y]^T$ consists of the coordinates of the feature point in the image plane expressed in pixels. Then the camera model is defined by

$$\begin{bmatrix} x \\ y \end{bmatrix} = \frac{f}{Z}\begin{bmatrix} X \\ Y \end{bmatrix} \tag{2.5}$$

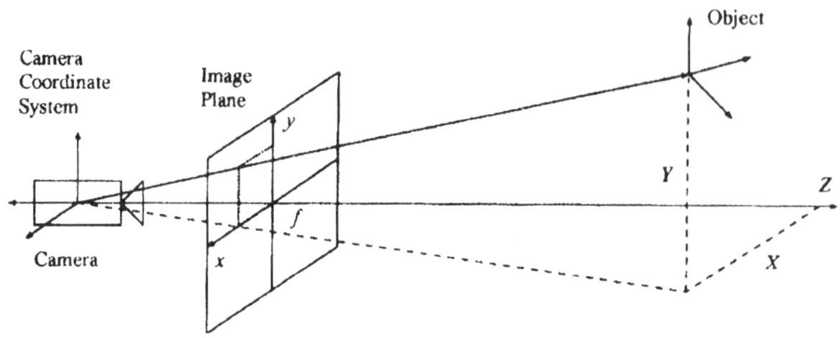

FIGURE 2.4
Perspective imaging model.

where f is the focal length of the lens (Figure 2.4) [26]. Note that $f < 0$ because the x and y coordinates of the image plane are aligned to the X and Y coordinates of the camera coordinate system. Thus $Z < 0$ for the object in the view area of the camera.

This expression is extended to multiple feature points. Suppose that there are n feature points on an object. Let $r_i = [X_i \; Y_i \; Z_i]^T$ $(i = 1, \ldots, n)$ be the positions (orientations are not included in r_i) of the feature points with respect to the camera coordinate system and the corresponding points mapped to the image plane have the coordinates $\xi_i = [x_i \; y_i]^T$ $(i = 1, \ldots, n)$. Figure 2.5 illustrates an example of the perspective transformation with four features ($n = 4$). Then the object image can be represented by a $2n$-dimensional feature vector $\xi \stackrel{\text{def}}{=} [\xi_1^T \cdots \xi_n^T]^T$, which is a function of the relative position between the object and the camera r. This relation is called the *camera model* and is expressed by the mapping $\iota: \mathbf{R}^6 \to \mathbf{R}^{2n}$ defined by

$$\xi \stackrel{\text{def}}{=} \iota(r), \quad \xi_d = \iota(r_d) \tag{2.6}$$

where r_d is the desired relative position between the camera and the object and ξ_d is the desired feature vector.

2.4 Task Feasibility

The robot configuration should avoid singular points while tracking. Thus we restrict the robot configuration in a region $\mathcal{V} \subset \mathbf{R}^m$ that does not contain the singular points. Also, we assume that for all $p \in \mathcal{U}$, where $\mathcal{U}$ is a subset of $\mathbf{R}^{m_{obj}}$ that contains all solutions of (2.3), the

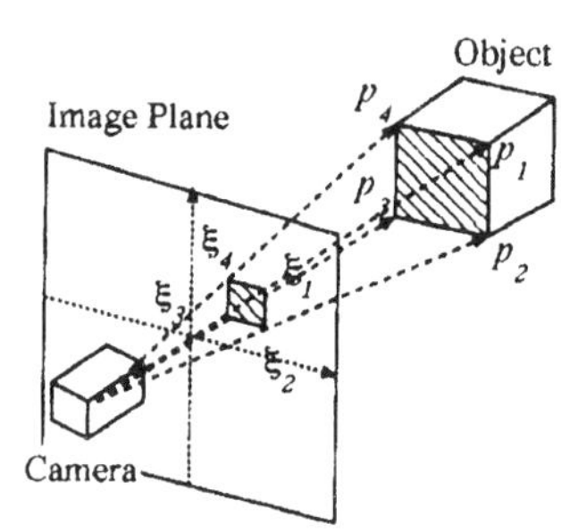

FIGURE 2.5
Example of perspective transformation with four features.

solution q^* of $s_{cam}(q^*) = s_{obj}(p) + r_d$ is in $\mathscr{V}$. This is a feasibility condition for object tracking. To satisfy this condition $m \geqslant m_{obj}$ is necessary. Also, it is useful to introduce the *feature manifold* $\mathscr{M}$, which is defined by

$$\mathscr{M} = \{\xi = \iota(r) : q \in \mathscr{V} \text{ and } p \in \mathscr{U}\} \tag{2.7}$$

The features on the feature manifold are called the *admissible features*.

3 JACOBIANS

The robot Jacobian that transforms the joint velocity to the hand velocity plays an important role in Cartesian space control. If one wants to control the robot in the tool frame, a Jacobian transforming the joint velocity to the tool velocity (expressed in the tool frame) will be needed. Similarly, we need a Jacobian that transforms the joint velocity to the feature velocity in the image coordinate system. Many important characteristics of the visual servo system are described by using the Jacobian. In this section, two Jacobians called the image Jacobian and the motion Jacobian are defined, and then degenerateness and redundancy are introduced.

3.1 Definitions

Differentiation of the camera model (2.6) yields

$$\dot{\xi} = J\dot{q} + L\dot{p} \tag{2.8}$$

where

$$J \stackrel{\text{def}}{=} \frac{\partial \iota}{\partial r}\frac{\partial r}{\partial q}, \quad L \stackrel{\text{def}}{=} \frac{\partial \iota}{\partial r}\frac{\partial r}{\partial p} \tag{2.9}$$

The matrices $J(2n \times m)$ and $L(2n \times m_o)$ are called the *image Jacobian* and *motion Jacobian*, respectively. The image Jacobian transform the joint velocity to the feature velocity, and the motion Jacobian transforms the target velocity to the feature velocity. Since the vector r is expressed in the camera coordinate system,

$${}^{c}J_{rob} \stackrel{\text{def}}{=} \frac{\partial r}{\partial q} = {}^{c}R_{w}\frac{\partial s_{cam}}{\partial q} \tag{2.10}$$

is the robot Jacobian expressed in the camera coordinate system. It is straightforward to see [19, 31] that

$$J_{img} \stackrel{\text{def}}{=} \frac{\partial \iota}{\partial r} = \begin{bmatrix} J_{img}^{(1)} \\ \vdots \\ J_{img}^{(n)} \end{bmatrix} \tag{2.11}$$

where

$$J_{img}^{(i)} = \begin{bmatrix} -\dfrac{f}{Z_i} & 0 & \dfrac{x_i}{Z_i} & \dfrac{x_i y_i}{f} & -\dfrac{x_i^2 + f^2}{f} & y_i \\ 0 & -\dfrac{f}{Z_i} & \dfrac{y_i}{Z_i} & \dfrac{y_i^2 + f^2}{f} & -\dfrac{x_i y_i}{f} & -x_i \end{bmatrix} \tag{2.12}$$

Note that the submatrix $J^{(i)} = J_{img}^{(i)} {}^cJ_{rob}$ expresses the infinitesimal change of the ith feature according to the infinitesimal change of the joint angles.

Similarly, we have

$$L = -J_{img} {}^cR_w J_{obj} \tag{2.13}$$

where

$$J_{obj} \stackrel{\text{def}}{=} \frac{\partial s_{obj}}{\partial p} \tag{2.14}$$

3.2 Degenerated Features

Consider a feature point that lies on the optical axis of the camera. When the point moves on the optical axis, the image does not change. Thus, this point is not useful for controlling the camera position in the $\mathbb{Z}$ axis. On the other hand, when the camera rotates in any direction around the object, the image does not change. Thus, a point feature is not useful for controlling the camera orientation. In this sense, a point feature is degenerated for six-degree-of-freedom control of the camera.

In general, the features that do not change when the robot's joint or the object itself moves are called *degenerated features*. A simple test for degenerateness is to check the rank of the Jacobian. If there is a direction of motion in joint space $\tilde{\theta}$ that does not change the features, then the following equation holds:

$$J\tilde{\theta} = 0 \tag{2.15}$$

Thus the Jacobian J is not full rank. If the Jacobian is full rank, then

$$J\delta\theta \neq 0 \tag{2.16}$$

for all $\delta\theta \in \mathbf{R}^m$. Thus the features change for all direction of motion in joint space. A similar discussion holds for object motion. Therefore, to avoid the degenerated features, we assume that for all $q \in \mathscr{V}$ and $p \in \mathscr{U}$ the Jacobians J and L have full column rank, that is,

$$\operatorname{rank} J = m \quad \text{and} \quad \operatorname{rank} L = m_o \qquad \text{for } \forall q \in \mathscr{V} \quad \text{and} \quad p \in \mathscr{U} \tag{2.17}$$

To satisfy this condition $2n \geqslant m$ is necessary,[1] but it is not sufficient. For a trivial example of a six-degree-of-freedom robot ($m = 6$), three points on a line are degenerated for a camera rotation around the line. A nontrivial example by Michel and Rives [33] is as follows: Suppose that there are three points on a plane that are not collinear. Then $\operatorname{rank} J < 6$ if the camera lies on the cylinder that includes the three points and whose axis is perpendicular to the plane containing these points (see Figure 2.6). For any attitude of the camera, J is singular.

3.3 Redundant Features

The example of Michel and Rives suggests using at least four feature points to control the six-degree-of-freedom robot. The features are called *redundant* if the number of features is larger than that of joints. For four feature points, the number of features is eight and they are redundant. A sufficient condition for the image Jacobian being full rank is given in the following lemma.

[1] $m \geqslant m_o$ is necessary to track all object pose.

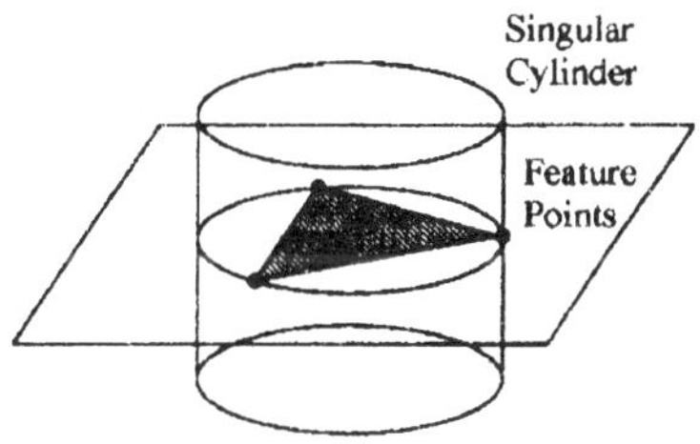

FIGURE 2.6
Singular cylinder.

Lemma 1 *Suppose that there are four points on a plane and the corresponding feature vector is admissible. Then the image Jacobian is full rank if any three of the feature points are not collinear in the image plane.*

Proof Let the plane on which the four points exist be $Z = pX + qY + r$. Then Z_i satisfies $Z_i = pX_i + qY_i + r$ for $i = 1, \ldots, 4$. Substituting (2.5) into this yields

$$\frac{f}{Z_i} = \frac{f - px_i - qy_i}{r} \tag{2.18}$$

and substituting this into (2.12) yields

$$J_{img}^{(i)} = \begin{bmatrix} -\dfrac{f - px_i - qy_i}{r} & 0 & \dfrac{x_i}{f}\dfrac{f - px_i - qy_i}{r} & \dfrac{x_i y_i}{f} & -\dfrac{x_i^2 + f^2}{f} & y_i \\ 0 & -\dfrac{f - px_i - qy_i}{r} & \dfrac{y_i}{f}\dfrac{f - px_i - qy_i}{r} & -\dfrac{x_i^2 + f^2}{f} & -\dfrac{x_i y_i}{f} & -x_i \end{bmatrix} \tag{2.19}$$

Define M_i and N as follows:

$$M_i = \begin{bmatrix} f & 0 & x_i & y_i & 0 & 0 & x_i^2/f & x_i y_i/f \\ 0 & f & 0 & 0 & x_i & y_i & x_i y_i/f & y_i^2/f \end{bmatrix}$$

$$N = \frac{1}{r}\begin{bmatrix} -1 & 0 & 0 & 0 & -r & 0 \\ 0 & -1 & 0 & r & 0 & 0 \\ p & 0 & 1 & 0 & 0 & 0 \\ q & 0 & 0 & 0 & 0 & r \\ 0 & p & 0 & 0 & 0 & -r \\ 0 & q & 1 & 0 & 0 & 0 \\ 0 & 0 & -p & 0 & -r & 0 \\ 0 & 0 & -q & r & 0 & 0 \end{bmatrix} \tag{2.20}$$

Then we have $J_{img}^{(i)} = M_i N$. Therefore,

$$J = MN^c J_{rob} \tag{2.21}$$

is obtained, where $M = [M_1^T \; M_2^T \; M_3^T \; M_4^T]^T$. Since M and ${}^cJ_{rob}$ are square, J becomes full rank if both M and ${}^cJ_{rob}$ are invertible and N is full rank. To check the singularity of M, it is straightforward to see that

$$\det M = \begin{vmatrix} 1 & x_1 & y_1 \\ 1 & x_2 & y_2 \\ 1 & x_3 & y_3 \end{vmatrix} \cdot \begin{vmatrix} 1 & x_2 & y_2 \\ 1 & x_3 & y_3 \\ 1 & x_4 & y_4 \end{vmatrix} \cdot \begin{vmatrix} 1 & x_3 & y_3 \\ 1 & x_4 & y_4 \\ 1 & x_1 & y_1 \end{vmatrix} \cdot \begin{vmatrix} 1 & x_4 & y_4 \\ 1 & x_1 & y_1 \\ 1 & x_2 & y_2 \end{vmatrix} \tag{2.22}$$

Thus M is invertible because any three feature points are not collinear. Since we are interested in the robot motion only in the nonsingular region, ${}^cJ_{rob}$ is invertible. On the other hand, if $p^2 + q^2 \neq 0$, the first six rows of N are linearly independent. If $p = q = 0$, the first four and the last two rows are linearly independent. Thus N is full rank. Therefore, the image Jacobian J is proved to be full rank.

Performance improvement by using redundant features is discussed in [17] and [25]. The smallest and the largest singular values of the image Jacobian play a central role for performance imprvement.

4 NONLINEAR CONTROL LAW

This section introduces a nonlinear controller and a nonlinear observer. An example of a two-link direct drive robot is also given.

4.1 Controlled Variable

Our goal is to track the object so as to keep the features of the object at the reference features. The models of robot, object motion, and camera are given by (2.2), (2.3), and (2.6), respectively. On the basis of these models, it is natural to adopt the features as the controlled variables, joint angles and joint velocities as the state, and the joint torque as the input. For a minimum set of features ($n = m/2$), this selection is appropriate.[2] However, for redundant features, the system becomes uncontrollable because the features cannot move in $\mathbf{R}^{2n}$ arbitrarily. To resolve this problem, one has to solve nonlinear geometric constraints on the features that represent the rigidness of the object. Since these constraints are difficult to solve, we linearize the constraints at the reference point and reduce the dimension of the feature vector to the dimension of the joint space.

Let $\mathcal{M}$ be a manifold, that is, the set of all admissible features (2.7), and consider a nominal point r_d that satisfies $\xi_d = \iota(r_d)$. Define a matrix B as follows:

$$B \overset{\text{def}}{=} \begin{cases} J(q^*, p^*) & \text{if} \quad n \geqslant m, \\ I & \text{if} \quad n = m \end{cases} \tag{2.23}$$

Note that J is a function of q and p; in this equation, p^* and q^* are a typical position of the object and a typical configuration of the robot that satisfy $s_{cam}(q^*) - s_{obj}(p^*) = r_d$. The matrix B is the image Jacobian at the nominal point if the features are redundant. If the features are minimum, then B is the identity matrix. The controlled variable is defined by

[2] There is no other choice.

using this B matrix in the following manner:

$$z \stackrel{\text{def}}{=} B^T(\xi - \xi_d), \quad \xi_d = \iota(r_d) \tag{2.24}$$

For J of full rank, $\xi \to \xi_d$ is guaranteed when $z \to 0$ [24, 25]. If the condition number (ratio of the largest and smallest singular values) of $J(q^*, p^*)$ is big, the variable z becomes very sensitive to numerical roundoff and the stability of the closed-loop system becomes very bad. To avoid this numerical instability problem, it is usual to define B as follows:

$$B = US^{-1}V^T \tag{2.25}$$

where the matrices U, S, V are obtained by singular value decomposition of $J(q^*, p^*)$, that is, $J(q^*, p^*) = USV$. However, we use (2.23) for notational simplicity.

4.2 Controller

Once the controlled variable is given, it is straightforward to compute a strictly linearizing controller. Taking the second derivative of z gives

$$\ddot{z} = B^T J M^{-1}(\tau - h) + \lambda + N\theta^* + \Phi\kappa(\theta^*) \tag{2.26}$$

where

$$\lambda = \begin{bmatrix} \lambda_1 \\ \vdots \\ \lambda_m \end{bmatrix}, \quad N = \begin{bmatrix} N_1 \\ \vdots \\ N_m \end{bmatrix}, \quad \Phi = \begin{bmatrix} \Phi_1 \\ \vdots \\ \Phi_m \end{bmatrix} \tag{2.27}$$

and

$$\begin{aligned} \lambda_i &\stackrel{\text{def}}{=} \dot{q}^T \frac{\partial^2 z_i}{\partial q^2} \dot{q}, \quad N_i \stackrel{\text{def}}{=} 2\dot{q}^T \frac{\partial^2}{\partial p \partial q} W \\ \Phi_{ijk} &\stackrel{\text{def}}{=} \left[\frac{\partial}{\partial p} \left\{ W^T \left(\frac{\partial z_i}{\partial p} \right)^T \right\} W \right]_{jk} \\ \Phi_i &\stackrel{\text{def}}{=} [\Phi_{i11} \quad \cdots \quad \Phi_{i1l} \quad \Phi_{i21} \quad \cdots \quad \Phi_{ill}] \\ \kappa(\theta^*) &\stackrel{\text{def}}{=} [\theta_1^{*2} \quad \cdots \quad \theta_1^* \theta_l^* \quad \theta_2^* \theta_1^* \quad \cdots \quad \theta_l^{*2}]^T \end{aligned} \tag{2.28}$$

Since B is full rank and J is a continuous function of p and q, $B^T J$ is invertible if p and q are close to p^* and q^*. Therefore, the actuator torque with new input v

$$\tau = M(B^T J)^{-1}(v - \lambda - N\theta^* - \Phi\kappa(\theta^*)) + h \tag{2.29}$$

yields a linear dynamics $\ddot{z} = v$. Thus we obtain the following theorem.

Theorem 1 *Define the new input v by*

$$v = -K_1 z - K_2 \dot{z} \tag{2.30}$$

where K_1 K_2 are positive definite gain matrices. Then the equilibrium point $(z, \dot{z}) = 0$ becomes exponentially stable by using the nonlinear input transformation (2.29).

4.3 Observer

The control law (2.29) and (2.30) requires $\dot{z}$ and θ^*, which are not usually known. Thus an estimator for these parameters is needed. Let the estimates of the parameter θ^* and controlled variable z be $\hat{\theta}$ and $\hat{z}$, respectively, and consider the following estimator [34, 43]:

$$\dot{\hat{z}} = B^T J\dot{q} + B^T LW\hat{\theta} + H(\hat{z} - z), \quad \dot{\hat{\theta}} = -W^T L^T BP(\hat{z} - z) \tag{2.31}$$

where H is any stable matrix and Q is any positive definite matrix. While P is selected to satisfy

$$H^T P + PH = -Q, \quad Q > 0 \tag{2.32}$$

Let the estimation error vectors be

$$\bar{z} \stackrel{\text{def}}{=} z - \hat{z}, \quad \bar{\theta} \stackrel{\text{def}}{=} \theta^* - \hat{\theta}, \quad e \stackrel{\text{def}}{=} \begin{bmatrix} \bar{z} \\ \bar{\theta} \end{bmatrix} \tag{2.33}$$

Then we obtain the following theorem.

Theorem 2 *For all $p \in \mathcal{U}$ and $q \in \mathcal{V}$ the estimator* (2.31) *makes the equilibrium point* $e = 0$ *asymptotically stable.*

It is easy to prove this theorem by taking the Lyapunov function candidate as follows:

$$V \stackrel{\text{def}}{=} e^T \tilde{P} e, \quad \tilde{P} \stackrel{\text{def}}{=} \begin{bmatrix} P & 0 \\ 0 & I \end{bmatrix} \tag{2.34}$$

This observer runs with the sampling rate of the joint servo. Thus the estimate of z is updated with the joint servo rate. Since new data z are not available during the vision sample interval, it is updated by using only the robot motion,

$$\dot{z} = J\dot{q} \tag{2.35}$$

4.4 Observer-Based Controller

Consider the following controller based on the estimated velocity of the feature vector $\hat{z}$:

$$\begin{aligned} \tau &= M(B^T J)^{-1}(v - \lambda - N\hat{\theta} - \Phi\kappa(\hat{\theta})) + h \\ v &= -K_1 z - K_2 B^T(J\dot{q} + LW(p)\hat{\theta}) \end{aligned} \tag{2.36}$$

Defining $\bar{\kappa} \stackrel{\text{def}}{=} \kappa(\theta^*) - \kappa(\hat{\theta})$ and substituting (2.36) into (2.26) yields

$$\ddot{z} = -K_1 z - K_2(\dot{z} - B^T LW\bar{\theta}) + N\bar{\theta} + \Phi\bar{\kappa} \tag{2.37}$$

Thus we obtain the following closed-loop dynamics:

$$\dot{x} = \bar{A}x + \bar{N}\bar{\theta} + \bar{\Phi}\bar{\kappa} \tag{2.38}$$

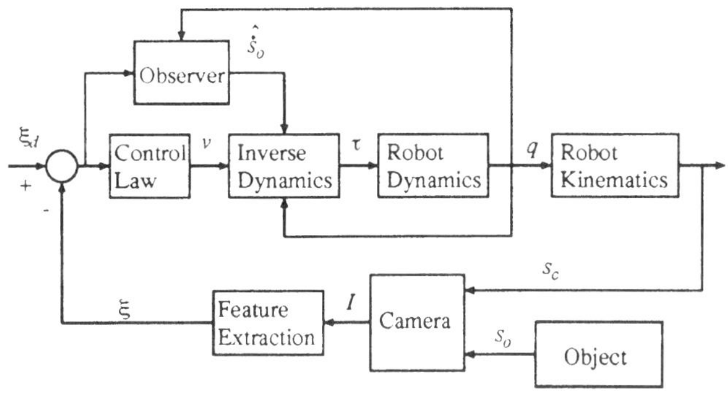

FIGURE 2.7
Observer-based controller.

where x, $\bar{A}$, $\bar{N}$, and $\bar{\Phi}$ are defined by

$$x \stackrel{\text{def}}{=} \begin{bmatrix} z \\ \dot{z} \end{bmatrix}, \quad \bar{A} \stackrel{\text{def}}{=} \begin{bmatrix} 0 & I \\ -K_1 & -K_2 \end{bmatrix}$$

$$\bar{N} \stackrel{\text{def}}{=} \begin{bmatrix} 0 \\ N + K_2 B^T L W \end{bmatrix}, \quad \bar{\Phi} \stackrel{\text{def}}{=} \begin{bmatrix} 0 \\ \Phi \end{bmatrix} \tag{2.39}$$

Finally, we have the main theorem.

Theorem 3 *For the system* (2.2), (2.3), *and* (2.6), *estimator* (2.31) *and controller* (2.36) *make the equilibrium point* $(x, e) = 0$ *asymptotically stable.*

The proof is omitted. It is tedious but straightforward based on Lyapunov's method. A block diagram of the observer-based controller is shown in Figure 2.7.

4.5 Example of Two-Link Robot

To see the procedure for designing the observer-based controller introduced in this section, an example of a planar two-link robot is given. Let us consider the robot shown in Figures 2.8 and 2.9. Figure 2.8 is the side view (from the $+Y_w$ direction). Figure 2.9 is the top view. The camera is mounted on the second link and looks upward. The object position is higher than the camera position. When the joint angle vector q equals zero, the robot is stretched out and the links are aligned with the X_w axis.

Robot Model

The robot model is given by (2.2) with

$$M = \begin{bmatrix} J_1 + J_2 + 2b\cos q_2 & J_2 + b\cos q_2 \\ J_2 + b\cos q_2 & J_2 \end{bmatrix}$$

$$h = \begin{bmatrix} -b(2\dot{q}_1\dot{q}_2 + \dot{q}_2^2)\sin q_2 + D_1\dot{q}_1 \\ b\dot{q}_1^2 \sin q_2 + D_2\dot{q}_2 \end{bmatrix} \tag{2.40}$$

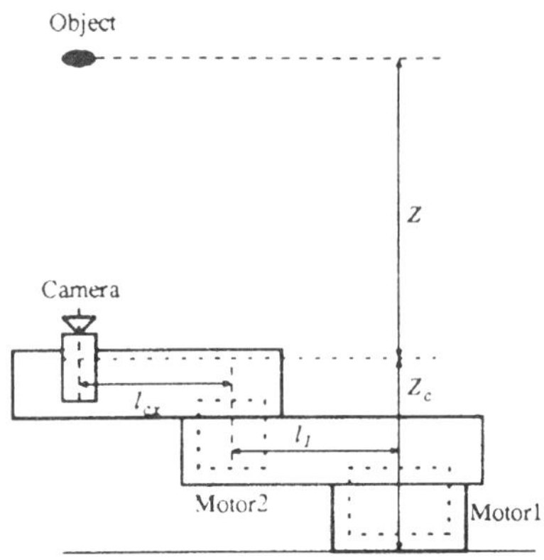

FIGURE 2.8
Two-link direct drive robot (side view).

where

$$
\begin{aligned}
J_1 &= I_1 + m_1 l_{g1} + m_2 + l_1 \\
J_2 &= I_2 + m_2 l_{g2} \\
b &= m_2 l_1 l_{g2}
\end{aligned}
\tag{2.41}
$$

In these equations, I_i and m_i are the moment of inertia and the mass of the ith link; l_{gi} is the length between the ith joint and the mass center of the ith link; D_i is the motor friction coefficient of the ith joint.

Camera Model

Let l_1 be the length of link 1 and $[l_{cx}\ l_{cy}]^T$ be the camera position with respect to link 2.

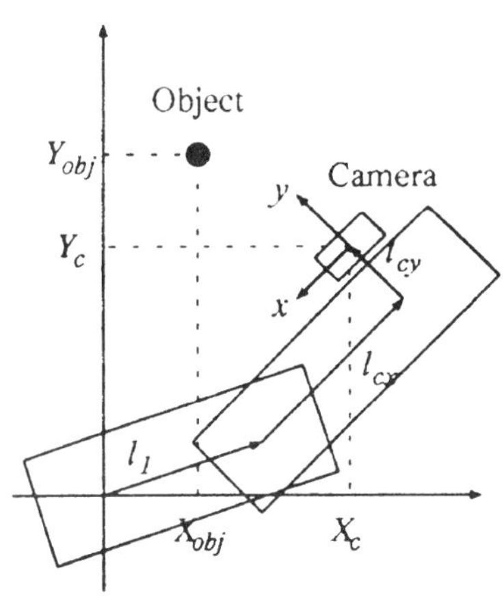

FIGURE 2.9
Two-link direct drive robot (overview).

Then we have the camera position and orientation

$$s_{cam} = \begin{bmatrix} l_1 c_1 + l_{cx} c_{12} - l_{cy} s_{12} \\ l_1 s_1 + l_{cx} s_{12} + l_{cy} c_{12} \\ Z_{cam} \\ 0 \\ 0 \\ \theta_1 + \theta_2 \end{bmatrix} \begin{matrix} \left.\begin{matrix} \\ \\ \\ \end{matrix}\right\} \text{position} \\ \left.\begin{matrix} \\ \\ \\ \end{matrix}\right\} \text{orientation} \end{matrix} \tag{2.42}$$

where $s_1 = \sin q_1$, $c_1 = \cos q_1$, $s_{12} = \sin(q_1 + q_2)$, $c_{12} = \cos(q_1 + q_2)$. Let the object position be $s_{obj} = [X_{obj} \ Y_{obj} \ Z_{obj} \ 0 \ 0 \ 0]^T$ (the orientation has no meaning because the object is a point). Assume that the depth $Z = Z_{obj} - Z_{cam}$ is known. Let the feature vector $\xi = [x \ y]^T$ be the position of the object in the image plane, f be the focal length of the lens, and r_i be the ith element of r. Then the object position with respect to the camera coordinate system is given by

$$\begin{bmatrix} X \\ Y \\ Z \end{bmatrix} = \begin{bmatrix} -\cos r_6 & -\sin r_6 & 0 \\ -\sin r_6 & -\cos r_6 & 0 \\ 0 & 0 & -1 \end{bmatrix} \begin{bmatrix} r_1 \\ r_2 \\ r_3 \end{bmatrix} \tag{2.43}$$

Therefore we have the camera model

$$\xi = \frac{f}{Z} \begin{bmatrix} X \\ Y \end{bmatrix} = \iota(r) \tag{2.44}$$

It is straightforward to compute (2.44), and we obtain

$$\xi = F \begin{bmatrix} l_{cx} + l_2 c_2 - X_{obj} c_{12} - Y_{obj} s_{12} \\ -l_{cy} + l_1 s_2 - X_{obj} s_{12} + Y_{obj} c_{12} \end{bmatrix} \tag{2.45}$$

where $F = \dfrac{f}{Z}$, $s_2 = \sin q_2$, $c_2 = \cos q_2$.

Jacobians

For this two-link robot example, the derivative of (2.6) is given by (2.8) with

$$J = F \begin{bmatrix} X_{obj} s_{12} - Y_{obj} c_{12} & -l_1 s_2 + X_{obj} s_{12} - Y_{obj} c_{12} \\ -X_{obj} c_{12} - Y_{obj} s_{12} & l_1 c_2 - X_{obj} c_{12} - Y_{obj} s_{12} \end{bmatrix}, \quad L = F \begin{bmatrix} -c_{12} & -s_{12} \\ -s_{12} & c_{12} \end{bmatrix} \tag{2.46}$$

It is easy to see that

$$\det J = F^2 l_1 (X_{obj} s_1 - Y_{obj} c_1), \quad \det L = -F^2 \tag{2.47}$$

Thus, J becomes singular only if the object is on the line connecting the first and second joints (just above the first link and its extension), and L is always nonsingular. To avoid the singular configuration, one may select more features than necessary. An example with redundant features may be found in [23].

Object Motion Model

In the previous example, the object height is constant. Thus the object degree of freedom is 2 and the object position is uniquely defined by $p = [X_{obj} \;\; Y_{obj}]^T$.

Straight Motion As shown in Figure 2.10(a), if the object motion is straight and the object velocities in the X and Y directions are v_X and v_Y, respectively, then we have (2.3) with

$$W(p) = I, \quad \theta^* = [v_X \quad v_Y]^T \tag{2.48}$$

Circular Motion If the object motion is circular with constant velocity ω as depicted in Figure 2.10(b), the object position and its time derivative are described by

$$\begin{bmatrix} X_{obj} \\ Y_{obj} \end{bmatrix} = \begin{bmatrix} r\cos\omega t \\ r\sin\omega t \end{bmatrix}, \quad \frac{d}{dt}\begin{bmatrix} X_{obj} \\ Y_{obj} \end{bmatrix} = \begin{bmatrix} -r\omega\sin\omega t \\ r\omega\cos\omega t \end{bmatrix} \tag{2.49}$$

where r is the radius of the circle. Thus the object velocity is given by (2.9) with

$$W(p) = \begin{bmatrix} -Y_{obj} \\ X_{obj} \end{bmatrix}, \quad \theta^* = \omega \tag{2.50}$$

For the case of an unknown center of the circle, let the center be (c_x, c_y). Then the object becomes

$$\begin{bmatrix} X_{obj} \\ Y_{obj} \end{bmatrix} = \begin{bmatrix} r\cos\omega t + c_x \\ r\sin\omega t + c_y \end{bmatrix} \tag{2.51}$$

Since the object velocity is the same as (2.49), we have the following parameterization:

$$W(p) = \begin{bmatrix} -Y_{obj} & 0 & 1 \\ X_{obj} & -1 & 0 \end{bmatrix}, \quad \theta^* = \begin{bmatrix} \omega \\ \omega c_x \\ \omega c_y \end{bmatrix} \tag{2.52}$$

Figure 8 Motion For "figure 8" motion as shown in Figure 2.10(c), the object position

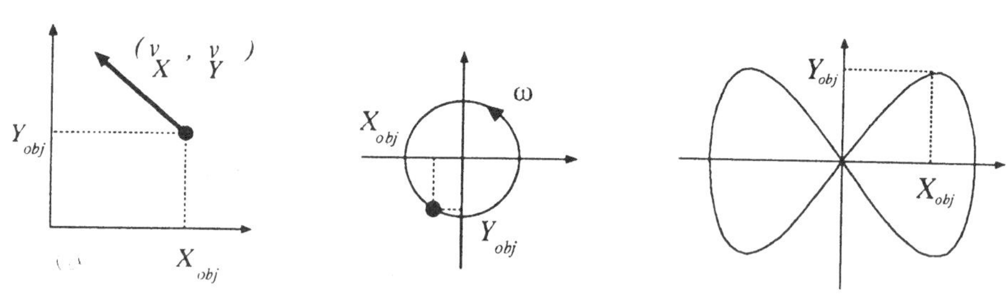

FIGURE 2.10
Object motions: (a) linear, (b) circular, (c) figure 8.

becomes

$$\begin{bmatrix} X_{obj} \\ Y_{obj} \end{bmatrix} = \begin{bmatrix} R_1 \cos \omega t \\ R_2 \sin 2\omega t \end{bmatrix} \tag{2.53}$$

Then the motion is modeled by (2.9) with

$$W(p) = \begin{bmatrix} 0 & -\dfrac{Y_{obj}}{X_{obj}} \\ X_{obj}^2 & -\dfrac{Y_{obj}^2}{X_{obj}^2} \end{bmatrix}, \quad \theta^* = \begin{bmatrix} \dfrac{1}{R}\omega \\ R\omega \end{bmatrix} \tag{2.54}$$

where $R = \dfrac{R_1^2}{2R_2}$.

Controller

Suppose that $\xi_d = [0 \ \ 0]^T$. If the distance from the object to the origin of the work space is close to l_1, the second joint angle is close to $\pi/2$ and the configuration is far from singular. For simplicity, we consider such a task and assume that J is invertible. Since $n = m = 2$, $B = I$ and $z = \xi$.

If the object motion is straight, the object motion model is (2.48) and we have

$$\begin{aligned} \ddot{z} &= J\ddot{q} + \lambda + N\begin{bmatrix} v_X \\ v_Y \end{bmatrix} \\ \lambda &= \left[(\dot{q}_1 + \dot{q}_2)\begin{bmatrix} 0 & -1 \\ 1 & 0 \end{bmatrix} J + Fl_1\dot{q}_1 \begin{bmatrix} 0 & c_2 \\ 0 & s_2 \end{bmatrix} \right] \dot{q} \\ N &= 2F(\dot{q}_1 + \dot{q}_2)\begin{bmatrix} s_{12} & -c_{12} \\ -c_{12} & -s_{12} \end{bmatrix} \end{aligned} \tag{2.55}$$

Thus we obtain the controller

$$\tau = MJ^{-1}\left(-K_1 z - K_2 \dot{z} - \lambda - N\begin{bmatrix} v_X \\ v_Y \end{bmatrix} \right) + h \tag{2.56}$$

If the object motion is circular, the object motion model is (2.49) and we have

$$\begin{aligned} \ddot{z} &= J\ddot{q} + \lambda + N\omega + \Phi\omega^2 \\ \Phi &= F\begin{bmatrix} -s_{12} & c_{12} \\ c_{12} & s_{12} \end{bmatrix}\begin{bmatrix} -Y_{obj} \\ X_{obj} \end{bmatrix} \\ N &= 2(\dot{q}_1 + \dot{q}_2)\Phi \end{aligned} \tag{2.57}$$

where λ is given by (2.55). Thus the controller is

$$\tau = MJ^{-1}(-K_1 z - K_2 \dot{z} - \lambda - N\omega - \Phi\omega^2) + h \tag{2.58}$$

5 LINEARIZED CONTROLLER

The controller given in the previous section is rigorous and useful for theoretical analysis. However, the computational burden of the inverse dynamics becomes very large for robots that have many degrees of freedom, and real-time implementation of inverse dynamics is not easy [20, 21].

On the other hand, many general-purpose industrial robots are the velocity control type and the robot joint contains high-ratio gears. Thus the terms of acceleration, centrifugal force, and Colioris force can be neglected. For these robots the controller can be linearized. Since the computational burden is reduced considerably by linearization, a linearized controller is useful for implementation on industrial robots.

The linearized dynamics of the visual servo system (2.24) at $(q, p) = (q^*, p^*)$ is given by

$$\dot{z} = B^T J\dot{q} + B^T LW\theta^* \tag{2.59}$$

where $J = J(q^*, p^*)$ and $L = L(q^*, p^*)$. Then the following control law:

$$\dot{q} = (B^T J)^{-1}(v - B^T LW\theta^*) \tag{2.60}$$

linearizes the dynamics from v to z as $\dot{z} = v$. Thus, with a positive definite gain matrix K, the feedback law $v = -Kz$ makes the equilibrium point $z = 0$ exponentially stable.

For the linearized system (2.59), a linearized observer similar to (2.31) can be used:

$$\dot{\hat{z}} = B^T J\dot{q} + B^T LW\hat{\theta} + H(\hat{z} - z), \quad \dot{\hat{\theta}} = -W^T L^T BP(\hat{z} - z) \tag{2.61}$$

Then the observer-based controller becomes as follows:

$$\dot{q} = (B^T J)^{-1}(-K\hat{z} - B^T LW\hat{\theta}) \tag{2.62}$$

Note that B, J, and L are constant matrices but W may depend on the object position p. Thus the observer-based controller is not strictly linear but the nonlinear dynamics of the robot are not considered. Some examples of the linearized controller with the PUMA robot will be described in the following section.

6 EXPERIMENTS

6.1 Two-Link Direct Drive Robot

To show the effectiveness of the nonlinear observer and controller, simulations and experiments are carried out. Figure 2.11 shows an experimental setup with a two-link direct drive robot.

A two-link planar robot illustrated in Figures 2.8 and 2.9 was designed and manufactured in our laboratory. Two direct drive motors (NSK RS0810, NSK AS0408) are used. Maximum rotational velocities are 1/3 and 1.5 (rpm), and maximum torques are 9000 and 1000 (kgf mm), respectively. The size and inertial parameters of this robot are $l_1 = 300$, $l_{cx} = 177$, $l_{cy} = 88$, $l_{g1} = 159$, $l_{g2} = 53.7$ (mm), $m_1 = 11.1$, $m_2 = 4.05$ (kg), $I_1 = 8.96 \times 10^5$, $I_2 = 5.20 \times 10^4$ (kg mm^2), $D_1 = 2050$, $D_2 = 159$ (kgf mm s).

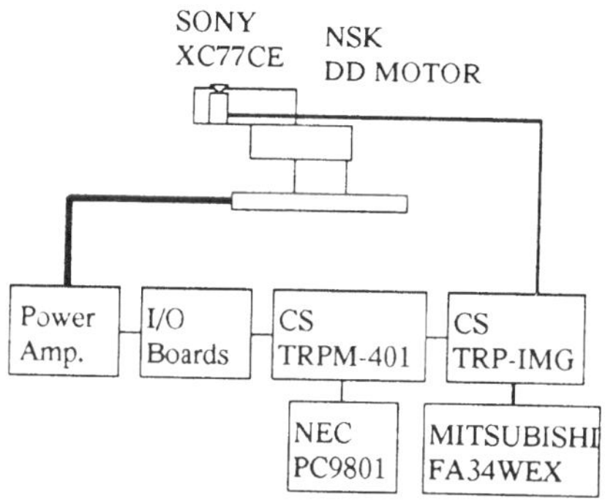

FIGURE 2.11
Visual feedback control system.

The host computer for the manipulator control is a personal computer (NEC PC9801). A transputer board TRPM-401, made by Concurrent Systems Inc., Japan, is added to the PC. The TRPM-401 board has 2 Mbytes of dynamic RAM and a transputer T800 [32], a 32-bit microprocessor with an on-chip floating-point processing unit. The vision processing board is Concurrent Systems TRP-IMG. The main processing unit of TRP-IMG is also a transputer. The TRP-IMG board has three digital signal processors IMS A110 connected in a pipeline. These DSPs do image convolution operation in real time. The board has 3 Mbytes of video RAM and 1 Mbyte of dynamic RAM. The video RAM is organized in 48 planes of 256×256 pixels, one byte per pixel. A TS68483 graphic controller chip is used as a video timing generator. The video digitizer chip AD9502 is used to sample a CCIR standard video input signal at a rate of 12.5 MHz and provides digital video information with a resolution of 8 bits. The I/O boards consist of an interface board (Concurrent Systems TRP98-2) for communication with the transputer with DA and counter boards. The DA boards are used to output the command torque, and the counter boards are used to input the encoder reading values. A CCD video camera SONY XC77CE is mounted on the end effector of the robot. The internal calibration of the camera and the external calibration of the geometrical relationships between the camera and the end effector are carried out on the basis of the calibration algorithm proposed by Tsai and Lenz [44, 45].

The sampling period of the vision system is 33 ms. The sampling period of the manipulator control is fixed to 1 ms. The feature vector (image coordinates of the center of the object) is computed in the TRP-IMG board and is passed to the TRPM-401 board every 33 ms, where the joint servo computations and observer update are carried out with the sampling period 1 ms.

Straight Motion

Simulations and experiments for object motion in a straight line are described. The initial object position is $s_{obj} = [282.5, -119, 1000]^T$ (mm), initial joint angle is $q = [-60, 80]^T$ (deg), initial object image is $\xi = [18, -20]^T$ (pixel), and reference image is $\xi_d = [0, 0]^T$ (pixel). The camera looks upward and the object moves above the robot. The task is to track the object so that the object image is kept at the reference point (origin of the image plane).

The object starts to move in the Y direction at 40 mm/sec at $t = 6$ sec and stops at $t = 10.5$; moves again in the X direction at 20 mm/sec at $t = 13.5$ and stops at $t = 17$. A stroboscopic plot of the robot motion is shown in Figure 2.12. Each triplet of line segments

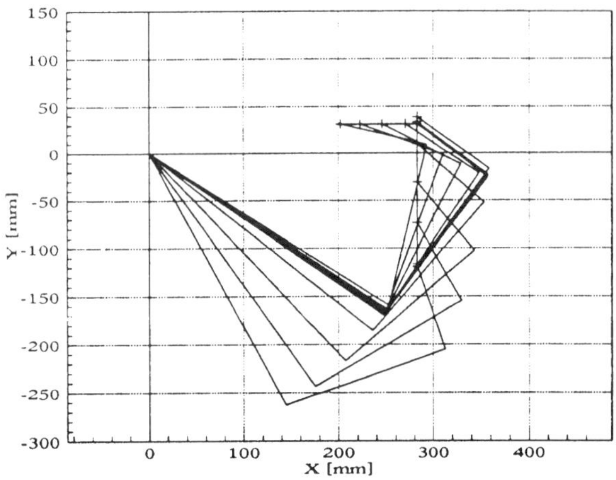

FIGURE 2.12
Straight motion.

shows a configuration of the robot and the camera. The base of the robot is at (0, 0) and the line segments passing through the base denote the first link. The second link is connected to the first link. The third line segment denotes the offset l_{cy} in Figure 2.9. The positions of the camera center are marked by $+s$.

Simulation The simulation results are shown in Figure 2.13. The horizontal axis is time; (a)

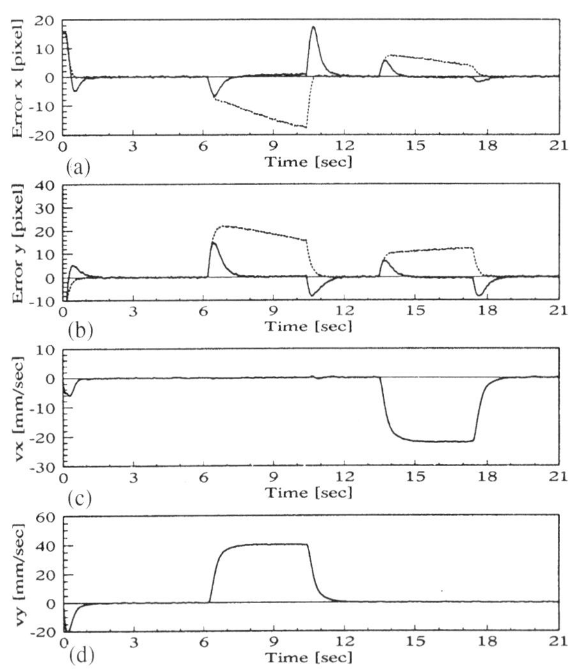

FIGURE 2.13
Simulation results (straight motion): (a) x error, (b) y error, (c) X estimated velocity, (d) Y estimated velocity; —, with observer; – –, without observer.

and (b) are feature errors in x and y coordinates in the image plane; (c) and (d) are estimates of the object velocity in X and Y coordinates in the base frame. A white Gaussian noise with variance 0.1 (pixel^2) was added to the feature position ξ. The solid lines in (a) and (b) are the results with the observer and the dotted lines are the results without the observer, that is, the controller (2.36) but $\hat{\theta} \equiv 0$. This controller

$$\begin{aligned} \tau &= M(B^T J)^{-1}(v - \lambda) + h \\ v &= -K_1 z - K_2 B^T J \dot{q} \end{aligned} \tag{2.63}$$

is referred to as *task-level inverse dynamics* in [22]. The same gains as with the observer-based control are used. Plots of the first 3 seconds in (a) and (b) show the response for a step change of the reference point. Since the robot moves to track the reference change, the observer is excited and generates some object velocity estimation. However, the object does not actually move. Thus the object velocity estimation has a bad effect on the tracking performance (i.e., overshoot). On the other hand, once the object is tracked at the reference position, the observer estimates the object velocity accurately. Note that the error in the image plane is reduced considerably while the object is moving with high velocity. Since the convergence time of the observer used in this simulation is almost 1 second, the overshoot at the moment of velocity change is inevitable.

Experiment Experiments are carried out on a direct drive robot. The parameters are the same as in the simulation. Results are shown in Figure 2.14. The errors in the image plane

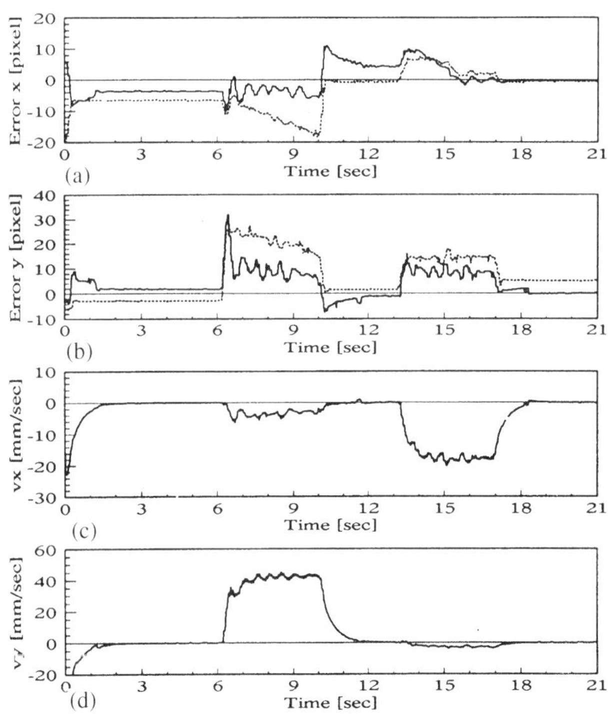

FIGURE 2.14

Experimental results (straight motion): (a) x error, (b) y error, (c) X estimated velocity, (d) Y estimated velocity; —, with observer; – –, without observer.

are due to friction, but they are not larger than 10 pixels. There are couplings in the estimation of velocity. The estimates are oscillatory compared with the simulation, and the effect of oscillation is found in the controlled error. However, the controlled error is fairly reduced while the object is moving.

Circular Motion

The second case is a circular motion with radius 20 mm. The center of the circle is at $(265, -120)$ in the base coordinate system. The object starts to move with angular velocity 2.8 rad/sec at $t = 6.5$, changes the speed to 1.4 rad/sec at $t = 17.5$, and stops at $t = 23.5$. The robot motion is shown in Figure 2.15. The line segments show the robot links and the + marks show the camera position.

Simulation The simulation results are shown in Figure 2.16. The horizontal axis is the time; (a) and (b) are the feature errors; (c) is the estimate of the angular velocity. The speed of convergence is slow, but the observer estimates the object velocity. When the observer is used, the error in the image plane decreases as the estimated value converges. On the other hand, there are steady-state errors for the controller without an observer.

Experiment Experimental results are given in Figure 2.17. The estimated value is accurate and the controlled error is reduced, but it does not converge to zero because the information on the center of the circle used in the design of the observer includes measurement error.

6.2 Six-Link PUMA Robot

Robot Control System Configuration

Real-time experiments are carried out on the visual feedback control system depicted in Figure 2.18. The host computer is an NEC PC9801 with TRPM-401 added in. A parallel computation scheme for joint level servo is implemented on a network of eight transputers (ADS TBE02). These transputers communicate with each other via a transputer link. The vision processing board is a TRP-IMG. The sampling periods of the vision system and the joint servo are 33 and 1 ms, respectively. See Section 6.1 for a detailed description of this equipment.

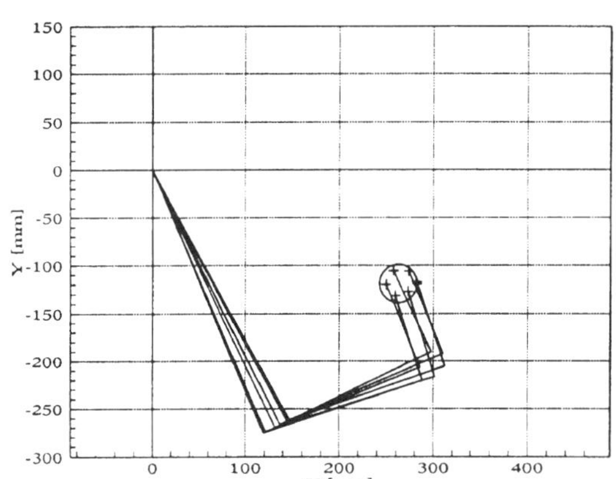

FIGURE 2.15
Circular motion.

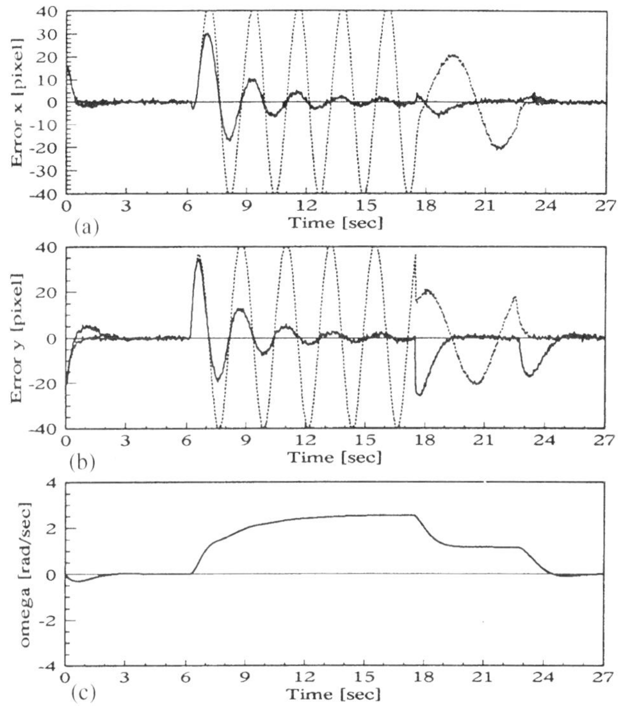

FIGURE 2.16

Simulation results (circular motion): (a) x error, (b) y error, (c) estimated angular velocity; —, with observer; – –, without observer.

Linear Motion

Feature Points and Image Jacobian We have tested the largest singular values of the image Jacobian for three objects shown in Figure 2.19. They are white boards with three, four, and five black marks. Three points are arranged to make a regular triangle with edge length 100 mm. Four points are on corners of a square with edge length 100 mm. For five points, an extra point with height 30 mm is added at the center of the square. The marks are on a plane except for the one at the center of the square. The features are the x and y coordinates of the image center of each mark. To avoid the singular cylinder mentioned in Section 3.2 and Figure 2.6, the reference camera position is located outside the cylinder. Computing the minimum singular values (σ_{min}) of the image Jacobians J_3, J_4, and J_5 (i.e., the image Jacobians for three, four, and five feature points, respectively) yields

$$\sigma_{min}(J_3) = 0.339, \quad \sigma_{min}(J_4) = 0.614, \quad \sigma_{min}(J_5) = 3.55 \tag{2.64}$$

Thus five features are desirable to obtain accurate position control of the camera in the 3-D work space. Therefore we carry out the next experiment with five feature points.

Experimental Setup The board with five features are attached to a PUMA 550. PUMA 550 and PUMA 560 are the same size but the former has five degrees of freedom. It does not have wrist rotation, which corresponds to joint 4 of the PUMA 560. The world coordinate system is at the base of the PUMA 560, which holds the camera. A nominal camera position

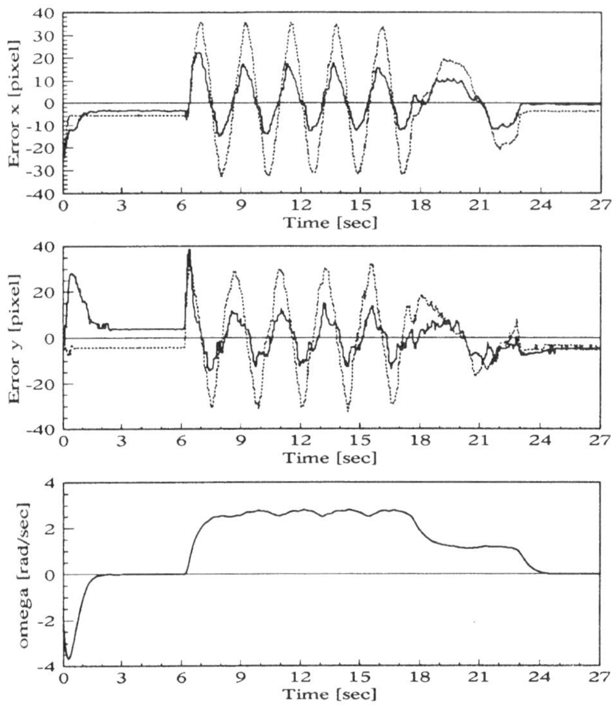

FIGURE 2.17
Experimental results (circular motion): (a) x error, (b) y error, (c) estimated angular velocity; —, with observer; – –, without observer.

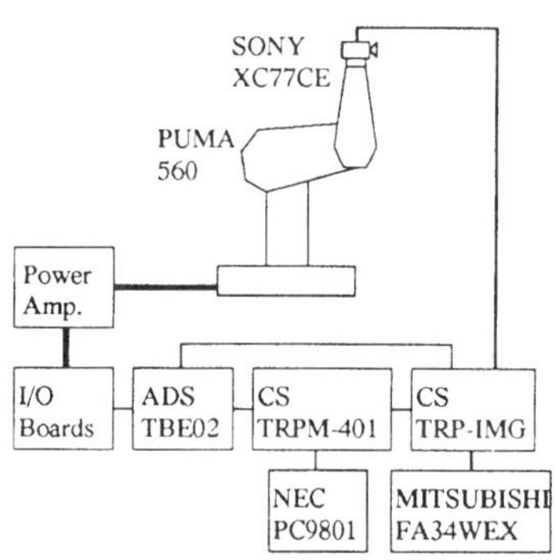

FIGURE 2.18
Visual feedback control system.

is in front of the marks and the distance is about 1000 mm. The nominal positions of the object and camera are shown in Figure 2.20. The X_w–Y_w–Z_w coordinate system is the world coordinate system.

Object Motion In this experiment, the object moves up and down; that is, the object motion is translational in the Z_w direction. Thus, $\dot{s}_o = [0\ 0\ v_z^*\ 0\ 0\ 0]^T$, where v_z^* is the object velocity in the vertical direction Z_w. Since the object motion is one dimensional, we can

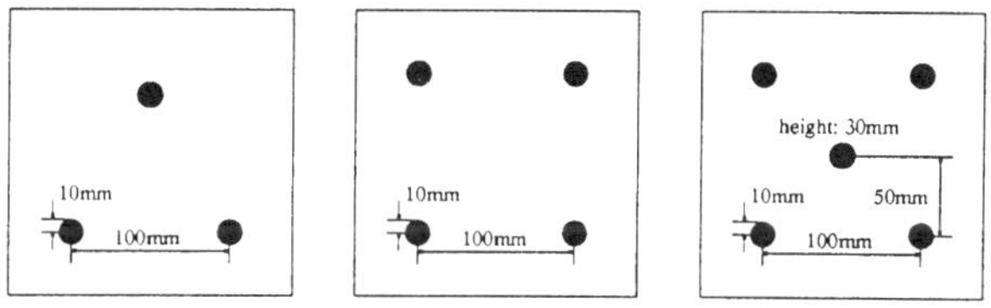

FIGURE 2.19
Configuration of feature points. Features are selected as the x and y coordinates in the image plane of the center of each circle. The center mark of five marks has height 30 mm.

choose the generalized coordinate as $p = Z_{obj}$, where Z_{obj} is the object height in the world coordinate system. Then we have $\dot{p} = v_z^*$ and the parameterization (2.3) is given by $W = 1$ and $\theta = v_z$. Since $\partial s_{obj}/\partial p = [0\ 0\ 1\ 0\ 0\ 0]^T$, the matrix L for the observer becomes as follows:

$$L = J_{img}\,{}^cR_w \begin{bmatrix} 0 \\ 0 \\ 1 \\ 0 \\ 0 \\ 0 \end{bmatrix} \tag{2.65}$$

The observer estimates v_z^*, that is, the object velocity, in the vertical direction. At $t = 10$ sec the object starts to move with velocity -20 mm/sec, that is, 20 mm/sec in the downward ($-Z_w$ direction), and stops at $t = 15$. After 10 seconds of pause, it moves upward with velocity 10 mm/sec and stops at $t = 35$.

Experiment The experimental results are shown in Figure 2.21(a) and (b). In Figure 2.21(a), the vertical axis shows the position of the camera in the world coordinate system. The solid and broken lines are the results with and without the observer, respectively. The dotted line is the reference trajectory of the camera. The control law with the observer is given by (2.62) and (2.61), and the control law without the observer is

$$\dot{q} = -(B^T J)^{-1} K z \tag{2.66}$$

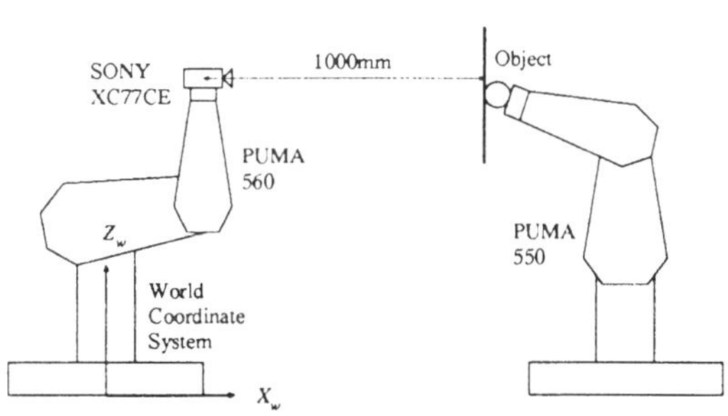

FIGURE 2.20
Robot configuration and object position.

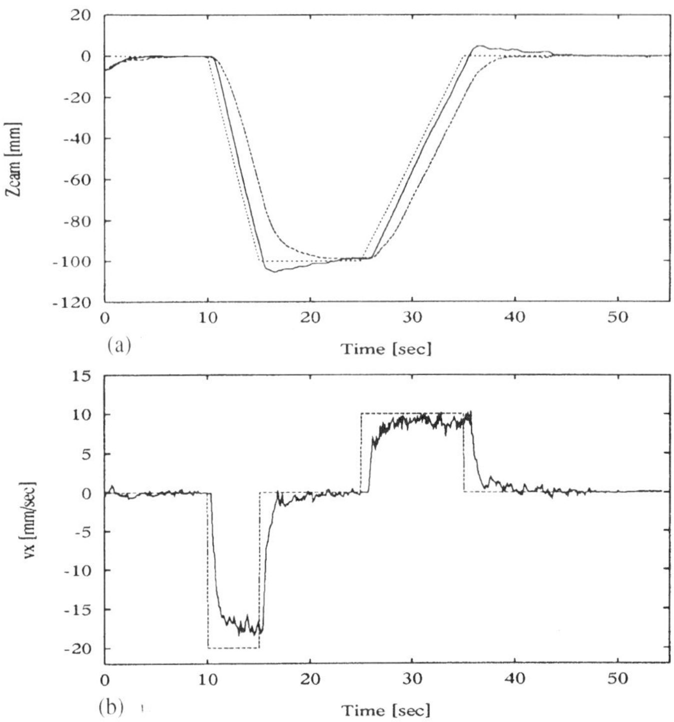

FIGURE 2.21

Step response. (a) Object position —, with observer; – –, without observer; ···, reference trajectory. (b) estimated velocity —, estimation; – –, true value.

The matrix K is the gain, which is the same as the observer-based controller. By using the observer the tracking speed is improved and the tracking error is reduced considerably. As shown in the nonlinear case, overshoot is a shortcoming of the observer-based scheme. Figure 2.21(b) shows the estimated velocity of the object. The broken line shows the true value and the solid line is the experimental result. The observer estimates the object velocity fairly accurately.

Another objective of the observer is interpolation of the visual data obtained with a very slow sampling rate. The values of z and its estimate $\hat{z}$ are plotted in Figure 2.22. The time $t = 0$ in Figure 2.22 corresponds to the time $t = 10$ in Figures 2.21. Delay of one visual sample is found in the period 0.1–0.4 because the velocity estimation is not correct during this period. Once the velocity estimation converges to the true value, the delay is canceled.

Tracking a Minirobot

Khepera Robot This section gives experimental results of visual tracking of Khepera, a mini two-wheeled mobile robot. Khepera was developed at the Laboratory of Micro Information, EPFL, Switzerland. Khepera has an MC68331 CPU and 256 Kbyte RAM and can be programmed in C. Figure 2.23 shows an overview of the Khepera robot. The wheel base is 53 mm.

Object Motion The camera tracks the Khepera robot as it moves on the floor. The motion is circular with radius 97 mm. We assume that the center of the circle is known. Two cases

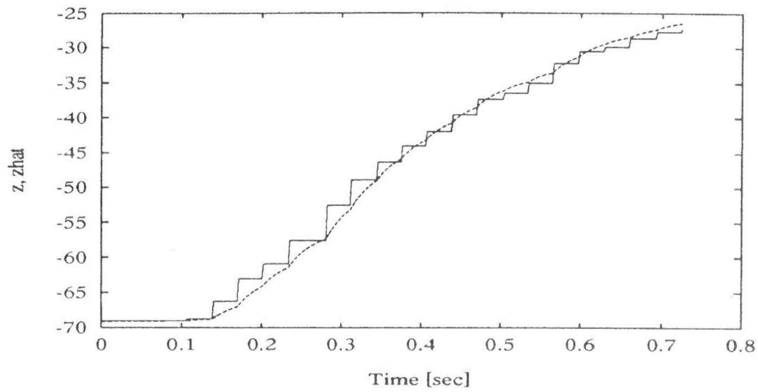

FIGURE 2.22
Estimated velocity ($\hat{v}_z$). —, measurements z; – –, interpolated value $\hat{z}$.

with velocities 0.9 and 1.8 rad/sec are examined. The experimental setup is depicted in Figure 2.24. The feature is the center of Khepera in the image. The camera tracks the object in a plane parallel to the floor; that is, the orientation and the height of the camera are kept constant.

Let the object position be $[X_{obj} \;\; Y_{obj} \;\; 0]^T$. Orientation is not considered in this case. Then the generalized coordinates of the object become $p = [X_{obj} \;\; Y_{obj}]^T$. Since the center position is assumed to be known, we can use (2.50) for the object motion model. Using $\partial s_{obj}/\partial p = [1 \;\; 1 \;\; 0 \;\; 0 \;\; 0 \;\; 0]^T$ yields

$$L = J_{img}\,{}^{c}R_{w}\begin{bmatrix}1\\1\\0\\0\\0\\0\end{bmatrix} \tag{2.67}$$

The observer estimates the rotational velocity of the object.

The Khepera robot starts to move counterclockwise at $t = 5$ sec with rotational velocity $\omega = 0.9$ rad/sec, changes its velocity to $\omega = 1.8$ at $t = 20$; and stops at $t = 35$. After 5 seconds of rest, the Khepera starts to move again clockwise with $\omega = -0.9$, changes the velocity to $\omega = -1.8$ at $t = 54$, and stops at $t = 69$.

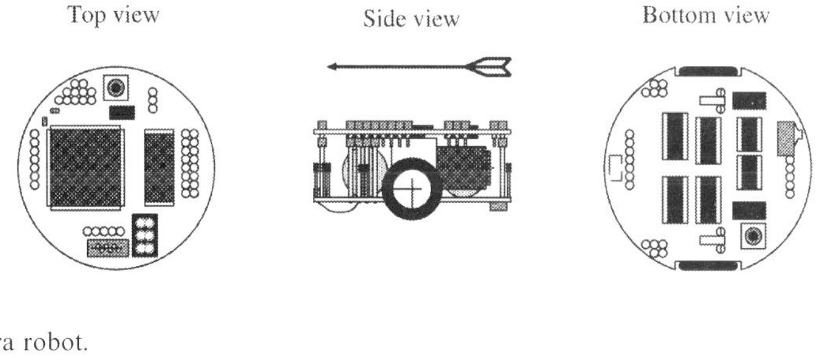

FIGURE 2.23
Overview of Khepera robot.

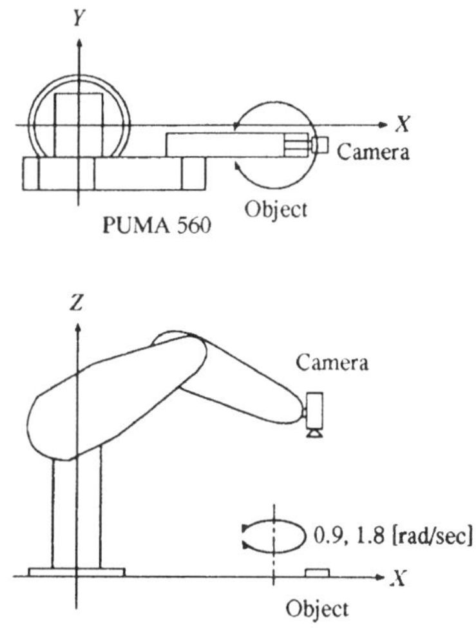

FIGURE 2.24
Experimental setup.

Control Law In this experiment, the controlled degree of freedom is two, that is, the X and Y coordinates of the camera position in the world coordinate system. Thus, to simplify the controller the controlled variable z is set to ξ. In other words, we choose $B = I$. Also, since the optical axis of the camera is aligned with the Z_w axis, which is orthogonal to the floor, the image Jacobian linearized at the reference position can be simplified to

$$J = -\frac{f}{Z_r} {}^cR_w J_{rob} \tag{2.68}$$

where Z_r is the camera height at the reference position. Then the control law is given by

$$\dot{q} = ({}^cR_w J_{rob})^{-1} \begin{bmatrix} {}^cv_{cam} \\ 0 \end{bmatrix}, \quad {}^cv_{cam} = \begin{bmatrix} -\frac{Z_r}{f}(-K\hat{\xi} - LW\hat{\theta}) \\ -K_z(Z - Z_r) \end{bmatrix}$$

$$\dot{\hat{\xi}} = J\dot{q} + LW\hat{\theta} + H(\hat{\xi} - \xi), \quad \dot{\hat{\theta}} = -W^T L^T P(\hat{\xi} - \xi) \tag{2.69}$$

where K and K_z are gain matrices and $\hat{\xi}$ is the observer estimation of the object features.

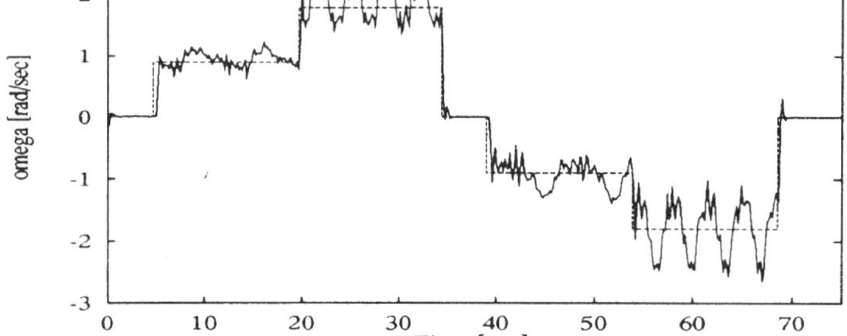

FIGURE 2.25
Estimated rotational velocity ($\hat{\omega}$). —, Estimated; – –, true.

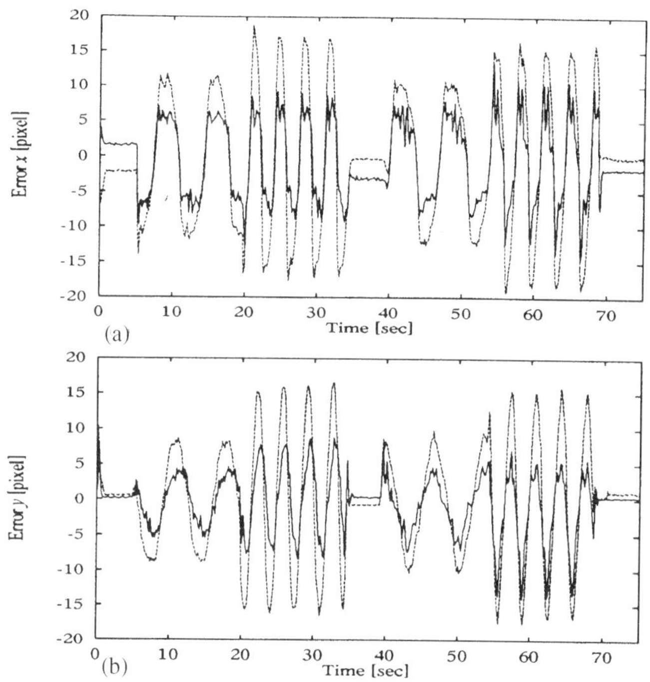

FIGURE 2.26
Controlled error. (a) x direction; (b) y direction. —, With observer; – –, without observer.

Experiment The initial joint angles are $q = [10.9 \ \ -42.1 \ \ -70.9 \ \ 2.01 \ \ 23.0 \ \ 0.01]^T$ and the initial camera position is $s_{cam} = [803.6 \ \ 1.588 \ \ 1103.0]$. Figure 2.26 shows the estimated value of ω. The solid line is the estimated value and the dashed line is the true value. The observer estimates the velocity fairly accurately, but oscillations are found. The oscillations are due to the calibration error of the rotation center. Also, the arm configuration becomes almost singular, that is, the arm is almost stretched out, when the object is at the farthest position. Thus the joint control accuracy is not very good around the farthest point. This singularity problem is another reason for oscillation.

Figure 2.26(a) and (b) show the error in the image plane for the x and y directions, respectively. The solid line is the result with the observer whose control law is given by (2.69). The broken line is the result without the observer. The control law without the observer is similar to (2.69) but the estimated velocity and the estimated feature are replaced by zero and the measured features, that is,

$$\dot{q} = ({}^cR_w J_{rob})^{-1} \begin{bmatrix} {}^v v_{cam} \\ 0 \end{bmatrix}, \quad {}^c v_{cam} = \begin{bmatrix} -\frac{f}{Z}(-K\xi) \\ -K_z(Z - Z_r) \end{bmatrix} \tag{2.70}$$

Note that the error is reduced for both directions by using the observer.

7 CONCLUSIONS

The vision sensor includes delay in its structure. Also, the sampling rate of the vision sensor is usually very slow. Thus, increasing the feedback gain yields oscillations. For these systems

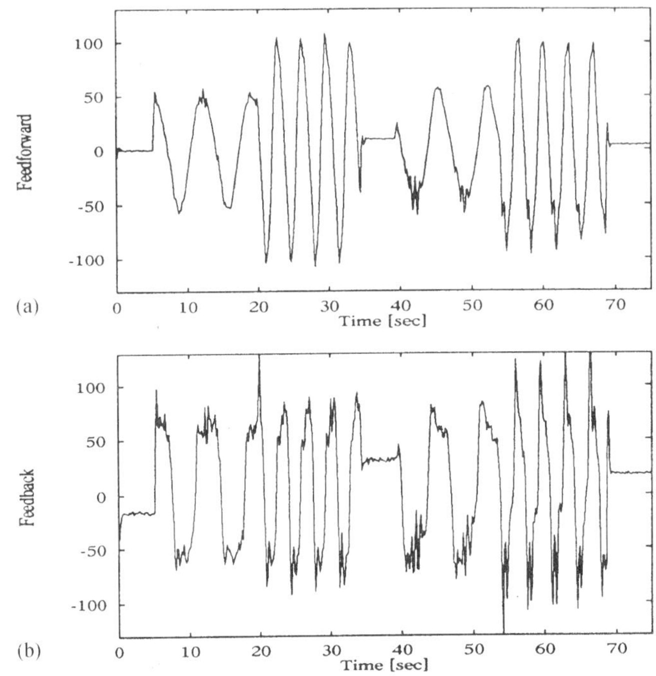

FIGURE 2.27
Control input. (a) Feedforward part $-LW\theta$; (b) feedback part $-K\hat{\xi}$.

a feedforward control scheme is effective. The results given in this chapter emphasize the usefulness of object velocity feedforward in the application of vision-based control.

We have introduced a visual feedback controller with a velocity observer. The observer compensates the delay by estimating the object velocity. Also, the observer provides intersample information to the joint servo by updating the visual information with the sampling rate of the joint servo. Thus the problems of slow sampling time and delay of the vision sensor are resolved. The tracking performance is improved by feedforwarding the object velocity. Stability of the observer-based control system is presented in a nonlinear form. Simulations and experiments with a two-link direct drive robot have exhibited the effectiveness of the observer-based control scheme. A linearized version suitable for industrial robot control is also presented. Experimental results with PUMP 560 have shown stable and accurate performance of the observer-based visual servo controller.

REFERENCES

[1] P. K. Allen, A. Timcenko, B. Yoshimi, and P. Michelman. Automated tracking and grasping of a moving object with a robotic hand-eye system. *IEEE Trans. Robotics and Automation*, 9(2):152–165, 1993.

[2] F. Chaumette and A. Santos. Tracking a moving object by visual servoing. In *12th IFAC World Congress, Vol. 9*, pages 409–414, Sydney, Australia, 1993.

[3] P. I. Corke. Video-rate robot visual servoing. In *Visual Servoing*, K. Hashimoto ed., World Scientific, pages 257–283, Singapore, 1993.

[4] P. I. Corke. Visual control of robot manipulators—a review. In *Visual Servoing*, K. Hashimoto ed., World Scientific, pages 1–32, Singapore, 1993.

[5] P. I. Corke. *Visual Control of Robots: High Performance Visual Servoing*. Research Studies Press, Somerset, England, 1997.
[6] P. I. Corke and M. C. Good. Dynamic effects in high-performance visual servoing. In *IEEE Int. Conf. Robotics and Automation*, pages 1838–1843, Nice, France, 1992.
[7] P. I. Corke and M. C. Good. Controller design for high speed-performance visual servoing. In *12th IFAC World Congress, Vol. 9*, pages 395–398, Sydney, Australia, 1993.
[8] P. I. Corke and M. C. Good. Dynamic effects in visual closed-loop systems. *Trans. on Robotics and Automation*, 12(5):671–683, 1996.
[9] C. Canudas de Wit, B. Siciliano, and G. Bastin, eds. *Theory of Robot Control*. Springer-Verlag, Berlin, 1996.
[10] B. Espiau, F. Chaumette, and P. Rives. A new approach to visual servoing in robotics. *IEEE Trans. Robotics and Automation*, 8(3):313–326, 1992.
[11] J. T. Feddema and O. R. Mitchell. Vision guided servoing with feature-based trajectory generation. *IEEE Trans. Robotics and Automation*, 5(5):691–700, 1989.
[12] M. A. Fischler and R. C. Bolles. Random sample consensus: A paradigm for model fitting with applications to image analysis and automated cartography. *Communications of the ACM*, 24(6):381–395, 1981.
[13] B. K. Ghosh, M. Jankovic, and Y. T. Yu. Perspective problems in system theory and its application to machine vision. *Journal of Mathematical Systems, Estimation and Control*, 4(1):3–38, 1994.
[14] B. K. Ghosh, M. Lei, and E. P. Loucks. Visionics: A new vision guided estimation of a dynamical system. In *12th IFAC World Congress, Vol. 9*, pages 415–418, Sydney, Australia, 1993.
[15] B. K. Ghosh and E. P. Loucks. A realization theory for perspective systems with application to parameter estimation problems in machine vision. *IEEE Trans. on Automatic Control*, 41(12):1706–1722, 1996.
[16] B. K. Ghosh, E. P. Loucks, and M. Jankovic. Multistage nonlinear estimation with application to image based parameter estimation. In *13th IFAC World Congress, Vol. F*, pages 447–452, San Francisco, 1996.
[17] K. Hashimoto, A. Aoki, and T. Noritsugu. Visual servoing with redundant features. In *35th IEEE Conf. on Decision and Control*, pages 2482–2484, Kobe, Japan, 1996.
[18] K. Hashimoto, T. Ebine, and K. Kimura. Visual servoing with hand-eye manipulator—optimal control approach. *IEEE Trans. on Robotics and Automation*, 12(5):766–774, 1996.
[19] K. Hashimoto et al. Manipulator control with image-based visual servo. In *IEEE Int. Conf. Robotics and Automation*, pages 2267–2272, Sacramento, CA, 1991.
[20] K. Hashimoto and H. Kimura. A new parallel algorithm for inverse dynamics. *Int. J. Robotics Research*, 8(1):63–76, 1989.
[21] K. Hashimoto and H. Kimura. An implementation of a parallel algorithm for real-time model-based control on a network of microprocessors. *Int. J. Robotics Research*, 9(4):37–47, 1990.
[22] K. Hashimoto and H. Kimura. Dynamic visual servoing with nonlinear model-based control. In *12th IFAC World Congress, Vol. 9*, pages 405–408, Sydney, Australia, 1993.
[23] K. Hashimoto and H. Kimura. LQ optimal and nonlinear approaches to visual servoing. In *Visual Servoing*, K. Hashimoto ed., World Scientific, pages 165–198, Singapore, 1993.
[24] K. Hashimoto and H. Kimura. Visual servoing with nonlinear observer. In *IEEE Int. Conf. Robotics and Automation*, pages 484–489, Nagoya, Japan, 1995.
[25] K. Hashimoto and T. Noritsugu. Observer-based control for visual servoing. In *13th IFAC World Congress, Vol. F*, pages 453–458, San Francisco, 1996.
[26] B. K. P. Horn. *Robot Vision*. MIT Press, Cambridge, MS, 1986.
[27] S. Hutchinson, G. D. Hager, and P. I. Corke. A tutorial on visual servo control. *Trans. on Robotics and Automation*, 12(5):651–670, 1996.
[28] W. Jang, K. Kim, M. Chung, and Z. Bien. Concepts of augmented image space and transformed feature space for efficient visual servoing of an "eye-in-hand robot." *Robotica*, 9:203–212, 1991.
[29] R. Kelly. Robust asymptotically stable visual servoing of planar robots. *Trans. on Robotics and Automation*, 12(5):759–766, 1996.
[30] A. J. Koivo and N. Houshangi. Real-time vision feedback for servoing robotic manipulator with self-tuning controller. *IEEE Trans. Systems, Man, and Cybernetics*, 21(1):134–142, 1991.
[31] H. Longuet-Higgins and K. Prazdny. The interpretation of a moving retinal image. *Proc. Royal Soc. London Ser. B*, 208:385–397, 1984.
[32] Inmos Ltd. *Transputer Reference Manual*. Prentice Hall, Herts, UK, 1988.
[33] H. Michel and P. Rives. Singularities in the determination of the situation of a robot effector from the perspective view of 3 points. Technical Report 1850, INRIA, 1993.
[34] A. P. Morgan and K. S. Narendra. On the stability of nonautonomous differential equations $\dot{x} = [a + b(t)]x$ with skew symmetric matrix $b(t)$. *SIAM J. Control*, 15:163–176, 1977.
[35] R. Nevatia. Depth measurement by motion stereo. *Computer Graphics and Image Processing*, 5:203–214, 1976.

[36] N. Papanikolopoulos, P. K. Khosla, and T. Kanade. Visual tracking of a moving target by a camera mounted on a robot: A combination of control and vision. *IEEE Trans. Robotics and Automation*, 9(1):14–35, 1993.

[37] A. A. Rizzi and D. E. Koditchek. A dynamical sensor for robot juggling. In *Visual Servoing*, K. Hashimoto ed., World Scientific, pages 229–256, Singapore, 1993.

[38] A. A. Rizzi and D. E. Koditchek. Further progress in robot juggling: The spatial two-juggle. In *IEEE Int. Conf. Robotics and Automation*, pages 919–924, Atlanta, GA, 1993.

[39] A. R. Rizzi and D. E. Koditchek. Further progress in robot juggling: solvable mirror laws. In *IEEE Int. Conf. Robotics and Automation*, pages 2935–2940, San Diego, CA, 1994.

[40] A. A. Rizzi and D. E. Koditchek. An active visual estimator for dexterous manipulation. *Trans. on Robotics and Automation*, 12(5):697–713, 1996.

[41] J. J. E. Slotine and W. Li. *Applied Nonlinear Control.* Prentice Hall, Englewood Cliffs, NJ, 1991.

[42] M. W. Spong and M. Vidyasagar. *Robot Dynamics and Control.* John Wiley & Sons, New York, 1989.

[43] A. Teel, R. Kadiyala, P. Kokotovic, and S. Sastry. Indirect techniques for adaptive input-output linearization of non-linear systems. *Int. J. Control*, 53(1):193–222, 1991.

[44] R. Y. Tsai. A versatile camera calibration technique for high-accuracy 3D machine vision metrology using off-the-shelf TV cameras and lenses. *IEEE J. Robotics and Automation*, RA-3(4):323–354, 1987.

[45] R. Y. Tsai and R. K. Lenz. A new technique for fully autonomous 3D robotic hand/eye calibration. *IEEE Trans. Robotics and Automation*, 5(3):345–358, 1989.

[46] L. E. Weiss, A. C. Sanderson, and C. P. Newman. Dynamic sensor-based control of robots with visual feedback. *IEEE J. Robotics and Automation*, RA-3(5):404–417, 1987.

[47] D. B. Westmore and W. J. Wilson. Direct dynamic control of a robot using an end-point mounted camera and Kalman filter position estimation. In *IEEE Int. Conf. Robotics and Automation*, pages 2376–2384, Sacramento, CA, 1991.

[48] W. J. Wilson. Visual servo control of robots using Kalman filter estimates of relative pose. In *12th IFAC World Congress, Vol. 9*, pages 399–404, Sydney, Australia, 1993.

[49] W. J. Wilson, C. C. Williams Hulls, and G. S. Bell. Relative end-effector control using Cartesian position based visual servoing. *Trans. on Robotics and Automation*, 12(5):684–696, 1996.

CHAPTER 3

Using Active Deformable Models in Robotic Visual Servoing

MICHAEL J. SULLIVAN, NIKOLAOS P. PAPANIKOLOPOULOS, RAHUL SINGH, and IOANNIS PAVLIDIS

Department of Computer Science and Engineering, University of Minnesota, Minneapolis, Minnesota

The potential of visual servoing systems using eye-in-hand cameras has been demonstrated by many research efforts. However, previous efforts have generally used one of only two approaches to the extraction of an error signal from the visual input: blob analysis or pixel-level feature tracking. In this chapter, we describe a third approach that combines some of the advantages of both previous methods. We use deformable active models to track image contours related to the object of interest. These contour models provide global information about the position of the object. In addition, by combining such models with *a priori* knowledge of the object shape, this approach may be extended to provide the orientation of the object in three dimensions.

We present a model-based approach for visual tracking and eye-in-hand robotic visual servoing. Our approach uses active deformable models to track a rigid or a semi-rigid object in the manipulator's work space. These deformable models (also known as "snakes") approximate the contour of the object boundary, defined by a set of control points. During tracking, the control points are updated at frame rates by minimizing an energy function involving the relative position of model points, image data, and the characteristics of figure pixels. When visual servoing is combined with the use of active deformable models, movement of the manipulator can compensate for translations and deformations of the object's image. To verify the potential of our approach, we run several experiments and present our findings.

1 INTRODUCTION

Robotic systems that operate in uncalibrated and/or uncontrolled environments must be able to react flexibly to changes in their environment. In the simplest case, such changes may be the result of the movement of a single object in the manipulator's work space. Previous work [1, 10, 13] has demonstrated that such changes can be handled effectively by incorporating information from eye-in-hand visual sensors into the manipulator's feedback loop. However, these systems sometimes have difficulty tracking targets that are semi-rigid or partially occluded. We propose to overcome these difficulties by incorporating active deformable models (commonly referred to as "snakes") into the visual system. Active deformable models attempt to conform to contours in the image as defined by the intensity gradient that correspond to the boundaries of the object being tracked. As the object contours translate or deform, the parameters of the active deformable model are adjusted. Simultaneously, the parameters of the active deformable model can be used as inputs to the manipulator controller.

The system presented in this chapter combines recent work from two different streams of research in the computer vision and robotics communities to improve the performance of eye-in-hand manipulators tracking moving objects. We have incorporated active deformable models into the established visual-servoing paradigm. By combining the two methods, we strengthen both. The use of visual servoing to produce compensating movements of the end-effector reduces the amount of deformation required from active deformable models to cope with object movements. At the same time, active deformable models allow manipulators controlled by visual servoing to deal with semi-rigid objects and motions (such as rotations) that challenge feature-based approaches to tracking.

Recent work in visual servoing has demonstrated the benefit of "closing the control loop" of a robotic manipulator guided by an "eye-in-hand" visual sensor. Generally, systems designed using this technique seek to hold an aspect of the visual input invariant through appropriate movements of the manipulator on which the camera is mounted. For example, Papanikolopoulos *et al.* [20] identify one or more features in the image and seek to maintain the features' locations in the image plane by producing compensating translations and rotations of the "end-effector." Visual servoing obviates the need to maintain a detailed, metric work space model. Rather than constructing a model of effector and target positions and computing a trajectory which matches that of the target, the system reacts directly to information provided by the sensor during the last control iteration.

In this chapter, we use active deformable models to provide the control signal to a visual servoing system. The organization of the chapter is as follows. First, Section 2 highlights the importance of the problem. Then Section 3 describes the issues and motivations for using this approach. Section 4 presents some previous work conducted in the area. Section 5 discusses the approach we propose. In this section, we also present an algorithm for the automatic selection of control points. In Section 6, we describe the hardware used to implement our experimental system. In Section 7, we present experimental results, and in Section 8 these results are discussed. Finally, in Section 10, we conclude and highlight the contributions made by this work.

2 IMPORTANCE OF THE VISUAL SERVOING PROBLEM

Vision-based control and active vision can have a significant impact on space applications, intelligent highways, manufacturing, and nuclear waste cleanup efforts. Vision-based control can enhance the performance of industrial robots in assembly lines, aid in better alignment of an object with the camera in automatic inspection systems, improve the automatic assembly of electronic devices (surface mount technology), assist in the realization of vehicle following (platooning), make possible autonomous satellite docking and recovery, and improve the efficiency of outdoor navigation techniques.

One area where robotic devices enhanced with sensing capabilities can have a significant impact is the area of nuclear waste cleanup. In particular, autonomous or semiautonomous robotic devices can participate in the inspection of waste storage tanks, detect and remove buried waste, automate the handling and analysis of contaminants, and help in decontamination and decommissioning operations. Moreover, the manipulator is a useful tool because it intervenes between the hazardous environment and the human operator. Since the human operator does not have any direct view of the environment where the task takes place, sensing devices must be used in order to provide some information about the status of the robotic task and the environment. Thus, in order to improve the efficiency of robotic devices in hazardous sites, it is important to augment them with sensing devices. Among the sensing devices, visual sensors play a critical role. The primary advantage of vision sensors is their ability to provide information on relatively large regions of the work space. This same ability, however, presents problems that must be overcome. Noise, time-consuming image processing and large amounts of visual information make their use challenging. Therefore, the modeling and design of robotic devices that include visual sensing has become a difficult and challenging task. These difficulties increase if we include many different sensing modules in the same feedback loop.

It is important to mention that currently no framework covers all the issues that are introduced by integrating the vision sensor or any sensor in the feedback loop of a robotic device. We think that there is a significant waste due to the fact that there is a trend to build systems that address only the use of specific sensing modules in the feedback loop. Small changes in the hardware or the software of a specific sensing module require significant redesign of the whole system, thereby increasing the cost and the development time. We firmly believe that the described system addresses some of these sensor-based control issues and provides a unified way of looking at problems of this type.

3 ISSUES

Many visual servoing systems use feature-based approaches or image attributes derived from image features such as optical flow. These systems typically find correspondences between features present in two images acquired at different instances in time from a camera mounted on the end-effector. Such systems may have difficulty handling cases in which object features become occluded or object motion or deformation alters the feature beyond recognition. For example, systems that define a feature as a template of pixels can fail when a feature rotates relative to the template used to match it.

To overcome these difficulties, the system proposed in this chapter incorporates contour tracking techniques. When a contour corresponding to the object boundary can be extracted from the image, it provides information about the object location in the environment. If prior information about the set of objects that may appear in the environment is available to the system, the contour might be used to recognize the object or to determine its distance from the camera. In other words, if a contour can be extracted from the image and this contour corresponds to an object boundary, the contour provides information useful to a visual servoing system. If additional, prior information about object shape and size can be combined with the contour information, the system could be extended to respond to object rotations and changes in depth.

For contour extraction, we have adopted the active deformable model methodology. Active deformable model techniques attempt to identify image contours by minimizing a contour energy function that includes terms representing regularization constraints such as contour smoothness and continuity as well as terms dependent on image attributes such as local contrast. A number of approaches have been proposed for formulating and finding a minimum for these functions, since the introduction of active deformable models to computer vision by Kass *et al.* [16]. The algorithm presented by Williams and Shah [29] and

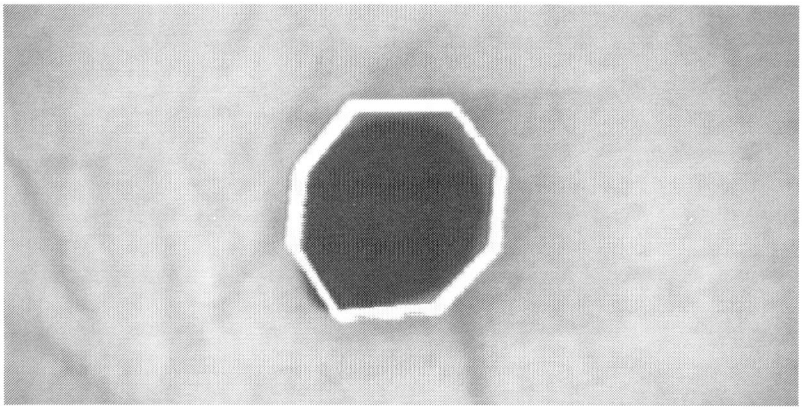

FIGURE 3.1
An active deformable contour tracking a balloon.

elaborated by Yoshimi and Allen [28] is particularly well suited for our application as it is both iterative and greedy. Because it is iterative, partial solutions are available during the minimization process; because it is greedy, the quality of these partial solutions tends to increase. We take advantage of these properties by running the control system and the contour extraction algorithm simultaneously. The controller issues commands to the arm based on the most recent partial solution. If the image were static, the system would tend to a steady state in which the object is centered in the image and the minimizing algorithm reaches a stable solution. However, since the object we are interested in is moving, the movements of the arm tend to change the image in ways that increase the energy of the current model configuration (because areas of high contrast in the image have moved), which forces the minimization algorithm to find a new configuration of minimum energy.

On the other hand, as long as object translations and deformations between frames are reasonably small, a minimum configuration of the active deformable model in one frame will be close to a minimum configuration for the model in the subsequent frame. The algorithm may find the minimum in a few iterations. In the best case, the minimization algorithm will be fast enough and the interframe displacements small enough that a minimum configuration can be found for each frame. This theoretical ideal may be impossible to achieve in practice—it would require a perfect model, signal processing system, and control law, but it illustrates the advantage of combining models and servoing. In short, tracking with active deformable models makes servoing possible and, in turn, the act of servoing simplifies the task of deforming the model to fit the contours of the image in the plane.

The center of the model—defined as the average location of the control points or as the two-dimensional center of mass—is used as the input to the manipulator's controller. Optimal estimation and control techniques (an LQG regulator) are used to deal with noise in this signal. We have conducted experiments that indicate the feasibility of this approach in dynamic but controlled environments, such as an automated factory floor.

When the speed of the minimization algorithm relative to the speed of the contour displacements and deformations in the image plane is sufficient, the system presented in this chapter tracks reliably. We have begun experimental work (see Section 7) in an effort to define and quantify the relevant factors (e.g., image displacement, image deformation, speed of the minimization algorithm, gains for energy function, number of control points). When the key factors have been identified, future work will focus on improving the critical elements of the system.

4 PREVIOUS WORK

This work draws on two streams of research in the computer vision and robotics communities. We have combined techniques developed by the visual servoing community with contour extraction techniques developed in the graphics and computer vision communities.

Several research efforts have focused on using vision information in the dynamic feedback loop [7, 10, 13, 14, 19, 24, 27]. Weiss *et al.* [24] have proposed a model reference adaptive control scheme for robotic visual servoing. In this work, servoing is performed with the goal of reducing the error between the desired image attributes (center of mass, first or second moment of the image) and the current image attributes. The verification of the proposed algorithms has been limited to simulation. Allen *et al.* [1] have proposed an approach that uses image-differencing techniques in order to track and grab a moving object. Dickmanns [11, 12] has presented methods (Kalman filters) for the integration of vision information in the feedback loop of various mechanical systems such as satellites and automobiles. Koivo and Houshangi [17] have proposed an adaptive scheme for visually servoing a manipulator based on the information obtained by a static sensor. Feddema and Lee [13] have proposed a MIMO adaptive controller for hand–eye visual tracking. Their work has been used as the basis for our approach. Several other researchers [3, 18] have proposed strategies for vision-based exploration. Finally, Ghosh [14] has addressed several vision-based robotic issues with the aid of new "Realization Theory" for perspective systems.

The concept of active deformable models, also called "snakes," was first introduced to the field of computer vision by Kass *et al.* [16]. Snakes have been used in a number of applications including image-based tracking of rigid and nonrigid objects. Using snakes requires a minimization process of an energy function. Several techniques have been used to solve this problem, including variational calculus [16], dynamic programming [2], and greedy methods using heuristics [28, 29]. The latter method has the advantage of being fast as well as numerically stable. Our method uses a greedy method similar to that used by Williams and Shah [29] and Yoshimi and Allen [28].

Other researchers have also combined elements of visual servoing and active deformable model techniques to approach different problems than the one presented in this chapter. Blake, Curwen, and Zisserman [4] have presented a different algorithm for contour estimation and used it in a system that tracks a contour in an image (it does not include a robotic component). Yoshimi and Allen [28] have used a greedy, iterative minimization algorithm to track a robotic finger with a static camera and detect contact between the finger and a stationary object. Finally, Colombo *et al.* [9] published a description of a system that uses a spline contour model to plan and execute a movement that positions an eye-in-hand robot so as to bring a known object into a canonical orientation relative to the camera. They report initial simulation results.

5 PROPOSED APPROACH

We describe a system that tracks a moving, deformable object in the work space of a robotic arm with an eye-in-hand camera. For these experiments, we have used a figure–ground approach to object detection and identification. The figure–ground methodology allows pixels to be identified as object or background pixels, a distinction that is useful during the initial placement of the active deformable model. However, this limits the applicability of our system to environments in which the background is uniform.

Once the active deformable model has been placed, two simultaneous processes commence. One process uses an iterative, greedy algorithm to find a minimum-energy configuration of the active deformable model. The second process issues control commands to the manipulator based on the current configuration of the active deformable model. Movements of the manipulator alter the position of the camera and, consequently, the image forces used by the minimization algorithm, closing the control loop.

5.1 Placing the Model

Movement in a scene can be detected by comparing two images acquired by a camera: a ground image taken before the movement occurred and the current image. This difference image is defined as (where x and y are image coordinates):

$$I_{diff}(x, y) = |I_{ground}(x, y) - I_{curr}(x, y)| \tag{3.1}$$

To enhance the boundary contours of the object's image in the difference image, we increase the contrast of the difference image with a simple thresholding operation, where:

$$I'_{diff}(x, y) = \begin{cases} 0 & \text{if } I_{diff}(x, y) < T \\ 255 & \text{otherwise} \end{cases} \tag{3.2}$$

for a threshold T. In this work, when we use the term difference image to refer to a specific image, we mean the binary image I'_{diff} that is obtained from these operations.

When the system begins operation, a background image of the scene is captured to be used as the ground image, I_{ground}, for the calculation of image differences. The object to be tracked is introduced to the scene and the tracking system is alerted by a signal from the operator. Either manually or automatically, a bounding box is placed around the region of differences created by the object.

After a bounding box has been selected, an initial configuration for the active deformable model is chosen by one of several algorithms. In the course of this research effort, we experimented initially with three significantly different techniques.

The initial placement algorithm simply placed control points along the bounding box by splitting each edge of the bounding box into a number of model edges. In other words, the four corners of the bounding box became control points in the initial configuration. If desired, one or more equidistant points were also chosen on each edge. When sufficiently large values were chosen for the ballon constraint (described in Section 5.3), this crude placement method was fairly successful. The model quickly collapsed upon the image contour. However, this method is ill suited for experiments using variations of active deformable models that incorporate task-specific constraints (as opposed to the generic "snake" constraints of equidistance and equal angles). It does not provide useful default values for the constraints. Indeed, the initial configuration of the active deformable model has little relation to the desired model configuration.

Current work has focused on the use of a more specific, but still fairly generic, determination of initial model configuration. First, a blob is chosen as the primary object of interest. Then the boundary of the connected blob is extracted by an edge-following algorithm (similar to the Boundary-Following Algorithm described in [15]). Then a predetermined number of control points is placed along the boundary. A configuration determined by this method is generally well suited for tracking using generic constraints and is a reasonable configuration for more specific constraints—at least for one orientation of the model.

There are at least three serious disadvantages of this approach:

- The connected blob chosen for boundary following may not correspond to the image contour that should be tracked. There may be blobs in the difference image that are unrelated to the object of interest. These blobs may be caused by noise or the presence of other objects. Alternatively, the object of interest may create more than one blob in the difference image because it does not create a uniform difference with the background. Both of these types of errors should be alleviated by the application of simple image processing techniques, such as blob size thresholding combined with morphological operators.
- Points that are equally spaced on the perimeter are not necessarily equidistant. For example, points on a serrated boundary will be much closer in image coordinates than in perimeter coordinates. An optimization algorithm could overcome this limitation in the general case, but at high computational cost. There are probably heuristic approaches that would find good configurations in the majority of cases, but we have not identified any.
- Points chosen by a perimeter walk may not be points of deformation on the object contour or points of high curvature on the contour. Ideally, the control points would be placed at points where deformation will occur, or at an object corner, where changing relative viewpoint changes the angle projected on the image plane. Since the perimeter walking scheme does not consider object characteristics or the curvature of the extracted boundary, it does not reflect these characteristics of the contour.

In order to address the last two of these weaknesses, we have undertaken preliminary investigation of an automatic placement technique discussed in the next section.

All three of the methods considered are suited for use only in conditions in which the contour model must be determined from a single example provided at run time. For situations in which the contour model is known *a priori*, it would be straightforward to apply a Generalized Hough Transform to the edge-detected image to determine the position, orientation, and any uncontrolled parameters of the model.

Once the active deformable model has been placed by any of the methods described here, its movements are controlled by the minimization of an energy function as described in Section 5.3.

5.2 Automatic Selection of Control Points Using the P & P Algorithm

We have developed an algorithm (P & P algorithm in [22]) that locates points of high curvature (corners) using a method similar to that in [6]. It also locates key in-between low curvatures points (key flat points) by employing a procedure conjugate to that for locating corners. We have tried the particular algorithm for the selection of control points.

Selection of Corner Points

The determination of corners is done in a way very similar to the method followed in Brault's algorithm [6]. The notable difference is that there is no need for parameter tuning. The basic mechanism is the same as that of Brault's algorithm [6]. Each point c of the curve is seen as a potential corner. The neighboring points from either side of point c contribute to the cornerness of c to a degree determined by certain conditions.

In more detail, the angles $\omega(c+i)$ and $\omega(c-i)$ (see Figure 3.2) are computed for each pair of neighbors $c \pm i$ $(i = 1, 2 \ldots)$. For a pair $c \pm i$ to belong to the corner domain of point c, the following inequalities must be satisfied:

$$\omega(c+i) < \frac{\pi}{2} \quad \text{and} \quad \omega(c-i) < \frac{\pi}{2} \tag{3.3}$$

The contribution CF (cornerness factor) of each pair $c \pm i$ to the making of the candidate corner c is computed by the formula

$$CF(c, i) = \cos(\omega(c+i)) * \cos(\omega(c-i)) \tag{3.4}$$

Using $\frac{\pi}{2}$ as a fixed upper limit in the inequalities, formula (3.3) is a departure from the method followed in Brault and Plamondon [6] and is what renders the corner determination parameterless. The first $M(c)$ points that satisfy the inequalities (3.3) constitute the corner domain of point c and their total contribution to the cornerness of point c is computed by

$$TCF(c) = \sum_{i=1}^{M(c)} CF(c, i) \tag{3.5}$$

The corner segmentation points are identified by searching the values of the function $TCF(c)$. The TCF values of the curve points present a very consistent pattern: strings of nonzero values spaced by strings of zero values. Each of the nonzero strings corresponds to a high-curvature segment, and the maximum value contained in each such string corresponds to a corner segmentation point.

Selection of Key Low-Curvature Points

While corners are the perceptually most important parts in a curve, corners alone provide insufficient data for an accurate reconstruction of a curve. The situation improves substantially if we provide some key points with rather low-curvature surroundings that lie between

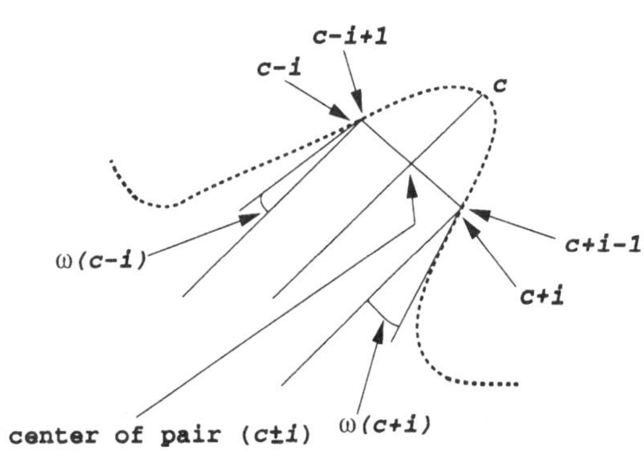

FIGURE 3.2
Geometric model for corner determination as proposed by Brault and Plamondon [6].

corners, as extra segmentation points. The way we find these key low-curvature points is conjugate to the way we find the corner points.

More precisely, a separate processing step is taking place for the location of the *key low-curvature points*. The geometric parameters shown in Figure 3.3 are the same as those in Figure 3.2 and are computed for each pair of neighbors $f \pm i$ $(i = 1, 2 \ldots)$ of every point f of the curve. This time, however, the larger the angles $\omega(f + i)$, and $\omega(f - i)$ are than $\frac{\pi}{2}$, the more the corresponding pair of neighboring points contributes to the *low curvatureness* of point f. As a result, by a suitable analysis of the angles $\omega(f + i)$ asnd $\omega(f - i)$, one can determine whether or not the pair of points $f \pm i$ is a part of the low-curvature domain of f and, in addition, can estimate the importance of the contribution of these points to the low curvatureness of point f.

The angles $\omega(f + i)$ and $\omega(f - i)$ must satisfy the following inequalities:

$$\omega(f + i) > \frac{\pi}{2} \quad \text{or} \quad \omega(f - i) > \frac{\pi}{2} \tag{3.6}$$

The contribution FF (low-curvatureness factor) of each pair $f \pm i$ to the making of the candidate key low-curvature point i is computed by the formula

$$FF(f, i) = |\cos(\omega(f + i))| * |\cos(\omega(f - i))| \tag{3.7}$$

In contrast to Eq. (3.5), Eq. (3.7) uses the absolute value of the trigonometric function *cos* since the range of the angles $\omega(f + i)$ and/or $\omega(f - i)$ now features $\frac{\pi}{2}$ as a lower and not as an upper limit. The total contribution of the first $M(f)$ points belonging to the low-curvature domain of f (the ones that satisfy the inequalities (3.6)) is computed by

$$TFF(f) = \sum_{i=1}^{M(f)} FF(f, i) \tag{3.8}$$

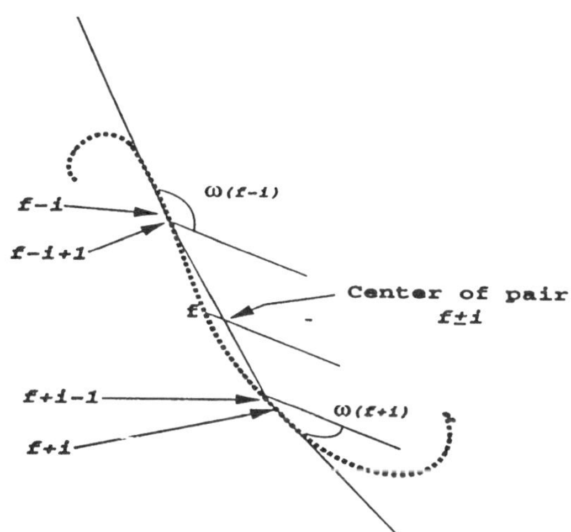

FIGURE 3.3
Geometric model for key low-curvature point determination.

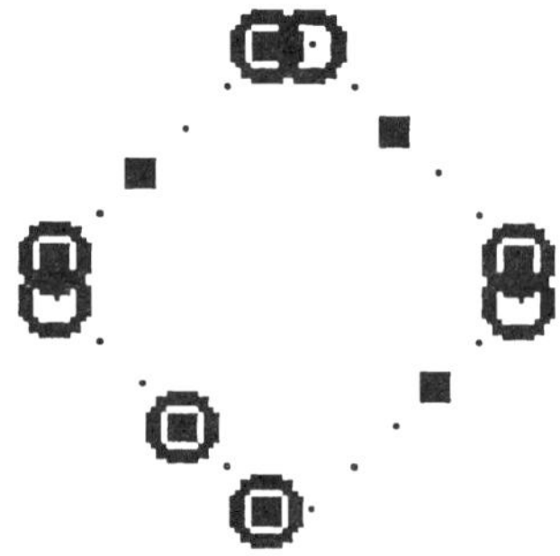

FIGURE 3.4
A square contour.

The identification of the key low-curvature segmentation points from the function $TFF(f)$ is done in a way analogous to the determination of corner points from the function $TCF(c)$.

Evaluation of the Proposed Algorithm

In order to get an indication of the goodness of the algorithmic selection of control points in terms of the accuracy of shape description, the following experiment was devised. Let a contour $\mathscr{C}$ of an arbitrary shape consist of N points ($\mathscr{C} = (\mathbf{P}_1, \mathbf{P}_2, \ldots, \mathbf{P}_N)$). Let the P & P algorithm select for the contour $\mathscr{C}$ a set $\mathscr{S}$ of m control points ($\mathscr{S} = (\mathbf{P}_{s1}, \mathbf{P}_{s2}, \ldots, \mathbf{P}_{sm})$). Also let a set $\mathscr{T}$ of m control points ($\mathscr{T} = (\mathbf{P}_{t1}, \mathbf{P}_{t2}, \ldots, \mathbf{P}_{tm})$) be chosen in such a way that an error norm is driven to minimum (optimal polygonal fit). The norm chosen for the purposes of the particular experiment was the Euclidean distance error of the polygonal fit represented by the point set. The set $\mathscr{T}$ was determined after an exhaustive search of all the $\binom{N}{m}$ combinations for the contour $\mathscr{C}$. It is interesting to compare the set of control points given by the P & P algorithm with the optimal polygonal fit point set for a variety of shapes (see Figures 3.4 through 3.7).

The small circles in these figures represent the points of the optimal polygonal fit set, while the points given by the P & P algorithm are represented by small squares. In all the shapes, the prominent corners are included in both the optimal polygonal fit set and the set of the P & P algorithm. Discrepancies arise only for the key flat points of the algorithm. The equivalent points of the optimal polygonal fit are mostly clustered in noisy areas of the shape.

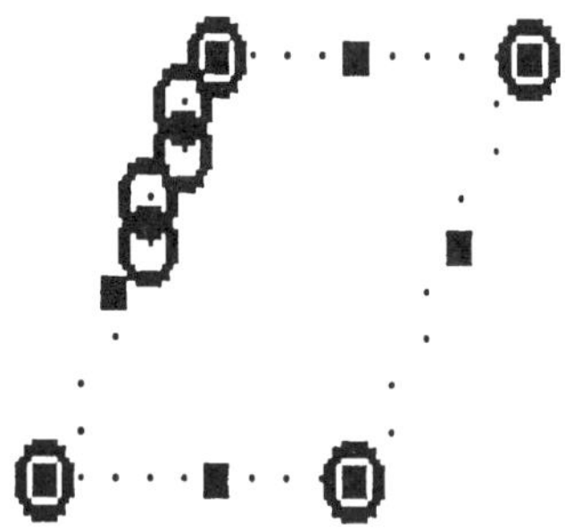

FIGURE 3.5
A parallelogram contour.

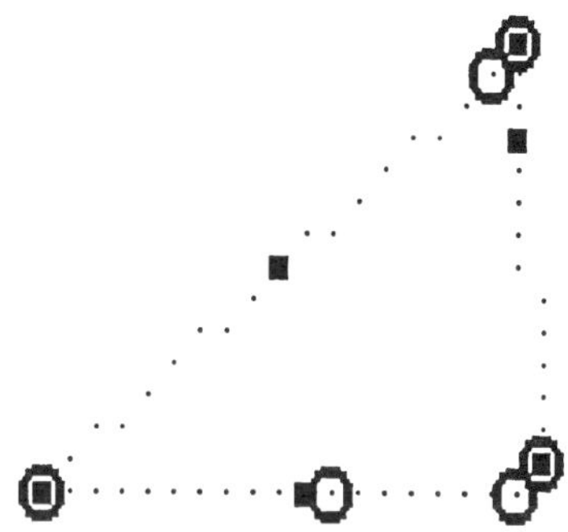

FIGURE 3.6
A triangular contour.

In contrast, the key flat points of the algorithm are uniformly distributed between the prominent corner points. This behavior is highly desirable, because the algorithm has not been designed specifically for a polygonal fit but for a more generic fit that may be even a spline fit. In fact, some model-based techniques use the control points for polygonal fits [25, 26, 28] and some others for spline fits [4]. The algorithm loses very little in terms of polygonal fit accuracy by placing the key flat points in a distributed instead of a clustered manner. For example, in the irregular contour case of Figure 3.7, the error of the optimal fit is 0.8189 pixels while the error of the P & P fit is 2.1701 pixels. The error of an arbitrary polygonal fit for this shape could run as high as 42.8378 pixels. The small compromise the algorithm concedes in the polygonal fit case pays off in the splint fit case, where a clustered distribution like the one favored by the optimal polygonal fit would give very poor results.

5.3 The Active Deformable Model

The formulation of active deformable models used in this work to approximate the object boundary draws on the work done in recent years by the computer vision community on active deformable models of contours, often referred to as "snakes." Given a continuous contour, described as a vector:

$$\mathbf{v}(s) = (x(s),\ y(s)) \tag{3.9}$$

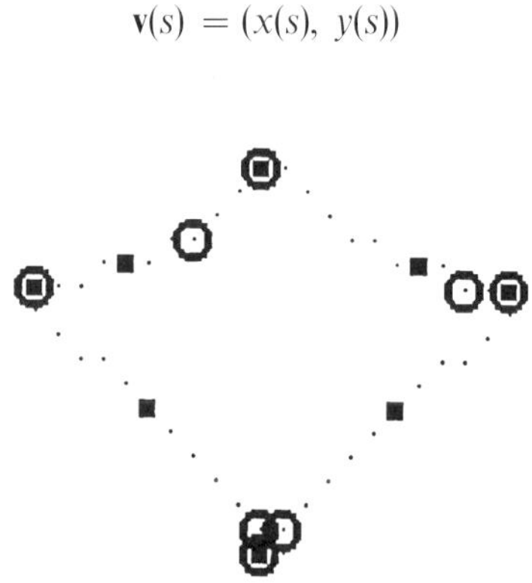

FIGURE 3.7
An irregular contour.

where s is the arc length, Kass *et al.* [16] related the task of finding a contour in an image to the minimization of an energy function (adopting the notation used in [29]):

$$E^*_{snake} = \int_0^1 E_{snake}(\mathbf{v}(s))\, ds = \int_0^1 [E_{int}(\mathbf{v}(s)) + E_{image}(\mathbf{v}(s)) + E_{con}(\mathbf{v}(s))]\, ds \tag{3.10}$$

In this function, E^*_{snake} is the total energy of the active deformable model, E_{int} is a measure of internal energy, such as that caused by curvature, and E_{image} is a function of image characteristics. The term E_{con} is derived from external constraints. When this continuous model is approximated in a discrete domain (e.g., a digital image) the equation becomes:

$$E^*_{snake} = \sum_{j=1}^{n} [\alpha E_{cont}(v_j) + \beta E_{curve}(v_j) + \gamma E_{image}(v_j)] \tag{3.11}$$

in which E_{cont} is derived from the distance between v_j and its neighbors, $v_{(j-1)\bmod n}$ and $v_{(j+1)\bmod n}$. E_{curve} is a function of the angle at point v_j. Again, E_{image} represents the image forces acting on the active deformable model. The terms α, β, and γ are weighting parameters that control the proportion of the active deformable model's energy derived from each of the three terms, which are assumed to be normalized.

Kass *et al.* [16] proposed that a minimum be found for this energy function with a variational calculus approach. Amini *et al.* [2] have proposed a method based on dynamic programming. We have chosen to adopt the greedy method developed by Williams and Shah [29]. In the greedy method, each point on the contour is considered in turn. An energy score is calculated for locations near the current location of the control point and the control point is moved to the location that results in the lowest energy.

The E_{curve}, E_{image}, and E_{cont} terms are usually sufficient to define an active deformable model approximation of an image contour when all terms vary significantly across the neighborhood of possible control point locations. However, using our current techniques, when the active deformable model is placed, it may have several control points that are far enough from the target's image that the image gradient is unvaryingly zero throughout the neighborhood of candidate locations. For these points, the term E_{image} plays no role at all

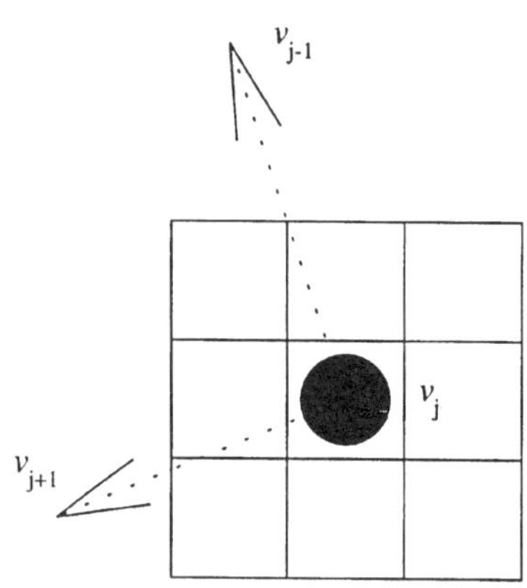

FIGURE 3.8
A single snake point in its window.

and they respond only to the internal energy and external constraints, rather than to a combination of image energy and constraints.

To facilitate the initial placement of the active deformable model, we have augmented the energy equation with an E_{model} term inspired by the "balloon factor" used by Yoshimi and Allen [28] to overcome a tendency toward implosion in their active deformable models.

The E_{model} term is calculated as follows. First, a neighborhood of the control point in the difference image is examined. If the percentage of difference pixels set within the neighborhood falls short of a predetermined level, the control point is defined as "outside" the object's image. To bias movement of the control point toward the object's image, the locations closest to the object's image are assigned the value -1 for E_{model}. Other locations are assigned the value 0. The locations closest to the object's image can be determined because the active deformable model control points are numbered counterclockwise around the closed active deformable model. A similar energy assignment is performed for control points that are "inside" the object's image. Besides aiding initial placement of the contour, this model energy also occasionally comes into play during later tracking stages when an object moves very quickly or has been temporarily lost for some other reason (e.g., occlusion).

5.4 The Control Signal Computation

Concurrently with the energy minimization process already described, a control signal is generated from the current configuration of the active deformable model by a process running on a separate processor. The purpose of this process is to determine the necessary camera translation to recenter the contour extracted by the active deformable model in the image plane.

It is necessary to choose a definition for the location of a contour. We have considered two options (1) the average location of the control points and (2) the centroid of the closed polygon defined by the contour. We have chosen to use the average location of the control points. This definition may be unsatisfying if the control points become bunched together on one side of the contour, but, in practice, this rarely occurs as the smoothness and continuity constraints penalize such configurations. Therefore, the slight improvement in these cases does not justify the additional processing time.

There is much more information in the configuration of the active deformable model than location. Future systems should be able to take use this information for three-dimensional (3-D) tracking and to overcome partial occlusions.

6 THE MINNESOTA ROBOTIC VISUAL TRACKER

The Minnesota Robotic Visual Tracker (MRVT) [5] that was used for these experiments consists of the Robot/Control Subsystem (RCS) and the Visual Processing System (VPS).

The RCS includes a PUMA 560 robotic arm, its Unimate computer–controller, and a VME-based Single Board Computer (SBC). The manipulator's trajectory is controlled by the Unimate controller as directed by path updates provided by an Ironics 68030 VME SBC running CHIMERA. A Sun Sparc-Station 330 hosts CHIMERA and shares its VME bus with the SBC via BIT-3 bus extenders. BIT-3 bus extenders also provide shared-memory communication between the RCS and VPS.

The VPS receives input from a video source such as a camera mounted on the end-effector of a robot arm, a static camera, or stored imagery played back through a Silicon Graphics Indigo or a videotape recorder (see Figure 3.9). The output of the VPS may be displayed in

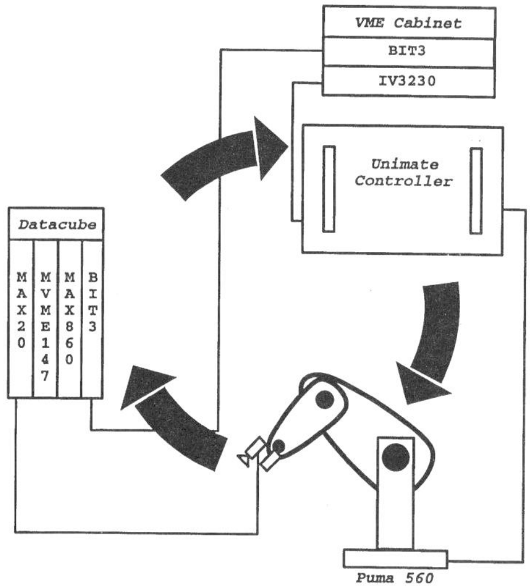

FIGURE 3.9
MRVT system architecture.

a readable format or be transferred to another system component and used as an input into a control subsystem. This flexibility offers a diversity of methods by which software can be developed and tested on our system. The main component of the VPS is a Datacube MaxTower system consisting of a Motorola MVME-147 single board computer running OS-9, a Datacube MaxVideo20 video processor, and a Datacube Max860 vector processor in a portable seven-slot VME chassis. The VPS performs the calculation of the difference

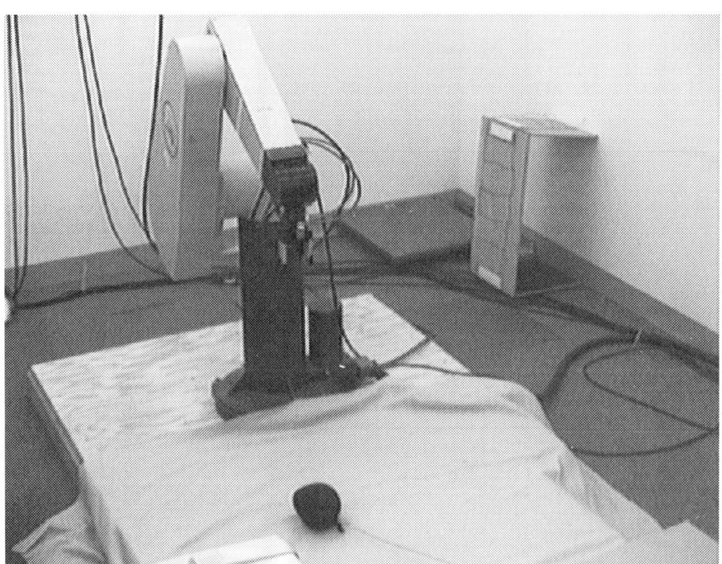

FIGURE 3.10
Experimental setup for balloon tracking.

image and the active deformable model energy minimization and calculates any desired control input. It can supply the data or the input via shared memory to an off-board processor via a Bit-3 bus extender for use as input to the RCS. The video processing and calculations required to produce the desired control input are performed under a pipeline programming model using Datacube's Imageflow libraries.

7 EXPERIMENTS

Initially, two types of experiments were run. In the first, a partially inflated balloon was moved by hand in the robot's work space. These runs were analyzed for timing information as well as qualitative information about system performance. Quantitative measures of tracking quality are not available from these runs, as the nature of the experiments denies access to "ground truth." To obtain quantitative data about system performance, a second set of experiments were conducted. In these trials, an SGI Indigo workstation was used to create a display of an object in motion along a circular path. While the MRVT tracked the object on the display, the control commands issued to the controller were collected. By comparing the control commands with the actual path of the object, tracking performance can be quantified.

In the balloon-tracking experiments, a black balloon attached to a stick was maneuvered in the manipulator's work space by an operator. The work space background was gray and fairly uniform, creating few distracting difference pixels (i.e., nonobject pixels that appear in the difference image). Empirically discovering gains that overcame this noise and resulted in good tracking performance was not difficult. The minimization algorithm performed approximately 2000 point updates per second (e.g., over eight trials, totaling 13 minutes and 9 seconds, eight-point snakes performed 251 updates per second). This update rate was seemed adequate for snakes with as many as 16 control points.

Informal testing did reveal one difficulty with the current implementation. The E_{model} term, which aids initial placement, interferes with tracking when the active deformable model is not a simple polygon. Various techniques to guarantee simplicity have been implemented and tested, but none has been effective without unacceptable performance penalties.

In the second set of experiments, a target was generated on an SGI Indigo and presented on a 27-inch monitor just outside the robot's workspace (see Figure 3.11). This target, a 7.3-cm square, repeatedly traveled in a circular path with a diameter of 25.7 cm or along a square path with sides of 27 cm. While traveling at about 8 cm/sec, deformation was introduced by rotating the square 360 degrees on its z-axis during each circuit. The position commands sent to the Unimate controller were collected. The first 1200 points from two sample runs are plotted in Figures 3.12 and 3.13. These figures contain data from a four-point and an eight-point model, respectively.

These plots demonstrate the trade-off between additional control points and system performance. In the four-point trials, the minimization algorithm performed 505 updates per second and the control loop sent 212 path instructions per second to the arm. In the eight-point trials, the minimization algorithm performed half as many updates per second (250) and only 146 control instructions were sent per second. Apparently, two iterations are not enough for the minimization algorithm to converge. Although the eight-point snake was able to track the target, the plots reveal many more oscillations in the path and a lack of consistency. One goal of future work should be to improve the performance of the minimization algorithm, so that better tracking can be obtained with more complex models.

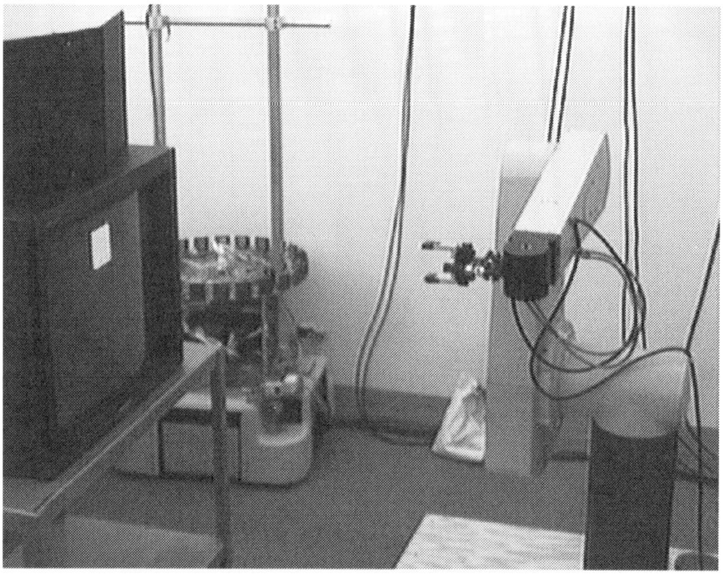

FIGURE 3.11
Experimental setup or quantitative trials.

For comparison, Figure 3.14 plots the path of a manipulator following the same target along a square path at similar speed, demonstrating how the controller handles discontinuities in the target path and acceleration and deceleration of the manipulator.

We also tried the P & P algorithm for the automatic selection of control points. Preliminary results of experiments incorporating the P & P algorithm for automatic control point selection in a model-based tracking scheme [26] suggest that this approach holds great promise. The P & P algorithm extends the previous version of our system in two important ways. It automates the selection of both the number and location of control points.

Experiments were conducted in which a target was presented on a 27-inch monitor located 1 meter from the end-effector mounted camera. The target, a 7.3-cm tall square or triangle, moved around a rectangular path of 100 cm at approximately 8 cm/sec. The position commands sent to the robotic arm were collected and are graphically illustrated in Figures 3.15–

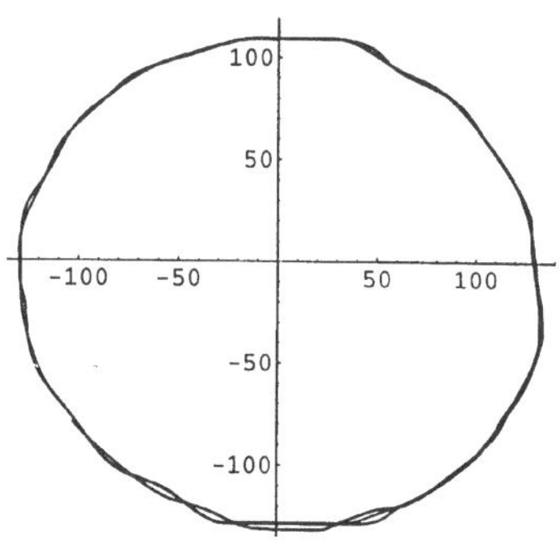

FIGURE 3.12
Tracking rotating square target with a four-point model (measurements in mm).

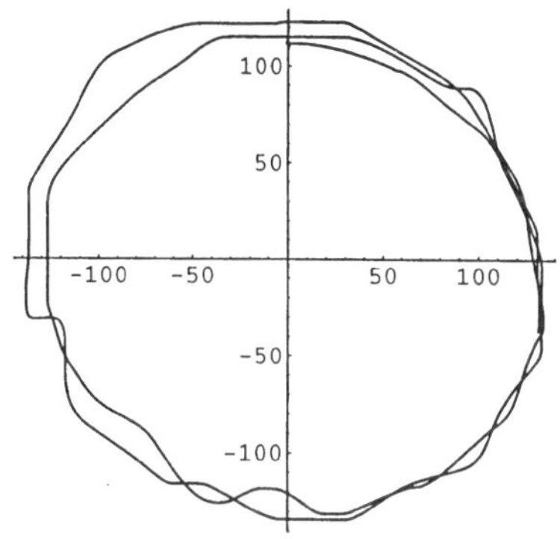

FIGURE 3.13
Tracking a rotating square target with an eight-point model (measurements in mm).

3.17. Previous results [26] (see Figure 3.16) were compared with results using the P & P Algorithm (see Figure 3.17).

The previous system used a *predetermined* number of control points irrespective of the target's shape. These points were *manually* placed near the object contour in a highly regular configuration. The generic constraints used by the tracking algorithm created a bias toward equidistant points and equal angles between edges. The new system uses the P & P algorithm to select control points automatically. Because the P & P algorithm does not choose equally spaced points, the constraints used during tracking were modified to reward configurations with angles close to the initial angles and distances close to the initial distances.

The model-based tracking scheme with the manual selection of control points worked well only when a small number of control points was selected and the points described the contour well. Because that system encouraged equidistance between control points and equal angles between edges, it performed best when the contour of the object being tracked could be approximated by an equilateral polygon (a highly regular shape) with as many vertices as the model had control points. For less regular shapes or control point configurations, performance degraded. For example, the system in [26] lost track of the square target after

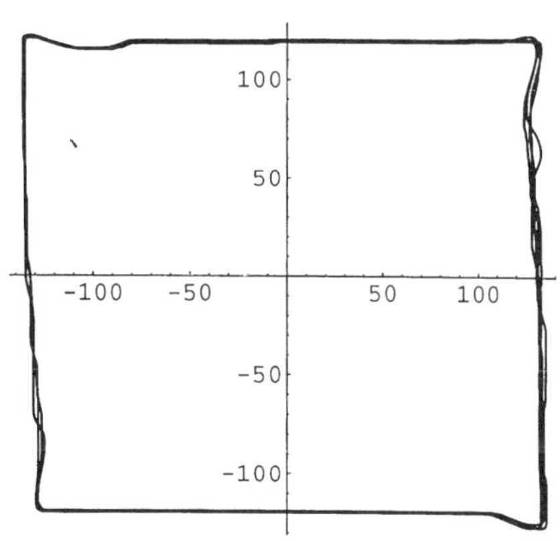

FIGURE 3.14
Tracking a square target with a four-point model (measurements in mm).

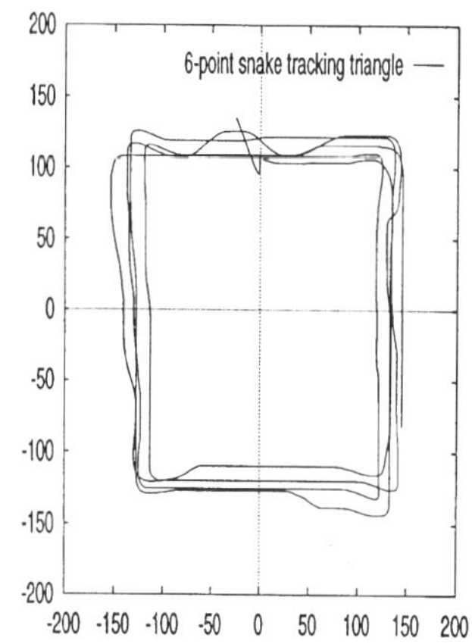

FIGURE 3.15
Tracking of a triangular target with the P & P algorithm (measurements in mm).

just one revolution when an eight-point model was used (see Figure 3.16). The old system was not tested with the (nonequilateral) triangular target, as this target is not a highly regular shape.

The system using the P & P algorithm for automatic point selection performed substantially better. Ten trials were measured. In the first five, the arm tracked the moving square. In the second five, the triangular target was tracked. Results from the first trial with each target are presented in Figures 3.15 and 3.17, respectively. The control point selection algorithm invariable selected 10 points for the square and six points for the triangle that appropriately described the shapes. The tracker maintained tracking of the objects for several revolutions. In this experiment, the P & P tracker exhibited its ability to maintain tracking of different target shapes (square, triangle) at fairly high speeds.

In order to show the generality of the approach, we used the method in another domain (pedestrian tracking). With exactly the same formulation as in the case of visual servoing, our system can successfully track motion of a walking pedestrian, even when the pedestrian's

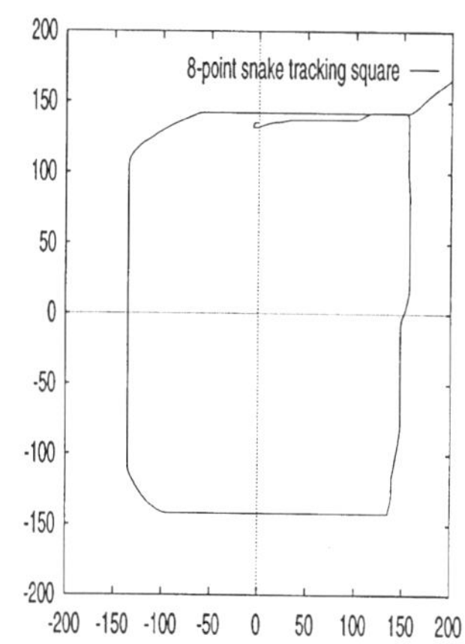

FIGURE 3.16
Tracking of a square target without the P & P algorithm. The target was lost after one revolution (measurements in mm).

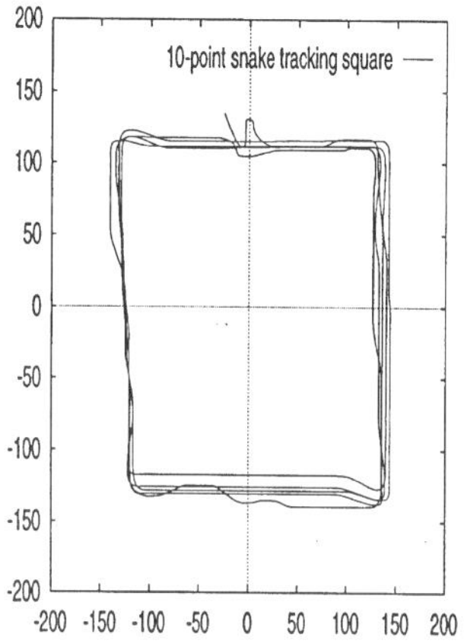

FIGURE 3.17
Tracking of a square target with the P & P algorithm (measurements in mm).

image deforms in unexpected ways such as those caused by thrusting out one's arms or kicking a leg forward in an exaggerated manner (Figures 3.18 and 3.19). It is also fairly robust with respect to occlusions, such as when two pedestrians pass in opposite directions or a single pedestrian passes behind a large tree. Potentially, more than one pedestrian could be tracked simultaneously. Although such a system should be equally robust with respect to occlusions caused by two tracked pedestrians passing one another, it would probably not be possible to tell whether the active deformable models had continued to track the same individual. Such a system might have difficulty distinguishing between two pedestrians approaching one another and then returning the way they came and two pedestrians walking past one another.

FIGURE 3.18
A six-point active deformable model tracking a pedestrian.

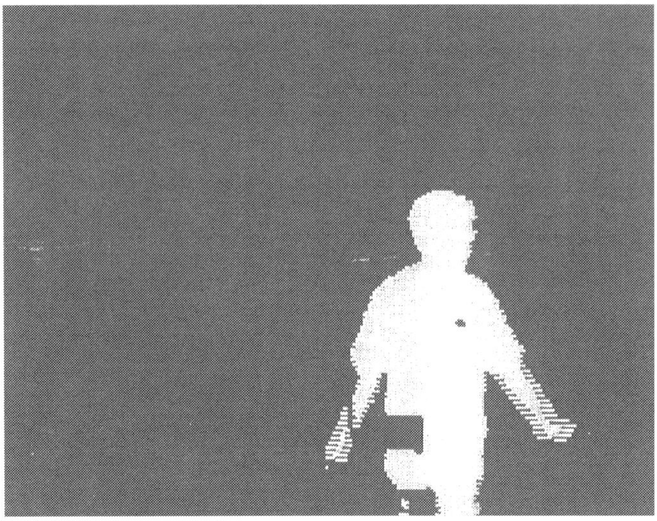

FIGURE 3.19
The difference image that provides image forces for the active deformable model.

Further development of the transportation-related system will require overcoming the inherent limitations of using a difference image to provide image forces for the active deformable model. These problems include short and long time-scale changes in the background caused by lighting changes or continuous regular movement of objects in the scene, for example, the rustling of leaves in the wind. The system is also vulnerable to the effects of camera self-motion. A slight jitter in the camera mount could cause many patches of noise in the difference image. Although these patches will generally be ignored once contour tracking has begun, they do disturb the initial placement of the snake. Richards *et al.* [23] describe two enhancements of the difference image framework to overcome these difficulties. First, by slowly modifying the ground image in a controlled way, changes in the background can be incorporated in the ground image. Second, to overcome the placement problem, additional processing of image regions can be done to identify portions of the image consistent with the appearance of a pedestrian. We plan to incorporate these improvements in our system. Consideration should also be given to methods that would make it possible to mount the camera in a moving vehicle.

8 DISCUSSION

Although the results of the experiments described in Section 7 demonstrate the promise of a system combining the active deformable models for visual tracking with a visual servoing system, they also illustrate drawbacks of the current implementation.

Two factors affect the quality of tracking which must be discriminated. The initial set of experiments conflates changes produced by the sheer number of control points with effects caused by the match between the number of control points and with the points of high curvature on the object boundary.

For example, performance degraded significantly when an eight-point model was used to track a four-sided figure. However, there are two reasonable explanations for this difference.

(1) The extra computation required to minimize an eight-point model reduced the total model update time by a factor large enough to create a qualitative drop-off in overall performance, or (2) the match between object shape and model was not good enough to achieve a stable minimum.

It should be noted that an important strength of the minimization algorithm (its local character) is also a weakness in this case. In no sense does the algorithm trade off higher curvature in one region to achieve lower curvature in another. It relentlessly attempts to reduce curvature (or approach a default angle) at every control point. Further, because the minimization considers only a small number of alternative positions for the control point, it cannot make dramatic changes in configuration to arrive at a globally optimal configuration.

The current system would also benefit from a theoretical basis for the selection of the gains applied to the different elements of the energy function. At present, these gains must be empirically determined for each application by observing the behavior of the active deformable model in action and adjusting parameters to overcome performance deficiencies.

Empirically determined gains have given satisfactory results, but a theoretical framework for gain selection would allow the automatic determination of gains, which will be necessary for deployment if such systems are to be used successfully in commercial manufacturing settings.

9 FUTURE WORK

There are number of promising areas for the further development of this system. These include further exploration of the performance of the algorithm described here and enhancements of the system. These enhancements may either increase the robustness of the system or extend its capabilities.

One issue that should be further explored is the necessity of using difference images as the input to the placement and energy minimization algorithms. If we can assume that more prior information is available about the shape, color, or texture of the object, then an alternative placement algorithm could be developed. If color or texture is known, then a different segmentation routine could be used. If shape is known, then a Generalized Hough Transform could be applied to an edge-detected image. The energy minimization algorithm relies on the difference image to provide image segmentation for the E_{model} term in Eq. (3.11). It is also used as an input to the edge detection process, but this design decision was made solely to increase ease of implementation. When new placement routines are available, the minimization algorithm should be tested with raw gray scale image data.

More experiments should also be done to determine whether the mean of control points is the most useful definition of the center of the active deformable model. Although the mean is very simple to compute, it directly refects the location of the control points—not the location of the entire shape. Consider that there are many sets of control points that define the same boundary (when control points are allowed to be collinear, which they frequently are). These sets of control points do not, however, have the same mean. If the location of a model is defined as the center of mass of the shape defined by its boundary, then the location of the model is invariant across these different sets of control points.

System robustness can be improved by arriving at a reliable measure for system failure. One such measure for the energy minimization technique described in this chapter is a "crossover" in the active deformable model. As mentioned previously, when the model is not a simple polygon, the E_{model} term no longer works in concert with the other energy terms, which frequently leads to uncontrollable expansion of the model. If a computationally inexpensive check could be devised for violation of this condition, the system could be stopped and new control points selected.

Finally, the ability of the system to move relative to the target object can be enhanced by making better use of the information available in the momentary configuration of the active deformable model. Currently, only the location of the mean of the model control points is recovered. By using the relative positions and distributions of the control points, the control input can be extended to take into account apparent scaling or skewing of the model points. For example, increases in the model scale should correlate inversely with decreases in the distance from the object to the camera. Theoretical groundwork for this extension exists in the previous work of Colombo [9] and Andrew Blake's group [8].

10 CONCLUSIONS

We have presented an approach to visual servoing using active deformable models to track image contours. We use these models to track the boundaries of the object's image in the difference image. By tracking the object's contour, we avoid some difficulties associated with visual servoing techniques that track features, such as the occlusion of features or changes in the features due to object deformations. Moreover, because we close the control loop by using partial solutions from an iterative technique, the movement of the manipulator actually simplifies the task of the process that tracks the object using active deformable models. To illustrate the potential of our algorithms, we implemented them on the MRVT system and presented a detailed description of their real-time performance.

Acknowledgments

This work has been supported by the National Science Foundation through Contracts IRI-9410003 and IRI-9502245, the Center for Transportation Studies through Contract USDOT/DTRS 93-G-0017-01, the Minnesota Department of Transportation through Contracts 71789-72983-169 and 71789-72447-159, the Department of Energy (Sandia National Laboratories) through Contracts AC-3752D and AL-3021, the 3M Corporation, the Army High Performance Computing Center and the Army Research Office through Contract DAAH04-95-C-0008, the McKnight Land-Grant Professorship Program, and the Department of Computer Science of the University of Minnesota. Michael Sullivan has also been supported by an NSF Graduate Fellowship in Visual Perception and Motor Control.

REFERENCES

[1] P. K. Allen, B. Yoshimi, and A. Timcenko. Real-time visual servoing. In *Proc. of the IEEE Int. Conf. on Robotics and Automation*, pp. 851–856, April 1991.

[2] A. A. Amini, T. E. Weymouth, and R. C. Jain. Using dynamic programming for solving variational problems in vision. *IEEE Transactions on Pattern Analysis and Machine Intelligence*, 12(9):211–218, 1990.

[3] A. Blake and A. Yuille. *Active Vision*. MIT Press, Cambridge, MA, 1992.

[4] A. Blake, R. Curwen, and A. Zisserman. A framework for spatiotemporal control in the tracking of visual contours. *Int. Journal of Computer Vision*, 11(2):127–145, 1993.

[5] S. A. Brandt, C. E. Smith, and N. P. Papanikolopoulos. The Minnesota Robotic Visual Tracker: A flexible testbed for vision-guided robotic research. In *Proc. of the 1994 IEEE Int. Conf. on Systems, Mean and Cybernetics*, pp. 1363–1368, San Antonio, 1994.

[6] J. J. Brault and R. Plamondon. Segmenting handwritten signatures at their perceptually important points. *IEEE Transactions on Pattern Analysis and Machine Intelligence*, Vol. 15, pp. 953–957, 1993.

[7] F. Chaumette and P. Rives. Vision-based-control for robotic tasks. In *Proc. of the IEEE Int. Workshop on Intelligent Motion Control*, pp. 395–400, 20–22 Aug. 1990.

[8] R. Cipolla and A. Blake. Surface orientation and time to contact from image divergence and deformation. In *ECCV '92*, pp. 187–202. Springer-Verlag, 1992.

[9] C. Colombo, B. Allotta, and P. Dario. Affine visual servoing: A framework for relative positioning with a robot. In *Proc. IEEE Int. Conf. on Robotics and Automation*, pp. 464–471, 1995.

[10] P. I. Corke and R. P. Paul. Video-rate visual servoing for robots. Technical Report MS-CIS-89-18, Grasp Lab, Department of Computer and Information Science, University of Pennsylvania, February 1989.

[11] E. D. Dickmanns and A. Zapp. Autonomous high speed road vehicle guidance by computer vision. In *Proc. of the 10th IFAC World Congress*, July 1987.

[12] E. D. Dickmanns. Computer vision for flight vehicles. *Z. Flugwiss. Weltraumforsch*, 12:71–79, 1988.

[13] J. T. Feddema and C. S. G Lee. Adaptive image feature prediction and control for visual tracking with a hand-eye coordinated camera. *IEEE Trans. on Systems, Man, and Cybernetics*, 20(5):1172-1183, 1990.

[14] B. K. Ghosh. Image based estimation problems in system theory: Motion and shape estimation of a planar textured surface undergoing a rigid flow. In *Proc. of the American Control Conf.*, pp. 1322–1326, 1993.

[15] R. Jain, R. Kasturi, and B. G. Schunck. *Machine Vision*. McGraw-Hill, New York, 1995.

[16] B. Kass, A. Witkin, and D. Terzopoulos. Snakes: Active contour models. *Int. Journal of Computer Vision*, 1(4):321–331, 1987.

[17] A. J. Koivo and N. Houshangi. Real-time vision feedback for servoing of a robotic manipulator with self-tuning controller. *IEEE Trans. Systems, Man, and Cybernetics*, 21(1):134–142, 1991.

[18] K. N. Kutulakos and C. R. Dyer. Recovering shape by purposive viewpoint adjustment. In *Proc. of the IEEE Conf. on Computer Vision and Pattern Recognition*, pp. 16–22, 1992.

[19] R. C. Luo, R. E. Mullen Jr., and D. E. Wessel. An adaptive robotic tracking system using optical flow. In *Proc. of the IEEE Int. Conf. on Robotics and Automation*, 568–573, 1988.

[20] N. P. Papanikolopoulos, P. K. Khosla, and T. Kanade. Visual tracking of a moving target by a camera mounted on a robot: A combination of control and vision. *IEEE Trans. on Robotics and Automation*, 9(1):14–35, February 1993.

[21] I. Pavlidis, M. J. Sullivan, R. Singh, and N. P. Papanikolopoulos. Improving the performance of model-based target tracking through automatic selection of control points. In *Proc. of the Int. Symposium on Robotics and Manufacturing*, pp. 711–716, 1996.

[22] I. Pavlidis and N. P. Papanikolopoulos. A curve segmentation algorithm that automates deformable-model-based target tracking. Technical Report TR 96-041, Department of Computer Science, University of Minnesota, 1996.

[23] C. A. Richards, C. E. Smith, and N. P. Papanikolopoulos. Detection and tracking of traffic objects in IVHS vision sensing modalities. In *Proc. Fifth Annual Meeting of ITS America*, pp. 453–461, 1995.

[24] A. C. Sanderson, L. E. Weiss, and C. P. Neuman. Dynamic sensor-based control of robots with visual feedback. *IEEE Journal of Robotics and Automation*, 3(5):404–417, Oct. 1987.

[25] M. J. Sullivan, C. A. Richards, C. E. Smith, O. Masoud, and N. P. Papanikolopoulos. Pedestrian tracking from a stationary camera using active deformable models. In *Proceedings of the Intelligent Vehicles '95 Symposium*, pp. 90–95, 1995.

[26] M. J. Sullivan and N. P. Papanikolopoulos. Using active deformable models to track deformable objects in robotic visual servoing experiments. In *Proceedings of the 1996 IEEE International Conference on Robotics and Automation*, pp. 2929–2934, Minneapolis, April 22–28, 1996.

[27] M. M. Trivedi, C. Chen, and S. B. Marapane. A vision system for robotic inspection and manipulation. *Computer*, 22(6):91–97, June 1989.

[28] B. H. Yoshimi and P. K. Allen. Visual control of grasping and manipulation tasks. In *Proc. of the 1994 IEEE Int. Conf. on Multisensor Fusion and Integration for Intelligent Systems*, pp. 575–582, Las Vegas, 1994.

[29] D. J. Williams and M. Shah. A fast algorithm for active contours and curvature estimation. *CVGIP: Image Understanding*, 55(1): 14–26, 1992.

CHAPTER 4

Visually Guided Tracking and Manipulation

MING LEI

RVSI Acuity CiMatrix, Canton, Massachusetts

BIJOY K. GHOSH

Department of Systems Science and Mathematics, Washington University, St. Louis, Missouri

1 INTRODUCTION

For a robot manipulator to operate properly in an unstructured environment, it is essential to employ a variety of sensors. A charge-coupled device (CCD) camera–based machine vision sensor is a typical noncontact sensor that provides feedback information to the manipulator controller. By including cameras inside the control servo loop of a manipulator, visual servoing can easily be achieved. There are three basic strategies of visual servoing. The first strategy uses a camera mounted on the end arm of a manipulator, which is commonly referred to as an eye-in-hand configuration. Because of the close proximity of the camera and the end effector to the workpiece, this technique is desirable for close inspection, gauging, and automated part recognition. The second strategy employs an overhead camera. This technique is usually implemented in a carefully designed and controlled environment in which the depth of a scene is known or fixed. For example, it can be used for servo control of a manipulator used to grasp an unoriented workpiece on a workbench or a conveyor. The third strategy is a natural extension of the second strategy; it uses multiple cameras whose pose and zoom may be controlled to improve the viewing conditions. Because cameras are not mounted on the arms more manipulators can be added to the system when needed for multirobot tasks without altering the overall system configuration. Visual servo control may assume different forms. Depending on the choice of feedback representation, a position-based or image-based scheme can be proposed. In a position-based scheme, the three-dimensional (3-D) position and orientation information of the environment is first inferred from a set of derived image features and then used in the manipulator controller. On the other hand, image-based visual servo control defines task reference configurations directly in the image space by using image features that are uniquely related to spatial position and orientation information.

Most of the research on using vision information in servoing robot, especially on visual tracking, has been conducted with the eye-in-hand configuraton [1–6]. Because of the inherent difficulty of recovering a 3-D scene from a single camera, it is natural to seek a task specification directly in the image space. In visual tracking, for example, such a task can be defined simply as "move the manipulator in such a way that the projection of a moving or static object is always at the desired location in the image" [4]. The "desired location" is usually described by some preselected feature points [4] or special marks [3] on the object or obtained through teach-by-showing techniques [2]. Starting from this point of view, the cited work discusses various aspects of the modeling methodology, feature estimation, and control law design. In the eye-in-hand configuration, visual tracking has been formulated as compensating for the instantaneous error of the desired location generated by the motion of the object. Typically, no attempt is made to model this motion, however. As a consequence, the tracking performs well only when the speed is low. Also, the rotational component of the object around an axis perpendicular to the optical axis of the camera must be kept minimal.

With the third strategy, however, the fact that no camera moves with the manipulator admits the possibility of estimating trajectories of a moving object. In doing so, motion prediction becomes possible and tracking performance can be improved. The efficacy of the motion modeling is demonstrated in [6], where an autoregressive (AR) discrete-time model is used to predict the location of features of a moving object for grasping. In a trajectory filtering and prediction technique for a moving object introduced in [7], a manipulator is guided with two cameras to track and grasp an object. The feedback scheme is position based, because the tracking is designed with respect to the centroid of the object, which is obtained by triangulation of the centroids of the 2-D images. In order to obtain the 3-D centroid accurately, the two image centroids must be correctly isolated to correspond to the same 3-D physical point. This is a difficult task, because the perspective projection of a 3-D centroid of an object does not necessarily correspond to the centroid of the image of the object.

Most approaches to motion analysis assume the camera-centered structure [8, 9]. The motion is usually considered as a rotation around the origin of the camera coordinate frame, followed by a translation. As different values for the motion parameters may be obtained for each new frame, these parameters have to be combined in order to produce the actual (natural) motion of the scene. Object-centered models for estimating the natural values for translation and the center of rotation are proposed in [10, 11]. The center is not necessarily the geometric center; it may be a point that has only a purely translational motion. The equations are developed with respect to the unknown center using multiframe features. In order to do this, special assumptions are usually made. For example, the motion is formulated as a constant rotation around a center followed by a constant translation of the center. It is shown that the solution of the simultaneous equations so obtained is less sensitive to noise than in the two-view case.

Motion parameters cannot be completely determined from a monocular sequence. The 3-D structure interpretation as well as the translation can be specified only up to an arbitrary scale factor. However, the problem can be overcome with binocular images obtained by stereo vision approaches [12]. In stereo vision algorithms, however, it is usually required to establish some correspondence, or matching features, between the two images taken at the same time. Because the matching process can be time consuming, alternative methods have been suggested, such as the correspondenceless approaches [13].

In this chapter, we present a new approach to the problem of tracking and grasping a moving object by a manipulator with multiple cameras. We propose a new object-centered model for the motion of the object by defining a 3-D reference point in the object. As the reference point is used to represent the translational motion of the object, the tracking can be designed with respect to the 3-D point. However, no attempt is made to recover 3-D feature points of the object on the basis of stereo matching. Instead, only the image of the reference point in each camera is needed, and it is determined from the 2-D features of the object within the same camera. By properly defining an error function between the image of the robot gripper and that of the reference point, the image-based tracking control law is obtained using the nonlinear regulator theory.

The chapter is organized as follows. Section 2 presents a complete modeling study of the multicamera hand–eye system. Section 3 proposes a new approach to the estimation of motion of the object. The problem of determining the reference point in the image planes is completely studied. A general framework for image-based robot tracking and grasping of a moving object is presented in Section 4. In Section 5 we report simulation results. Section 6 concludes the chapter.

2 MODELING OF THE TRACKING AND GRASPING SYSTEM

2.1 System Configuration

Figure 4.1 illustrates the configuration of the tracking and grasping system, where $\mathcal{I}$ ($\mathcal{I} \geqslant 2$) fixed cameras are directed toward the work space of a robot manipulator. The robot's base coordinate frame is chosen as the world coordinate frame (WCF) of the system, which is denoted by $o_w x_w y_w z_w$. For $i = 1, \ldots, \mathcal{I}$, let $o_{ci} x_{ci} y_{ci} z_{ci}$ denote the ith camera coordinate frame (CCFi), where o_{ci} and $o_{ci} z_{ci}$ are respectively the center of the lens and the optical axis of the ith camera. For any point p in the work space, its coordinates in the WCF and CCFi are denoted respectively by (x_w^p, y_w^p, z_w^p) and $(x_{ci}^p, y_{ci}^p, z_{ci}^p)$ $(i = 1, \ldots, \mathcal{I})$.

The coordinates of point p in the WCF and the CCFi can be related by

$$\begin{bmatrix} x_{ci}^p \\ y_{ci}^p \\ z_{ci}^p \end{bmatrix} = R_i \begin{bmatrix} x_w^p \\ y_w^p \\ z_w^p \end{bmatrix} + d_i \qquad (i = 1, \ldots, \mathcal{I}) \tag{4.1}$$

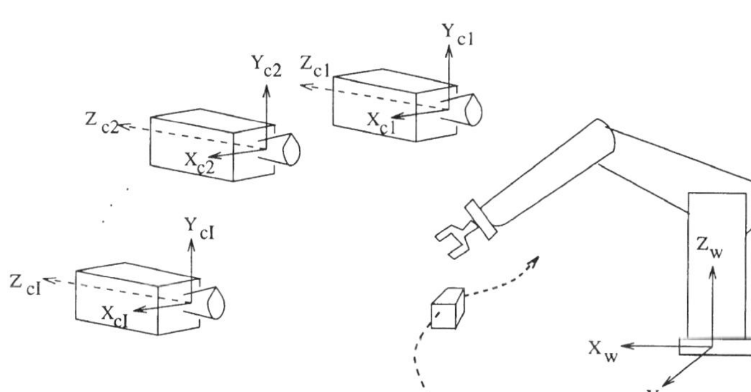

FIGURE 4.1
Configuration of the tracking and grasping system.

where

$$R_i = \begin{bmatrix} r_{i1} & r_{i2} & r_{i3} \\ r_{i4} & r_{i5} & r_{i6} \\ \tau_{i7} & r_{i8} & r_{i9} \end{bmatrix}, \qquad d_i = \begin{bmatrix} d_{i1} \\ d_{i2} \\ d_{i3} \end{bmatrix}$$

are known 3×3 orthonormal and 3×1 matrices, respectively.

2.2 Model of the Robot Manipulator

For the task of tracking and grasping a moving object, a six degree-of-freedom PUMA 560 manipulator is used with a parallel jaw gripper mounted on the end effector. The dynamics of the robot is described by

$$D(q)\ddot{q} + C(q, \dot{q}) + G(q) = \tau \tag{4.2}$$

where $q = [q_1\ q_2\ q_3\ q_4\ q_5\ q_6]^{\mathrm{T}}$, $\tau = [\tau_1\ \tau_2\ \tau_3\ \tau_4\ \tau_5\ \tau_6]^{\mathrm{T}}$, q_i, $\dot{q}_i$, $\ddot{q}_i$ $(i = 1, 2, \ldots, 6)$ are the position, velocity, and acceleration of joint i, respectively; τ_i is the torque acting at joint i; and D, C, and G are respectively the 6×6 inertia matrix, 6×1 vector of centripetal and Coriolis terms, and 6×1 vector of gravity terms. The position and orientation of the gripper, denoted by $[x_w^g\ y_w^g\ z_w^g]^{\mathrm{T}}$ and $[\mathbf{n}^g\ \mathbf{s}^g\ \mathbf{a}^g]$, respectively, can be represented as trigonometric functions of q by the forward kinematic equations. Alternatively, using the OAT representation [14], the orientation of the gripper can also be described by three independent Euler angles O^g, A^g, and T^g. Clearly, O^g, A^g, T^9 are also functions of the joint displacement q.

On introducing state variables $x_1 = q$, $x_2 = \dot{q}$, and $x^{\mathrm{T}} = [x_1^{\mathrm{T}}\ x_2^{\mathrm{T}}]$, (4.2) has the form

$$\begin{bmatrix} \dot{x}_1 \\ \dot{x}_2 \end{bmatrix} = \begin{bmatrix} x_2 \\ D^{-1}(C(x) + G(x_1)) \end{bmatrix} + \begin{bmatrix} 0 \\ D^{-1}(x_1) \end{bmatrix} \tau \tag{4.3}$$

$$\triangleq f(x) + g(x)\tau \tag{4.4}$$

The pose (position and orientation) of the gripper can be described by

$$\begin{bmatrix} x_w^g \\ y_w^g \\ z_w^g \\ O^g \\ A^g \\ T^g \end{bmatrix} = \begin{bmatrix} h_1(x_1) \\ h_2(x_1) \\ h_3(x_1) \\ h_4(x_1) \\ h_5(x_1) \\ h_6(x_1) \end{bmatrix} \triangleq h(x_1) \tag{4.5}$$

It is well known [15] that for system (4.4) with output equation (4.5), a state feedback

$$\tau = \alpha(x) + \beta(x)v \tag{4.6}$$

and change of coordinates

$$[\xi_1\quad \xi_2\quad \cdots\quad \xi_{12}]^{\mathrm{T}} = \Phi(x) \tag{4.7}$$

with $\alpha(x)$, $\beta(x)$, and $\Phi(x)$ appropriately computed from vector fields f, g, and h will render the system into two decoupled subsystems:

$$\underline{\text{Plant (I):}} \qquad \dot{\xi}^{(1)} = A\xi^{(1)} + Bv^{(1)} \tag{4.8}$$

$$[x_w^g \quad y_w^g \quad z_w^g]^{\mathrm{T}} = C\xi^{(1)} \tag{4.9}$$

$$\underline{\text{Plant (II):}} \qquad \dot{\xi}^{(2)} = A\xi^{(2)} + Bv^{(2)} \tag{4.10}$$

$$[O^g \quad A^g \quad T^g]^{\mathrm{T}} = C\xi^{(2)} \tag{4.11}$$

where

$$\xi^{(1)} = [\xi_1 \quad \xi_2 \quad \cdots \quad \xi_6)^{\mathrm{T}}, \qquad \xi^{(2)} = [\xi_7 \quad \xi_8 \quad \cdots \quad \xi_{12}]^{\mathrm{T}}$$

$$v^{(1)} = [v_1 \quad v_2 \quad v_3]^{\mathrm{T}}, \qquad v^{(2)} = [v_4 \quad v_5 \quad v_6]^{\mathrm{T}}$$

$$A = \begin{bmatrix} 0_{3\times 3} & I_{3\times 3} \\ 0_{3\times 3} & 0_{3\times 3} \end{bmatrix}, \qquad B = \begin{bmatrix} 0_{3\times 3} \\ I_{3\times 3} \end{bmatrix}, \qquad C = [I_{3\times 3} \quad 0_{3\times 3}]$$

The linearized and decoupled plants (4.8), (4.9), (4.10), and (4.11) will be used in the design of the tracking and grasping controllers.

2.3 Modeling the System of Cameras

Each camera is modeled by the perspective projection. For a point p in the robot work space with world coordinates (x_w^p, y_w^p, z_w^p) and camera coordinates $(x_{ci}^p, y_{ci}^p, z_{ci}^p)$ in the CCF_i $(i = 1, \ldots, \mathscr{I})$, its image in image plane $O_i X_i Y_i$ can be given by

$$X_i^p = \frac{x_{ci}^p}{z_{ci}^p} f_i, \qquad Y_i^p = \frac{y_{ci}^p}{z_{ci}^p} f_i \qquad (i = 1, \ldots, \mathscr{I}) \tag{4.12}$$

where f_i is the focal length of the ith camera.

For convenience, we denote

$$\mathbf{x}^p = [x_w^p \quad y_w^p \quad z_w^p]^{\mathrm{T}}$$

$$\mathbf{X}^p = [X_1^p \quad Y_1^p \quad X_2^p \quad Y_2^p \quad \cdots \quad X_{\mathscr{I}}^p \quad Y_{\mathscr{I}}^p]^{\mathrm{T}}$$

Combining (4.1) and (4.12) yields

$$X_i^p = \frac{r_{i1}x_w^p + r_{i2}y_w^p + r_{i3}z_w^p + d_{i1}}{r_{i7}x_w^p + r_{i8}y_w^p + r_{i9}z_w^p + d_{i3}} f_i \tag{4.13}$$

$$Y_i^p = \frac{r_{i4}x_w^p + r_{i5}y_w^p + r_{i6}z_w^p + d_{i2}}{r_{i7}x_w^p + r_{i8}y_w^p + r_{i9}z_w^p + d_{i3}} f_i \tag{4.14}$$

Denote

$$Q_{i1} = [q_{i1} \quad q_{i2} \quad q_{i3}]$$

$$Q_{i2} = [q_{i4} \quad q_{i5} \quad q_{i6}]$$

$$Q_{i3} = [q_{i7} \quad q_{i8} \quad q_{i9}]$$
$$q_{ih} = r_{ih} f_i / d_{i3} \qquad (i = 1, \ldots, \mathcal{I}; h = 1, 2, \ldots, 6)$$
$$q_{ih} = r_{ih} / d_{i3} \qquad (i = 1, \ldots, \mathcal{I}; h = 7, 8, 9)$$
$$p_{ij} = d_{ij} f_i / d_{i3} \qquad (i = 1, \ldots, \mathcal{I}; j = 1, 2)$$

Then (4.13) and (4.14) can be written as

$$X_i^p = \frac{Q_{i1} C\xi^{(1)} + p_{i1}}{Q_{3i} C\xi^{(1)} + 1} \triangleq p_{i1}(\mathbf{x}^p) \tag{4.15}$$

$$Y_i^p = \frac{Q_{i2} C\xi^{(1)} + p_{i2}}{Q_{3i} C\xi^{(1)} + 1} \triangleq p_{i2}(\mathbf{x}^p) \tag{4.16}$$

Define a mapping P by

$$\mathbf{X}^p = P(\mathbf{x}^p) \triangleq [p_{11}(\mathbf{x}^p) \quad p_{12}(\mathbf{x}^p) \quad \cdots \quad p_{\mathcal{I}1}(\mathbf{x}^p) \quad p_{\mathcal{I}2}(\mathbf{x}^p)]^{\mathrm{T}} \tag{4.17}$$

From (4.17), we have

$$M(\mathbf{X}^p)\mathbf{x}^p = N(\mathbf{X}^p) \tag{4.18}$$

where

$$M(\mathbf{X}^p) = \begin{bmatrix} Q_{11} - X_1^p Q_{13} \\ Q_{12} - Y_1^p Q_{13} \\ \vdots \\ Q_{\mathcal{I}1} - X_{\mathcal{I}}^p Q_{\mathcal{I}3} \\ Q_{\mathcal{T}2} - Y_{\mathcal{I}}^p Q_{\mathcal{I}3} \end{bmatrix}, \qquad N(\mathbf{X}^p) = \begin{bmatrix} X_1^p - p_{11} \\ Y_1^p - p_{12} \\ \vdots \\ X_{\mathcal{I}}^p - p_{\mathcal{I}1} \\ Y_{\mathcal{I}}^p - p_{\mathcal{I}1} \end{bmatrix}$$

Because the number of equations in (4.18) is larger than the number of unknowns, the least squares solution of $\mathbf{x}^p$ can be used to define the inverse map of P:

$$\mathbf{x}^p = P^{-1}(\mathbf{X}^p) \triangleq (M^{\mathrm{T}}M)^{-1}M^{\mathrm{T}}N \tag{4.19}$$

Therefore, the position vector $\mathbf{x}^p$ and the image vector $\mathbf{X}^p$ of point p can be related by transformations (4.17) and (4.19).

2.4 Modeling a Moving Rigid Object

We have discussed the estimation of motion using a camera-centered or object-centered model. For the task of motion tracking, where prediction of the position of the object is important, the object-centered model may be more suitable. In the following, we model the general motion of a rigid object by decomposing it into a translation of some point, which we shall call a reference point, and a rotation of the object around the point. For ease of conveying our idea of the tracking and grasping design, we assume that the moving object is a regular hexahedron.

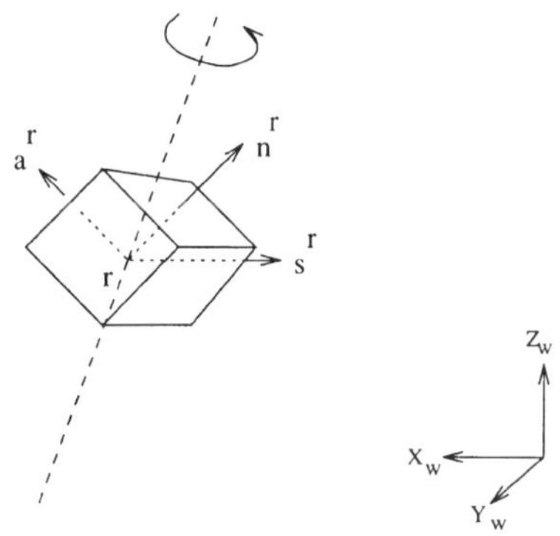

FIGURE 4.2
Motion of a rigid object.

A reference point of a moving object is defined as a 3-D point such that

1. Its position relative to the object does not change over time.
2. The point has a purely translational motion and no rotation.
3. With the constraint that the translational motion of the reference point is smooth, its distance from the centroid of the object is kept as small as possible.

The definition of the reference point consists of two important aspects. First, the choice of the reference point should be such that its motion is smooth and therefore predictable. Second, because the tracking and grasping will be designed with respect to the reference point, the point should be kept close to the centroid of the object in order to achieve stable grasping.

The rotational motion of the object can be described as follows. Let $r\mathbf{n}^r\mathbf{s}^r\mathbf{a}^r$ be a coordinate frame attached to the object centered at the reference point with each axis being parallel to the corresponding sides of the hexahedral object (Figure 4.2). When the object moves, the orientation of the coordinate frame changes (with respect to the WCF) according to (see also [16])

$$[\dot{\mathbf{n}}^r \quad \dot{\mathbf{s}}^r \quad \dot{\mathbf{a}}^r] = \Omega[\mathbf{n}^r \quad \mathbf{s}^r \quad \mathbf{a}^r] \tag{4.20}$$

with

$$\Omega = \begin{bmatrix} 0 & -\omega_3 & \omega_2 \\ \omega_3 & 0 & -\omega_1 \\ -\omega_2 & \omega_1 & 0 \end{bmatrix} \tag{4.21}$$

where $\omega = [\omega_1 \ \omega_2 \ \omega_3]^T$ is the angular velocity.

3 ESTIMATION OF THE MOTION FIELD OF THE REFERENCE POINT

3.1 Formulation of the 3-D Reference Point Problem

The three assumptions that a reference point should satisfy can be described mathematically as follows. Let $(x_w^{uj}, y_w^{uj}, z_w^{uj})$ $(j = 1, 2, \ldots, \mathcal{J})$ be the coordinates of 3-D feature points in the objects in the WCF. We assume these points can be observed by each camera. Since $r\mathbf{n}^r\mathbf{s}^r\mathbf{a}^r$ is the coordinate frame whose origin is at the reference point $\mathbf{x}^r = [x_w^r \ y_w^r \ z_w^r]^T$, by assumption (1), there exist $k_{j1}, k_{j2}, k_{j3} \in \mathbb{R}^1$ such that

$$\begin{bmatrix} x_w^{uj} \\ y_w^{uj} \\ z_w^{uj} \end{bmatrix} - \begin{bmatrix} x_w^r \\ y_w^r \\ z_w^r \end{bmatrix} = k_{j1}\mathbf{n}^r + k_{j2}\mathbf{s}^r + k_{j3}\mathbf{a}^r = [\mathbf{n}^r \quad \mathbf{s}^r \quad \mathbf{a}^r] \begin{bmatrix} k_{j1} \\ k_{j2} \\ k_{j3} \end{bmatrix} \tag{4.22}$$

holds for any time t. Differentiating (4.22) and using (4.20), the 3-D motion of the feature point is described by

$$\begin{bmatrix} \dot{x}_w^{uj} \\ \dot{y}_w^{uj} \\ \dot{z}_w^{uj} \end{bmatrix} = \begin{bmatrix} \dot{x}_w^r \\ \dot{y}_w^r \\ \dot{z}_w^r \end{bmatrix} + \Omega \left(\begin{bmatrix} x_w^{uj} \\ y_w^{uj} \\ z_w^{uj} \end{bmatrix} - \begin{bmatrix} x_w^r \\ y_w^r \\ z_w^r \end{bmatrix} \right) \tag{4.23}$$

Using (4.1), Eq. (4.2) can be written in the CCFi as

$$\begin{bmatrix} \dot{x}_{ci}^{uj} \\ \dot{y}_{ci}^{uj} \\ \dot{z}_{ci}^{uj} \end{bmatrix} = \begin{bmatrix} \dot{x}_{ci}^r \\ \dot{y}_{ci}^r \\ \dot{z}_{ci}^r \end{bmatrix} + \bar{\Omega} \left(\begin{bmatrix} x_{ci}^{uj} \\ y_{ci}^{uj} \\ z_{ci}^{uj} \end{bmatrix} - \begin{bmatrix} x_{ci}^r \\ y_{ci}^r \\ z_{ci}^r \end{bmatrix} \right) \tag{4.24}$$

where $\bar{\Omega} = R_i R_i^{\mathrm{T}}$. Clearly, $\bar{\Omega}$ is skew symmetric. Hence

$$\bar{\Omega} = \begin{bmatrix} 0 & -\bar{\omega}_3 & \bar{\omega}_2 \\ \bar{\omega}_3 & 0 & -\bar{\omega}_1 \\ -\bar{\omega}_2 & \bar{\omega}_1 & 0 \end{bmatrix} \tag{4.25}$$

In assumption (3), the smoothness of the motion is interpreted as the 3-D velocity of the reference point changing slowly over any two adjacent frames. Let any two adjacent frames be denoted by $k = 1, 2$. Then it is required that

$$\dot{z}_{ci}^r(1) = \dot{z}_{ci}^r(2) \tag{4.26}$$

and the objective function

$$J_2 = \frac{f_i^2}{(\dot{z}_{ci}^r(1))^2} [(\dot{x}_{ci}^r(1) - \dot{x}_{ci}^r(2))^2 + (\dot{y}_{ci}^r(1) - \dot{y}_{ci}^r(2))^2] \tag{4.27}$$

is minimized. Furthermore, the requirement that the reference point be as close to the centroid as possible can be satisfied by minimizing

$$J_1 = \sum_{k=1}^{2} \frac{f_i^2}{(z_{ci}^r(k))^2} [(x_{ci}^r(k) - x_{ci}^c(k))^2 + (y_{ci}^r(k) - y_{ci}^c(k))^2 + (z_{ci}^r(k) - z_{ci}^c(k))^2] \tag{4.28}$$

where weighting factors $z_{ci}^r(1)$ and $z_{ci}^r(2)$ indicate that the object is paid greater attention when it moves closer to the ith camera.

Finally, by assumption (2), we have

$$\begin{bmatrix} x^r_{ci}(2) \\ y^r_{ci}(2) \\ z^r_{ci}(2) \end{bmatrix} = \begin{bmatrix} x^r_{ci}(1) \\ y^r_{ci}(1) \\ z^r_{ci}(1) \end{bmatrix} + T_v \begin{bmatrix} \dot{x}^r_{ci}(1) \\ \dot{y}^r_{ci}(1) \\ \dot{z}^r_{ci}(1) \end{bmatrix} \tag{4.29}$$

where T_v is the vision sampling period.

Let

$$J = J_1 + \varepsilon J_2 \tag{4.30}$$

where $\varepsilon \geqslant 0$ is a weighting factor. We have

Definition 1 *The 3-D reference point problem is to find a 3-D point* $(x^r_{ci}(k),\ y^r_{ci}(k),\ z^r_{ci}(k))$ *for* $k = 1, 2$ *that minimizes the objective function* J in (4.30) *subject to the constraint* (4.29) *and the constraints given by Eq.* (4.24) *for* $t = kT_v$, $k = 1, 2$ *and* $j = 1, 2, \ldots, \mathcal{J}$. *The point obtained is called the 3-D reference point.*

Suppose that $(x^{uj}_{ci}(k),\ y^{uj}_{ci}(k),\ z^{uj}_{ci}(k))$ and $(\dot{x}^{uj}_{ci}(k),\ \dot{y}^{uj}_{ci}(k),\ \dot{z}^{uj}_{ci}(k))$ are known for $j = 1, 2, \ldots, \mathcal{J}$, $k = 1, 2$. Assume further the $\bar{\Omega}$ can be found from another source [17, 18]. Then the reference point $(x^r_{ci}(k),\ y^r_{ci}(k),\ z^r_{ci}(k))$ and its 3-D velocity field $(\dot{x}^r_{ci}(k),\ \dot{y}^r_{ci}(k),\ \dot{z}^r_{ci}(k))$ can be determined by solving the 3-D reference problem. However, in order to find $(x^{uj}_{ci}(k),\ y^{uj}_{ci}(k),\ z^{uj}_{ci}(k))$ from images of (at least two) different cameras, the problem of matching feature points among the cameras has to be solved. This is a difficult task, especially when real-time implementation is required. To avoid the point-matching problem, it is essential that the image $(X^r_i(k),\ Y^r_i(k))$ of the reference point and its optical flow $(\dot{X}^r_i(k),\ \dot{Y}^r_i(k))$ in the ith camera be determined directly from the measurement $(X^{uj}_i(k),\ Y^{uj}_i(k))$ and $(\dot{X}^{uj}_i(k),\ \dot{Y}^{uj}_i(k))$ of feature points from the ith camera itself, where the related quantities are defined as

$$X^r_i = \frac{x^r_{ci}}{z^r_{ci}} f_i, \qquad Y^r_i = \frac{y^r_{ci}}{z^r_{ci}} f_i \tag{4.31}$$

$$X^{uj}_i = \frac{x^{uj}_{ci}}{z^{uj}_{ci}} f_i, \qquad Y^{uj}_i = \frac{y^{uj}_{ci}}{z^{uj}_{ci}} f_i \tag{4.32}$$

3.2 Formulation of the 2-D Reference Point Problem

We want to determine (X^r_i, Y^r_i) and $(\dot{X}^r_i, \dot{Y}^r_i)$ from (X^{uj}_i, Y^{uj}_i) and $(\dot{X}^{uj}_i, \dot{Y}^{uj}_i)$, the measurement obtained purely from the ith camera. In this section, since all the derivations are with respect to the same camera, the subscripts "i" and "ci" can be dropped for notational brevity. We further simplify the notation for the reference point by using (X, Y) and (x, y, z) in place of (X^r, Y^r) and (x^r, y^r, z^r), respectively. Moreover, we denote T_v *by* T.

To reformulate the reference point problem in the 2-D image space, we want to represent the equations involved in Definition 1 in the 2-D image coordinates. To this end, we can rewrite J_2 in (4.27) as follows. By (4.31),

$$xf = Xz, \qquad yf = Yz \tag{4.33}$$

Differentiating these equations yields

$$\dot{x}f = \dot{X}z + X\dot{z} \tag{4.34}$$

$$\dot{y}f = \dot{Y}z + Y\dot{z} \tag{4.35}$$

Noticing that $\dot{z}(1) = \dot{z}(2)$ in (4.26), we have

$$\begin{aligned} J_2 &= \left(\frac{\dot{x}(1)f - \dot{x}(2)f}{\dot{z}(1)}\right)^2 + \left(\frac{\dot{y}(1)f - \dot{y}(2)f}{\dot{z}(1)}\right)^2 \\ &= \left(\frac{\dot{X}(1)z(1) + X(1)\dot{z}(1)}{\dot{z}(1)} - \frac{\dot{X}(2)z(2) + X(2)\dot{z}(2)}{\dot{z}(2)}\right)^2 \\ &\quad + \left(\frac{\dot{Y}(1)z(1) + Y(1)\dot{z}(1)}{\dot{z}(1)} - \frac{\dot{Y}(2)z(2) + Y(2)\dot{z}(2)}{\dot{z}(2)}\right)^2 \end{aligned}$$

Denoting $\rho = \dot{z}/z$, J_2 can be written as

$$J_2 = \left(X(1) - X(2) + \frac{\dot{X}(1)}{\rho(1)} - \frac{\dot{X}(2)}{\rho(2)}\right)^2 + \left(Y(1) - Y(2) + \frac{\dot{Y}(1)}{\rho(1)} - \frac{\dot{Y}(2)}{\rho(2)}\right)^2 \tag{4.36}$$

Here $\rho(1)$ and $\rho(2)$ are assumed to be nonzero. The treatment of the degenerate cases when $\rho(1) = \rho(2) = 0$ can be found in [18].

Consider J_1 in (4.28). The location of the centroid of an object is generally not known. The centroid can be approximated by

$$x^c = \frac{1}{\mathscr{J}} \sum_{j=1}^{\mathscr{J}} x^{uj}, \qquad y^c = \frac{1}{\mathscr{J}} \sum_{j=1}^{\mathscr{J}} y^{uj}, \qquad z^c = \frac{1}{\mathscr{J}} \sum_{j=1}^{\mathscr{J}} z^{uj} \tag{4.37}$$

Denote $d^{uj} = z^{uj}/z$. By (4.31),

$$\frac{x^c f}{z} = \frac{1}{\mathscr{J}} \sum_{j=1}^{\mathscr{J}} \frac{x^{uj} f}{z^{uj}} \frac{z^{uj}}{z} = \frac{1}{\mathscr{J}} \sum_{j=1}^{\mathscr{J}} X^{uj} d^{uj} \tag{4.38}$$

$$\frac{y^c f}{z} = \frac{1}{\mathscr{J}} \sum_{j=1}^{\mathscr{J}} \frac{y^{uj} f}{z^{uj}} \frac{z^{uj}}{z} = \frac{1}{\mathscr{J}} \sum_{j=1}^{\mathscr{J}} Y^{uj} d^{uj} \tag{4.39}$$

$$\frac{z^c f}{z} = \frac{1}{\mathscr{J}} \sum_{j=1}^{\mathscr{J}} \frac{z^{uj} f}{z} = f \frac{1}{\mathscr{J}} \sum_{j=1}^{\mathscr{J}} d^{uj} \tag{4.40}$$

Therefore, J_1 can be rewritten as

$$J_1 = \sum_{k=1}^{2} \left[\left(\frac{x(k) - x^c(k)}{z(k)}\right)^2 + \left(\frac{y(k) - y^c(k)}{z(k)}\right)^2 + \left(\frac{z(k) - z^c(k)}{z(k)}\right)^2\right] f^2 \tag{4.41}$$

or, in the 2-D image coordinates,

$$J_1 = \sum_{k=1}^{2} \left[(X(k) - \frac{1}{\mathscr{J}} \sum_{j=1}^{\mathscr{J}} X^{uj}(k) d^{uj}(k))^2 + (Y(k) - \frac{1}{\mathscr{J}} \sum_{j=1}^{\mathscr{J}} Y^{uj}(k) d^{uj}(k))^2 + f^2 \left(1 - \frac{1}{\mathscr{J}} \sum_{j=1}^{\mathscr{J}} d^{uj}(k)\right)^2 \right] \tag{4.42}$$

Equation (4.29) can be written in the 2-D image coordinates as follows. From Eq. (4.26), multiplying both sides of (4.29) by $f/\dot{z}(2)$ and $f/\dot{z}(1)$, respectively, yields

$$\begin{bmatrix} x(2)\,f/\dot{z}(2) \\ y(2)\,f/\dot{z}(2) \\ z(2)\,f/\dot{z}(2) \end{bmatrix} = \begin{bmatrix} x(1)f/\dot{z}(1) \\ y(1)f/\dot{z}(1) \\ z(1)f/\dot{z}(1) \end{bmatrix} + T \begin{bmatrix} \dot{x}(1)f/\dot{z}(1) \\ \dot{y}(1)f/\dot{z}(1) \\ \dot{z}(1)f/\dot{z}(1) \end{bmatrix} \tag{4.43}$$

Using (4.31), (4.34), and (4.35), Eq. (4.43) becomes

$$\begin{bmatrix} X(2)/\rho(2) \\ Y(2)/\rho(2) \\ f/\rho(2) \end{bmatrix} = \begin{bmatrix} X(1)/\rho(1) \\ Y(1)/\rho(1) \\ f/\rho(1) \end{bmatrix} + T \begin{bmatrix} \dot{X}(1)/\rho(1) + X(1) \\ \dot{Y}(1)/\rho(1) + Y(1) \\ f \end{bmatrix} \tag{4.44}$$

Finally, we want to represent Eq. (4.24) in the 2-D image coordinates. To this end, multiplying both sides of (4.24) by f/z^{uj} yields

$$\frac{\dot{x}^{uj}}{z^{uj}} f = \frac{\dot{x}}{z} f/d^{uj} - \bar{\omega}_3(Y^{uj} - Y/d^{uj}) + \bar{\omega}_2(1 - 1/d^{uj}) f \tag{4.45}$$

$$\frac{\dot{y}^{uj}}{z^{uj}} f = \frac{\dot{y}}{z} f/d^{uj} - \bar{\omega}_3(X^{uj} - X/d^{uj}) + \bar{\omega}_1(1 - 1/d^{uj}) f \tag{4.46}$$

$$\frac{\dot{z}^{uj}}{z^{uj}} f = \frac{\dot{z}}{z} f/d^{uj} - \bar{\omega}_2(X^{uj} - X/d^{uj}) + \bar{\omega}_1(Y^{uj} - Y/d^{uj}) f \tag{4.47}$$

Differentiating (4.32) and using (4.31), (4.45), (4.46), and 4.47), we have

$$\begin{aligned} \dot{X}^{uj} &= \frac{\dot{x}^{uj} f}{z^{uj}} - X^{uj}\frac{\dot{z}^{uj}}{z^{uj}} \\ &= \left[\dot{X} + (X - X^{uj})\frac{\dot{z}}{z} - \frac{\bar{\omega}_2}{f} X^{uj} X + \left(\frac{\bar{\omega}_1}{f} X^{uj} + \bar{\omega}_3\right) Y - \bar{\omega}_2 f \right] \Big/ d^{uj} \\ &\quad + \frac{\bar{\omega}_2}{f}(X^{uj})^2 - \left(\frac{\bar{\omega}_1}{f} X^{uj} + \bar{\omega}_3\right) Y^{uj} + \bar{\omega}_2 f \end{aligned} \tag{4.48}$$

$$\begin{aligned} \dot{Y}^{uj} &= \frac{\dot{y}^{uj} f}{z^{uj}} - Y^{uj}\frac{\dot{z}^{uj}}{z^{u}} \\ &= \left[\dot{Y} + (Y - Y^{uj})\frac{\dot{z}}{z} + \frac{\bar{\omega}_1}{f} Y^{uj} Y - \left(\frac{\bar{\omega}_2}{f} Y^{uj} + \bar{\omega}_3\right) X + \bar{\omega}_1 f \right] \Big/ d^{uj} \\ &\quad + \left(\frac{\bar{\omega}_2}{f} Y^{uj} + \bar{\omega}_3\right) X^{uj} - \frac{\bar{\omega}_1}{f}(Y^{uj})^2 - \bar{\omega}_1 f \end{aligned} \tag{4.49}$$

Rewriting (4.48) and (4.49) yields

$$\dot{X} + X\frac{\dot{z}}{z} - X^{uj}\left(\frac{\dot{z}}{z} + \frac{\bar{\omega}_2}{f} X - \frac{\bar{\omega}_1}{f} Y\right) + \bar{\omega}_3 Y + c_X^{uj} d^{uj} = \bar{\omega}_2 f \tag{4.50}$$

$$\dot{Y} + Y\frac{\dot{z}}{z} - Y^{uj}\left(\frac{\dot{z}}{z} + \frac{\bar{\omega}_2}{f}X - \frac{\bar{\omega}_1}{f}Y\right) - \bar{\omega}_3 X + c_Y^{uj} d^{uj} = -\bar{\omega}_1 f \tag{4.51}$$

where

$$c_X^{uj} \triangleq -\dot{X}^{uj} + \frac{\bar{\omega}_2}{f}(X^{uj})^2 - \left(\frac{\bar{\omega}_1}{f}x^{uj} + \bar{\omega}_3\right)Y^{uj} + \bar{\omega}_2 f$$

$$c_Y^{uj} \triangleq -\dot{Y}^{uj} + \left(\frac{\bar{\omega}_2}{f}Y^{uj} + \bar{\omega}_3\right)X^{uj} - \frac{\bar{\omega}_1}{f}(Y^{uj})^2 - \bar{\omega}_1 f$$

Note that Eqs. (4.50) and (4.51) hold for any time, particularly for $k = 1, 2$. Because $(X^{uj}(k),\ Y^{uj}(k))$ and $(\dot{X}^{uj}(k),\ \dot{Y}^{uj}(k))$ are quantities that can be measured, Eqs. (4.50) and (4.51) for $j = 1, 2, \ldots, \mathscr{J}$ are constraints on the image of the reference point. In the following, time k is added to the equations whenever it is needed to represent the two adjacent frames. We have

Definition 2 *Consider the minimization problem with the objective function* (3.9) *subject to the constraints given by* (4.44), (4.50), *and* (4.51), *where J_1 and J_2 are given by* (3.21) *and* (3.15) *respectively. The unknown variables are $\dot{X}(k)$, $\dot{Y}(k)$, $X(k)$, $Y(k)$, $\rho(k)$, and $d^{uj}(k)$, for $k = 1, 2$. The 2-D reference point problem is to find the 2-D images $(X(k),\ Y(k))$ and $(\dot{X}(k),\ \dot{Y}(k))$ for $k = 1, 2$ of a 3-D point such that the minimization problem is solved. $(X(k),\ Y(k))$ is called the 2-D reference point.*

The following lemma is useful for solving the 2-D reference problem.

Lemma 1 *Consider the feature points uj ($j = 1, 2, \ldots, \mathscr{J}$) in the object. If $c_X^{u1}c_Y^{uj} - c_Y^{u1}c_X^{uj} \neq 0$ for $j \neq 1$, then*

$$d^{uj} = a_j\left(\rho + \frac{\bar{\omega}_2}{f}X - \frac{\bar{\omega}_1}{f}Y\right) \tag{4.52}$$

for $j = 1, 2, \ldots, \mathscr{J}$, where $a_j \neq 0$ can be computed from the image of feature uj.

Proof Subtracting Eq. (4.50) with feature uj ($j \neq 1$) from Eq. (4.50) with feature $u1$ yields

$$c_X^{u1}d^{u1} - c_X^{uj}d^{uj} = (X^{u1} - X^{uj})\left(\rho + \frac{\bar{\omega}_2}{f}X - \frac{\bar{\omega}_1}{f}Y\right) \tag{4.53}$$

Similarly, from (4.51),

$$c_Y^{u1}d^{u1} - c_Y^{uj}d^{uj} = (Y^{u1} - Y^{uj})\left(\rho + \frac{\bar{\omega}_2}{f}X - \frac{\bar{\omega}_1}{f}Y\right) \tag{4.54}$$

If $c_X^{u1}c_Y^{uj} - c_Y^{u1}c_X^{uj} \neq 0$, then d^{uj} ($j = 1, 2, \ldots, \mathscr{J}$) can be uniquely represented by (4.52), where

$$a_1 = \frac{c_Y^{uj}(X^{u1} - X^{uj}) - c_X^{uj}(Y^{u1} - Y^{uj})}{c_X^{u1}c_Y^{uj} - c_Y^{u1}c_X^{uj}}$$

$$a_j = \frac{c_Y^{u1}(X^{u1} - X^{uj}) - c_X^{u1}(Y^{u1} - Y^{uj})}{c_X^{u1}c_Y^{uj} - c_Y^{u1}c_X^{uj}} \qquad (j \neq 1)$$

d^{uj}, by definition, is a positive quantity. Hence $a_j \neq 0$. □

Remark A robust way to compute a_j is to use the least squares solution of the following equations:

$$\begin{bmatrix} c_X^{u1} & -c_X^{u2} & 0 & \cdots & 0 \\ c_Y^{u1} & -c_Y^{u2} & 0 & \cdots & 0 \\ c_X^{u1} & 0 & -c_X^{u3} & \cdots & 0 \\ c_Y^{u1} & 0 & -c_Y^{u3} & \cdots & 0 \\ \vdots & \vdots & \vdots & \vdots & \vdots \\ c_X^{u1} & 0 & 0 & \cdots & -c_X^{u\mathscr{J}} \\ c_Y^{u1} & 0 & 0 & \cdots & -c_Y^{u\mathscr{J}} \end{bmatrix} \begin{bmatrix} a_1 \\ a_2 \\ a_3 \\ a_4 \\ \vdots \\ a_{\mathscr{J}} \end{bmatrix} = \begin{bmatrix} X^{u1} - X^{u2} \\ Y^{u1} - Y^{u2} \\ X^{u1} - X^{u3} \\ Y^{u1} - Y^{u3} \\ \vdots \\ X^{u1} - X^{u\mathscr{J}} \\ Y^{u1} - Y^{u\mathscr{J}} \end{bmatrix} \tag{4.55}$$

Theorem 1 *Assume that the angular velocity of the moving object is known. Then the 2-D reference point $(X(k),\ Y(k))$ together with $\rho(k)$ for $(k = 1, 2)$ satisfies*

$$\alpha_{11}(1)X(1) + \alpha_{12}(1)Y(1) + \alpha_{13}(1)\rho(1) - \bar{\omega}_2 f b_1(1) + \varepsilon\left(\frac{\gamma_{11}}{\rho(1)} g_1 + \frac{\gamma_{21}}{\rho(1)} g_2\right) - \lambda_1(1 + T\beta_{11}(1)) - \lambda_2 T\beta_{21}(1) = 0 \tag{4.56}$$

$$\alpha_{12}(1)X(1) + \alpha_{22}(1)Y(1) + \alpha_{23}(1)\rho(1) + \bar{\omega}_1 f b_1(1) + \varepsilon\left(\frac{\gamma_{12}}{\rho(1)} g_1 + \frac{\gamma_{22}}{\rho(1)} g_2\right) - \lambda_1 T\beta_{21}(1) - \lambda_2(1 + T\beta_{22}(1)) = 0 \tag{4.57}$$

$$\begin{aligned}\alpha_{13}(1)X(1) &+ \alpha_{23}(1)Y(1) + \alpha_{33}(1)\rho(1) - f^2 b_1(1) \\ &- \varepsilon\left(\frac{\gamma_{11}X(1) + \gamma_{12}Y(1) + \gamma_{13}}{\rho^2(1)} g_1 + \frac{\gamma_{21}X(1) + \gamma_{22}Y(1) + \gamma_{23}}{\rho^2(1)} g_2\right) \\ &+ \lambda_1 T(X(2) - \beta_{13}(1)) + \lambda_2 T(Y(2) - \beta_{23}(1)) + \lambda_3(T\rho(2) - 1) = 0\end{aligned} \tag{4.58}$$

$$\alpha_{11}(2)X(2) + \alpha_{12}(2)Y(2) + \alpha_{13}(2)\rho(2) - \bar{\omega}_2 f b_1(2) + \lambda_1(1 + T\rho(1)) = 0 \tag{4.59}$$

$$\alpha_{12}(2)X(2) + \alpha_{22}(2)Y(2) + \alpha_{23}(2)\rho(2) + \bar{\omega}_1 f b_1(2) + \lambda_2(1 + T\rho(1)) = 0 \tag{4.60}$$

$$\alpha_{13}(2)X(2) + \alpha_{23}(2)Y(2) + \alpha_{33}(2)\rho(2) - f^2 b_1(2) + \lambda_3(1 + T\rho(1)) = 0 \tag{4.61}$$

$$h_1(X(1),\ Y(1),\ \rho(1),\ X(2)) = 0 \tag{4.62}$$

$$h_2(X(1),\ Y(1),\ \rho(1),\ Y(2)) = 0 \tag{4.63}$$

$$h_3(\rho(1),\ \rho(2)) = 0 \tag{4.64}$$

where

$$b_X(k) = \frac{1}{\mathscr{J}} \sum_{j=1}^{\mathscr{J}} X^{uj}(k) a_j(k)$$

$$b_Y(k) = \frac{1}{\mathscr{J}} \sum_{j=1}^{\mathscr{J}} Y^{uj}(k) a_j(k)$$

$$b_1(k) = \frac{1}{\mathscr{J}} \sum_{j=1}^{\mathscr{J}} a_j(k)$$

$$\alpha_{11}(k) = (1 - \bar{\omega}_2/fb_X(k))^2 + (\bar{\omega}_2/fb_Y(k))^2 + (\bar{\omega}_2 b_1(k))^2$$

$$\alpha_{12}(k) = \bar{\omega}_1/f(1 - \bar{\omega}_2/fb_X(k))b_X(k) - \bar{\omega}_2/f(1 + \bar{\omega}_1/fb_Y(k))b_Y(k) - \bar{\omega}_1\bar{\omega}_2 b_1^2(k)$$

$$\alpha_{13}(k) = -(1 - \bar{\omega}_2/fb_X(k))b_X(k) + \bar{\omega}_2/fb_Y^2(k) + f\bar{\omega}_2 b_1^2(k)$$

$$\alpha_{22}(k) = (\bar{\omega}_1/fb_X(k))^2 + (1 + \bar{\omega}_1/fb_Y(k))^2 + (\bar{\omega}_1 b_1(k))^2$$

$$\alpha_{23}(k) = -\bar{\omega}_1/fb_X^2(k) - (1 + \bar{\omega}_1/fb_Y(k))b_Y(k) - f\bar{\omega}_1 b_1^2(k)$$

$$\alpha_{33}(k) = b_X^2(k) + b_Y^2(k) + f^2 b_1^2(k)$$

$$\beta_{13}(k) = \frac{1}{\mathcal{J}} \sum_{j=1}^{\mathcal{J}} (X^{uj}(k) - a_j(k)c_X^{uj}(k))$$

$$\beta_{11}(k) = \bar{\omega}_2/f\beta_{13}(k)$$

$$\beta_{12}(k) = -\bar{\omega}_1/f\beta_{13}(k) - \bar{\omega}_3$$

$$\beta_{23}(k)\frac{1}{\mathcal{J}} \sum_{j=1}^{\mathcal{J}} (Y^{uj}(k) - a_j(k)c_Y^{uj}(k))$$

$$\beta_{21}(k) = \bar{\omega}_2/f\beta_{23}(k) + \bar{\omega}_3$$

$$\beta_{22}(k) = -\bar{\omega}_1/f\beta_{23}(k)$$

$$\gamma_{n1} = \beta_{n1}(1) - \beta_{n1}(2)(1 + T\beta_{11}(1)) - \beta_{n2}(2)T\beta_{21}(1)$$

$$\gamma_{n2} = \beta_{n2}(1) - \beta_{n1}(2)T\beta_{12}(1) - \beta_{n2}(2)(1 + T\beta_{22}(1))$$

$$\gamma_{n3} = -\beta_{n1}(2)T\bar{\omega}_2 f + \beta_{n2}(2)T\bar{\omega}_1 f$$

$$\gamma_{n4} = -\beta_{n1}(2)T\beta_{13}(1) - \beta_{n2}(2)T\beta_{23}(1) + \beta_{n3}(1) - \beta_{n3}(2)$$

where $n = 1, 2$, and the functions g_1, g_2, h_1, h_2, and h_3 are defined by

$$g_1 = (\gamma_{11}X(1) + \gamma_{12}Y(1) + \gamma_{13})/\rho(1) + \gamma_{14} - T\bar{\omega}_2 f \tag{4.65}$$

$$g_2 = (\gamma_{21}X(1) + \gamma_{22}Y(1) + \gamma_{23})/\rho(1) + \gamma_{24} + T\omega_1 f \tag{4.66}$$

$$h_1 = (1 + T\rho(1))X(2) - (1 + T\beta_{11}(1))X(1) - T\beta_{12}(1)Y(1) - T\beta_{13}(1)\rho(1) - T\bar{\omega}_2 f \tag{4.67}$$

$$h_2 = (1 + T\rho(1))Y(2) - T\beta_{21}(1)X(1) - (1 + T\beta_{22}(1)Y(1) - T\beta_{23}(1)\rho(1) + T\bar{\omega}_1 f \tag{4.68}$$

$$h_3 = (1 + T\rho(1))\rho(2) - \rho(1) \tag{4.69}$$

Proof The proof consists of three parts.

(1) Picking independent equations from (4.44), (4.50), and (4.51).

There are $2\mathcal{J}$ equations in (4.50) and (4.51) for $j = 1, 2, \ldots, \mathcal{J}$. However, these equations are not independent. From Lemma 1, we have $\mathcal{J}$ independent equations given by (4.52) for $j = 1, 2, \ldots, \mathcal{J}$. Substitution of these equations into (4.50) and (4.51) yields

$$\dot{X} + X\rho = (X^{uj} - a_j c_X^{uj})\left(\rho + \frac{\bar{\omega}_2}{f}X - \frac{\bar{\omega}_1}{f}Y\right) - \bar{\omega}_3 Y + \bar{\omega}_2 f \tag{4.70}$$

$$\dot{Y} + Y\rho = (Y^{uj} - a_j c_Y^{uj})\left(\rho + \frac{\bar{\omega}_2}{f}X - \frac{\bar{\omega}_1}{f}Y\right) + \bar{\omega}_3 X - \bar{\omega}_1 f \tag{4.71}$$

Summing (4.70) for each j from 1 to $\mathscr{J}$ and then dividing the summation by $\mathscr{J}$ yields

$$\dot{X} + X\rho = \left(\rho + \frac{\bar{\omega}_2}{f} X - \frac{\bar{\omega}_1}{f} Y\right)\beta_{13} - \bar{\omega}_3 Y + \bar{\omega}_2 f \tag{4.72}$$

Similarly by (4.71),

$$\dot{Y} + Y\rho = \left(\rho + \frac{\bar{\omega}_2}{f} X - \frac{\bar{\omega}_1}{f} Y\right)\beta_{23} + \bar{\omega}_3 X - \bar{\omega}_1 f \tag{4.73}$$

Since ρ is assumed nonzero, dividing (4.72) and (4.73), respectively, by ρ yields

$$\frac{\dot{X}}{\rho} + X = \frac{\beta_{11}X + \beta_{12}Y + \bar{\omega}_2 f}{\rho} + \beta_{13} \tag{4.74}$$

$$\frac{\dot{Y}}{\rho} + Y = \frac{\beta_{21}X + \beta_{22}Y - \bar{\omega}_1 f}{\rho} + \beta_{23} \tag{4.75}$$

So the equations in (4.50) and (4.51) are equivalent to those in (4.52), (4.74), and (4.75). Finally, using (4.74) and (4.75), the equations in (4.44) can be written as

$$\frac{X(2)}{\rho(2)} = \frac{(1 + T\beta_{11}(1))X(1) + T\beta_{12}(1)Y(1) + T\bar{\omega}_2 f}{\rho(1)} + T\beta_{13}(1) \tag{4.76}$$

$$\frac{Y(2)}{\rho(2)} = \frac{T\beta_{21}(1)X(1) + (1 + T\beta_{22}(1))Y(1) - T\bar{\omega}_1 f}{\rho(1)} + T\beta_{23}(1) \tag{4.77}$$

$$\frac{1}{\rho(2)} = \frac{1}{\rho(1)} + T \tag{4.78}$$

Therefore, the constraints in the 2-D reference point problem can also be equivalently given by Eqs. (4.52), (4.74), (4.75), (4.76), (4.77), and (4.78).

(2) Eliminating variables $\dot{X}$, $\dot{Y}$, d^{uj} ($j = 1, 2, \ldots, \mathscr{J}$) from the 2-D reference point problem. Using (4.52), J_1 in (4.42) can be written as

$$\begin{aligned} J_1 = \sum_{k=1}^{2} \Bigg[& \left(1 - \frac{\bar{\omega}_2}{f} b_X(k)\right) X(k) + \frac{\omega_1}{f} b_X(k) Y(k) - b_X(k)\rho(k) \Bigg]^2 \\ & + \left[-\frac{\bar{\omega}_2}{f} b_Y(k) X(k) + \left(1 + \frac{\bar{\omega}_1}{f} b_Y(k)\right) Y(k) - b_Y(k)\rho(k) \right]^2 \\ & + f^2 \left[-\frac{\bar{\omega}_2}{f} b_1(k) X(k) + \frac{\bar{\omega}_1}{f} b_1(k) Y(k) - b_1(k)\rho(k) + 1 \right]^2 \end{aligned} \tag{4.79}$$

Substitution of (4.74) and (4.75) into J_2 in (4.36) yields

$$J_2 = g_1^2 + g_2^2 \tag{4.80}$$

where

$$g_1 = \frac{\beta_{11}(1)X(1) + \beta_{12}(1)Y(1) + \bar{\omega}_2 f}{\rho(1)} - \frac{\beta_{11}(2)X(2) + \beta_{12}(2)Y(2) + \bar{\omega}_2 f}{\rho(2)} + \beta_{13}(1) - \beta_{13}(2) \tag{4.81}$$

$$g_2 = \frac{\beta_{21}(1)X(1) + \beta_{22}(1)Y(1) + \bar{\omega}_1 f}{\rho(1)} - \frac{\beta_{21}(2)X(2) + \beta_{22}(2)Y(2) - \bar{\omega}_1 f}{\rho(2)} + \beta_{23}(1) - \beta_{23}(2) \tag{4.82}$$

It is now clear that the 2-D reference point problem becomes the minimization problem with the objective function (4.30), with J_1 and J_2 being defined in (4.79) and (4.80), respectively, subject to the constraints given by Eqs. (4.76), (4.77), and (4.78).

(3) Solving the constrained minimization problem.

Functions g_1 and g_2 in (4.81) and (4.82) can be simplified by substituting Eqs. (4.76), (4.77), and (4.78) into them. The results are given by Eqs. (3.65) and (3.66). Furthermore, from (4.78), we have

$$\frac{1}{\rho(2)} = \frac{1 + T\rho(1)}{\rho(1)} \tag{4.83}$$

Substitution of (4.83) into (4.76) and (4.77) yields

$$h_1 = 0 \tag{4.84}$$

$$h_2 = 0 \tag{4.85}$$

where h_1 and h_2 are defined by Eqs. (4.67) and (4.68), respectively. Equation (4.83) can also be written as

$$h_3 = 0 \tag{4.86}$$

where h_3 is given by (4.69).

Define Lagrangian

$$L = J_1 + \varepsilon J_2 + 2\lambda_1 h_1 + 2\lambda_2 h_2 + 2\lambda_3 h_3 \tag{4.87}$$

Then Eqs. (4.56), (4.57), (4.58), (4.59), (4.60), and (4.61) can be obtained by setting

$$\frac{\partial L}{\partial([X(1)Y(1)\rho(1)X(2)Y(2)\rho(2)]^{\mathrm{T}})} = 0 \tag{4.88}$$

4 THE CONTROL DESIGN FOR TRACKING AND GRASPING

We have modeled the motion of the object as the motion of the reference point coupled with the change in orientation of the object. Because of the decoupled nature of the robot dynamics, the control for tracking and grasping the object can be designed in two separate steps. To grasp the moving object, the center of the robot gripper is commanded to track the motion of the reference point. In the meantime, aligning the orientation of the gripper with that of the object can also be posed as an orientation tracking problem. However, since a dynamic model for the motion of the reference point is required in the tracking control design, we will first discuss the problem of fitting the data of the reference point to a dynamic system.

4.1 Model Fitting for the Reference Point

For most real-time tracking applications, the motion of the object is generally not known. Therefore, the best one can hope for is to approximate the motion trajectory locally with simple models. In the following, 3-D circular paths with varying radii are used to approximate the 3-D trajectory of the reference point.

Let the reference point move along some arbitrary circle in 3-D space. Ordinary differential equations for the motion can be given by

$$\begin{bmatrix} \dot{x}^r_w \\ \dot{y}^r_w \\ \dot{z}^r_w \end{bmatrix} = \Omega' \left(\begin{bmatrix} x^r_w \\ y^r_w \\ z^r_w \end{bmatrix} - \begin{bmatrix} \delta_x \\ \delta_y \\ \delta_z \end{bmatrix} \right) \tag{4.89}$$

where Ω' still takes the form of (4.21) with its elements ω'_1, ω'_2, and ω'_3 being some unknown parameters. Note that the Ω' here describes the circular path the reference point follows and has nothing to do with the Ω in (4.21). Define

$$\begin{bmatrix} \delta'_x \\ \delta'_y \\ \delta'_z \end{bmatrix} = -\Omega' \begin{bmatrix} \delta_x \\ \delta_y \\ \delta_z \end{bmatrix}$$

Then (4.89) can be written as

$$\dot{\mathbf{x}}^r \triangleq \begin{bmatrix} \dot{x}^r_w \\ \dot{y}^r_w \\ \dot{z}^r_w \end{bmatrix} = \Omega' \begin{bmatrix} x^r_w \\ y^r_w \\ z^r_w \end{bmatrix} + \begin{bmatrix} \delta'_x \\ \delta'_y \\ \delta'_z \end{bmatrix} \triangleq \phi(\mathbf{x}^r; \theta_P) \tag{4.90}$$

where

$$\theta_P = [\omega'_1 \quad \omega'_2 \quad \omega'_3 \quad \delta'_x \quad \delta'_y \quad \delta'_z]^{\mathrm{T}}$$

We call ω'_1, ω'_2, ω'_3, δ'_x, δ'_y, and δ'_z the 3-D motion parameters.

The 3-D circular trajectory can be projected into 2-D image planes. To do this, first observe that the time trajectory of (4.89) starting from any fixed initial 3-D position stays in the plane

$$\omega'_1 x^r_w + \omega'_2 y^r_w + \omega'_3 z^r_w + s = 0 \tag{4.91}$$

where s is a scalar related to the initial position. Using (4.1), Eq. (4.91) can be written in the coordinates of the CCFi as

$$[p_i \quad q_i \quad r_i] \begin{bmatrix} x^r_{ci} \\ y^r_{ci} \\ z^r_{ci} \end{bmatrix} + s_i = 0 \tag{4.92}$$

where

$$[p_i \quad q_i \quad r_i] = [\omega'_1 \quad \omega'_2 \quad \omega'_3] R_i^{-1}$$
$$s_i = s - [\omega'_1 \quad \omega'_2 \quad \omega'_3] R_i^{-1} d_i$$

Using (4.12) and (4.92) yields

$$\frac{1}{z^r_{ci}} = -\frac{p_i x^r_{ci} + q_i y^r_{ci} + r_i z^r_{ci}}{s_i z^r_{ci}} = -\frac{p_i X^r_i + q_i Y^r_i + r_i f_i}{f_i s_i} \tag{4.93}$$

Differentiating both sides of the equations in (4.12) and using (4.90) yields

$$\begin{aligned} \dot{X}^r_i &= (\dot{x}^r_{ci} f_i - X^r_i \dot{z}^r_{ci})/z^r_{ci} \\ &= -\omega'_3 Y^r_i + \omega'_2 f_i - \frac{X^r_i}{f_i}(-\omega'_2 X^r_i + \omega'_1 Y^r_i) + (\delta'_x f_i - X_i \delta'_z)\frac{1}{z^r_{ci}} \end{aligned} \tag{4.94}$$

Substituting (4.93) into (4.94), we have

$$\dot{X}^r_i = S_{i1}(X^r_i, Y^r_i) \tag{4.95}$$

where

$$S_{i1} = a_{i11} + a_{i12} X^r_i + a_{i13} Y^r_i + X^r_i(a_{i14} X^r_i + a_{i15} Y^r_i)$$

and a_{ijk} can be appropriately defined. Similarly, we have

$$\dot{Y}^r_i = S_{i2}(X^r_i, Y^r_i) \tag{4.96}$$

where

$$S_{i2} = a_{i21} + a_{i22} X^r_i + a_{i23} Y^r_i + Y^r_i(a_{i14} X^r_i + a_{i15} Y^r_i)$$

We call a_{ijk} the 2-D motion parameters. They are to be estimated from the image of the reference point. For notational simplicity, we write

$$\dot{\mathbf{X}}^r = S(\mathbf{X}^r; \theta_I) \triangleq \begin{bmatrix} S_{11}(X^r_1, Y^r_1; \theta_I) \\ S_{12}(X^r_1, Y^r_1; \theta_I) \\ \vdots \\ S_{\mathcal{I}1}(X^r_{\mathcal{I}}, Y^r_{\mathcal{I}}; \theta_I) \\ S_{\mathcal{I}2}(X^r_{\mathcal{I}}, Y^r_{\mathcal{I}}; \theta_I) \end{bmatrix} \tag{4.97}$$

where θ_I is the column vector consisting of the parameters a_{ijk}.

4.2 Parameter Estimation: Recursive Least Squares

We first show how to estimate 2-D motion parameters. From (4.95), we have

$$\dot{X}_i^r(k) = \Phi(k)^{\mathrm{T}}\theta_{I,i1} \tag{4.98}$$

where $\Phi(k)$ and $\theta_{I,i1}$ are defined, respectively, as

$$\Phi(k) = [1 \quad X_i^r(k) \quad Y_i^r(k) \quad (X_i^r(k))^2 \quad X_i^r(k)Y_i^r(k)]^{\mathrm{T}}$$
$$\theta_{I,i1} = [a_{i11} \quad a_{i12} \quad a_{i13} \quad a_{i14} \quad a_{i15}]^{\mathrm{T}}$$

Since $\dot{X}_i^r(k)$ and $\Phi(k)$ can be obtained from the image of the reference point in the ith camera, the unknown parameters $\theta_{I,i1}$ in (4.98) can be recursively estimated [19]. In the same way, motion parameters a_{i21}, a_{i22}, and a_{i23} can be estimated. One important observation is that the estimation of 2-D motion parameters of the ith camera requires only the knowledge of the image of the reference point in the same camera. Therefore, the estimation scheme can be implemented in parallel for multiple cameras.

4.3 Image-Based Tracking of the Reference Point

Let the 2-D motion dynamics (4.97) of the reference point be an exosystem that generates signals for the images of the gripper of the robot plant (4.8) to follow. The image vector of the center of the gripper is given by

$$\mathbf{X}^g = P(\mathbf{x}^g) = P(C\xi^{(1)}) \tag{4.99}$$

where P is defined in (4.17). Hence, the difference between the image of the reference point and the image of the gripper can be measured by an error output defined by

$$e = \mathbf{X}^g - \mathbf{X}^r \tag{4.100}$$

The motion tracking can be posed as a state feedback regulator problem (SFRP) [20] as follows.

Definition 3 *The image-based tracking problem is to solve the SFRP with plant* (4.8), *exosystem* (4.97), *and error output* (4.100). *If the problem is solvable, then the resulting state feedback is called the image-based controller.*

Note that the 3-D pose of the gripper can be obtained by measuring the robot joint displacement vector. Therefore, assuming matrices R_i and d_i in (4.1) are accurate, we can compute the image vector $\mathbf{X}^g$ from (4.99). On the other hand, we can also measure the image vector $\mathbf{X}^g$ as the center of the gripper from the image planes. Clearly, if there is error in measuring R_i or d_i, the computed image vector of the gripper will not coincide with the measured one. To obtain matrices R_i and d_i precisely, calibration is necessary. Traditionally, calibration is accomplished by observing with the cameras a number of known 3-D points in the work space of the robot. The calibration parameters so obtained give a good approximation for all the points within the work space. For the task of motion tracking, however, the calibration method is not efficient. In fact, a precise calibration in the neighborhood of the motion trajectory suffices.

Assume that the image vector of the gripper is also processed, in addition to that of the moving object. Then the local calibration can be achieved as follows.

Let (X_i^g, Y_i^g) be the measured image of the center of the gripper from camera i. Furthermore, x_w^g, y_w^g, z_w^g can be computed from the forward kinematic equations. By (4.9), it can be seen that Eq. (4.15) can be written as

$$X_i^g(kT_v) = \Phi(k)^{\mathrm{T}}\theta$$

where

$$\Phi(k) = \begin{bmatrix} 1 \\ \mathbf{x}^g(kT_v) \\ -X_i^g(kT_v)\mathbf{x}^g(kT_v) \end{bmatrix}, \qquad \theta = \begin{bmatrix} p_{i1} \\ Q_{i1}^{\mathrm{T}} \\ Q_{i3}^{\mathrm{T}} \end{bmatrix}$$

Therefore, when new images are taken at the next sampling time, the parameters in θ can be updated on line from the recursive least squares method. Similarly, Eq. (4.16) can be used for parameter estimation.

Using the newly estimated parameters p_{i1}, p_{i2}, and q_{ih} in Eq (4.99), we can design the image-based controller according to the following theorem.

Theorem 2 *The image-based controller is given by*

$$v^{(1)} = c(\mathbf{X}^r) + K(\xi^{(1)} - \pi(\mathbf{X}^r)) \tag{4.101}$$

where K satisfies $\sigma(A + BK) \subset \mathbf{C}^-$, and $\pi(\mathbf{X}^r) = [\pi_1 \ \pi_2 \ \cdots \ \pi_6]^{\mathrm{T}}$ and $c(\mathbf{X}^r) = [c_1 \ c_2 \ c_3]^{\mathrm{T}}$ are obtained, respectively, from

$$[\pi_1 \quad \pi_2 \quad \pi_3]^{\mathrm{T}} = P^{-1}(\mathbf{X}^r) \tag{4.102}$$

$$[\pi_4 \quad \pi_5 \quad \pi_6 \quad c_1 \quad c_2 \quad c_3]^{\mathrm{T}} = \frac{\partial \pi}{\partial \mathbf{X}^r} S(\mathbf{X}^r; \theta_I) \tag{4.103}$$

Proof Denote $\bar{\mathbf{x}}^r = [\delta_x \ \delta_y \ \delta_z]^{\mathrm{T}}$. First we note that $\bar{\mathbf{X}}^r = P(\bar{\mathbf{x}}^r)$ is an equilibrium of the system (4.97). Using coordinate transformations,

$$\tilde{\xi}^{(1)} = \xi^{(1)} - [(\bar{\mathbf{x}}^r)^{\mathrm{T}} \quad 0 \quad 0 \quad 0]^{\mathrm{T}} \tag{4.104}$$

$$\tilde{\mathbf{X}}^r = \mathbf{X}^r - \bar{\mathbf{X}}^r \tag{4.105}$$

plant (4.8) and exosystem (4.97) read, respectively,

$$\dot{\tilde{\xi}}^{(1)} = A\tilde{\xi}^{(1)} + Bv^{(1)} \tag{4.106}$$

$$\dot{\tilde{\mathbf{X}}}^r = \tilde{S}(\tilde{\mathbf{X}}^r; \theta_I) \triangleq S(\tilde{\mathbf{X}}^r + \bar{\mathbf{X}}^r; \theta_I) \tag{4.107}$$

The error output (4.100) is

$$e = P(C\tilde{\xi}^{(1)} + \bar{\mathbf{x}}^r) - \bar{\mathbf{X}}^r - \tilde{\mathbf{X}}^r \triangleq \tilde{h}(\tilde{\xi}^{(1)}) - \tilde{\mathbf{X}}^r \tag{4.108}$$

Letting

$$\alpha_i = Q_{i3}(C\tilde{\xi}^{(1)} + \bar{\mathbf{x}}^r) + 1 \qquad (i = 1, \ldots, \mathscr{I})$$

the first component of $\tilde{h}(\tilde{\xi}^{(1)})$ is given by

$$\frac{Q_{11}(C\tilde{\xi}^{(1)} + \bar{\mathbf{x}}^r) + p_{11}}{\alpha_1} - \bar{X}_1^r = \frac{(Q_{11} - \bar{X}_1^r Q_{13})C\tilde{\xi}^{(1)} + (Q_{11} - \bar{X}_1^r Q_{13})\bar{\mathbf{x}}^r - (\bar{X}_1^r - p_{11})}{\alpha_1}$$
$$= \frac{(Q_{11} - \bar{X}_1^r Q_{13})C\tilde{\xi}^{(1)}}{\alpha_1}$$

The last term is obtained using Eq. (4.18). The rest of components can be similarly computed. Hence,

$$\tilde{h}(\tilde{\xi}^{(1)}) = \begin{bmatrix} (Q_{11} - \bar{X}_1^r Q_{13})C\tilde{\xi}^{(1)}/\alpha_1 \\ (Q_{12} - \bar{Y}_1^r Q_{13})C\tilde{\xi}^{(1)}/\alpha_1 \\ \vdots \\ (Q_{\mathscr{I}1} - \bar{X}_{\mathscr{I}}^r Q_{\mathscr{I}3})C\tilde{\xi}^{(1)}/\alpha_{\mathscr{I}} \\ (Q_{\mathscr{I}2} - \bar{Y}_{\mathscr{I}}^r Q_{\mathscr{I}3})C\tilde{\xi}^{(1)}/\alpha_{\mathscr{I}} \end{bmatrix}$$

Clearly, the origin is an equilibrium of the system with plant (4.106), exosystem (4.107), and error output (4.108). Using the regulator theory, the analytical mappings $\tilde{\pi}(\tilde{\mathbf{X}}^r) = [\tilde{\pi}_1 \; \tilde{\pi}_2 \; \cdots \; = \tilde{\pi}_6]^{\mathrm{T}}$ and $\tilde{c}(\tilde{\mathbf{X}}^r) = [\tilde{c}_1 \; \tilde{c}_2 \; \tilde{c}_3]^{\mathrm{T}}$ should satisfy

$$\frac{\partial \tilde{\pi}(\tilde{\mathbf{X}}^r)}{\partial(\tilde{\mathbf{X}}^r)} \tilde{S}(\tilde{\mathbf{X}}^r; \theta_I) = A\tilde{\pi} + B\tilde{c} \tag{4.109}$$

$$\tilde{h}(\tilde{\pi}(\tilde{\mathbf{X}}^r)) - \tilde{\mathbf{X}}^r = 0 \tag{4.110}$$

The first component of (4.110) gives

$$(Q_{11} - \bar{X}_1^r Q_{13})C\tilde{\pi} = \tilde{X}_1^r Q_{13}(C\tilde{\pi} + \bar{\mathbf{x}}^r) + \tilde{X}_1^r$$

which is equivalent to

$$(Q_{11} - X_1^r Q_{13})C\tilde{\pi} = (X_1^r - \bar{X}_1^r)(1 + Q_{13}\bar{\mathbf{x}}^r) \tag{4.111}$$

From (4.18),

$$(Q_{11} - \bar{X}_1^r Q_{13})\bar{\mathbf{x}}^r = \bar{X}_1^r - p_{11} \tag{4.112}$$

Substituting (4.112) in the right-hand side of (4.111) yields

$$(Q_{11} - X_1^r Q_{13})(C\tilde{\pi} + \bar{\mathbf{x}}^r) = X_1^r - p_{11}$$

Therefore, Eq. (4.110) is equivalent to

$$M(\mathbf{X}^r)(C\tilde{\pi} + \bar{\mathbf{x}}^r) = N(\mathbf{X}^r)$$

So

$$\begin{bmatrix} \tilde{\pi}_1 \\ \tilde{\pi}_2 \\ \tilde{\pi}_3 \end{bmatrix} = P^{-1}(\mathbf{X}^r) - \bar{\mathbf{X}}^r \tag{4.113}$$

Having obtained $\tilde{\pi}_1$, $\tilde{\pi}_2$, and $\tilde{\pi}_3$, we can solve (4.109) for $\tilde{\pi}_4$, $\tilde{\pi}_5$, $\tilde{\pi}_6$ and $\tilde{c}_1$, $\tilde{c}_2$, $\tilde{c}_3$ in a successive fashion. The image-based controller is thus given by

$$v^{(1)} = \tilde{c}(\tilde{\mathbf{X}}^r) + K(\tilde{\xi}^{(1)} - \tilde{\pi}(\tilde{\mathbf{X}}^r)) \tag{4.114}$$

with $\sigma(A + BK_I) \subset \mathbf{C}^-$. Now define

$$\pi(\mathbf{X}^r) = \tilde{\pi}(\mathbf{X}^r - \bar{\mathbf{X}}^r) + \bar{\xi}^{(1)}$$
$$c(\mathbf{X}^r) = \tilde{c}(\mathbf{X}^r - \bar{\mathbf{X}}^r)$$

Then it is trivial to verify (4.102), (4.103), and (4.101) from (4.113), (4.109), and (4.114), respectively. □

Remark For real-time implementation of control law (4.101), one needs to find the closed-form solutions for π_4, π_5, π_6 and c_1, c_2, c_3 as functions of $\mathbf{X}^r$. This can be achieved as follows. Denote $\pi^{(1)} = [\pi_1 \;\; \pi_2 \;\; \pi_3]^{\mathrm{T}}$, $\pi^{(2)} = [\pi_4 \;\; \pi_5 \;\; \pi_6]^{\mathrm{T}}$. Using (4.18), (4.19) and (4.102) yields the identity

$$M(\mathbf{X}^r)\pi^{(1)}(\mathbf{X}^r) = N(\mathbf{X}^r) \tag{4.115}$$

Differentiating (4.115) with respect to $\mathbf{X}^r$ yields

$$M(\mathbf{X}^r)\frac{\partial \pi^{(1)}}{\partial \mathbf{X}^r} = L_0 + L_1\pi_1 + L_2\pi_1 + L_3\pi_3 \triangleq L \tag{4.116}$$

where L_0 and L_i are diagonal matrices, that is,

$$L_0 = \mathrm{diag}(d_{13}, d_{13}, \ldots, d_{\mathcal{J}3}, d_{\mathcal{J}3})$$
$$L_i = \mathrm{diag}(r_{1i+6}, r_{1i+6}, \ldots, r_{\mathcal{J}i+6}, r_{\mathcal{J}i+6}), \qquad i = 1, 2, 3$$

Postmultiplying (4.116) by $S(\mathbf{X}^r; \theta_I)$ yields

$$M(\mathbf{X}^r)\pi^{(2)} = LS \tag{4.117}$$

So $\pi^{(2)}$ can be solved. Differentiating (4.117) and postmultiplying the resulting equation by

$S(\mathbf{X}^r; \theta_I)$ yields

$$M(\mathbf{X}^r)\frac{\partial \pi^{(2)}}{\partial \mathbf{X}^r} S = 2(L_1\pi_4 + L_2\pi_5 + L_3\pi_6)S + L\frac{\partial S}{\partial \mathbf{X}^r} S \tag{4.118}$$

So

$$M(\mathbf{X}^r)c(\mathbf{X}^r) = \left(2(L_1\pi_4 + L_2\pi_5 + L_3\pi_6) + L\frac{\partial S}{\partial \mathbf{X}^r}\right) S \tag{4.119}$$

4.4 Tracking the Change in Orientation for Grasping

The grasping design here is to find a feedback control law $v^{(2)}$ so that the closed-loop system of plant (4.10) is asymptotically stable and both plates of the gripper are properly aligned with the moving object so that grasping may occur at any point. Clearly, this can be formulated as a tracking problem.

Consider Eq. (4.20), which describes the change in orientation of the object. Because the orientation of the gripper is described by OAT angles, we want to represent Eq. (4.20) in the OAT coordinates so that the two orientations can be easily compared with each other.

Equation (4.20) can be locally described in OAT space using the following relationship (see [14]):

$$\begin{bmatrix} n_x & s_x & a_x \\ n_y & s_y & a_y \\ n_z & s_z & a_z \end{bmatrix} = \begin{bmatrix} SOSACT - COST & -SOSAST - COCT & -SOCA \\ -COSACT - SOST & COSAST - SOCT & COCA \\ -CACT & CAST & -SA \end{bmatrix} \tag{4.120}$$

where $S \triangleq \sin$, $C \triangleq \cos$, and the superscript r is temporarily dropped for brevity. Differentiating $SA = -a_z$ yields

$$CA\dot{A} = -\dot{a}_z = -\omega_1 a_y + \omega_2 a_x$$

Eliminating a_x, a_y by (4.120) yields

$$\dot{A} = -\omega_1 CO - \omega_2 SO \tag{4.121}$$

Differentiating $-SOCA = -a_x$ yields

$$-CO\dot{O}CA + SOSA\dot{A} = \dot{a}_x = \omega_2 a_z - \omega_3 a_y$$

By (4.120) and (4.121), we have

$$\dot{O} = (-\omega_1 SO + \omega_2 CO)\tan A + \omega_3 \tag{4.122}$$

Finally, differentiating $CAST = s_z$ yields

$$-SA\dot{A}ST + CACT\dot{T} = \dot{s}_z = \omega_1 s_y - \omega_2 s_x$$

By (4.120) and (4.121), we have

$$\dot{T} = (-\omega_1 SO + \omega_2 CO)/CA \tag{4.123}$$

Therefore, if we denote

$$\eta^r = \begin{bmatrix} O^r \\ A^r \\ T^r \end{bmatrix}$$

Eq. (4.20) can be written as

$$\dot{\eta}^r = \zeta(\eta^r) \triangleq \begin{bmatrix} (\omega_2 \cos O^r - \omega_1 \sin O^r) \tan A^r + \omega_3 \\ -\omega_2 \sin O^r - \omega_1 \cos O^r \\ (\omega_2 \cos O^r - \omega_1 \sin O^r)/\cos A^r \end{bmatrix} \tag{4.124}$$

Assume that the angular velocity has been estimated. Using equation (4.124), the control law for tracking the change in orientation of the object can be given by

$$v^2 = \frac{\partial \zeta}{\partial \eta^r} \zeta + K_{O,p}(\eta^g - \eta^r) + K_{O,v}(\dot{\eta}^g - \zeta) \tag{4.125}$$

where $K_{O,p}$ and $K_{O,v}$ are 3×3 gain matrices satisfying

$$\sigma\left(\begin{bmatrix} 0 & I_{3\times 3} \\ K_{O,v} & K_{O,p} \end{bmatrix}\right) \in \mathbf{C}^-$$

4.5 Implementation Issues

In the recursive estimation of the 2-D motion parameters from Eq. (4.97), we find that the estimation results are less sensitive to noise when the second-order terms are neglected. In other words, using affine equations yields robust estimation results. For example, with two cameras (e.g., $\mathscr{I} = 2$), the projected motion (4.97) can be approximated by

$$\begin{bmatrix} \dot{X}_1^r \\ \dot{Y}_1^r \\ \dot{X}_2^r \\ \dot{Y}_2^r \end{bmatrix} = \begin{bmatrix} a_{111} \\ a_{121} \\ a_{211} \\ a_{221} \end{bmatrix} + \begin{bmatrix} a_{112} & a_{113} & 0 & 0 \\ a_{122} & a_{123} & 0 & 0 \\ 0 & 0 & a_{212} & a_{213} \\ 0 & 0 & a_{222} & a_{223} \end{bmatrix} \begin{bmatrix} X_1^r \\ Y_1^r \\ X_2^r \\ Y_2^r \end{bmatrix} \tag{4.126}$$

In general, due to the processing delay, the vision sampling rate $1/T_v$ can be achieved around 5 to 10 Hz. This is rather slow compared with the robot servoing rate. Therefore, important issues such as multirate sampling and significant time delay in computing the reference point and the orientation of the object have to be dealt with.

Let the robot servo sampling periods be T_s. For convenience, we assume that $T_v/T_s = N$, where N is an integer, and that the vision sampler and the servo sampler are synchronized. Figure 4.3 illustrates the relation between the two sampling periods on the time axis. Because

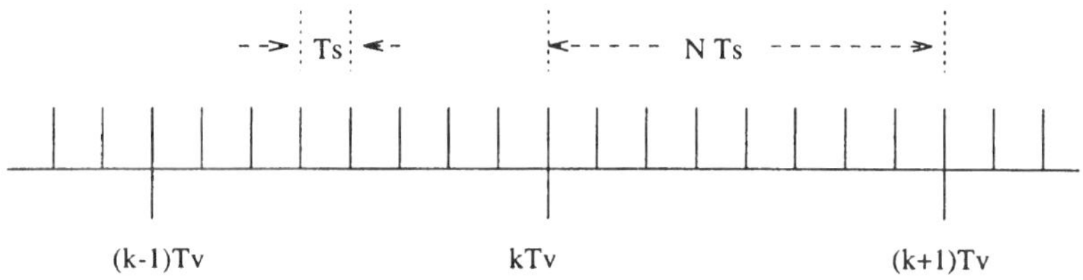

FIGURE 4.3
Vision and robot servo sampling periods on the time axis.

the information needed at sampling instant $(k-1)T_v$ for the on-line controller will not become available until a later time kT_v, prediction is necessary.

For example, consider

$$v(t) = \begin{bmatrix} v^{(1)}(t) \\ v^{(2)}(t) \end{bmatrix}$$

and $v^{(1)}$, $v^{(2)}$ are given by (4.101) and (4.125), respectively. At time instant

$$t_1 = kT_v + mT_s, \qquad m \in \{0, 1, \ldots, N-1\}$$

the following prediction scheme can be used in computing $v(t_1)$.

1. The image vector of the reference point $\mathbf{X}^r(t_1)$ is predicted from

$$\hat{\mathbf{X}}^r((N+m)T_s)$$

where $\hat{\mathbf{X}}^r(t)$ is the solution of the differential equation (4.126) with initial condition

$$\hat{\mathbf{X}}^r(0) = \mathbf{X}^r((k-1)T_v)$$

and a_{ijk} updated by the recursive estimation at time instant $(k-1)T_v$.
2. The orientation $\eta^r(t_1)$ is predicted in a similar way to $\mathbf{X}^r(t_1)$.

5 SIMULATION RESULTS AND DISCUSSION

Simulation is done on the basis of the following system configuration. Two CCD cameras are separated by 1 m and are 2 m away from the origin of the robot's base frame with the orientation matrices and translation vectors in (4.1) chosen as follows:

$$R_1 = \begin{bmatrix} 0.5070 & 0.8619 & 0 \\ -0.4480 & 0.2635 & 0.8543 \\ 0.7364 & -0.4331 & 0.5198 \end{bmatrix}, \qquad d_1 = \begin{bmatrix} -0.1521 \\ 0.1344 \\ -2.5296 \end{bmatrix}$$

$$R_2 = \begin{bmatrix} 0 & 1.0000 & 0 \\ -0.5767 & 0 & 0.8170 \\ 0.8170 & 0 & 0.5767 \end{bmatrix}, \quad d_2 = \begin{bmatrix} 0 \\ 0.1730 \\ -2.3260 \end{bmatrix}$$

We assume that the focal length of each camera is 25 mm and the vision sampling rate is 5 Hz.

5.1 The 2-D Reference Point Problem

In Figure 4.4, let u_j ($j = 1, 2, 3, 4$) be the 3-D feature points of a moving object. Let C be the centroid of the four feature points; B of feature points $u1$, $u2$, and $u3$; and A of feature points $u1$ and $u2$. The effectiveness of the method of determining the 2-D reference point from Theorem 1 can be demonstrated by the following simulations.

1. Suppose the object moves in such a way that point C translates constantly in 3-D and the object rotates around point C. Then, from the observations of the 2-D images of points u_j, the 2-D reference point can be solved that is exactly the image of point C no matter what $\varepsilon \geqslant 0$ is.
2. Suppose the object moves in such a way that point A translates in 3-D and the object rotates around point A. Let u_j ($j = 1, 2, 3, 4$) be the observable features. For $\varepsilon = 10$, the X coordinates of the images of points A, C and the reference point r versus time are plotted in Figure 4.5(a). Similarly, their Y-coordinates are plotted in Figure 4.5(b). The trajectories of these points in image plane $X-Y$ are plotted in Figure 4.5(c). It can be seen from Figure 4.5(c) that the 2-D reference point represents a balance between the smoothness of its trajectory (for prediction) and the nearness of the point to the centroid of the object (for stable grasping).
3. Suppose the object undergoes the same motion as in (1). However, we assume only that 3-D feature points $u1$, $u2$, and $u3$ are observable. The centroid of the object is thus represented by point B. For $\varepsilon = 5$, the trajectories of the 2-D reference point and the images of point B and C are plotted in the $X-Y$ plane in Figure 4.5(d).

5.2 Image-Based Tracking

Assume that the robot sampling rate is 100 Hz. The maximum velocity of the gripper is 0.25 m/sec. A moving cube with average speed 0.1 m/sec is required to be tracked and grasped. We use the sinusoidal curve (which was proposed in [6]) to demonstrate the

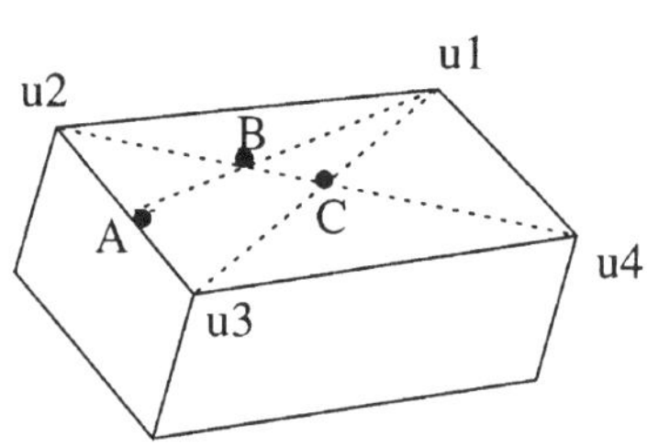

FIGURE 4.4
Illustration of the points used in the simulation.

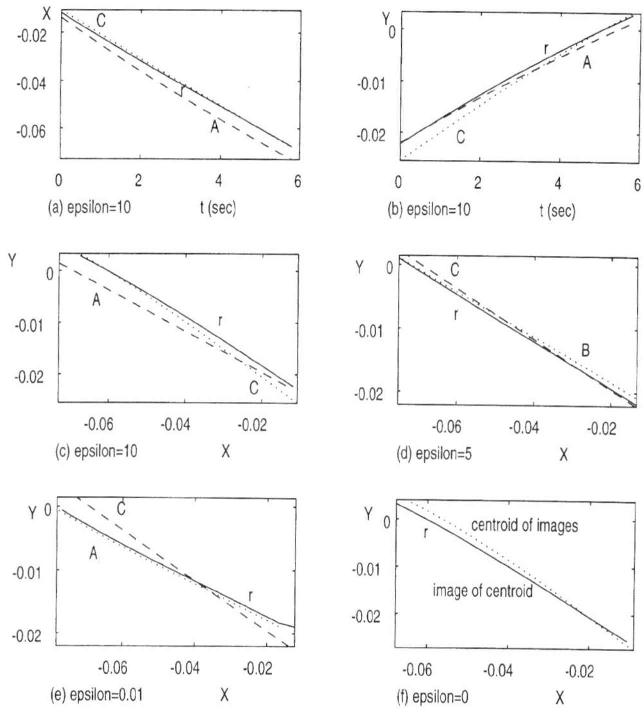

FIGURE 4.5
Estimation of the 2-D reference point.

tracking performance of the proposed schemes. The trajectory is given by

$$
\begin{aligned}
\text{Curve:}\quad x_w^r(t) &= 0.4572 + 0.508\sin(0.08\pi t) \quad (\text{m})\\
y_w^r(t) &= -0.0508 + -0.0508\sin(0.16\pi t) \quad (\text{m})\\
z_w^r(t) &= 0.05 \quad (\text{m})\\
O(t) &= 1.7708 - \pi\sin(0.08\pi t)/3 \quad (\text{rad})\\
A(t) &= 0 \quad (\text{rad})\\
T(t) &= 0 \quad (\text{rad})
\end{aligned}
$$

To simulate calibration errors, two new matrices are generated by multiplying R_1 and R_2 by some rotation matrix that represents a small rotation of angle β around a certain axis. Using projection (4.15), (4.16), the image location and velocity are generated. Uniformly distributed random variables with mean zero and different variances (noise level) are then added to them to simulate the quantization error effect and measurement noise. The resulting quantities are used as (X_i^r, Y_i^r) and $(\dot{X}_i^r, \dot{Y}_i^r)$ $(i = 1, 2)$. The image location (X_i^g, Y_i^g) of the gripper is similarly generated. Each image plane is digitized into 512×512 in pixel size.

Figure 4.6 shows the tracking performance of the image-based scheme for the sinusoidal trajectory with a noise level of 1 pixel and no calibration error. The initial pose of the gripper is $[0.508\ 0.0762\ 0.058]^{\mathrm{T}}$ m and $[\pi/2\ 0\ 0]^{\mathrm{T}}$ radian. Since the initial position error is large,

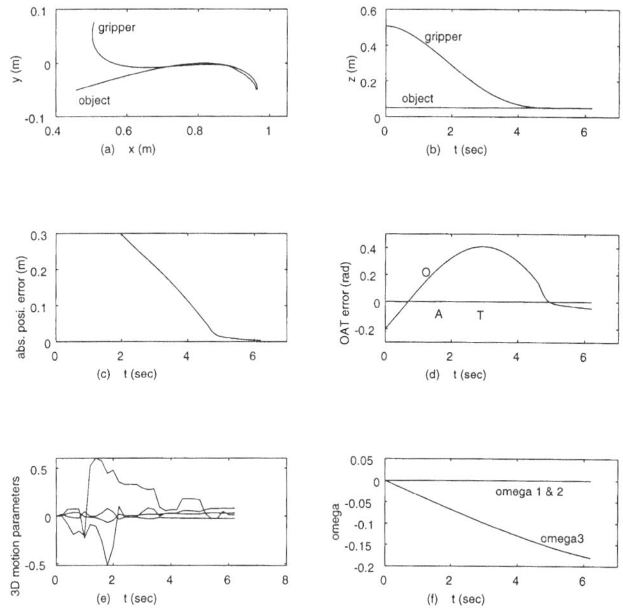

FIGURE 4.6
Tracking of the sinusoidal trajectory (noise level 1 pixel). (a) Gripper position versus cube position on xy plane. (b) Corresponding position in z direction. (c) Absolute position error between gripper and cube. (d) OAT angle tracking error versus time. (e) Part of estimated 3-D motion parameters. (f) Estimated angular velocity.

on-line polynomial trajectory planning is used to drive the robot into the vicinity (0.05 m) of the moving cube. The image-based controller is then switched on for fast and stable tracking. It can be seen that the trajectory of the gripper is quite smooth. The grasping occurs when the distance (tolerance) between the gripper and the cube is stable for three vision samples (here tolerance = 0.005 m is used).

The effect of calibration errors is simulated with $\beta = 0.02$ radian. The image-based tracking results with noise level 2 pixels are plotted in Figure 4.7. The estimated calibration parameters in (4.15) and (4.16) are plotted in Figure 4.7(g).

6 CONCLUSIONS

A complete modeling study for the multicamera robot hand–eye system has been presented. Assuming a rigid manipulator whose dynamic model is exactly known, the feedback linearized and decoupled model is used. The rigid motion of the object is modeled as a translation of the reference point and a rotation of the object around the point. For each camera, a complete solution is obtained to compute the motion of the reference point in the image plane, utilizing 2-D image features of the object only from the same image plane. In so doing, the traditional stereo vision approach is avoided.

Nonlinear regulator theory is applied to the control law design. Based on the estimated 2-D image reference point of the object, dynamic models in 2-D are obtained and used as the exosystem in the regulator design. The image-based tracking scheme is proposed. On the

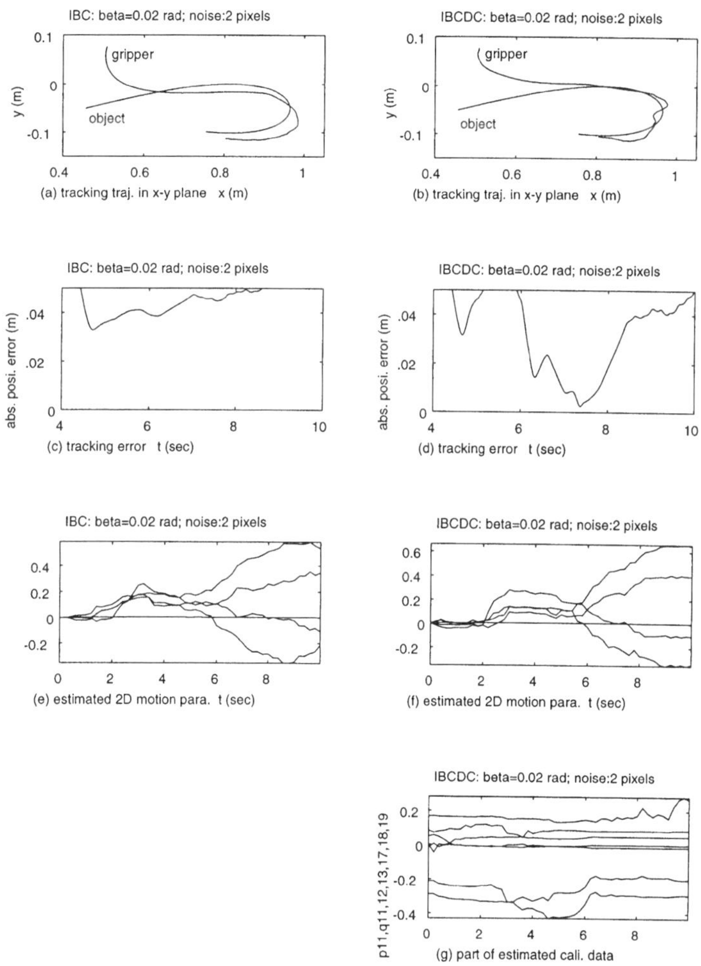

FIGURE 4.7
Performance comparison of IBC and IBCDC.

other hand, using the OAT orientation representation of the object, the control design for grasping is solved as an orientation tracking problem.

In our analysis of tracking and grasping, regular polyhedral objects are assumed. The approach needs to be extended to deal with objects of arbitrary shape in order to improve its applicability. Furthermore, although it has been demonstrated by simulations that the tracking and grasping algorithms have good performance in terms of speed and tracking accuracy, we believe that further effort should be made to utilize video-rate pipelined hardware or parallel processor arrays if we are ever to achieve real-time capabilities.

REFERENCES

[1] L. E. Weiss, A. C. Sanderson, and C. P. Neuman, Dynamic sensor-based control of robots with visual feedback. *IEEE J. Robot. Automat.* RA-3(5):404–417, 1987.

[2] J. T. Feddema and C. S. G. Lee, Adaptive image feature prediction and control for visual tracking with a hand–eye coordinated camera. *IEEE Trans. Syst. Man Cybernet.* 20:1172–1183, 1990.
[3] K. Hashimoto, T. Kimoto, T. Ebine, and H. Kimura, Manipulator control with image-based visual servo. *Proceedings IEEE International Conference on Robotics and Automation*, Sacramento, CA, 1991, pp. 2267–2272.
[4] N. Papanikolopoulos and P. K. Khosla, Adaptive robotic visual tracking: Theory and experiments. *IEEE Trans. Automat. Control* 38:429–445, 1993.
[5] B. Espiau, F. Francois, and P. Rives, A new approach to visual servoing in robotics. *IEEE Trans. Robot. Automat.* 8:313–326, 1992.
[6] A. J. Koivo and N. Houshangi, Real-time vision feedback for servoing robotic manipulator with self-tuning controller. *IEEE Trans. Syst. Man Cybernet.* 21(1):134–142, 1991.
[7] P. K. Allen, A. Timcenko, B. Yoshimi, and P. Michelman, Trajectory filtering and prediction for automated tracking and grasping of a moving object. *Proceedings IEEE Conference on Robotics and Automation*, Nice, France, 1992, pp. 1850–1856.
[8] R. Y. Tsai and T. S. Huang, Uniqueness and estimation of three-dimensional motion parameters of rigid objects with curved surfaces. *IEEE Trans. PAMI*, 6(1):13–26, 1984.
[9] J. K. Aggarwal and A. Mitiche, Structured motion from images. *Proceedings Image Understanding Workshop DARPA*, Miami Beach, FL, Dec. 1985, pp. 89–97.
[10] T. J. Broida and R. Chellappa, Estimating the kinematics and structure of a rigid object from a sequence of monocular images. *IEEE Trans. PAMI* 13(6):497–513, 1991.
[11] H. Shariat and K. Price, Motion estimation with more than two frames. *IEEE Trans. PAMI* 12(5):417–434, 1990.
[12] A. M. Waxman and J. H. Duncan, Binocular image flows: Step toward stereo-motion fusion. *IEEE Trans. PAMI* 8(6): 1986.
[13] J. Aloimons and A. Basu, Shape and 3D Motions from contour without point-to-point correspondences: General principles. *Proceedings IEEE Conference on Computer Vision Pattern Recognition*, Miami Beach, FL, 1986, pp. 518–527.
[14] K. S. Fu, R. C. Gonzalez, and C. S. G. Lee, *Robotics: Control, Sensing, Vision, and Intelligence.* McGraw-Hill, New York, 1987.
[15] T. J. Tarn, A. K. Bejczy, A. Isidori, and Y. Chen, Nonlinear feedback in robot arm control. *Proceedings 23rd IEEE Conference on Decision and Control*, Las Vegas, Dec. 1984, pp. 736–751.
[16] M. W. Spong and M. Vidyasagar, *Robot Dynamics and Control.* John Wiley & Sons, 1989.
[17] R. M. Haralick, H. Joo, C. Lee, X. Zhuang, V. Vaidya, and M. Kim, Pose estimation from corresponding point data. *IEEE Trans. Syst. Man Cybernet*, 19(6):1426–1446, 1989.
[18] M. Lei, Vision based robot tracking and manipulation. Ph.D. dissertation, Department of Systems Science and Mathematics, Washington University, St. Louis, May 1994.
[19] K. J. Astrom and B. Wittenmark, *Adaptive Control.* Addison-Wesley, 1989.
[20] A. Isidori and C. I. Byrnes, Output regulation of nonlinear systems. *IEEE Trans. AC* 35(2):131–140, 1990.
[21] Z. Lin, V. Zeman, R. V. Patel, On-line robot trajectory planning for catching a moving object. *Proceedings IEEE International Conference on Robotics and Automation,* 1989, pp. 1726–1731.
[22] D. J. Fleet, *Measurement of Image Velocity.* Kluwer Academic Publishers, 1992.

SECTION III

Multiple Sensor Fusion in Planning and Control

CHAPTER 5

Complementary Sensor Fusion in Robotic Manipulation

BIJOY K. GHOSH, ZHENYU YU, DI XIAO, NING XI and TZYH-JONG TARN
Washington University, St. Louis, Missouri

ABSTRACT

In this chapter we analyze problems in robotic manipulation in an uncalibrated environment. The environment consists of a PUMA 560 robotic manipulator, a turntable rotating around a vertical axis equipped with an encoder that records the instantaneous angular displacement with respect to an axis chosen *a priori*, and a CCD camera–based vision sensor that is fixed permanently on the ceiling. It is assumed that a part with a known shape but unknown orientation is placed on the turntable, which is rotating with unknown motion dynamics. Furthermore, the relative positions of the manipulator, turntable, and camera (the calibration parameters) are assumed to be unknown. In spite of our lack of knowledge of the orientation of the part and the calibration parameters, the objective is to track the rotating part (with an *a priori* specified relative orientation) and grasp the part with the end effector of the manipulator.

In addition, we consider planning and control of a robot manipulator for a class of constrained motions. Here the task under consideration is to control a robot so that a tool grasped by its end effector follows a path on an unknown surface with the aid of a single-camera vision system. To accomplish the task, we propose a new planning and control strategy based on multisensor fusion. Three different sensors—joint encoders, a wrist force–torque sensor, and a vision system with a single camera fixed above a work space—are employed. First, based on sensory information, we decouple control variables into two subspaces: one for force control and the other for control of constrained motion. This decoupling allows one to design control schemes for regulation of force and for constrained motion separately. Second, we develop a new scheme by means of sensor fusion to handle the uncertainties in an uncalibrated work space. The contact surface is assumed to be unknown but the trajectory to be followed is visible to the vision system, and the precise position and orientation of the camera with respect to the robot are also assumed to be unknown. Overall, the contributions described in this chapter are the following: (1) multisensor fusion used for both force–torque and visual sensors with complementary observed data, as opposed to many sensor fusion schemes in the literature with redundant data; (2) intelligent manipulation of a robot that can work in an uncalibrated work space with a camera that is not calibrated with respect to the robot.

1 INTRODUCTION

In this chapter we discuss two important problems in robotic manipulation. The first problem deals with manipulation in an uncalibrated environment. The second problem deals with motion planning with multisensor fusion. A third related problem that is not the main emphasis of this chapter is servoing. Of course, in each of these problems, "vision" plays an important role. This chapter emphasizes that in many instances, vision alone is not sufficient, and one has to combine visual information with one or more additional sensory inputs. This leads to many multisensor fusion–based algorithms, which are discussed in this chapter. Before we elaborate on these algorithms, we make a few background and somewhat historical remarks.

Control of robot manipulators with vision in the feedback loop has an exciting history starting with the pioneering work of Hill and Park [1] and Weiss, Sanderson, and Neuman [2]. Subsequent work in this area has focused on visual servoing, wherein the emphasis is on visually locating the position and orientation of a part and controlling a robot manipulator to grasp and manipulate the part. If the part is not stationary, then the process of locating the part and repositioning the robot must be performed by utilizing feedback control, which has been studied in [3–7]. Using vision in the feedback loop has many advantages over the more direct "look and go" approach. Some of the advantages are that a visually guided robot is more flexible and robust and has the potential to perform satisfactorily even under structural uncertainty. This is evidenced by the "controlled active vision" scheme introduced by Papanikolopoulos *et al.* [8], where the goal is to accomplish a task in spite of environmental and target-related unknown and possibly changing factors. Other instances of visual guidance have been evidenced by the work of Allen *et al.* [3], when the objective is to grasp a toy train undergoing a planar circular motion. The position of the train is observed visually and the orientation is automatically specified by the position.

The concept of multisensor fusion is to combine data from multiple sensors to obtain inferences that may not be possible from a single sensor alone [9]. Without going into the details of the specific reason why a single sensor, for example, the visual sensor, cannot be used reliably for all the different tasks that we propose to perform, we note that the main purpose of using multisensor fusion is to compensate for the speed of computation. The vision system we use is neither fast nor accurate—hence the need for "sensor fusion."

There are many other multisensor fusion schemes in the literature [10–21]. For example, Allen and Bajcsy [17] used stereo edges to match objects to a fixed world model and then adopted a tactile sensor to investigate the occluded parts. Flynn [18] has combined a sonar and an infrared proximity sensor in order to reduce errors inherent in both sensor domains. Magee *et al.* [19] presented a new method for combining data from intensity and range sensors. Algorithms based on fusing static thermal and visual images obtained from outdoor scenes have been reported by Nandhakumar and Aggarwal [20] and Mitiche and Aggarwal [21].

As opposed to multisensor fusion–based servoing, where the sensory information automatically generates the feedback control, in this chapter we propose multisensor fusion–based planning, where the sensory information automatically feeds the planner. The motion planning schedule is generated autonomously as a result, and the robot controller simply follows the motion plan. This simplifies the problem of controller synthesis while relieving the computational burden of the controller. On the other hand, the planner has an additional structure, since it now receives (multi) sensory input. Because "planning" is not performed in real time, it suffices to use a planner with a somewhat slower computational capability.

It has long been recognized that sensor-based control is an important issue in robotics.

As a robot is expected to accompish more complex tasks, the need to take advantage of multiple sensors in a system becomes stronger. The growing use of multiple sensors in robotics greatly extends the application domain of robots. In practice, the environment in which robots are employed is often poorly structured, especially in flexible manuacturing systems, where manufacturing lines are not well prepared due to limited time and expense. To compensate for the uncertainties in the environment, robots need rich and reliable information from the sensors.

In this chapter, we also propose a new sensor-based control strategy for a class of constrained motion of a robot in an uncalibrated work space. Our attempt it to provide a robot with certain intelligence in the sense that it can handle uncertainties in the environment without explicit intervention of reprogramming. Concretely, the task under our consideration is to control a robot such that the tip of a tool grasped by the end effector of the robot follows a curve on an unknown surface (both the shape and the exact location of the surface are unknown), as shown in Figure 5.1. Many tasks in manufacturing engineering, such as welding, cutting materials along a curve, and scribing parts, are in this category. To accomplish a task of this kind, it is natural to locate the precise position of the curve. In practice, however, it may be very hard to describe the curve exactly, especially in the case that the contact surface is unknown. At first glance, the task we consider looks relatively simple, because guided by their vision system, humans can easily accomplish it without any apparent difficulty. As a matter of fact, as is also the case with perception, the elementary operative intelligence that people use unconsciously to interact with their environment (e.g., assembling a device) turns out to be extremely difficult to duplicate using a computer-controlled robot. To deal with uncertainties in the work space, in our research three different sensors have been used. They are encoders mounted at each joint of a robot with six degrees of freedom, a force–torque sensor mounted at the wrist of the robot, and a vision system with a single camera fixed above an uncalibrated work space, respectively.

How to integrate information from different sensors is a topic in the field of multisensor fusion. A number of results have been obtained in the area [22–26]. However, to our best knowledge of the literature in the field, most researchers focus on fusion in order to sense data by means of redundancy of information obtained from the sensors. As is well known, a

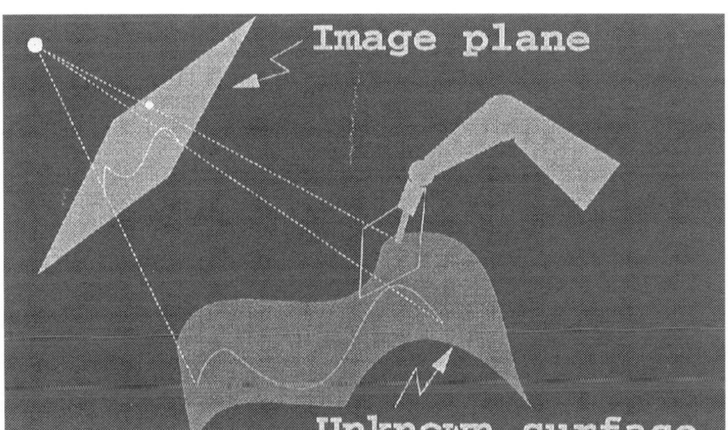

FIGURE 5.1
A typical trajectory-following task with vision and force–torque sensor.

force–torque sensor and a vision system are disparate sensors. Nelson and Khosla [23] proposed a resolvability concept for assimilating these two sensors and utilized them in different stages separately during contact transitions based on their resolvabilities. In this chapter, we integrate information from the force–torque sensor and the vision system simultaneously to ensure the completion of the task. Actually, we employ the force–torque sensor not only for maintaining contact between the tool and the surface but also for determining the normal vector of the tangent plane of the surface at the contact point. The vision system is used to monitor the difference between the tool and the path. A control signal is created by projecting the error in the image plane onto the tangent plane. In this way, the vision system is implemented in the closed loop. The proposed control has a hierarchical structure. The lower level is a hybrid position–force control with nonlinear feedback. The upper level is a planner that generates a desired motion based on the information from multiple sensors. The relation between the motion of the robot on the tangent plane and its projection on the image plane is continuously updated in time.

The study of control of a robot manipulator with visual information in a feedback loop is often referred to as visual servoing. Roughly speaking, the approaches used in visual servoing can be classified into two categories: position-based and image-based methods [27]. The majority of research work on visual servoing has emphasized dynamic visual tracking with an eye-in-hand configuration. The reason is partially that with the camera rigidly mounted on the robot arm, positioning against a static scene and tracking a moving object can be easily posed as a feature-tracking problem. However, the eye-in-hand configuration limits the work space of the robot arm and therefore is not suitable for some cases. In our work we choose an alternative configuration — one camera fixed above the uncalibrated work space. Our method can easily be extended to a multicamera system with a potential for better results. To complete our task, we need to control the contact force and the motion constrained on the unknown surface simultaneously without the possibility of a conflict. Here, servoing either on the image plane or in the task space is not suitable, if not impossible. It is evident that motion servoing on the tangent plane of the surface gives us a better way to design control. In order to utilize visual information, most of the previous work on visual servoing needed precise calibration of the camera with respect to the robot. Calibration is a time-consuming procedure and makes the implementation expensive or even impossible in some cases. In this chapter, we introduce a new image-based motion-planning approach for the constrained motion of the robot that is robust against uncertainties in the pose of the camera.

The study of the control problem for constrained motion of a robot can be dated back to the 1950s. The past few decades have witnessed much progress in this field (see [28] for a summary of the control strategies). Among them two approaches, hybrid position–force control [29–34] and impedance control [28, 35–39], have attracted much attention from researchers. The hybrid position–force control approach has a clear physical meaning, and impedance control unifies the planning and control as a whole procedure. However, in order to use impedance control, one has to translate a task into a desired impedance, a relation between motion and force. In some cases, this is very difficult. In this chapter, we adopt a hybrid position–force control approach by decoupling control variables in two perpendicular subspaces: tangent space and the normal direction. We conclude that with this decoupling, one can design control laws for constrained motion and force regulation separately. Our method proposed in this chapter differs from previous work in the force control community. In our case, the contact surface is assumed to be unknown. Therefore, the desired trajectory for constrained motion in the hybrid position–force control scheme is unknown *a priori* and has to be planned in real time.

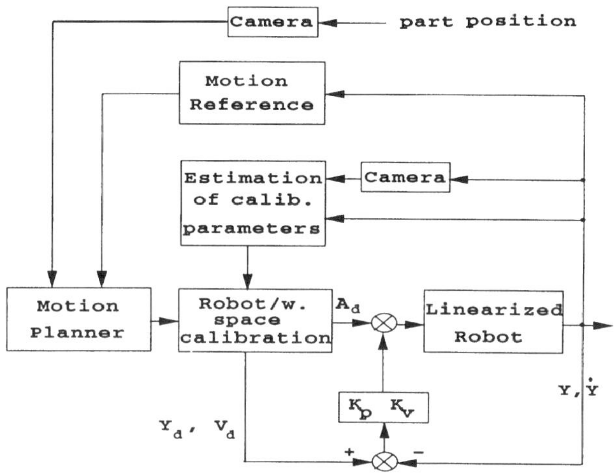

FIGURE 5.2
A block diagram showing the planning and control structure.

The planning and control structure employed in this chapter is shown in Figure 5.2. The sensory data are fed back to a planner that can generate a desired trajectory in real time for the robot, instead of being fed directly back to the controller as in most previous work in robotics. We also show in Figure 5.3 how control is implemented by using information from the vision and force–torque sensor.

2 GRASPING

2.1 Experimental Setup

We consider a manufacturing work cell as shown in Figure 5.4. The work cell is equipped with a rotating conveyor (equipped with encoders that measure the rotation angle), a robotic

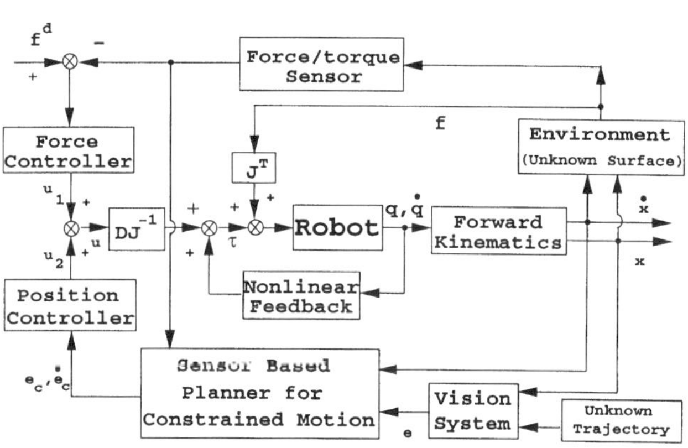

FIGURE 5.3
Block diagram of the control system with vision and force–torque sensor.

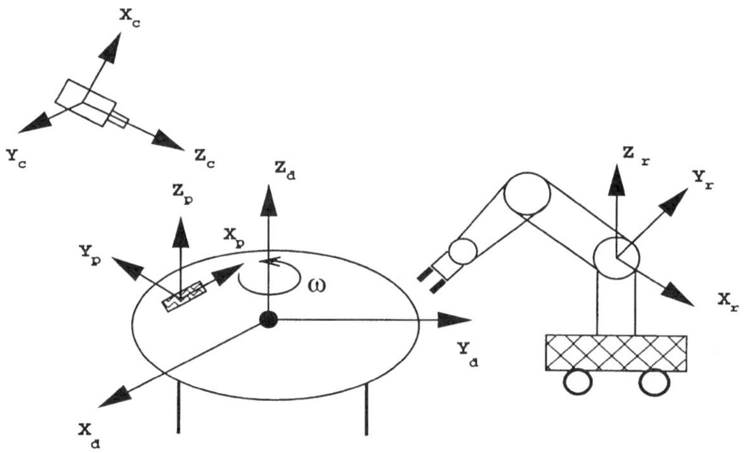

FIGURE 5.4
A typical manufacturing work cell.

manipulator, and a computer vision system with a single charge-coupled device (CCD) camera. The precise relative positions of the camera, robot, and conveyor are assumed to be unknown. In spite of the lack of calibration data, the objective is to compute the instantaneous position and orientation of a part placed on the turntable with respect to the coordinate system attached to the base frame of the robot. A second objective is to feed this information to a motion planner, which operates in either a time base or an event base. The planner computes the relevant position, velocity, and acceleration profile that the end effector needs to follow in order to achieve the desired task, which in our experiment is to pick up a part from the rotating conveyor. The following assumptions are made about or work cell.

A1. The precise position and orientation of the camera with respect to the robot coordinate frame are unknown. In addition, the precise position and orientation of the conveyor with respect to the robot coordinate frame are unknown.

A2. The plane of the conveyor and the XY plane of the base frame of the robot are assumed to be parallel.

A3. The part has a known simple shape. In particular, we assume that observing feature points placed on the top surface of the part enables one to determine the orientation of the part.

A4. The entire work cell is in the view field of the camera. The center of the conveyor and a reference point on the conveyor are also assumed to be observed by the camera.

A5. The intrinsic parameters (the focal length etc.) of the camera are known.

The technical contents of this section are now summarized. Because the camera has not been selectively placed at any specific known position in the work cell, we propose a virtual rotation algorithm that would virtually rotate the camera in a vertical position with respect to the disc conveyor. This is summarized in Section 2.2. In Section 2.2 we also describe how the position and orientation of the part are computed. This is first done assuming that the height of the part is negligible compared with its dimension. Subsequently, we consider parts with feature points that are a certain distance above the disc conveyor.

In the next phase of this chapter, described in Section 2.3, we consider the problem of robot calibration. *A priori*, we do not assume that the position of the robot is known with respect to the coordinate frame attached to the disc conveyor. We propose to compute the associated parameters by observing feature points on the end effector. The underlying problem that we outline in Section 2.3 is that of computing the coordinates of the feature points with respect to the frame attached to the conveyor.

Our final problem, described in Section 2.4, is to derive a control law for tracking and grasping. The problem we consider is to make a plane for a robot to track a target whose position, orientation, and velocity are estimated continuously. In doing so, we implement a "parallel guidance" controller. We propose that an error reduction term be added to the position and velocity of the target to form a desired position, velocity, and acceleration profile for the robot. The error reduction term can be carefully planned using both time-based and event-based approaches. In Section 2.5, a description of various experimental implementations have been provided.

2.2 Estimation and Calibration

Camera Self-Calibration

As described before, we assume that the camera has been placed at an unknown position in the work cell. In this section we describe a virtual rotation algorithm to ascertain the relative position of the camera with respect to the coordinate frame of the conveyor. Note that any point on the conveyor undergoes a circular trajectory as the conveyor rotates. The image of such a circular trajectory is an ellipse on the image plane of the camera. The shape of the ellipse depends on the relative orientation of the camera with respect to the normal vector of the plane of the conveyor. In order to rotate virtually and obtain the corresponding image from the top view of the camera, we transform the observed image in such a way that the projected ellipse is transformed to a circle. The details of the steps are described as follows.

Let $(x_r, y_r)_i$ *be the coordinates of the ith reference point on the image plane. Let* (x_c, y_c) *be the coordinates of the image of the center of the turntable.*

1. *A rotation* R_k *of the camera around its optical center is applied to transform the image of the center of the turntable to the center of the image plane. Kanatani's standard rotation* [40] *can be used in this step.*
2. *Obtain parameters that describe the ellipse traced out by the ith reference point on the image plane. A recursive least squares fitting algorithm is used for this purpose.*
3. *A rotation* R_Z *around the z axis of the camera is applied to transform the major axis of the elipse on the image plane into a position parallel to the y axis of the image plane. Note that this would automatically place the minor axis along the x axis.*
4. *A rotation* R_Y *of the camera around the y axis of the image plane is applied to transform the ellipse to a circle. Note that this circle would automatically have its center on the x axis of the image plane.*

The entire process of the camera rotations is illustrated in Figure 5.5. Note that the rotation R_k is necessary to align the center of the conveyor with the center of the image plane. It is explained in Appendix 1 that this step is necessary in order to choose the correct sign of $\tan \alpha$ in (5.54). In general, the camera sees the original circular trajectory of the reference point up to an elliptic cone. The angle of rotation in step 4 is determined by the plane that intersects the elliptic cone on a circle. The details are described in Appendix 1.

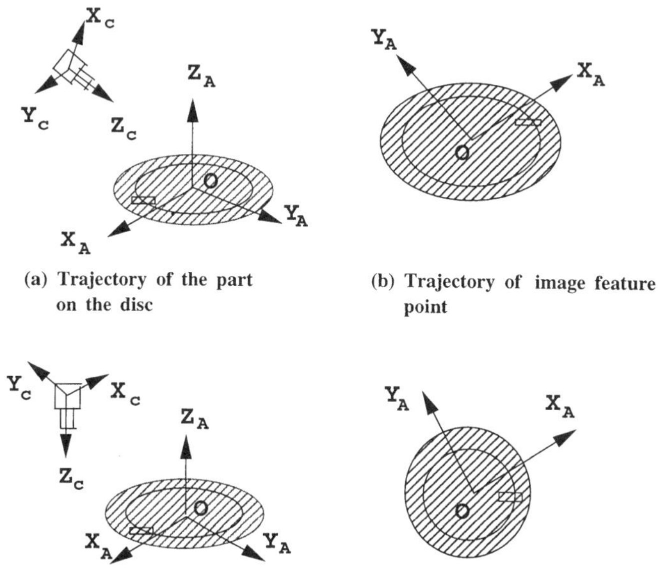

(a) Trajectory of the part on the disc (b) Trajectory of image feature point

(c) Virtual rotation of the camera (d) Transformed image

FIGURE 5.5
The virtual rotation scheme.

Estimating Points on the Part in Three Dimensions

In this subsection we assume that a part of nonnegligible height has been placed on the conveyor. As shown in Figure 5.6, a feature point on the part is at a height h from the surface of the conveyor. Let $oxyz$ be any Cartesian coordinate frame with its origin o at the center of the conveyor and its z axis perpendicular to the conveyor. After virtual rotation of the camera, the position of a point p in three-dimensional space can be represented by

$$x_p = \frac{ax_{ref} - by_{ref}}{l^2} + \frac{(\bar{a} - a)x_{ref} - (\bar{b} - b)y_{ref}}{lLf} h \tag{5.1}$$

$$y_p = \frac{bx_{ref} + ay_{ref}}{l^2} + \frac{(\bar{b} - b)x_{ref} + (\bar{a} - a)y_{ref}}{lLf} h \tag{5.2}$$

where

$$a = \overrightarrow{C_oP_o} \cdot \overrightarrow{C_oR_o}; \qquad b = \overrightarrow{C_oP_o} \times \overrightarrow{C_oR_o} \tag{5.3}$$

$$\bar{a} = \overrightarrow{C_oO_o} \cdot \overrightarrow{C_oR_o}; \qquad \bar{b} = \overrightarrow{C_oO_o} \times \overrightarrow{C_oR_o} \tag{5.4}$$

where $\overrightarrow{AB}$ denotes the vector from A to B and $\cdot$ and $\times$ are operations of dot product and cross product, respectively. The points C_o, R_o, and P_o are the transformed images (after virtual rotation) of the disc center, the reference point, and the point p via perspective projection, respectively. The point O_o is the intersection of the optical axis and the transformed image plane and h stands for the distance between the point p and the plane of the disc conveyor. The coordinates $(x_{ref}, y_{ref}, 0)$ are the coordinates of the reference point R in the frame $oxyz$ and (x_p, y_p, h) the coordinates of the point p in the same frame. The scalar L denotes the length of the reference line segment $\overrightarrow{CR}$, l is the length of the transformed

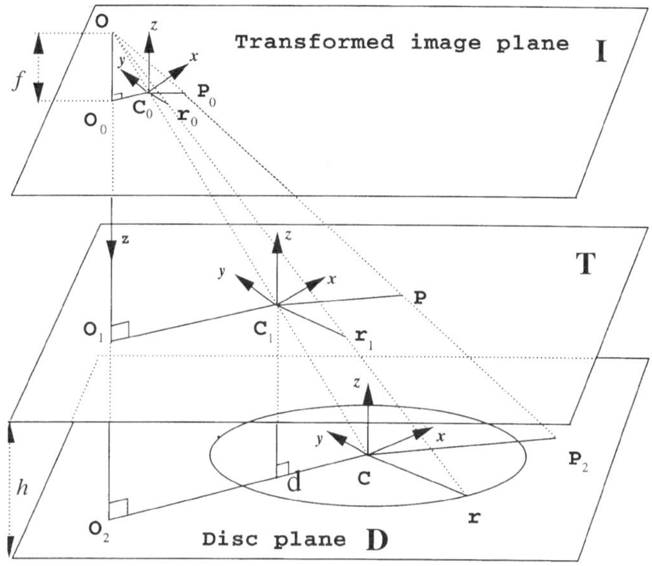

FIGURE 5.6
Determining the position of a point from its image using virtual rotation.

image $\overrightarrow{C_o R_o}$ of the reference line segment, and f is the focal length of the camera. Note that a, b, $\bar{a}$, $\bar{b}$, and l can easily be computed in terms of the image coordinates of C_o, R_o, and P_o and the parameters L and f are known constants. For a given camera, f is also a constant. Hence, if the cordinates (x_{ref}, y_{ref}) of the reference point are given in the frame $oxyz$, the position of the point p in the same frame is an affine function of h, the height of the point from the conveyor.

Equations (5.1) and (5.2) describe an affine line in the coordinate frame $oxyz$. The affine line passes through the optical center of the camera and the feature point (x_p, y_p, h). Consider a feature point on the part with coordinates (x_p, y_p, h) and suppose that two different images are taken at two different times t_1 and t_2. Applying the virtual rotation algorithm to each of the two images, it follows that we can write

$$x_p = \theta_1(t_i) + \theta_2(t_i)h$$

$$y_p = \theta_3(t_i) + \theta_4(t_i)h$$

for $i = 1, 2$. Note that these equations describe two different affine lines described by the feature point p at two different instants of time t_1 and t_2. The parameters θ_1, θ_2, θ_3, θ_4 can be correspondingly determined (5.1) and (5.2) and can take different values for different instants of time. However, because the part does not move in the rotating coordinate frame, it follows that (x_p, y_p, h) remain invariant at t_1 and t_2 and we have

$$\theta_1(t_1) + \theta_2(t_1)h = \theta_1(t_2) + \theta_2(t_2)h$$

$$\theta_3(t_1) + \theta_4(t_1)h = \theta_3(t_2) + \theta_4(t_2)h$$

which can be solved for h using any standard least squares method.

Sensor Integration for Estimation in Real Time

In Figure 5.4, a part placed on the disc conveyor is shown schematically. In order to describe the process of sensor integration, the following three coordinate frames have been defined. These are the fixed disc frame $x_D y_D z_D o_D$, which is assumed not to rotate with the conveyor; the attached disc frame $x_A y_A z_A o_A$, which is attached to the conveyor and rotates with it; and the part coordinate frame $x_p y_p z_p o_p$, which is attached to the centroid of the part to represent its position and orientation.

The z axes of both fixed and attached disc frames have been assumed to coincide with the axis of rotation of the conveyor. The position of the turntable can be described by $\theta(t)$, which is the angle between the x axes of the two coordinate frames. The angle $\theta(t)$ can be measured by an encoder sensor. The origin of the part coordinate frame is attached to the centroid of the part to represent the position of the part. The orientation of the part coordinate represents the orientation of the part. Let the position of the centroid of a part with respect to the attached disc frame (i.e., the relative position of the part) be represented by x_a, which is assumed to be obtainable from the vision system. The position $x_d(t)$ of the part with respect to the fixed coordinate frame (i.e., the absolute position of the part) is given by

$$x_d(t) = R_z(\theta(t))x_a \tag{5.5}$$

where

$$R_z(\theta(t)) = \begin{bmatrix} \cos\theta(t) & -\sin\theta(t) & 0 \\ \sin\theta(t) & \cos\theta(t) & 0 \\ 0 & 0 & 1 \end{bmatrix} \tag{5.6}$$

represents a rotational transformation around the z axis.

Let n_a, s_a, and r_a be the unit vectors of a coordinate frame of a part expressed in the attached disc frame. The matrix $R_a = [n_a \; s_a \; r_a]$ representing the orientation of the part in the attached disc coordinate frame (i.e., the relative orientation of the part) can likewise be obtained from the vision system. Let $R_d(t) = [n_d(t) \; s_d(t) \; r_d(t)]$ be the orientation of the part expressed in the fixed disc frame (i.e., the absolute orientation of the part), where $n_d(t)$, $s_d(t)$, and $r_d(t)$ are the unit vectors of the part coordinate frame expressed in the fixed disc frame. The matrices $R_d(t)$ and R_a are related as follows:

$$R_d(t) = R_z(\theta(t))R_a \tag{5.7}$$

Equations (5.5) and (5.7) reflect how absolute information $x_d(t)$ and $R_d(t)$ is obtained through fusing the relative information x_a and R_a with the encoder measurement $\theta(t)$. Differentiating both sides of (5.5) yields

$$\dot{x}_d(t) = \Omega(\dot{\theta})R_z(\theta)x_a \tag{5.8}$$

where $\Omega(\dot{\theta})$ is given by

$$\Omega(\dot{\theta}) = \begin{bmatrix} 0 & -\dot{\theta} & 0 \\ \dot{\theta} & 0 & 0 \\ 0 & 0 & 0 \end{bmatrix} \tag{5.9}$$

Equation (5.8) shows that the velocity of a part on the conveyor can be obtained by fusing relative information x_a with encoder measurement $\theta(t)$ and $\dot{\theta}(t)$, the speed of rotation of the conveyor. In fact, $\dot{\theta}(t)$ can also be measured by the encoder.

Because the relative position of a part with respect to the conveyor has been assumed not to change with time, the absolute position, orientation, and velocity of the part are updated at the rate at which the encoder measurements are taken. Since the encoder measurements are updated at a very high frequency, absolute information is assumed to be obtained in real time, without requiring that the relative information be updated with high frequency by the vision system. The computational burden on the vision system is therefore greatly reduced.

2.3 Robot Calibration

In order to determine the relation between the fixed disc frame and the base frame of the robot, we need to describe a set of points in both frames, as the two frames are related by

$$^{b}P = {}^{b}R_{d}{}^{d}P + {}^{b}T_{d}$$

where ^{b}P and ^{d}P are the coordinates of a point in the base frame and the fixed disc frame, respectively. From assumption A2 in Section 2.1, it is seen that there is one unknown in the rotation matrix $^{b}R_{d}$. $^{b}T_{d}$ has three unknown elements. For a point q, if we know its coordinates in both frames (i.e., $^{b}P_{q}$ and $^{d}P_{q}$), then from the last equality we have three equations for the four unknown variables. Obviously, in order to get a unique solution to the relation, we need to know at least two points in both frames. Fortunately, from reading encoders of the robot, the coordinates of points on the end effector with respect to the base frame of the robot can be readily obtained. In what follows, we describe the points on the end effector in the fixed disc frame with the aid of the single camera.

Suppose that an image was taken at time t. At the very moment the coordinates of the reference point in the fixed disc frame are

$$({}^{d}X_{ref}, {}^{d}Y_{ref}, {}^{d}Z_{ref}) = (L\cos(\theta(t)),\ L\sin(\theta(t)),\ 0)$$

After computing the corresponding a, b, $\bar{a}$, and $\bar{b}$ via the image data, the coordinates of a point on the end effector in the fixed disc frame can be easily obtained by substituting the results into (5.1) and (5.2), as long as the distance of the point from the turntable is known.

Case 1: The Plane of the Turntable Is Known

Assume that the plane of the turntable in the base frame of the robot is known; that is, the distance between the plane of the turntable and the XY plane of the base frame is known. It follows that we can compute the z coordinate of a point on the end effector in the base frame of the robot from reading the encoders on the robot. Therefore the height of the point on the end effector from the turntable can be computed. Consequently, the coordinates of the point in th fixed disc frame are obtained.

Case 2: The Plane of the Turntable Is Unknown

Suppose that the distance between the plane of the turntable and the XY plane of the base frame is unknown *a priori*, although we continue to assume that they are parallel. We now describe how to determine the height of the point from the turntable in order to obtain coordinates of the point in the fixed disc frame. We consider the following two subcases:

Two Points on the End Effector Let (x_i, y_i, z_i) $(i = 1, 2)$ be coordinates of the ith point on the end effector with the associated coordinates on the image plane being given by (X_i, Y_i). The coordinates of the ith point on the end effector with respect to the camera frame can be represented by

$$(x_i, y_i, z_i) = \left(\frac{X_i}{f} z_i, \frac{Y_i}{f} z_i, z_i\right) \tag{5.10}$$

Note that the z coordinates of the two points in the base frame of the robot are known and so is their difference, say d. After virtually rotating the camera, the difference of the z coordinates of the two points in the camera frame has the same value with opposite sign, since z axes of the base frame and the camera frame are parallel but point in opposite directions. Without loss of generality, we assume $z_1 - z_2 = d$, which implies, if $d > 0$, that the second point on the end effector is farther away from the turntable. Also, in the base frame of the robot, the distance between the two points on the end effector is known, say s, via encoders of the robot. Of course, s is still preserved in the camera frame. Hence, we have

$$(x_1 - x_2)^2 + (y_1 - y_2)^2 + (z_1 - z_2)^2 = s^2$$

that is,

$$(X_1 z_1 - X_2 z_2)^2 + (Y_1 z_1 - Y_2 z_2)^2 + f^2(z_1 - z_2)^2 = f^2 s^2 \tag{5.11}$$

Solving these equations for z_2, we have

$$z_2 = \frac{-F \pm \sqrt{F^2 - EG}}{E}$$

In practice, the problem itself guarantees that one real solution exists. Hence, the quadratic equation z_2 must have two real solutions (i.e., $F^2 - EG \geqslant 0$). In other words, we can determine z_2 up to two solutions if $F < 0$ and $G > 0$. However, in many cases we can recover z_2 uniquely, since the point should be in front of the camera (i.e., $z_2 > 0$). For instance, if $d = 0$ (i.e., the line segment joining the two points is parallel to the turntable) then $G < 0$. As a result,

$$z_2 = \frac{-F + \sqrt{F^2 - EG}}{E}$$

is the unique solution. By continuity, for small enough d, the solution for z_2 is unique. Generally speaking, if $G < 0$ then the unique solution exists. In such cases, we can determine the distances of the two points on the end effector from the turntable, respectively,

$$h_1 = H - z_1 = H - \frac{-F + \sqrt{F^2 - EG}}{E} - d$$

$$h_2 = H - z_2 = H - \frac{-F + \sqrt{F^2 - EG}}{E}$$

Three Points on the End Effector Similarly to case A, without loss of generality, we assume that

$$z_1 - z_2 = d_1, \qquad z_2 - z_3 = d_2 \tag{5.12}$$

and that

$$\begin{aligned}(x_1 - x_2)^2 + (y_1 - y_2)^2 + (z_1 - z_2)^2 &= s_1^2\\(x_2 - x_3)^2 + (y_2 - y_3)^2 + (z_2 - z_3)^2 &= s_2^2\\(x_3 - x_1)^2 + (y_3 - y_1)^2 + (z_3 - z_1)^2 &= s_3^2\end{aligned}$$

where the d_i and s_j $(i = 1, 2, j = 1, 2, 3)$ can be obtained by reading the encoders on the robot. Eliminating z_1 and z_2 by employing Eq. (5.12), it turns out that

$$\begin{aligned}E_{12}z_3^2 + 2(E_{12}d_2 + F_{12}d_1)z_3 + E_{12}d_2^2 + 2F_{12}d_1d_2 + G_1d_1^2 - f^2s_1^2 &= 0\\E_{23}z_3^2 + 2F_{23}d_2z_3 + G_2d_2^2 - f^2s_2^2 &= 0\\E_{13}z_3^2 + 2F_{13}(d_1 + d_2)z_3 + G_3(d_1 + d_2)^2 - f^2s_3^2 &= 0\end{aligned}$$

where

$$\begin{aligned}E_{ij} &= (X_i - X_j)^2 + (Y_i - Y_j)^2\\F_{ij} &= X_i(X_i - X_j) + Y_i(Y_i - Y_j)\\G_i &= (X_i^2 + Y_i^2 + f^2)\\i &= 1, 2, 3, j = 1, 2, 3\end{aligned}$$

Therefore, z_3 satisfies three quadratic equations. In fact, there is a real common solution to the equations in practice. Generically, the equations have only one common solution and hence give rise to a unique solution for z_i $(i = 1, 2, 3)$. As a result, the distance of the ith point from the turntable is $h_i = H - z_i$ $(i = 1, 2, 3)$.

Computation of the Relation between the Frames

Having described at least two points in both the fixed disc frame and the base frame of the robot, we obtain

$$\begin{bmatrix}{}^bx_i\\{}^by_i\\{}^bz_i\end{bmatrix} = {}^bR_d\begin{bmatrix}{}^dx_i\\{}^dy_i\\{}^dz_i\end{bmatrix} + {}^bT_d, \qquad i = 1, 2, \ldots$$

where $({}^bx_i, {}^by_i, {}^bz_i)$ and $({}^dx_i, {}^dy_i, {}^dz_i)$ are the coordinates of the ith point with respect to the base frame and the fixed disc frame, respectively. Recalling assumption A2, we know that ${}^bR_d$

has the following structure:

$$
{}^{b}R_{d} = \begin{bmatrix} r_{11} & r_{12} & 0 \\ -r_{12} & r_{11} & 0 \\ 0 & 0 & 1 \end{bmatrix} \tag{5.13}
$$

In this case, we can determine ${}^{b}R_{d}$ and ${}^{b}T_{d}$ with knowing two points in the two frames. As a matter of fact, it is seen that

$$
\begin{bmatrix} {}^{b}x_{2} - {}^{b}x_{1} \\ {}^{b}y_{2} - {}^{b}y_{1} \\ {}^{b}z_{2} - {}^{b}z_{1} \end{bmatrix} = {}^{b}R_{d} \begin{bmatrix} {}^{d}x_{2} - {}^{d}x_{1} \\ {}^{d}y_{2} - {}^{d}y_{1} \\ {}^{d}z_{2} - {}^{d}z_{1} \end{bmatrix}
$$

or

$$
\begin{bmatrix} {}^{b}e_{1} \\ {}^{b}e_{2} \\ {}^{b}e_{3} \end{bmatrix} = {}^{b}R_{d} \begin{bmatrix} {}^{d}e_{1} \\ {}^{d}e_{2} \\ {}^{d}e_{3} \end{bmatrix}, \quad \begin{bmatrix} {}^{b}e_{1} \\ {}^{b}e_{2} \\ {}^{b}e_{3} \end{bmatrix} = \begin{bmatrix} {}^{b}x_{2} - {}^{b}x_{1} \\ {}^{b}y_{2} - {}^{b}y_{1} \\ {}^{b}z_{2} - {}^{b}z_{1} \end{bmatrix}, \quad \begin{bmatrix} {}^{d}e_{1} \\ {}^{d}e_{2} \\ {}^{d}e_{3} \end{bmatrix} = \begin{bmatrix} {}^{d}x_{2} - {}^{d}x_{1} \\ {}^{d}y_{2} - {}^{d}y_{1} \\ {}^{d}z_{2} - {}^{d}z_{1} \end{bmatrix}
$$

which are linear equations in r_{11} and r_{12}. As long as the line joining the two points is not parallel to the z axis of the base frame, the equations always have a unique solution. However, such a solution may not satisfy the constraint $r_{11}^{2} + r_{12}^{2} = 1$ due to the possible noise in the observed data and computation errors. In other words, ${}^{b}R_{d}$ obtained in this way may not be orthogonal and may therefore be meaningless.

Actually, the problem of determining ${}^{b}R_{d}$ can be viewed as the optimization problem of determining ${}^{b}R_{d}$ with the structure in (5.13) such that

$$
\left\| \begin{bmatrix} {}^{b}e_{1} \\ {}^{b}e_{2} \\ {}^{b}e_{3} \end{bmatrix} - {}^{b}R_{d} \begin{bmatrix} {}^{d}e_{1} \\ {}^{d}e_{2} \\ {}^{d}e_{3} \end{bmatrix} \right\|_{2}
$$

is minimized. Solving this optimization problem yields

$$
\begin{bmatrix} r_{11} \\ r_{12} \end{bmatrix} = \frac{1}{\sqrt{{}^{b}e_{1}^{2} + {}^{b}e_{2}^{2}}\sqrt{{}^{d}e_{1}^{2} + {}^{d}e_{2}^{2}}} \begin{bmatrix} {}^{b}e_{1}{}^{d}e_{1} + {}^{b}e_{2}{}^{d}e_{2} \\ {}^{b}e_{1}{}^{d}e_{2} - {}^{b}e_{2}{}^{d}e_{1} \end{bmatrix}
$$

Knowing ${}^{b}R_{d}$, we have

$$
{}^{b}T_{d} = \begin{bmatrix} {}^{b}x_{1} \\ {}^{b}x_{2} \\ {}^{b}x_{3} \end{bmatrix} - {}^{b}R_{d} \begin{bmatrix} {}^{d}x_{1} \\ {}^{d}x_{2} \\ {}^{d}x_{3} \end{bmatrix}
$$

Now we can determine any observed point on the part with respect to the base frame of the robot. Based on the recovered coordinates of the points on the part, we can easily know the position of the centroid of the part and the orientation of the part.

2.4 Robot Planning and Control

In this section we discuss the problem of multiple sensor–based robot motion planning and control involved in performing tasks such as tracking and grasping a moving part. The problem we consider is to make a plan for a robot to track a target whose position, orientation, and velocity are measured by the estimation and calibration schemes discussed in Section 3. In doing so, we use the concept of *parallel guidance*. We propose that an error reduction term be added to the position and velocity of the target to form a desired position, velocity, and acceleration pofile for the robot. When the error reduction term is carefully planned, it guarantees a time optimal and robust robot motion with given bounded control. The planner, we show, can be implemented as both time based and event based. This leads to a new event-based tracking scheme for a robot.

Robot Control

The dynamic model of a robot arm is given by

$$D(q)\ddot{q} + C(q, \dot{q}) + G(q) = \tau \tag{5.14}$$

where q, $D(q)$, $C(q, \dot{q})$, $G(q)$ and τ are respectively the joint angle vector, inertia matrix, load related to centripetal and Coriolis forces, load related to gravity, and joint torque vector. The joint torque has to satisfy the following constraints:

$$\tau_{i,min}(q, \dot{q}) \leqslant \tau_i \leqslant \tau_{i,max}(q, \dot{q}) \qquad i = 1, 2, \ldots m \tag{5.15}$$

where m is the number of joints. The output is given by $Y = H(q) = (X, \theta)^t$ where $X \in \mathbb{R}^3$ and $\theta \in \mathbb{R}^3$ represent the position and orientation of robot end effector. Let us consider the nonlinear feedback control law [41] given by

$$\tau = D(q)J^{-1}(q)(A_d + K_v(V_d - \dot{Y}) + K_p(Y_d - Y) - \dot{J}(q)\dot{q}) + C(q, \dot{q}) + G(q) \tag{5.16}$$

where $J(q)$ is the Jacobian of $H(q)$ with respect to q; K_v and K_p are the velocity and position feedback gains; A_d, V_d, and Y_d are the desired acceleration, velocity, and position vectors respectively. It is assumed that the robot has six joints, for otherwise a pseudoinverse is to be used in (5.16). It can be shown that if A_d, V_d, and Y_d satisfy the constraints $\dot{Y}_d = V_d$ and $\dot{V}_d = A_d$, then with proper choice of K_v and K_p, the closed-loop system is stable and the tracking error will vanish asymptotically provided the induced joint torque is within the limit specified by (5.15). The problem of robot motion planning considered here is to design A_d, V_d, and Y_d so that the robot end effector will track the motion of a part.

Robot Tracking with Planned Error Reduction: Parallel Tracking

The main problem in robot tracking is to eliminate the position and velocity error between the robot end effector and the part by controlling the robot. An obvious tracking plan would be to choose

$$\begin{cases} Y_d = Y_p(t) \\ V_d = V_p(t) \\ A_d = \dot{V}_d \end{cases} \tag{5.17}$$

which is to let the desired motion to be the motion of the part. However, if such a plan is used to control the motion of a robot, the required control will be large when the initial error between Y_d, the position of the part, and Y, the actual position of the robot, is large.

A large value of the position error can easily cause the command torque to exceed the limit and the lower level controller to shut down. Usually, in a global tracking problem the initial value of the position error is very large. The tracking problem [42–44] studied in the literature has always been a local tracking problem, wherein it is assumed that the initial error is either zero or remains very small.

In order to circumvent the problem of dealing with a large value of the initial error, a new tracking plan is proposed in the form of

$$\begin{cases} Y_d = Y_p(t) + Y_e(t) \\ V_d = \dot{Y}_d = V_p(t) + \dot{Y}_e(t) \\ A_d = \dot{V}_d = A_p(t) + \ddot{Y}_e(t) \end{cases} \tag{5.18}$$

where Y_e is an error reduction term that is set to $Y(0) - Y_p(0)$ at the time tracking starts and is gradually reduced to zero according to a plan. The proposed plan is based on using optimal control where the bounds on the control have also been taken into consideration. The proposed plan controls the robot to move at the same speed as the part, while the position error is gradually reduced to zero according to an error reduction plan. This scheme is motivated by the parallel guidance in missile pursuit, hence the name parallel tracking. The difference, however, is that the velocity of the part also needs to be tracked for stable grasping.

After the error reduction term is added to the target position, the initial positional error is set to be zero to guarantee that the control will not be out of range when tracking starts. The error reduction term will be planned so that the initial positional error will be reduced with a feasible control command.

The torque demand of a planned motion depends on the planned path, speed, and acceleration. It is usually difficult to obtain. Another method of motion planning often adopted is to specify a conservative constant speed and acceleration limit based on the off-line kinematic and dynamic work space analysis. In this way the motion planning problem is greatly simplified, making such planning schemes as time-optimal and minimum-energy planning possible [45, 46].

Assume the constraints on the desired motion are given by $\|V_d\| \leqslant v_m$ and $\|A_d\| \leqslant a_m$. We also assume that by prior knowledge, the motion of the target satisfies the constraints $\|V_p\| \leqslant v_{mp}$ and $\|A_p\| \leqslant a_{mp}$. It follows from (5.18) that a conservative constraint on the error reduction term planned error would be $\|\dot{Y}_e\| \leqslant v_m - v_{mp}$ and $\|\ddot{Y}_e\| \leqslant a_m - a_{mp}$. It is reasonable to assume $v_{mp} \leqslant v_m$ and $a_{mp} \leqslant v_m$ for the robot to be able to keep track of the target.

Another side effect of adding the error reduction term is that the resultant small tracking control error makes it possible for us to ignore the fact that the space of orientation angles is not Euclidean and to treat the orientation tracking in the same way as position tracking is treated.

Time-Based Optimal Parallel Tracking

When the error reduction term is planned to reduce the initial error in a time-optimal way, the robot tracking plan in (5.18) is considered time optimal. The time-optimal planning problem is stated as follows.

$$\min T = \int_0^T dt$$
$$\begin{cases} \dot{Y}_e = V \\ \dot{V} = A \end{cases} \tag{5.19}$$
$$\|V\| \leqslant v_{mye},\ \|A\| \leqslant a_{mye},\ V(0) = V(T) = 0,\ Y_e(0) = Y(0) - Y_p(0),\ Y_e(T) = 0$$

Let us define $s = \|Y_e(t)\|$. The preceding time optimal problem can be stated as follows:

$$\min T = \int_0^T dt$$
$$\begin{cases} \dot{s} = v \\ \dot{v} = a \end{cases} \tag{5.20}$$
$$|v| \leqslant v_{mye},\ |a| \leqslant a_{mye},\ v(0) = v(T) = 0,\ s(0) = \|Y(0) - Y_p(0)\|,\ s(T) = 0$$

This is a classical bang–bang time optimal control problem [47] and the solution is given by

$$a = \begin{cases} -a_{mye}, & \dfrac{v_{mye}}{a_{mye}} > t \geqslant 0 \\ 0, & \dfrac{s(0)}{v_{mye}} > t \geqslant \dfrac{v_{mye}}{a_{mye}} \\ a_{mye}, & t \geqslant \dfrac{s(0)}{v_{mye}} \end{cases} \tag{5.21}$$

Let $(m, n, p)^T$ be a unit vector along the direction of the vector $Y - Y_p$. The error reduction term Y_e and its two derivatives $\dot{Y}_e$ and $\ddot{Y}_e$ are given by

$$\begin{cases} \ddot{Y}_e = (m, n, p)^T a(t) \\ \dot{Y}_e = (m, n, p)^T v(t) \\ Y_e = (m, n, p)^T s(t) \end{cases} \tag{5.22}$$

where $v(t)$ and $s(t)$ are solution of (5.20) with $a(t)$ as input. Note in particular that the vector $(m, n, p)^T$ is independent of t.

The initial velocity error between the robot and the target can sometimes cause the control to be out of range too. A similar technique can be used to overcome this problem by adding an error reduction term V_e in the velocity. The robot tracking plan in (5.18) can be rewritten as

$$\begin{cases} Y_d = Y_p(t) + Y_e(t) + \displaystyle\int_0^t V_e\, dt \\ V_d = \dot{Y}_d = V_p(t) + \dot{Y}_e(t) + V_e \\ A_d = \dot{V}_d = A_p(t) + \ddot{Y}_e(t) + \dot{V}_e \end{cases} \tag{5.23}$$

where V_e is an added term to cancel the initial velocity error between the robot and the target.

V_e is set to $V(0) - V_p(0)$ at the time tracking starts and will be gradually reduced to zero based on a plan. The plan of V_e can be designed in the same way as the plan of Y_e.

Event-Based Optimal Parallel Tracking

Event-Based Planning Traditionally, a motion plan for a robot is expressed in terms of time and is driven by time during on-line execution. That is, the control command is stored as a function of time and is indexed on the basis of the time passed during on-line execution. The problem with this kind of planning and control is that the progress of planned motion is not affected even when the robot fails to follow the planned motion for unexpected reasons such as the existence of an obstacle or malfunction of a device. Therefore the control error will become very big and cause the control to be out of range and the controller to shut down.

In order to overcome this problem, the concept of event-based planning and control for a robot following a given path has been introduced by Tarn, Bejczy, and Xi [46]. The problem they tried to solve was to design a motion plan for a robot following a given path which constant constraints assigned on the velocity and acceleration of the robot. Such a motion plan can be time optimal or of minimum energy [45].

The event-based planning and control scheme tries to parameterize the motion plan in terms of curve length s traveled along the prespecified path. Let $v = \dot{s}$, $a = \dot{v}$, and $w = v^2$ and define $u = \dfrac{da}{ds}$. The motion-planning problem is now stated as follows.

$$\begin{aligned}
&\min T = \int_0^T dt = \int_0^S \frac{1}{v}\, ds \\
&\begin{cases} \dfrac{dw}{ds} = 2a \\ \dfrac{da}{ds} = u \end{cases} \\
&|w| \leqslant w_m,\ |a| \leqslant a_m,\ |u| \leqslant u_m,\ w(0) = w(S) = a(0) = a(S) = 0
\end{aligned} \tag{5.24}$$

where w_m and u_m are, respectively, an equivalent speed limit and the jerk constraint to guarantee smoother robot motion. This is again a classical time-optimal bang–bang control problem. The solution of this problem is given in the form of $u(s)$, $a(s)$, and $w(s)$ or $v(s) = \sqrt{w(s)}$ [46].

For a straight-line path the vector motion plan is given by

$$\begin{cases} \ddot{Y}_d = (m, n, p)^T a(s) \\ \dot{Y}_d = (m, n, p)^T v(s) \\ Y_d = (m, n, p)^T s + Y_0 \end{cases} \tag{5.25}$$

where Y_0 is the beginning point of the path and $(m, n, p)^T$ a unit vector along the path.

The motion reference s and desired position Y_d can be determined during actual execution by projecting the actual position of the robot onto the given path and then choosing the projection as Y_d. The minimum positional control error is generated when Y_d is chosen in this way. The parameter s is thus the curve length between this y_d and the beginning point Y_0. It is easy to observe that the motion reference s reflects the actual execution of a planned motion. The basic concept of this entire approach is actually the well-established feedback control principle.

The main advantage of using event-based robot motion planning and control is that the motion planning becomes a closed-loop process. The plan is realized during execution based on sensory measurements. It provides an efficient closed-loop algorithm for implementing the tracking control in real time, since the location and motion of the part as a function of time are unknown prior to execution. An important contribution of this chapter is to develop such an event-based tracking scheme based on the multiple sensory integration to achieve (1) stable global tracking and (2) robust grasping control of a part with unknown orientation.

Event-Based Tracking The concept of event-based planning and control can be employed to plan the error correction Y_e in (5.18). A straight line is chosen as the path for Y_e, which is the same as in Section 3.3.

The robot tracking plan in Section 3.3 can be reparameterized in terms of s, which is the distance between the part and the robot. This new choice of the event-based motion reference reflects the special characteristic of the tracking problem; that is, the target path is unknown prior to the execution.

Using the new motion reference, the event-based error correction term Y_e can be obtained as

$$a = \begin{cases} -a_{mye}, & s(0) - \dfrac{1}{2}\dfrac{v_{mye}^2}{a_{mye}} < s \leqslant s(0) \\ 0, & \dfrac{1}{2}\dfrac{v_{mye}^2}{a_{mye}} < s \leqslant s(0) - \dfrac{1}{2}\dfrac{v_{mye}^2}{a_{mye}} \\ a_{mye}, & s \leqslant \dfrac{1}{2}\dfrac{v_{mye}^2}{a_{mye}} \end{cases} \tag{5.26}$$

and

$$v = \begin{cases} -\sqrt{s a_{mye}(s(0) - s)}, & s(0) - \dfrac{1}{v_{mye}^2 < s \leqslant s(0)} \\ -v_{mye}, & \dfrac{1}{2}\dfrac{v_{mye}^2}{a_{mye}} < s \leqslant s(0) - \dfrac{1}{2}\dfrac{v_{mye}^2}{a_{mye}} \\ -\sqrt{s a_{mye} s}, & s \leqslant \dfrac{1}{2}\dfrac{v_{mye}^2}{a_{mye}} \end{cases} \tag{5.27}$$

The vector motion plan is obtained as in (5.22) except that a and v are now functions of s. The motion reference s is generated by calculating the magnitude of $Y - Y_p$.

Since the motion plan is now driven by an event-related or state of execution–related motion reference rather than time, this robot tracking scheme has all the advantages of original event-based motion planning and control. This new scheme is called *event-based robot tracking*. It is a new extension of the event-based robot motion planning and control strategy.

2.5 Experimental Results

Our experimental setup consists of three components that constitute the components of a work cell. They are the disc conveyor, the camera system, and the robotic manipulator. The manipulator is one of two PUMA 560 robotic manipulators separately controlled by two UMC controllers from Motion Tek. The two controllers interface with the main computer, a four-processor SGI IRIS 4D/VGX workstation, through shared memory. High-level planning and control algorithms such as the multiple sensor integration algorithm, the vision algorithm with an implicit calibration, and the parallel tracking algorithm discussed in this chapter are all implemented on an SGI workstation, making full utilization of its computing power. Schunk grippers are mounted on the manipulators to grasp parts to be manipulated. The disc conveyor rotates around a fixed axis. There is an encoder attached to the motor to generate measurement of the angle and speed of rotation of the conveyor. The resolution of the encoder is $(576 \times 10^3)/2\pi$ lines per radian. The rotation of the disc conveyor is independently controlled by a spare channel on one of the Motion Tek controllers. Two markers are placed on the disc conveyor, with one of them at the center of rotation. These two markers represent the x axis of the attached disc coordinate frame. They are identified as reference points by the vision algorithm in determining the relative position and orientation of a part placed on the conveyor. The vision system consists of two cameras in a stereo setup, an Intelledex IntelleVue vision processor, a black-and-white monitor, and a development system on a PC. Only one camera is used in this experiment. In order to allow communication between the Intelledex vision processor and the SGI IRIS 4D/VGX workstation, a parallel interface between them was developed. The task of the vision system is to identify the image coordinates of the markers representing the reference points and the outline of a part. Note that the outline of the part is not marked with markers. The results are then transferred to the SGI workstation to be further processed or directly used in the guidance of robot motion. The Intelldex vision processor is based on a 16-MHz Intel 80386 CPU with an Intel 80287 coprocessor. There are six $256 \times 256 \times 8$ bit image buffers, two of which can be used as image grabbers for direct image acquisition. Image resolution is 256×240 (although the cameras have a pixel array of 422×490). The display has 256 gray levels, while the digitizer can distingush only 64 gray levels.

In Figures 5.7–5.12, results of our experiments are shown. These experiments were carried out under the assumption that the part on the conveyor is of nonnegligible height and the robot position is not calibrated. The set of experiments has been further subdivided into two possible subcases. In subcase 1, we assume that the base frame of the robot is parallel to the

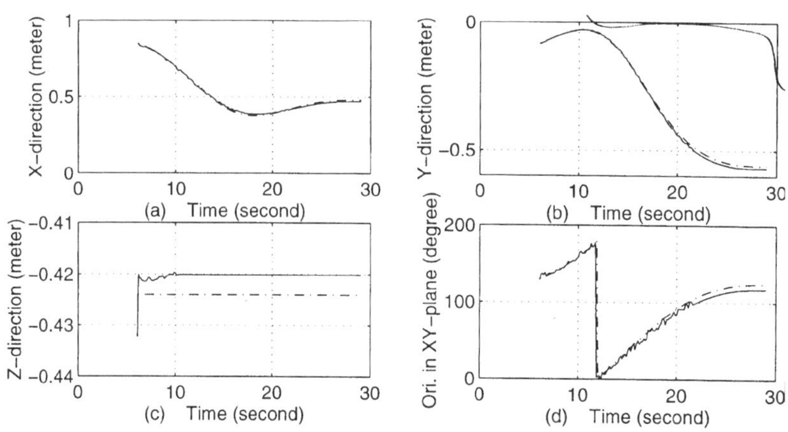

FIGURE 5.7
Estimated and actual pose of the part for the case in which the distance of the end effector from the disc conveyor is unknown.

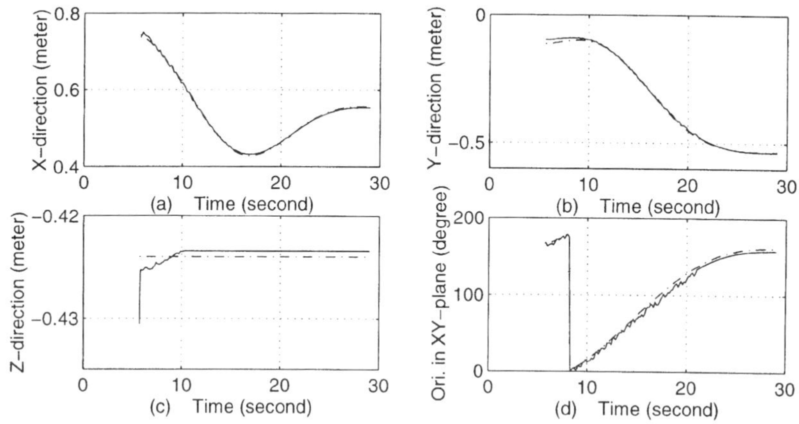

FIGURE 5.8
Errors in pose estimation for the case in which the distance of the end-effector from the disc conveyor is unknown.

plane of the conveyor and that the separation between the two planes is unknown. The height of the part is assumed to be unknown. In Figure 5.7, we show the estimated and the actual pose of the part. Figure 5.7(a)–(c) indicate the estimated and actual trajectory of the part in the X direction, Y direction, and Z direction of the base frame of the robot, respectively. In Figure 5.7(d), the estimated and actual orientations of the part in the base frame of the robot are shown. The dotted lines denote the actual values and the solid lines the estimated ones. In Figure 5.8, we show the estimation errors of the pose of the part. Figure 5.8(a)–(c) show the position errors in the X direction, Y direction, and Z direction, respectively. The error in orientation is provided in Figure 5.8(d). The position errors in the X direction and Y direction are around 1 cm and the orientation error is less than 5 degrees.

In subcase 2, we assume that the base frame of the robot is parallel to the plane of the conveyor and that the separation between the two planes is known. As in subcase 1, we

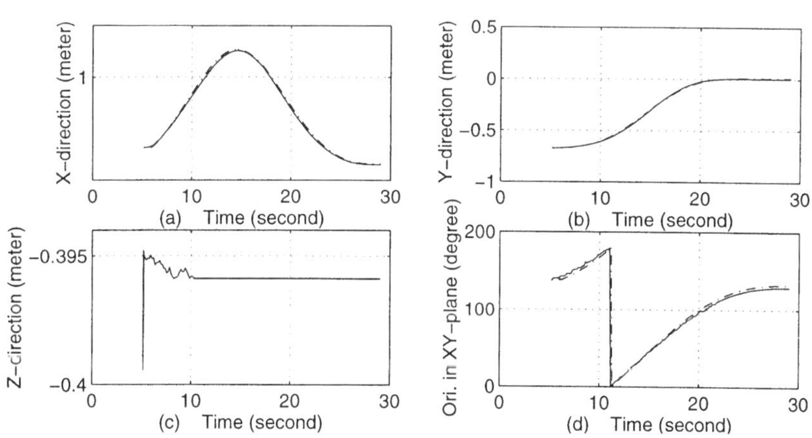

FIGURE 5.9
Estimated and actual pose of the part for the case where the distance of the end-effector from the disc conveyor is known and height of the part is 4 cm.

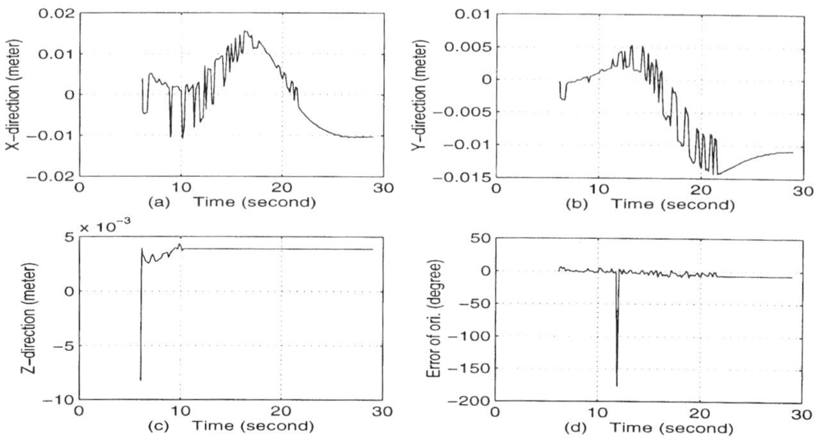

FIGURE 5.10
Errors in pose estimation for the case in which the distance of the end effector from the disc conveyor is known and the height of the part is 4 cm.

continue to assume that the height of the part is unknown. In Figures 5.9 and 5.10, we show our results when the true height of the part is 4 cm. The estimated and actual poses of the part are presented in Figure 5.9, where (a), (b), and (c) depict the estimated and actual trajectories of the part in the X direction, Y direction, and Z direction of the base frame of the robot. In Figure 5.9, the estimated trajectory of the part is denoted by solid lines and the actual trajectory of the part by dotted lines. In Figure 5.10, we show the estimation error of the pose of the part, where (a), (b), and (c) present the error in the X direction, Y direction, and Z direction, respectively. Estimated and actual orientations are given in Figure 5.9(d), whereas the error in orientation is given in Figure 5.10(d). The experiment was repeated assuming the unknown height of the part to be 7 cm, and the results are shown in Figures 5.11 and 5.12. As is evident from the figures, the position errors are within 1 cm and the orientation errors are less than 5 degrees.

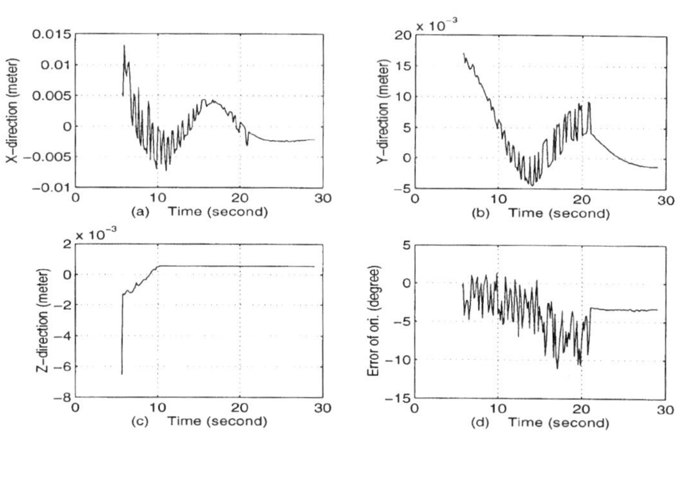

FIGURE 5.11
Estimated and actual pose of the part for the case in which the distance of the end effector from the disc conveyor is known and the height of the part is 7 cm.

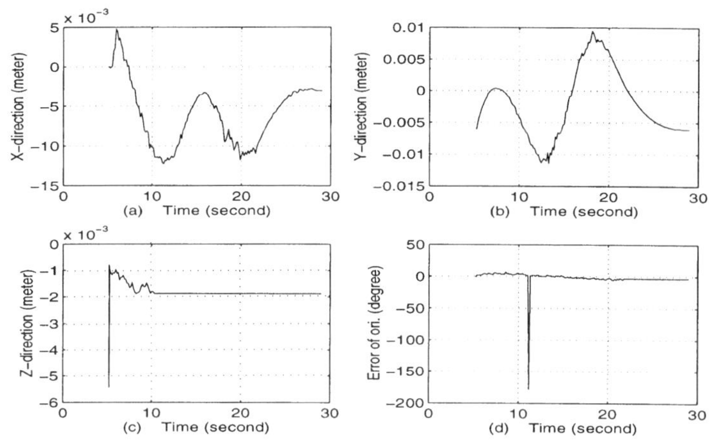

FIGURE 5.12
Errors in pose estimation for the case in which the distance of the end effector from the disc conveyor is known and the height of the part is 7 cm.

In comparing the results for subcases 1 and 2, the errors in position and orientation are somewhat comparable. In repeated trials, it is observed that the success rate (in terms of successfully grasping the part) in subcase 1 is lower than that in subcase 2. A quantitative comparison has not been made.

3 TRACKING AN UNKNOWN TRAJECTORY ON A SURFACE

3.1 Hybrid Control Design

Let us consider a nonredundant rigid robot with six degrees of freedom. As is well known, the dynamics of the robot in joint space can be written as

$$D(\mathbf{q})\ddot{\mathbf{q}} + c(\mathbf{q}, \dot{\mathbf{q}}) + p(\mathbf{q}) = \tau$$

where $\mathbf{q}$, $\dot{\mathbf{q}}$, and $\ddot{\mathbf{q}} \in \mathbb{R}^6$ are the joint angle vector, joint velocity vector, and joint acceleration vector, respectively. $D(\mathbf{q}) \in \mathbb{R}^{6\times 6}$ is the inertia matrix of the robot, and $c(\mathbf{q}, \dot{\mathbf{q}})$ stands for Coriolis and Centrifugal terms. $p(\mathbf{q})$ is the term caused by gravity. $\tau \in \mathbb{R}^6$ represents the joint torque vector.

When the end effector of the robot makes contact with an environment, the dynamics of the robot becomes

$$D(\mathbf{q})\ddot{\mathbf{q}} + c(\mathbf{q}, \dot{\mathbf{q}}) + p(\mathbf{q}) = \tau + \mathbf{F} \tag{5.28}$$

where $\mathbf{F}$ is the constraint force–torque vector in the joint space.

The pose of the end effector of the robot in the task space and the joint vector of the robot are related by

$$(x, y, z, O, A, T)^T = h(\mathbf{q})$$

from which we derive

$$(\dot{x}, \dot{y}, \dot{z}, \dot{O}, \dot{A}, \dot{T})^T = J\dot{\mathbf{q}}$$

where $J \in \mathbb{R}^{6\times 6}$ is the Jacobian of the robot. Let $\mathbf{x} = (x, y, z, O, A, T)^T$. Thus,

$$\dot{\mathbf{x}} = J\dot{\mathbf{q}} \tag{5.29}$$

$$\ddot{\mathbf{x}} = J\ddot{\mathbf{q}} + \dot{J}\dot{\mathbf{q}} \tag{5.30}$$

Throughout this section, we assume that the dynamics and kinematics of the robot are known.

Constrained Motion

Herein, we consider the constrained motion of the robot on a rigid surface. Also, the contact between the robot and the surface is assumed to be frictionless and a point contact. Suppose that the surface in the task space can be written as

$$z = z(x, y)$$

where (x, y, z) are coordinates in the task space and $z(\cdot, \cdot)$ is assumed to be smooth enough. If the end effector of the robot is kept in touch with the surface, the contrained motion of the robot is given by

$$\dot{z} = \frac{\partial z}{\partial x}\dot{x} + \frac{\partial z}{\partial y}\dot{y}$$

that is,

$$\left(\frac{\partial z}{\partial x}, \frac{\partial z}{\partial y}, -1\right)\begin{bmatrix}\dot{x}\\ \dot{y}\\ \dot{z}\end{bmatrix} = 0$$

Let

$$G = \left(\frac{\partial z}{\partial x}, \frac{\partial z}{\partial y}, -1, 0, 0, 0\right)$$

Then we have

$$G\dot{\mathbf{x}} = 0 \tag{5.31}$$

Furthermore, we obtain

$$\dot{G}\dot{\mathbf{x}} + G\ddot{\mathbf{x}} = 0 \tag{5.32}$$

On the other hand, under the assumption that the contact is frictionless, it turns out that

$$\mathbf{f}^T\delta\mathbf{x} = \mathbf{F}^T\delta\mathbf{q} = 0$$

where $\delta\mathbf{x}$ and $\delta\mathbf{q}$ represent the virtual displacements of the robot in the task space and the joint space, respectively, and $\mathbf{f}$ is the constraint force exerted on the end effector in the task space. Combination of the last equations and Eq. (5.29) gives rise to

$$\mathbf{F} = J^T \mathbf{f}$$

We assume that the contact occurs at a point of the end effector. It is easily seen that

$$\mathbf{f} = (f_x, f_y, f_z, 0, 0, 0)^T$$

Furthermore, it is true that

$$\mathbf{f} = \lambda G^T \tag{5.33}$$

for some scalar $\lambda \in \mathbb{R}$.

Decoupling of Control Variables

Let us utilize a nonlinear feedback control law, namely

$$\tau = c(\mathbf{q}, \dot{\mathbf{q}}) + p(\mathbf{q}) + DJ^{-1}\mathbf{u} - DJ^{-1}\dot{J}\dot{\mathbf{q}} \tag{5.34}$$

where $\mathbf{u}$ is a new control vector to be determined. Substituting the preceding control torque into the dynamics (5.28) of the robot yields

$$D(\mathbf{q})J^{-1}\ddot{\mathbf{x}} = DJ^{-1}\mathbf{u} + \mathbf{F}$$

Recall that from (5.30) and (5.33), we have

$$\ddot{\mathbf{x}} = \mathbf{u} + JD^{-1}J^T G^T \lambda = \mathbf{u} + M^T \lambda \tag{5.35}$$

where $M = GJD^{-1}J^T$. Let us denote $\Phi = GM^T$. Clearly, Φ is a nonzero scalar during the constrained motion since D is nonsingular. Define two subspaces as follows:

$$S_1 = \{\mathbf{y} \mid (I - M^T \Phi^{-1} G)\mathbf{y} = 0, \mathbf{y} \in \mathbb{R}^6\} \tag{5.36}$$

$$S_2 = \{\mathbf{y} \mid M^T \Phi^{-1} G \mathbf{y} = 0, \mathbf{y} \in \mathbb{R}^6\} \tag{5.37}$$

It is seen that S_1 and S_2 are subspaces of $\mathbb{R}^6$ at the contact point during the constrained motion and that

$$\mathbb{R}^6 = S_1 \oplus S_2$$

where $\oplus$ stands for the direct sum. Moreover, $\dim(S_1) = 1$ and $\dim(S_2) = 5$.

Premultiplying both sides of Eq. (5.35) by G, from constraint condition (5.32), it is seen that

$$\lambda = -\Phi^{-1}(G\mathbf{u} + \dot{G}\dot{x}) \tag{5.38}$$

Now, we represent the control vector $\mathbf{u}$ as $\mathbf{u} = u_1 + u_2$, $\mathbf{u}_i \in S_i$, $(i = 1, 2)$. Note that M^T is of full column rank. Then (5.38) immediately reduces to

$$\lambda = -\Phi^{-1}(G\mathbf{u}_1 + \dot{G}\dot{\mathbf{x}}), \quad \mathbf{u}_1 \in S_1 \tag{5.39}$$

which implies that the constraint force is controlled only by $\mathbf{u}_1(\in S_1)$. The force control law $\mathbf{u}_1 \in S_1$ is designed from (5.39). Premultiplying both sides of (5.35) by $(I - M^T\Phi^{-1}G)$ leads to

$$\left.\begin{aligned} &(I - M^T\Phi^{-1}G)\ddot{\mathbf{x}} = \mathbf{u}_2, \quad \mathbf{u}_2 \in S_2 \\ &G\dot{\mathbf{x}} = 0 \end{aligned}\right\} \tag{5.40}$$

which implies that, when constrained, the motion of the robot manipulator is restricted and its degrees of freedom are reduced from six to five and that the motion is controlled only by $\mathbf{u}_2 \in S_2$. In conclusion, as a hybrid control strategy, one may design the motion control law based on (5.40) and the force control law from (5.39), respectively.

Hybrid Control Scheme

In this section, we choose the force control law in the form

$$\mathbf{u}_1 = -M^T(\lambda^d + K_f \int_{t_0}^{t} (\lambda^d - \lambda)\,dt) - M^T\Phi^{-1}\dot{G}\dot{\mathbf{x}} \tag{5.4}$$

where, clearly, $\mathbf{u}_1 \in S_1$. Applying the control law to Eq. (5.39) gives at once

$$(\lambda - \lambda^d) + K_f \int_{t_0}^{t} (\lambda - \lambda^d)\,dt = 0$$

which will result in $\lambda \to \lambda^d$ as $t \to \infty$ with a proper choice of K_f.

During the constrained motion, (5.31) and (5.32) hold. Note that $z(x, y)$ is unknown in our case and so is G. However, $\mathbf{f}$ can be measured and from (5.33) we have

$$G = \left(-\frac{f_z}{f_z}, -\frac{f_y}{f_x}, -1, 0, 0, 0\right) = (\alpha, \beta, -1, 0, 0, 0)$$

where

$$\alpha = -\frac{f_x}{f_z}, \qquad \beta = -\frac{f_y}{f_z}$$

It should be pointed out that α and β depend only on the direction of $\mathbf{f}$, instead of its magnitude. From (5.31) and the last equation, during the constrained motion we have

$$\dot{\mathbf{x}} = \underbrace{\begin{bmatrix} 1 & 0 & 0 & 0 & 0 \\ 0 & 1 & 0 & 0 & 0 \\ \alpha & \beta & 0 & 0 & 0 \\ 0 & 0 & 1 & 0 & 0 \\ 0 & 0 & 0 & 1 & 0 \\ 0 & 0 & 0 & 0 & 1 \end{bmatrix}}_{T} \underbrace{\begin{pmatrix} \dot{x} \\ \dot{y} \\ \dot{O} \\ \dot{A} \\ \dot{T} \end{pmatrix}}_{\dot{\mathbf{x}}_c} \tag{5.42}$$

Furthermore, it is true that

$$\ddot{\mathbf{x}} = T\ddot{\mathbf{x}}_c + \dot{T}\dot{\mathbf{x}}_c \tag{5.43}$$

Since $GT = 0$, (5.40) becomes

$$T\ddot{\mathbf{x}}_c + (I - M^T\Phi^{-1}G)\dot{T}\dot{\mathbf{x}}_c = \mathbf{u}_2, \quad \mathbf{u}_2 \in S_2 \tag{5.44}$$

from which we can design the motion control law as

$$\mathbf{u}_2 = T[\ddot{\mathbf{x}}_c^d + K_v(\dot{\mathbf{x}}_c^d - \dot{\mathbf{x}}_c) + K_p(\mathbf{x}_c^d - \mathbf{x}_c)] + (I - M^T\Phi^{-1}G)\dot{T}\dot{\mathbf{x}}_c \tag{5.45}$$

Clearly, $\mathbf{u}_2 \in S_2$. Note that T is a matrix of full column rank, and substitution of (5.45) in (5.44) gives rise to

$$\ddot{\mathbf{e}}_c + K_v\dot{\mathbf{e}}_c + K_p\mathbf{e}_c = 0$$

where $\mathbf{e}_c = \mathbf{x}_c^d - \mathbf{x}_c$. With proper choices of K_v and K_p, $\mathbf{e}_c \to 0$ exponentially as $t \to \infty$.

Thus, the hybrid control scheme is given by

$$\mathbf{u} = \mathbf{u}_1 + \mathbf{u}_2 = T[\ddot{\mathbf{x}}_c^d + K_v\dot{\mathbf{e}}_c + K_p\mathbf{e}_c] + \dot{T}\dot{\mathbf{x}}_c - M^T\left(\lambda^d + K_f\int_{t_0}^{t} e_f\,dt\right)$$

where $e_f = \lambda^d - \lambda$. The matrix T depends only on the orientation of the tangent plane of the contact surface at the contact point, which can be measured via the force–torque sensor. $\dot{T}$ reflects the change of the orientation. The matrix M can also be known from the measurements of the force–torque sensor and the encoders of the robot. The proposed control scheme is simple and consists of three parts: one is for constrained motion, one is used to compensate the change of the orientation, and one is for force regulation. Although it is easy to design $\mathbf{u}_1$ and $\mathbf{u}_2$ separately, we do not have to compute them separately. Instead, the last equation can be directly used to compute the control law. The corresponding joint control torque can be obtained by substituting the last equation into (5.34).

3.2 Image-Based Motion Planning

In the previous section, it was shown that if the robot is given the desired trajectory, hybrid control can be used to complete the constrained motion. Since we assume that the constrained surface is unknown, it is reasonable to assume that the trajectory to be followed is also unknown. In this section, we shall propose a planning scheme for the constrained motion of the robot based on fusion of the visual data from an uncalibrated camera and additional information from encoders and force–torque sensor.

Relation between the Constrained Motion and Its Image

Suppose that $O_cX_cY_cZ_c$ is the camera frame with the origin being at the optical center of the camera and the Z axis perpendicular to the image plane. Let $OXYZ$ be the base frame of the robot. Then we have

$$\begin{pmatrix} x^c \\ y^c \\ z^c \end{pmatrix} = R \begin{pmatrix} x \\ y \\ z \end{pmatrix} + T \tag{5.46}$$

where (x^c, y^c, z^c) and (x, y, z) are the coordinates of a point p in the camera frame and the base frame of the robot, respectively. The parameter R is the rotation matix and $T \in \mathbb{R}^3$ is a translation vector.

From a pinhole model of the camera with focal length f, the coordinates of the image of the point p are given by

$$X = f\frac{x^c}{z^c c} \tag{5.47}$$

$$Y = f\frac{y^c}{z^c} \tag{5.48}$$

Suppose that we observe a moving point. Differentiating both sides of (5.48) with respect to time t and using (5.46) give rise to the following:

$$\begin{pmatrix} \dot{X} \\ \dot{Y} \end{pmatrix} = \frac{1}{z^c}\begin{bmatrix} f & 0 & -X \\ 0 & f & -Y \end{bmatrix} R \begin{pmatrix} \dot{x} \\ \dot{y} \\ \dot{z} \end{pmatrix} \tag{5.49}$$

When the motion of the end effector of the robot is constrained on a surface $z = z(x, y)$, (5.31) holds. We assume that the surface is unknown, that is, $z(x, y)$ is unknown. However, from the force–torque sensor mounted on the end effector, the robot can locally estimate the surface. More precisely, the tangent plane to the unknown surface at the contact point is computed by using the force–torque sensor. It is easily seen that

$$\mathbf{f}^T \dot{\mathbf{x}} = 0$$

and

$$\dot{z} = -\frac{f_x}{f_z}\dot{x} - \frac{f_y}{f_z}\dot{y}$$

Therefore, the relation between the motion of the end effector on the constrained surface and its perspective projection on the image plane of camera is given by

$$\begin{pmatrix} \dot{X} \\ \dot{Y} \end{pmatrix} = \underbrace{\frac{1}{z^c}\begin{bmatrix} f & 0 & -X \\ 0 & f & -Y \end{bmatrix} R \begin{bmatrix} 1 & 0 \\ 0 & 1 \\ \alpha & \beta \end{bmatrix}}_{Q} \begin{pmatrix} \dot{x} \\ \dot{y} \end{pmatrix}$$

$$= \frac{1}{z^c} Q \begin{pmatrix} \dot{x} \\ \dot{y} \end{pmatrix} \tag{5.50}$$

In (5.50), we do not know z^c and R exactly, while we know α and β via measurement of the force–torque sensor. However, it is easy to know the range of z^c (in our experimental setup,

$1 < z^c < 1.5\,\mathrm{m}$). Since both the camera and the base of the robot are fixed, R is constant. Once R is recovered, Q can be known via sensory information from the vision system and the force–torque sensor. It is easy to develop a recursive least squares algorithm based on Eq. (5.50) to recover R up to a scalar multiple. This procedure provides an adaptation scheme, wherein we do not have to determine the rotation matrix R exactly.

Motion Planning

After recovering R, as described in Section 3.1 and utilizing the force–torque sensor together with the vision system, we compute Q as defined in (5.50). With an uncalibrated camera, it is hard to recover exactly the trajectory on the unknown surface. We assume that the trajectory and the end effector of the robot are both visible. Hence it is easy to know the difference between the end effector and the trajectory on the image plane. We denote the difference on the image plane by $\mathbf{e}$. The desired motion of the robot can be written as

$$\begin{pmatrix} \dot{x}^d \\ \dot{y}^d \end{pmatrix} = -kQ^{-1}\mathbf{e} \tag{5.51}$$

where k is a positive scalar. It may be noticed that the direction of the desired velocity is crucial for the completion of the task. The magnitude of the velocity affects only the time period during which the task is completed, and different k values could be designed by using different approaches. For example, k could be determined by considering the limitation of control torque of the robot.

We can adopt the control strategy proposed in the previous sections to control the robot such that $\dot{x}$ tends to $\dot{x}^d$ and $\dot{y}$ tends to $\dot{y}^d$, exponentially in t. From the relation (5.50), it turns out that

$$\begin{pmatrix} \dot{X} \\ \dot{Y} \end{pmatrix} \to \frac{1}{z^c} QkQ^{-1}\mathbf{e} = -\frac{k}{z^c}\mathbf{e}$$

that is,

$$\dot{\mathbf{e}} \to -\frac{k}{z^c}\mathbf{e}$$

exponentially. It can be proved that the system $\dot{\mathbf{e}} = -\frac{k}{z^c}\mathbf{e}$ is globally asymptotically stable. Furthermore, by using the proposed hybrid control scheme and motion planning, $\mathbf{e} \to 0$ as $t \to \infty$. Because force control will maintain the contact between the end effector and the surface, the task can be accomplished under the proposed control law.

It should be pointed out that the control is robust against perturbations in Q. Suppose that $\bar{Q}$ is an estimate of Q. Let $\bar{Q}^{-1} = Q^{-1} + \Delta$. As long as the system $\dot{\mathbf{e}} = -(I + Q\Delta)\mathbf{e}$ is asymptotically stable, the proposed planning and control scheme guarantees that $\mathbf{e} \to 0$ as $t \to \infty$.

3.3 Experimental Results

An experimental system has been set up in the Center for Robotics and Automation at Washington University, as shown in Figure 5.13. It consists of one PUMA 560 manipulator,

FIGURE 5.13
Experimental setup for trajectory following on an unknown surface.

a vision system, and a table on which there is a curved unknown surface. The computer vision system consists of a CCD camera (fixed above the work space) with digital image resolution of 256 × 256 and the Intelledex Vision processor based on a 16-MHz Intel 80386 CPU. The focal length of the camera is 0.0125 m. The vision system interfaces to the host computer, an SGI 4D/340 VGX. Visual measurements are sent to the SGI by a parallel interface. The robot is controlled by a UMC controller, which also interfaces with the SGI via memory mapping.

In the experiments, we assume that the relative poses between the camera, table, and robot are unknown and an unknown surface is a curved surface that is randomly placed on the robot work space. The trajectory to be followed by the end effector of the robot is characterized by five markers. We adopt our proposed control strategy and control the robot successfully so that the trajectory-following task is completed in a robust manner. In the experiments, we choose $k = 0.5$. We assume that the end effector initially makes contact with the unknown curved surface (manually move the end effector of the robot to touch the surface). Force control is used to maintain the contact. First of all, the robot automatically moves in a predesigned pattern (i.e., straight line motion along the x direction and y direction, respectively) with the force control loop, while a recursive least squares algorithm

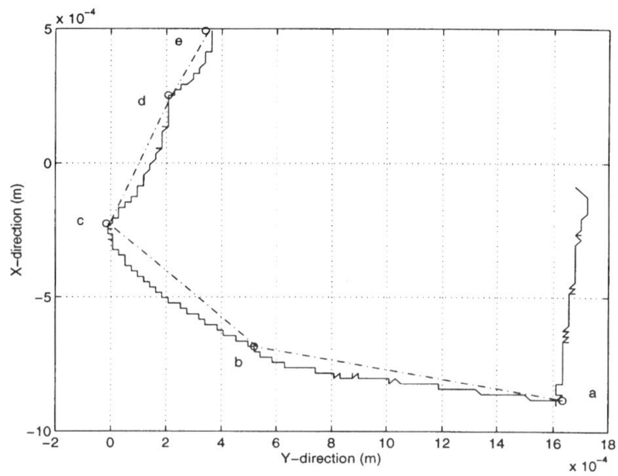

FIGURE 5.14
Projections of the desired and actual trajectories of the end effector on the image plane for slope plane.

is employed to recover the motion relation between the tangent plane and the image plane (in our experiments, during this procedure about 60 images are taken in 6 seconds). Then the proposed planning scheme is used to generate desired motion for the robot in real time and the robot is automatically controlled to follow the trajectory on the unknown surface without knowing the trajectory in the base frame of the robot.

The experimental results are illustrated in Figure 5.14–5.16. In Figure 5.14 the dotted and solid lines denote the projections of the trajectory to be followed on an unknown surface and the trajectory of the end effector of the robot on the image plane, respectively. Figure 5.15 shows the actual contact force with the desired force being -0.5 kg, while the actual trajectory of the end effector of the robot in the task space is given in Figure 5.16. In Figures 5.14 and 5.16, the points labeled by uppercase and lowercase letters are the markers on the

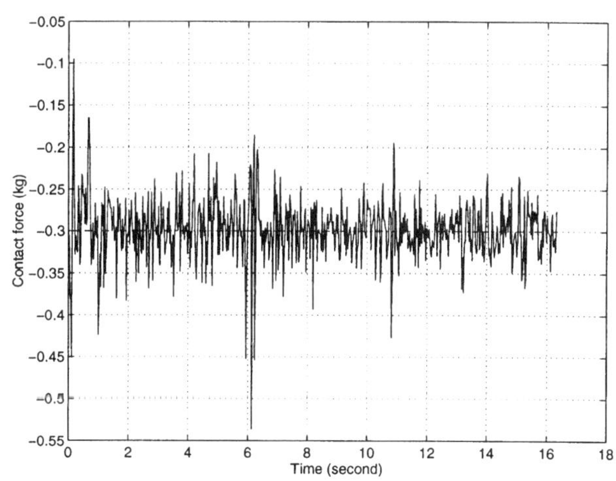

FIGURE 5.15
Desired and actual contact forces for slope plane.

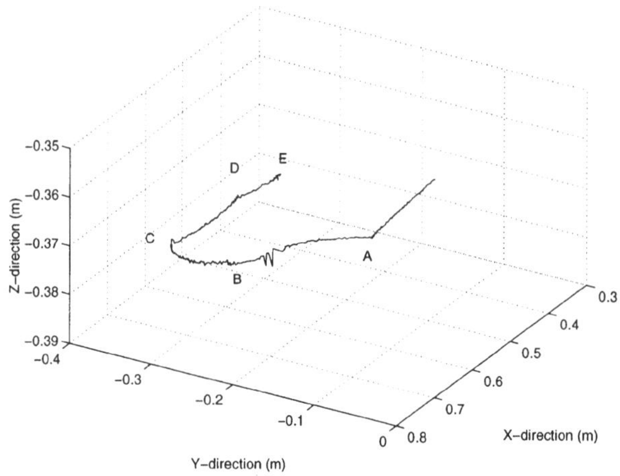

FIGURE 5.16
Actual trajectory of the end effector of the robot in the task space for the slope plane.

surface and their corresponding images on the image plane, respectively. Although the curved surface is deformable (see Figures 5.15 and 5.16), the task is still accomplished well.

4 CONCLUSION

In this chapter we demonstrate the advantage of using visual sensing in a multiple sensor integration mode in order to manipulate parts in a typical manufacturing work cell that is composed of a robot manipulator, a disc conveyor, and a camera system. Even though the visual computations are performed at a low rate, part position and orientation information can still be updated at the same rate as the feedback loop using an additional encoder sensor. We also demonstrate a practical tracking algorithm that pays attention to the fact that the torque that the robot control system can supply is bounded. The proposed algorithm is primarily based on error feedback, with an extra error reduction term added to force the torque requirement to remain within acceptable bounds. The proposed scheme is implemented on both time and event bases. The experimental results clearly demonstrate the advantages of the proposed scheme.

We have also developed a sensor fusion scheme for the control problem of constrained motions of a manipulator. At the contact point, the force–torque sensor mounted on the wrist of the robot provides local information about an unknown surface. The vision system with only one camera is used to reveal global information about the deviation of the end effector of the robot from the trajectory to be followed. In order to take advantage of multisensor fusion, we have designed a new planning and control scheme. The measurements of the sensors have been fed back to a planner, which generates a desired motion on the tangent plane to the unknown surface at the contact point for the robot with a force control loop, using a hierarchical structure. The force control ensures contact between the tool grasped by the end effector and the unknown surface, while the visual servoing guarantees that the difference between the images of the end effector and the trajectory vanishes. Therefore, the trajectory-following task is guaranteed to be accomplished.

The proposed method is simple to implement and robust against uncertainty in the configuration of the camera, the robot and the surface and utilizes an adaptive scheme. The experiments have successfully demonstrated the feasibility of the method. Because the proposed image-based motion planning scheme is implemented on the tangent plane of the contact surface, we can easily develop a fusion algorithm to utilize the scheme for the case where multiple uncalibrated cameras are available.

Acknowledgments

Partially supported by the Department of Energy under grant DE-FG02-90ER14140 and by the National Science Foundation under grant CAD-9404949 and Sandia National Laboratories contract AC 3752-C.

REFERENCES

[1] J. Hill and W. T. Park. Real time control of a robot with a mobile camera *Proceedings of the 9th ISIR*, Washington, DC, March 1979, pp. 233–246.

[2] L. Weiss, A. Sanderson, and C. Neuman. Dynamic sensor based control of robots with visual feedback. *IEEE J. Robot. Automat.* RA-3(5):404–417, 1987.

[3] P. Allen, B. Yoshimi, and A. Timcenko. Real-time visual servoing. In *Proceedings IEEE International Conference on Robotics and Automation*, 1991, pp. 851–856.

[4] E. D. Dickmanns and V. Graefe. Applications of dynamic monocular machine vision. *Machine Vision Appl.* 1:241–261, 1988.

[5] E. D. Dickmanns and V. Graefe. Dynamic monocular machine vision. *Machine Vision Appl.* 1:223–240, 1988.

[6] J. T. Feddema and O. R. Mitchell. Vision-guided servoing with feature-based trajectory generation. *IEEE Trans. Robot. Automat.* 5(5):691–700, 1989.

[7] P. I. Corke. Visual control of robot manipulators—a review. In *Visual Servoing*, K. Hashimoto, ed., pp. 1–32. World Scientific, 1994.

[8] N. Papanikolopoulos, P. K. Khosla, and T. Kanade. Visual tracking of a moving target by a camera mounted on a robot: A combinaton of control and vision. *IEEE Trans. Robot. Automat.* 9(1): 14–35, 1993.

[9] D. L. Hall. *Mathematical Techniques in Multisensor Data Fusion*. Artech House, Boston, 1992.

[10] T. C. Henderson and E. Shilcrat. Logical sensor systems. *J. Robot. Syst.* 1(2): 1984.

[11] T. C. Henderson, W. S. Fai, and C. Hansen. MKS: A multisensor kernel system. *IEEE Trans. Syst. Man Cybernet.* SMC-14(5):784–791, 1984.

[12] H. F. Durrant-Whyte. Integrating distributed sensor information, an application to a robot system coordinator. *Proceedings of 1985 IEEE International Conference on Systems, Man and Cybernetics*, 1985, pp. 415–419.

[13] H. F. Durrant-Whyte. Consistent integration and propagation of disparate sensor observations. *Proceedings 1986 IEEE International Conference on Robotics and Automation*, San Francisco, April 1986.

[14] R. C. Luo and M. Kav. Multisensor integration and fusion. *SPIE 1988 Conference on Multisensor Fusion*, Orlando, FL, April 1988.

[15] R. C. Luo, M. H. Lin, and R. S. Scherp. The issues and approaches of a robot multi-sensor integration. *Proceedings 1987 IEEE International Conference on Robotics and Automation*, Raleigh, NC, 1987, pp. 1941–1946.

[16] R. C. Luo and M.-H. Lin. Intelligent robot multi-sensor data fusion for flexible manufacturing systems. *Advances in Manufacturing Systems Integration and Processes, Proceedings of the 15th Conference on Production Research and Technology*, University of California, Berkeley, Jan. 1989.

[17] P. Allen and R. Bajcsy. Object recognition using vision and touching. *Proceedings of the 9th Joint Conference on Artificial Intelligence*, Los Angeles, Aug. 1985.

[18] A. Flynn. Redundant sensors for mobile robot navigation. Dept. of EECS, M.S. thesis, MIT, Cambridge, MA, July 1985.

[19] M. J. Magee, B. A. Boyter, C.-H. Chien, and J. K. Aggarwal. Experiments in Intensity Guided Range Sensing Recognition of Three-Dimensional Objects. *IEEE Trans. PAMI* PAMI-7(6):629–637, 1985.

[20] N. Nandhakumar and J. K. Aggarwal. Synergetic analysis of thermal and visual images for scene perception.

Proceedings of the Platinum Jubilee Conference on Systems and Signal Processing, Indian Institute of Science, Banglore, Dec. 1986.

[21] A. Mitiche and J. K. Aggarwal. Multiple sensor integration/fusion through image processing: A review. *Opt. Eng.* 25:380–386, March 1996.

[22] T. Ishikawa, S. Sakane, T. Sato, and H. Tsukune. Estimation of contact position between a grasped object and the environment based on sensor fusion of vision and force. *1996 IEEE/SICE/RSJ International Conference on Multisensor Fusion and Integration for Intelligent Systems*, Washington, DC, Dec. 8–11, 1996, pp. 116–123.

[23] B. J. Nelson and P. K. Khosla. Force and vision resolvability for assimilating disparate sensory feedback. *IEEE Trans. Robot. Automat.* 12:714–731, 1996.

[24] P. K. Allen. Integrating vision and touch for object recognition tasks. *Int. J. Robot. Res.* 7(6):15–33, 1988.

[25] S. A. Stansfield. A robotic perceptual system utilizing passive vision and active touch. *Int. J. Robot. Res.* 7(6):138–161, 1988.

[26] A. Castano and S. Hutchinson. Visual compliance: Task-directed visual servo control. *IEEE Trans. Robot. Automat.* 10:334–342, 1994.

[27] S. Hutchinson, G. D. Hager, and P. I. Corke. A tutorial on visual servo control. *IEEE Trans. Robot. Automat.* 12(5):651–670, 1996.

[28] D. E. Whitney. Force feedback control of manipulator fine motions. *ASME J. Dynam. Syst. Meas. Control* 99(2):91–97, 1977.

[29] O. Khatib. Dynamic control of manipulators in operation space. *6th CISM-IFTMM Congress on Theory of Machines and Mechanism*, New Delhi, India, 1983, pp. 1128–1131.

[30] T. Yoshikawa. Dynamic hybrid position/force control of robot manipulators — description on hand constraint and calculation of joint driving force. *Proceedings IEEE International Conference on Robotics and Automation*, 1986, pp. 1393–1398.

[31] O. Khatib and J. Burdick. Motion and force control of robot manipulators. *Proceedings IEEE International Conference on Robotics and Automation*, 1986, pp. 1381–1386.

[32] M. T. Mason. Compliance and force control for computer controlled manipulators. *IEEE Trans. Syst. Man Cybernet.* 11:418–431, 1981.

[33] M. H. Raibert and J. J. Craig. Hybrid position/force control of manipulators. *J. Dynam. Syst. Meas. Control* 103(2):126–133, 1981.

[34] R. P. Paul and B. Shimano. Compliance and control. *Proceedings of Joint Automatic Control Conference*, 1976, pp. 694–699.

[35] M. H. Ozaki and M. Takata. On the force feedback control of a manipulator with a compliant wrist force sensor. *Mech. Machine Theory* 18(1):57–62, 1986.

[36] H. Van Brussel nd J. Simons. Robot assembly by active force feedback asccommodation. *Manuf. Technol. CIRP Ann.* 28(1):397–401, 1979.

[37] J. K. Salisburg. Active stiffness control of a manipulator in Cartesian coordinates. *Proceedings of 19th IEEE Conference on Decision and Control*, 1980, pp. 95–100.

[38] R. K. Roberts, R. P. Paul, and B. M. Hilliberry. The effect of wrist force sensor stiffness on the control of robot manipulators. *Proceedings IEEE International Conference on Robotics and Automation*, 1985.

[39] N. Hogan. Programmable impedance industrial manipulators. *Proc. Conf. CAD/CAM Tech. in Mech. Eng.*, MIT, Cambridge, MA, 1982.

[40] K. Kanatani. *Group-Theoretical Methods in Image Understanding*. Springer-Verlag, 1990.

[41] T. J. Tarn, A. K. Bejczy, A. Isidori, and Y. L. Chen. Nonlinear feedback in robot arm control. *Proceedings of 23rd IEEE CDC*, Las Vegas, Dec. 1984.

[42] K. Hashimoto *et al.* Manipulator control with image-based visual servo. *IEEE International Conference on Robotics and Automation*, Sacramento, CA, 1991, pp. 2267–2272.

[43] N. Papanikolopoulos, P. K. Khosla, and T. Kanade. Adaptive robotic visual tracking. *Proceedings of 1991 American Control Conference*, Boston, June 26–28, 1991.

[44] B. H. Yoshimi and P. K. Allen. Active, uncalibrated visual servoing. *Proceedings of 1994 IEEE International Conference on Robotics and Automation*, San Diego, CA, May 8–13, 1994.

[45] O. Dahl. Path constrained robot control, Doctoral dissertation, Ound Institute of Technology, 1992.

[46] T.-J. Tarn, A. K. Bejczy and N. Xi. Motion planning in phase space for intelligent robot arm control. *Proceedings of IROS '92*, Raleigh, NC, 1992, vol. 3, pp. 1507–1514.

[47] A. E. Bryson, Jr. and Y. C. Ho. *Applied Optimal Control: Optimization, Estimation and Control*. John Wiley & Sons, 1975.

[48] R. Safaee-Rad, K. C. Smith, and B. Benhabib. Three-dimensional location estimation of circular features for machine vision. *IEEE Trans. Robot. Automat.* 8:624–640, 1992.

[49] B. H. Yoshimi and P. K. Allen. Alignment using an uncalibrated camera system. *IEEE Trans. Robot. Automat.* 11:516–521, 1995.

[50] J. T. Feddema. Visual servoing: A technology in search of an application. Workshop M-5 on Visual Servoing: Achievements, Applications and Open Problems, *1994 IEEE International Conference on Robotics and Automation.*

[51] Zhenyu Yu, B. K. Ghosh, N. Xi, and T. J. Tarn. Temporal and spatial sensor fusion in a robotic manufacturing workcell. *Proceedings of the 1995 IEEE International Conference on Robotics and Automation*, Nagoya, Japan, May 21, 1995.

[52] Zhenyu Yu, B. K. Ghosh, N. Xi, and T. J. Tarn. Multi-sensor based planning and control for robotics manufacturing systems. *Proceedings of the 1995 IEEE/RSJ International Conference on Intelligent Robotics and Systems*, Aug. 5–9, 1995.

[53] Zhenyu Yu. Vision-guided robot motion planning and control. Doctoral dissertation, Washington University, St. Louis, MO, Aug. 1995.

5 APPENDIX

Consider an ellipse in the image plane whose major axis is parallel to the y axis of the image plane. Let $(x_{ez}, 0)$ be its center and let a and b be the half-lengths of its minor and major axes, respectively.

Assume that the image plane is located at $z = f$, the camera is a pinhole camera, and the projection center is at the origin. Then the elliptic cone formed by the projection lines emitting from the projection center and passing through the ellipse can be described by

$$\frac{\left(x - x_{ez}\dfrac{z}{f}\right)^2}{a^2} + \frac{y^2}{b^2} = \frac{z^2}{f^2} \tag{5.52}$$

where $a^2 \leqslant b^2$ as a result of step 3 of the "Virtual rotation" algorithm in Section 2.1. We now describe the following theorem.

Theorem The planes that intersect the elliptic cone (5.52) on a circle are parallel to

$$z = -x \tan \alpha \tag{5.53}$$

where α is determined by

$$\tan \alpha = \frac{f(x_{ez}b^2 \pm a\sqrt{b^4 + b^2f^2 + b^2x_{ez}^2 - a^2b^2 - a^2f^2})}{b^2x_{ez}^2 - a^2b^2 - a^2f^2} \tag{5.54}$$

Proof It is well known [55] that planes that intersect an elliptic cone on a circle have at most two solutions that are different in orientation. In fact, it can be shown that up to parallel translations, the two planes can be made to pass through the minor axis of the elliptic cross section. In particular, for the elliptic cone (5.52), since the minor axis passes through the y axis, the two planes are of the form (5.53) for a suitable parameter α. We now proceed to show that indeed (5.54) is satisfied.

Assume that the plane (5.53) intersects the elliptic cone (5.52) in a circle. It follows that if the coordinate frame is rotated by an angle of α about the y axis and shifted to the intersection between the plane containing the circle and the z axis, the circle of intersection would be of the form

$$(\tilde{x} - \tilde{x}_0)^2 + (\tilde{y} - \tilde{y}_0)^2 = r^2, \qquad \tilde{z} = 0 \tag{5.55}$$

in the new coordinate frame $(\tilde{x}.\tilde{y}, z)$. The required transformation is given by

$$\begin{pmatrix} x \\ y \\ z \end{pmatrix} = \begin{pmatrix} \cos\alpha & 0 & \sin\alpha \\ 0 & 1 & 0 \\ -\sin\alpha & 0 & \cos\alpha \end{pmatrix} \begin{pmatrix} \tilde{x} \\ \tilde{y} \\ \tilde{z} \end{pmatrix} + \begin{pmatrix} 0 \\ 0 \\ \tilde{z}_0 \end{pmatrix} \tag{5.56}$$

where $\tilde{z}_0$ is the required shift in the z coordinate.

Describing the elliptic cone (5.52) in terms of the new coordinates $(\tilde{x}, \tilde{y}, \tilde{z})$ given by (5.56) and setting $\tilde{z} = 0$, we obtain the intersection between the elliptic cone and the plane given by

$$[\tilde{x}\cos\alpha - x_{ex}(-\tilde{\tau}\sin\alpha + \tilde{z}_0)/f]^2/a^2 + \tilde{y}^2/b^2 = (-\tilde{x}\sin\alpha + \tilde{z}_0)^2/f^2 \tag{5.57}$$

Since (5.57) is an equation of a circle, the following condition must be satisfied.

$$(\cos\alpha + x_{ez}\sin\alpha/f)^2/a^2 - \sin^2\alpha/f^2 = 1/b^2 \tag{5.8}$$

Solving (5.58) for α, one obtains (5.54). □

In order to discriminate between the two values of α given by (5.54), we proceed as follows. Assume that z axis (i.e., the optical axis of the camera) passes through the center of the circle. The correct solution of α can be determined by considering a circle centered at $(0, 0, z_0)$ contained in the plane parallel to the image plane rotated by an angle of α. The circle is projected as an ellipse on the image plane with the major axis along the x axis. It follows that one can compute

$$l_1 = f\frac{r\cos\alpha}{z_0 - r\sin\alpha} \tag{5.59}$$

and

$$l_2 = f\frac{r\cos\alpha}{z_0 + r\sin\alpha} \tag{5.60}$$

It is easy to see that l_1 is greater than l_2 and the sign of the x coordinate of the center of the elliptic image is positive. On the other hand, the sign of $-\tan\alpha$ is also positive. It follows that α can be determined by noting the sign of the center of the elliptic image.

CHAPTER 6

Feedback Control with Force and Visual Sensor Fusion

B. J. NELSON
Department of Mechanical Engineering, University of Minnesota, Minneapolis, Minnesota

P. K. KHOSLA
Department of Electrical and Computer Engineering, Carnegie Mellon University, Pittsburgh, Pennsylvania

ABSTRACT

Within the domain of robotic manipulation, roboticists have traditionally considered force feedback to be the most relevant sensing modality. This is because of the need for highly accurate information on the relative positions of objects and on the nature of contact forces between objects being manipulated. More recently, many researchers have realized the benefits of using visual servoing techniques to reduce alignment uncertainties between objects using imprecisely calibrated camera–lens–manipulator systems. These two sensing modalities, force and vision, are complementary in the sense that they are useful during different stages of task execution: vision brings parts into alignment; force ensures that reasonable contact forces are maintained as parts mating occurs. However, when one considers the integration of force and vision feedback from the traditional sensor integration viewpoint, the deficiencies in conventional approaches become apparent. In this chapter, we describe the concept of vision and force sensor resolvability as a means of comparing the ability of the two sensing modes to provide useful information during robotic manipulation tasks. By monitoring the resolvability of the two sensing modes with respect to the task, the information provided by the disparate sensors can be seamlessly assimilated during task execution. A nonlinear force–vision servoing algorithm that uses force and vision resolvability to switch manipulator control between sensors is experimentally analyzed. The advantages of the assimilation technique are demonstrated during contact transitions between a stiff manipulator and a rigid environment, a system configuration that easily becomes unstable when force control alone is used. Experimental results demonstrate robust contact transitions by the proposed nonlinear controller while the conflicting task requirements of fast approach velocities, maintaining stability, minimizing impact forces, and suppressing bounce between contact surfaces are simultaneously satisfied.

1 INTRODUCTION

Sensor integration has long been considered a key research topic in robotics research. When one considers the integration of force and vision feedback from the traditional sensor integration viewpoint, however, the deficiencies in conventional approaches become apparent. Force and vision sensors are disparate sensing modalities by nature and typically produce fundamentally different measurements, force and position. The output of a manipulator wrist force sensor yields measurements of force and torque along and about the three Cartesian axes. Properly tracked features in a visual sensor's sensing space yield measurements of the relative positions and orientations of objects in the world. This presents an inherent problem when attempting to integrate information from these two sensors. Conventional techniques for sensor integration operate in some common space closely related to the particular sensors used in the system, often using a probabilistic weighting method for combining information from different sensors [1–4]. Figure 6.1 illustrates the conventional technique schematically. Force and vision sensors do not share a common data representation. Conventional sensor integration also assumes that a temporally accurate cross-coupling between sensors can be modeled in sensor space. Vision and force sensing modes are appropriate during different stages of the task, making a temporal comparison of the two data sets meaningless during most of the task. These two facts indicate that traditional sensory integration techniques are not appropriate for the integration of force and vision feedback.

Instead, we advocate a task-oriented approach for assimilating information from force and vision sensors. We believe in the importance of the task model for combining information from disparate sensors, much as Jain [5] argues for the importance of environment models for the assimilation of information from disparate sensors in a mobile robot domain. It makes little sense to combine the measurements of force and position using, for example, a Kalman filter, because of the disparate nature of the feedback. A model of the task is required that has the capability of dynamically assimilating information from the two disparate sensing modes. As the task occurs, the task model determines when vision is appropriate and when force is appropriate by considering the nearness of contact surfaces and the resolution with which each sensor can sense object locations. In Figure 6.2 this is illustrated by the close relationship between the sensor feedback assimilation module and the environment model that describes the task.

Throughout a manipulation task, the data from both sensors must be compared in order to ascertain which sensing mode is more relevant to the task at the given time instant. Our

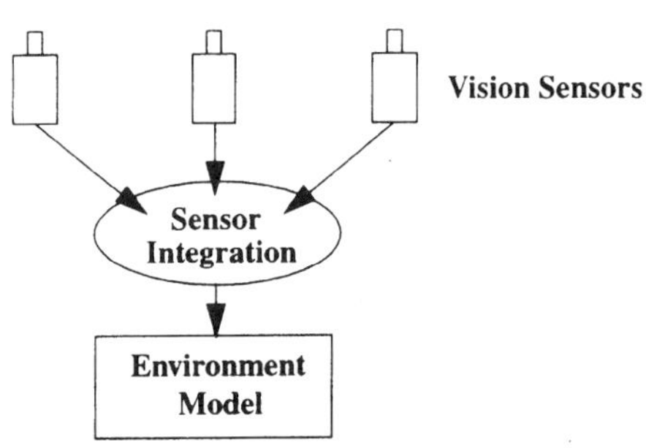

FIGURE 6.1
Conventional sensor integration.

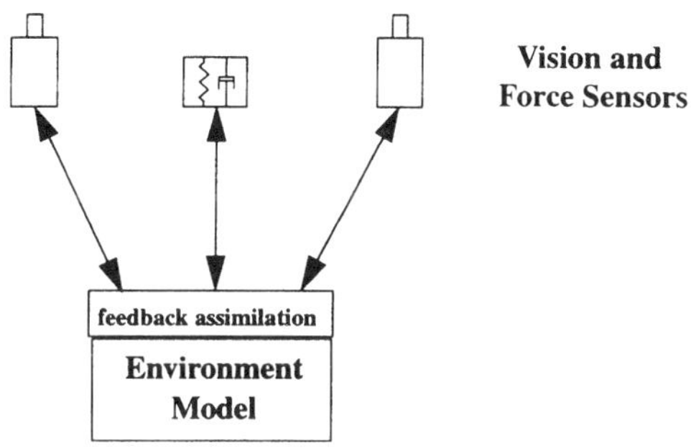

FIGURE 6.2
Task-oriented assimilation of disparate sensory feedback.

previous work in *resolvability* [6, 7] showed how various visual sensing configurations can be compared in terms of the resolution with which they can visually servo an object held by a manipulator. The resolvability measure has been used to guide active camera–lens systems throughout a manipulation task [8]. The concept of resolvability can be extended to force sensors in order to determine the resolution with which a force sensor can detect infinitesimal task displacements in the environment. Force resolvability is dependent not only on the force sensor but also on the stiffness of the entire system with respect to task displacements. This extension of resolvability provides a common measure for both sensors for evaluating when visual servoing or force servoing strategies are appropriate.

During transitions between force and vision sensing, a nonlinear force–vision control law compensates for the uncertain world until it becomes clear when a new sensing mode has been achieved or whether the system should return to the prior sensing mode. Both force and vision sensory data are considered simultaneously; however, control of the manipulator is achieved using only one of these modalities at any instant in time. In order to illustrate the advantages of assimilating disparate sensor feedback using our proposed method, we experimentally demonstrate the performance of the technique during contact transitions. Many researchers have studied the impact problem, and various impact strategies have been proposed. However, the fundamental problem of using force feedback alone to minimize impact forces while quickly achieving contact stably within imprecisely calibrated environments still exists. By combining vision feedback with force feedback using the concept of resolvability and our proposed nonlinear control strategy, we demonstrate how fast stable contact transitions with a stiff manipulator in a rigid environment can be achieved.

In this chapter, we demonstrate the use of force and vision resolvability for assimilating high-bandwidth visual feedback (30 Hz) and high-bandwidth force feedback (100 Hz) within the same manipulator feedback loop. After reviewing related work, we discuss the concept of vision and force resolvability and how they can be used in the sensor assimilation process. Next, a visual servoing control law is derived, followed by a description of the vision–force servoing control strategy. An important contribution of this work is that we show how vision can be used to simplify the stability problem greatly by allowing the effective use of low-gain force control with stiff manipulators (a Puma 560). As the stability of low-gain force control is much easier to maintain, the use of force feedback during manipulator fine motion is more easily realized because simple force control strategies can be used without the need for high-order models of the arm, sensor, and environment for choosing stable controller gains. The proper combination of force and vision feedback is the key to the success of this strategy.

2 PREVIOUS WORK

2.1 Force Control

Robotic force control has been studied since the 1950s. A survey of the history of force control can be found in Whitney [9]. Active impedance control has been suggested as the most general form of force control [10], but difficulties in programming impedance-controlled manipulators have resulted in very limited use of this strategy. Hybrid control [11] separates position control and force control into two separate control loops that operate in orthogonal directions, as shown in Figure 6.3, and has been extended to include manipulator dynamics [12].

One of the most important issues in force control is maintaining manipulator stability [9]. Force controllers must be properly formulated and tuned in order to maintain stability, and this can be difficult, particularly during initial contact between stiff surfaces. During impact, another important issue is the generation of large impact forces. An effective impact strategy based on a proportional gain explicit force controller with a feedforward signal and negative gains has been demonstrated experimentally in Volpe and Khosla [13]. The gains for the controller were chosen using a fourth-order model of the arm, sensor, and environment in which a frictionless arm was assumed (experiments were conducted with the CMU Direct-Drive Arm II). Although extremely high impact velocities were achieved (75 cm/s), large impact forces were also generated (90 N). This illustrates a typical problem exhibited by all force control strategies during impact with rigid objects; for example [14–21], high impact velocities, manipulator stability, low impact forces, and quickly achieving the desired force are all contradictory system requirements.

2.2 Visual Servoing

Visual servoing has a less extensive history than force control, mainly because of the lack of computational resources available for processing the large amounts of data contained in an image. Although previous researchers had considered fast visual feedback for guiding manipulator motion, for example [22], the visual servoing field was first well defined in Weiss [23]. Since this work, two types of visual servoing configurations have emerged, eye-in-hand

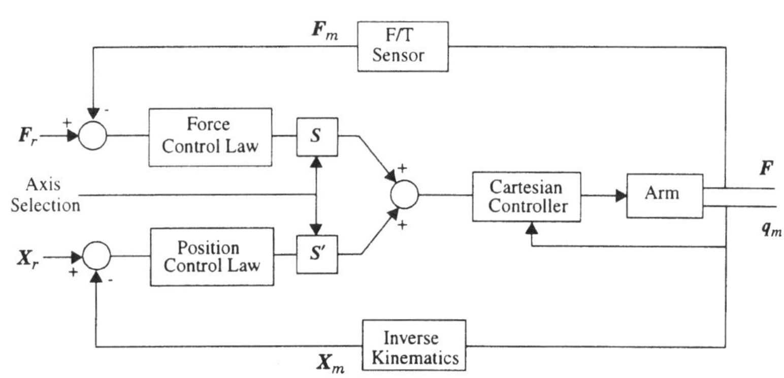

FIGURE 6.3

Hybrid force–position control loop, wher $\mathbf{F}_r$ is the reference force vector, $\mathbf{F}_m$ is the measured force vector, $\mathbf{X}_r$ is the reference position, $\mathbf{X}_m$ is the measured position, $\mathbf{S}$ and $\mathbf{S}'$ are the orthogonal selection matrices, $\mathbf{F}$ is the appied force, and q_m is a vector of measured joint positions.

configurations and static camera configurations. Eye-in-hand visual servoing tracks objects of interest with a camera mounted on a manipulator's end effector [24–32]. Static camera visual servoing guides manipulator motion based on feedback from a camera observing the end effector [33–36]. Most of this past work has been with monocular systems, although stereo systems have been used for visual servoing [36–38].

2.3 Sensor Resolution

The concept of sensor resolution plays an important role in the assimilation of force and vision feedback. In order to use visual feedback effectively to perform robotic tasks, many researchers have recognized that the placement of the sensor relative to the task is an important consideration, and sensor resolution has been considered in the past as a criterion for sensor planning [39–41]. These efforts concern static camera systems in which a required spatial resolution is known and a single camera placement is desired. In Das and Ahuja [42], a study of stereo, vergence, and focus cues for determining range is described in which the performance of each cue for determining range accuracy is characterized. This characterization can be used to control camera parameters in order to improve the accuracy of range estimates. Our resolvability approach can be used for determining the ability of a visually servoed manipulator to resolve positions and orientations of objects accurately along all six degrees of freedom. Resolvability provides a technique for estimating the relative ability of various visual sensor systems, including single camera systems, stereo pairs, multibaseline stereo systems, and three-dimensional (3-D) rangefinders, to control visually manipulated objects accurately and to provide spatially accurate data on objects of interest. Camera–lens intrinsic and extrinsic parameters can be actively controlled using a resolvability measure in conjunction with other sensor placement criteria so that the accuracy of visual control can be improved [6]. The concept can also be used for static sensor placement for either object recognition or visual servoing.

A measure similar to resolvability called *observability* (unrelated to observability in the controls sense) has been introduced [43a] and was extended to include manipulator configuration in [43b]. The concepts of resolvability and observability are both inspired by the concept of *manipulability* [44]. The proposed algorithm based on observability attempts to maximize $\sqrt{\det(\mathbf{J}\mathbf{J}^T)}$ throughout a camera trajectory, where $\mathbf{J}$ is the image Jacobian. The extension of observability to *motion perceptibility* includes the manipulator Jacobian as a component of the latter measure. Our emphasis on resolvability has been on its directional nature, as determined from the singular values of $\mathbf{J}$ and the eigenvectors of $\mathbf{J}^T\mathbf{J}$, and on using this decomposition to guide camera–lens motion actively during task execution [6]. In this chapter, we extend resolvability to include force sensors. This measure is then used to assimilate force and vision information within high-bandwidth manipulator feedback loops.

For force sensor design, strain gauge sensitivity, force sensitivity, and minimum sensor stiffness are three critical design parameters. In Nakamura *et al.* [45] and Uchiyama *et al.* [46], force sensor design techniques are based on measures derived in part from a singular value decomposition of the force sensor calibration matrix. A measure of force sensor resolution to be presented in Section 3.2 uses this past work as a foundation and extends the analysis to include system stiffness as well.

2.4 Fusing Force and Vision Feedback

Many researchers have considered the fusion of force and vision feedback, although very few have done this within high-bandwidth feedback control loops. One of the first papers to mention the benefits of integrating high-bandwidth visual and force feedback is that of Shirai and Inoue [22]. A 0.1-Hz visual servoing scheme was implemented and the use of force servoing was referred to, but a lack of computational resources hampered their effort, and many of the issues of combining the two sensing modalities went unnoticed. In Ishikawa *et al.* [47], visual servoing of 2 Hz was used to align a wrench with a bolt before a compliant wrenching operation is performed. Again, vision and force were not explicitly combined, and the issues concerning their integration remained unaddressed.

A Bayesian framework for combining visual observations and tactile data for grasping is described in Durrant-Whyte [48]. The nondynamic nature of the task made the use of the tactile feedback of questionable value for the experiments described. Rule-based approaches for combining vision and touch feedback for object recognition are described in Allen [49] and Stansfield [50]. An important observation made in Allen [49] is that touch feedback is successful for object recognition because vision provides cues to the tactile sensor. This is also important for object manipulation and is a primary reason why fast, stable contact transitions can be more easily realized with manipulators servoed under both vision and force control, rather than force control alone. The assimilation technique proposed in this chapter employs a nonlinear control strategy that is a combination of quantitative and rule-based approaches of combining force and vision sensing. The general idea behind a quantitative multisensor integration framework [2] is used to determine the appropriate sensor at any given instant. Rule-based methods are used to perform mode switching and to eliminate false force sensor readings.

3 SENSOR RESOLVABILITY

Sensor fusion techniques that use multiple sensors along the same task dimensions require that the system compare the characteristics of the feedback from each individual sensor at some point during the task. For sensors that have similar data representations, for example, multiple cameras that provide positional information, this is straightforward. For sensors that provide fundamentally different measurements, for example, force and vision sensors, this presents a problem. *Vision resolvability* compares the effects of camera–lens–task configurations on the ability to resolve accurately, and therefore visually servo, object positions and orientations [6, 7]. The more general concept of *sensor resolvability* described in this chapter provides a measure of the ability of both force and vision sensors to resolve positions and orientations in task space, thus providing a method for assimilating the data from the two sensors.

Resolvability is a function of the Jacobian of the mapping from task space to sensor space. A matrix form of the Jacobian is desired that contains both intrinsic and extrinsic sensor parameters in order to analyze the effects of these parameters on the structure of the Jacobian. For any sensor system, an equation of the form

$$\delta \mathbf{x}_S = \mathbf{J}(\phi)\delta \mathbf{X}_T \tag{6.1}$$

where $\delta \mathbf{x}_S$ is an infinitesimal displacement vector in sensor space; $\mathbf{J}(\phi)$ is the Jacobian matrix and is a function of the extrinsic and intrinsic parameters of the visual sensor as well as the number of features tracked and their locations on the image plane; and $\delta \mathbf{X}_T$ is an infinitesimal displacement vector in task space.

The ability of a vision or force sensor to resolve task positions and forces is directionally dependent. By performing a singular value decomposition on the task space to sensor space Jacobian and analyzing the singular values and the eigenvectors of $\mathbf{J}^T\mathbf{J}$ that result from the decomposition, the directional properties of the ability of the sensor to resolve positions and orientations become apparent.

Singular value decomposition is a common technique used for analyzing the range and null space of a matrix transformation. Details concerning the SVD can be found in Klema and Laub [51]. The SVD of a matrix A is given by

$$\mathbf{A} = \mathbf{U}\Sigma\mathbf{V}^T \tag{6.2}$$

where Σ is a diagonal matrix containing the square roots of the eigenvalues of $\mathbf{A}^T\mathbf{A}$ and $\mathbf{A}\mathbf{A}^T$. These eigenvalues are also the singular values of $\mathbf{A}$; $\mathbf{U}$ contains the eigenvectors of $\mathbf{A}\mathbf{A}^T$; and $\mathbf{V}$ contains the eigenvectors of $\mathbf{A}^T\mathbf{A}$. For resolvability, the eigenvectors of $\mathbf{J}^T\mathbf{J}$ are those in which we are interested, because these eigenvectors give us a set of basis vectors for the row space of $\mathbf{J}$, which is also the vector space described by $\Re(\mathbf{J}^T)$, the range of $\mathbf{J}^T$. Therefore, singular values of $\mathbf{J}$ and the corresponding eigenvectors of $\mathbf{J}^T\mathbf{J}$ indicate, in task space, how sensor space displacements will be affected by task displacements. Conversely, the eigenvectors of $\mathbf{J}\mathbf{J}^T$ tell us the affect of task space displacements in sensor space. For example, large singular values along a particular eigenvector of $\mathbf{J}^T\mathbf{J}$ tell us that small task displacements in the direction of the particular eigenvector will result in relatively large sensor space displacements. Thus, the task can be accurately observed and controlled along these directions. Small singular values indicate that task displacements in those directions are difficult to observe, and therefore difficult to control, because large task displacements cause small displacements in sensor space.

All of the eigenvectors of $\mathbf{J}^T\mathbf{J}$ combined with the corresponding singular values of $\mathbf{J}$ indicate the directional ability of the sensor system to resolve displacements in task space. It is this combination of singular values and eigenvectors that defines resolvability and is used in plotting the resolvability ellipsoid in the section on vision resolvability ellipsoids. In order to use an ellipsoidal representation of resolvability, we assume that the object of interest has an equal ability to translate and rotate about all of its Cartesian axes. Furthermore, we assume that the velocity of the object is constrained to fall within some six-dimensional spheroid, such that

$$\|\dot{\mathbf{X}}_T\| = (\dot{X}_T^2 + \dot{Y}_T^2 + \dot{Z}_T^2 + \omega_{x_T}^2 + \omega_{y_T}^2 + \omega_{z_T}^2)^{1/2} \tag{6.3}$$

and $\|\dot{\mathbf{X}}_T\| < 1$. Under these assumptions, the principal axes of the ellipsoid representing the ability of $\mathbf{J}$ to resolve positions and orientations in task space are given by $\sigma_1\mathbf{v}_1$, $\sigma_2\mathbf{v}_2$, $\sigma_3\mathbf{v}_3$, $\sigma_4\mathbf{v}_4$, $\sigma_5\mathbf{v}_5$, and $\sigma_6\mathbf{v}_6$, where σ_i is the ith singular value of $\mathbf{J}$ and $\mathbf{v}_i$ is the ith eigenvector of $\mathbf{J}^T\mathbf{J}$. The following sections derive Jacobians for force and vision sensors, present ellipsoids for various camera–lens–object configurations, and analyze the Jacobian for various force sensor configurations.

3.1 Vision Resolvability

Monocular Systems

Camera Model The mapping from task space to sensor space for any system using a camera as the visual sensor requires a camera–lens model in order to represent the projection of task objects onto the CCD image plane. For visual servoing, a simple pinhole camera model has proved adequate for visual tracking using our experimental setup. If we place the camera coordinate frame {C} at the focal point of the lens as shown in Figure 6.4, a feature on an object at ${}^C\mathbf{P}$ with coordinates (X_C, Y_C, Z_C) in the camera frame projects onto the camera's image plane at

$$x_i = \frac{fX_C}{s_x Z_C} + x_p \tag{6.4}$$

$$y_i = \frac{fY_C}{s_y Z_C} + y_p \tag{6.5}$$

where (x_i, y_i) are the image coordinates of the feature, f is the focal length of the lens, s_x and s_y are the horizontal and vertical dimensions of the pixels on the CCD array, and (x_p, y_p) is the piercing point of the optical axis on the CCD. This model assumes that $|Z_C| \gg |f|$, thus simplifying the terms in the denominator of (6.4) and (6.5).

The mapping from camera frame feature velocity to image plane optical flow, or sensor space velocity, can be obtained by differentiating (6.4) and (6.5). This yields the following equations:

$$\dot{x}_S = \frac{f\dot{X}_C}{s_x Z_C} - \frac{fX_C\dot{Z}_C}{s_x Z_C^2} = \frac{f\dot{X}_C}{s_x Z_C} - x_S\frac{\dot{Z}_C}{Z_C} \tag{6.6}$$

$$\dot{y}_S = \frac{f\dot{Y}_C}{s_y Z_C} - \frac{fY_C\dot{Z}_C}{s_x Z_C^2} = \frac{f\dot{Y}_C}{s_y Z_C} - y_S\frac{\dot{Z}_C}{Z_C} \tag{6.7}$$

where $x_S = x_i - x_p$ and $y_S = y_i - y_p$. This defines the mapping of object velocity with respect to the camera frame onto the image plane. The next step is to transform task space velocities into the camera frame and then project these camera frame velocities onto the sensor space to obtain the mapping from task space velocity to sensor space velocity.

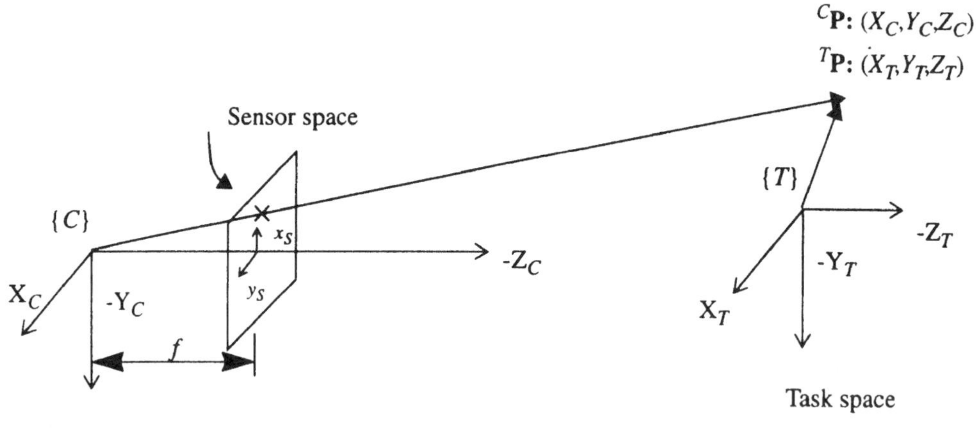

FIGURE 6.4
Task frame–camera frame definitions.

Objects Defined in a Task Frame For visually servoing a manipulator holding an object, the objective is to move the image coordinates of ${}^C\mathbf{P}$ to some location on the image plane by controlling the motion of ${}^C\mathbf{P}$. Typically, ${}^C\mathbf{P}$ is some feature on an object being held by a manipulator. Thus, the motion of ${}^C\mathbf{P}$ is induced relative to the tool frame of the manipulator being observed. Figure 6.4 shows the coordinate systems used to define the mapping from task space to sensor space, where ${}^C\mathbf{P}$ is defined with respect to the task frame as ${}^T\mathbf{P}$ with coordinates (X_T, Y_T, Z_T). For now, we assume that the rotation of the task frame $\{T\}$ with respect to $\{C\}$ is known. The velocity of ${}^T\mathbf{P}$ can be written as

$$ {}^C\dot{\mathbf{P}} = {}^C_T\mathbf{R}({}^T\mathbf{V} + {}^T\dot{\mathbf{P}} + {}^T\Omega \times {}^T\mathbf{P}) \tag{6.8} $$

where ${}^T\mathbf{V} = [\dot{X}_T \ \dot{Y}_T \ \dot{Z}_T]^T$ and ${}^T\Omega = [\omega_{X_T} \ \omega_{Y_T} \ \omega_{Z_T}]^T$ are the translational and rotational velocities of the task frame with respect to itself. These are manipulator end-effector velocities that can be commanded. Since we assume that the object being servoed or observed is rigidly attached to the task frame, ${}^T\dot{\mathbf{P}} = 0$, and (6.6) becomes

$$ {}^C\dot{\mathbf{P}} = {}^C_T\mathbf{R}({}^T\mathbf{V} + {}^T\Omega \times {}^T\mathbf{P}) \tag{6.9} $$

Furthermore, if we assume that $\{C\}$ and $\{T\}$ are aligned, as shown in Figure 6.4, then ${}^C_T\mathbf{R} = \mathbf{I}$ and the elements of ${}^C\mathbf{P}$ can be written as

$$ \begin{aligned} \frac{dX_C}{dt} &= \dot{X}_T + Z_T\omega_{Y_T} - Y_T\omega_{Z_T} \\ \frac{dY_C}{dt} &= \dot{Y}_T - Z_T\omega_{X_T} + X_T\omega_{Z_T} \\ \frac{dZ_C}{dt} &= \dot{Z}_T + Y_T\omega_{X_T} - X_T\omega_{Y_T} \end{aligned} \tag{6.10} $$

The assumption that $\{C\}$ and $\{T\}$ are aligned is used only in formulating the Jacobian from task space to sensor space. If the transformation from task space to sensor space is initially known and the commanded task frame velocity is known, then the coordinates (X_T, Y_T, Z_T) can be appropriately updated while visual servoing. It will also be necessary to account for task frame rotations when determining the velocity to command the task frame based on ${}^T\mathbf{V} = [\dot{X}_T \ \dot{Y}_T \ \dot{Z}_T]^T$ and ${}^T\Omega = [\omega_{X_T} \ \omega_{Y_T} \ \omega_{Z_T}]^T$. It would have been possible to include the terms of ${}^C_T\mathbf{R}$ in (6.10); however, the assumption made simplifies the derivation and does not affect the end result.

By combining (6.10) with (6.6) and (6.7), the entire Jacobian transformation for a single feature from task space to sensor space can now be written in the form

$$ \begin{bmatrix} \dot{x}_S \\ \dot{y}_S \end{bmatrix} = \begin{bmatrix} \dfrac{f}{s_x Z_C} & 0 & \dfrac{-x_S}{Z_C} & \dfrac{-Y_T x_S}{Z_C} & \dfrac{f Z_T}{s_x Z_C} + \dfrac{X_T x_S}{Z_C} & \dfrac{-f Y_T}{s_x Z_C} \\ 0 & \dfrac{f}{s_y Z_C} & \dfrac{-y_S}{Z_C} & \dfrac{-f Z_T}{s_y Z_C} - \dfrac{Y_T y_S}{Z_C} & \dfrac{X_T y_S}{Z_C} & \dfrac{f X_T}{s_y Z_C} \end{bmatrix} \begin{bmatrix} \dot{X}_T \\ \dot{Y}_T \\ \dot{Z}_T \\ \omega_{X_T} \\ \omega_{Y_T} \\ \omega_{Z_T} \end{bmatrix} \tag{6.11} $$

For this form of the Jacobian, the parameters of the Jacobian are given by

$$\phi = [f, s_x, s_y, x_S, y_S, Z_C, X_T, Y_T, Z_T]^T$$

Alternatively, the sensor coordinates may be omitted and replaced with camera frame coordinates to arrive at a Jacobian of the form

$$\begin{bmatrix} \dot{x}_S \\ \dot{y}_S \end{bmatrix} = \begin{bmatrix} \frac{f}{s_x Z_C} & 0 & \frac{-fX_C}{s_x Z_C^2} & \frac{-fX_C Y_T}{s_x Z_C^2} & \frac{fZ_T}{s_x Z_C} + \frac{fX_C X_T}{s_x Z_C^2} & \frac{-fY_T}{s_x Z_C} \\ 0 & \frac{f}{s_y Z_C} & \frac{-fY_C}{s_y Z_C^2} & \frac{-fZ_T}{s_y Z_C} - \frac{fY_C Y_T}{s_y Z_C^2} & \frac{fY_C X_T}{s_y Z_C^2} & \frac{fX_T}{s_y Z_C} \end{bmatrix} \begin{bmatrix} \dot{X}_T \\ \dot{Y}_T \\ \dot{Z}_T \\ \omega_{X_T} \\ \omega_{Y_T} \\ \omega_{Z_T} \end{bmatrix} \tag{6.12}$$

where the parameters are $\phi = [f, s_x, s_y, X_C, Y_C, Z_C, X_T, Y_T, Z_T]^T$. Either form may be desirable, depending on the design parameters desired for determining sensor placement.

Generally, several features on an object are tracked. For n feature points, the Jacobian is of the form

$$\mathbf{J}_v(k) = \begin{bmatrix} \mathbf{J}_1(k) \\ \vdots \\ \mathbf{J}_n(k) \end{bmatrix} \tag{6.13}$$

where $\mathbf{J}_i(k)$ is the Jacobian matrix for each feature given by the 2×6 matrix in (6.11) or (6.12). $\mathbf{J}_v(\phi)$ has been rewritten as $\mathbf{J}_v(k)$ in order to emphasize its time-varying nature due to the variance of x_S, y_S, and Z_C with time.

Binocular Systems with Parallel Optical Axes

In this section, the Jacobian for a stereo pair with parallel optical axes observing an object described relative to a task frame is derived. The derivation is based on equations for a stereo eye-in-hand system given in [37]. The term b represents the length of the baseline of the cameras, which is the line segment between camera focal points. The origin of the camera frame lies on the baseline midway between focal points, with the $-Z$ axis pointing toward the object task frame, as shown in Figure 6.5. The camera model is represented by

$$x_{Sl} = \frac{fX_C + b/2}{s_x Z_C} \qquad y_{Sl} = \frac{fY_C}{s_y Z_C} \tag{6.14}$$

$$x_{Sr} = \frac{fX_C - b/2}{s_x Z_C} \qquad y_{Sr} = \frac{fY_C}{s_y Z_C} \tag{6.15}$$

where it is assumed that f, s_x, and s_y are the same for both cameras. Through a derivation similar to that in Section 3.1.1, the mapping from task space velocity to sensor space velocity

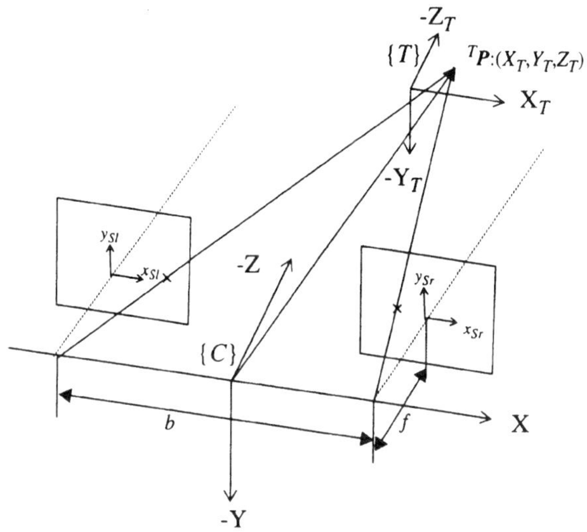

FIGURE 6.5
Task frame–camera frame definitions.

can be written as

$$
\begin{bmatrix} \dot{x}_{Sl} \\ \dot{y}_{Sl} \\ \dot{x}_{Sr} \\ \dot{y}_{Sr} \end{bmatrix} =
\begin{bmatrix}
\frac{f}{s_x Z_C} & 0 & \frac{-f(X_C+b/2)}{s_x Z_C^2} & \frac{-f(X_C+b/2)Y_T}{s_x Z_C^2} & \frac{fZ_T}{s_x Z_C} + \frac{f(X_C+b/2)(X_T+b/2)}{s_x Z_C^2} & \frac{-fY_T}{s_x Z_C} \\
0 & \frac{f}{s_y Z_C} & \frac{-fY_C}{s_y Z_C^2} & \frac{-fZ_T}{s_y Z_C} - \frac{fY_C Y_T}{s_y Z_C^2} & \frac{fY_C(X_T+b/2)}{s_y Z_C^2} & \frac{f(X_T+b/2)}{s_y Z_C} \\
\frac{f}{s_x Z_C} & 0 & \frac{-f(X_C-b/2)}{s_x Z_C^2} & \frac{-f(X_C-b/2)Y_T}{s_x Z_C^2} & \frac{fZ_T}{s_x Z_C} + \frac{f(X_C-b/2)(X_T-b/2)}{s_x Z_C^2} & \frac{-fY_T}{s_x Z_C} \\
0 & \frac{f}{s_y Z_C} & \frac{-fY_C}{s_y Z_C^2} & \frac{-fZ_T}{s_y Z_C} - \frac{fY_C Y_T}{s_y Z_C^2} & \frac{fY_C(X_T-b/2)}{s_y Z_C^2} & \frac{f(X_T-b/2)}{s_y Z_C}
\end{bmatrix}
\times \begin{bmatrix} \dot{X}_T \\ \dot{Y}_T \\ \dot{Z}_T \\ \omega_{X_T} \\ \omega_{Y_T} \\ \omega_{Z_T} \end{bmatrix} \tag{6.16}
$$

where $d = x_{Sl} - x_{Sr}$ is the disparity of each corresponding feature point. The sensor space vector contains four terms representing the optical flow of the feature in both the left and right images.

Binocular Systems with Perpendicular Optical Axes

An orthogonal stereo pair is shown in Figure 6.6. If the axes are aligned as shown in the figure, the Jacobian mapping from task space to sensor space can be written as

$$\begin{bmatrix} \dot{x}_{Sl} \\ \dot{y}_{Sl} \\ \dot{x}_{Sr} \\ \dot{y}_{Sr} \end{bmatrix} = \begin{bmatrix} \frac{f}{s_x Z_{Cl}} & 0 & \frac{-fX_{Cl}}{s_x Z_{Cl}^2} & \frac{-fX_{Cl}Y_T}{s_x Z_{Cl}^2} & \frac{fZ_T}{s_x Z_{Cl}} + \frac{fX_{Cl}X_T}{s_x Z_{Cl}^2} & \frac{-fY_T}{s_x Z_{Cl}} \\ 0 & \frac{f}{s_y Z_{Cl}} & \frac{-fY_{Cl}}{s_y Z_{Cl}^2} & \frac{-fZ_T}{s_y Z_{Cl}} - \frac{fY_{Cl}Y_T}{s_y Z_{Cl}^2} & \frac{fY_{Cl}X_T}{s_y Z_{Cl}^2} & \frac{fX_T}{s_y Z_{Cl}} \\ \frac{fX_{Cr}}{s_x Z_{Cr}^2} & 0 & \frac{f}{s_x Z_{Cr}} & \frac{fY_T}{s_x Z_{Cr}} & \frac{fX_{Cr}Z_T}{s_x Z_{Cr}^2} - \frac{fX_T}{s_x Z_{Cr}} & \frac{fX_{Cr}Y_T}{s_x Z_{Cr}^2} \\ \frac{fY_{Cr}}{s_y Z_{Cr}^2} & \frac{f}{s_y Z_{Cr}} & 0 & \frac{-fZ_T}{s_y Z_{Cr}} & \frac{fY_{Cr}Z_T}{s_y Z_{Cr}^2} & \frac{fY_{Cr}Y_T}{s_y Z_{Cr}^2} - \frac{fX_T}{s_y Z_{Cr}} \end{bmatrix} \begin{bmatrix} \dot{X}_T \\ \dot{Y}_T \\ \dot{Z}_T \\ \omega_{X_T} \\ \omega_{Y_T} \\ \omega_{Z_T} \end{bmatrix} \tag{6.17}$$

Vision Resolvability Ellipsoids

Vision resolvability measures the ability of a visual sensor to resolve object positions and orientations. For example, a typical single camera system has the ability to resolve accurately object locations that lie in a plane parallel to the image plane, but can resolve less accurately object depth based on the projection of object features on the image plane. Similarly, rotations within planes parallel to the image plane can be more accurately resolved than

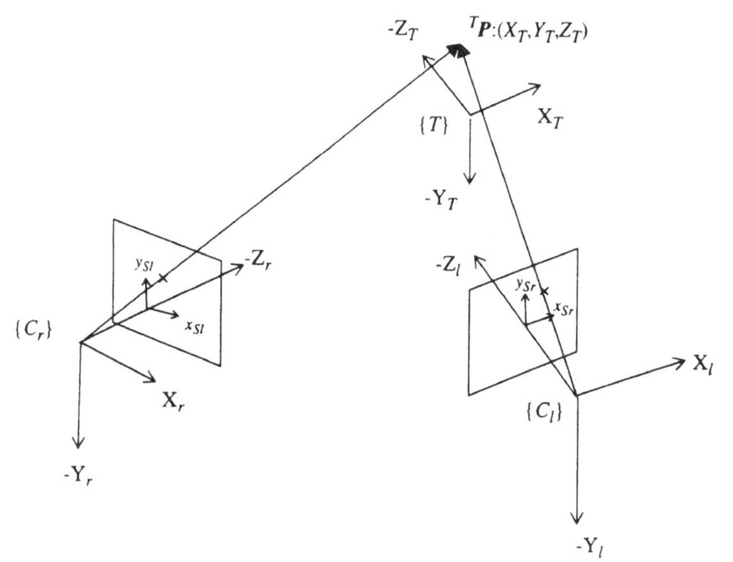

FIGURE 6.6
Task frame–camera frame definitions.

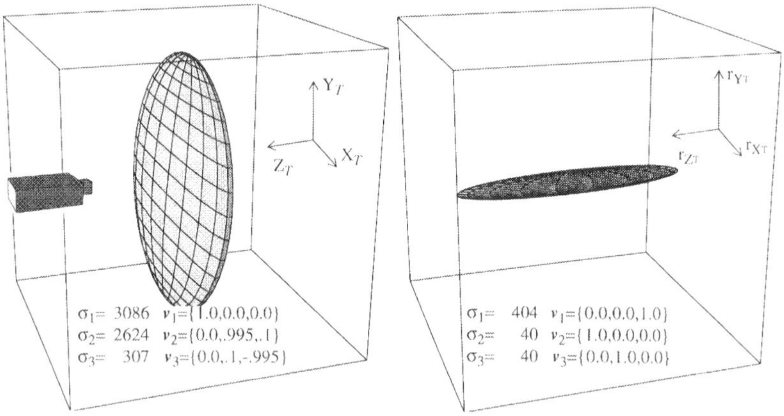

FIGURE 6.7
Resolvability ellipsoids: monocular system, $f = 24$ mm, depth = 1.0 m, two features located in the task frame at (0.1 m, 0.1 m, 0) and (−0.1 m, 0.1 m, 0).

rotations in planes perpendicular to the image plane. The degree of vision resolvability is dependent on many factors. For example, depth, focal length, number of features tracked and their image plane coordinates, position and orientation of the camera, and relative positions and orientations of multiple cameras all affect the magnitudes and directions of resolvability. Because of the difficulty in understanding the multidimensional nature of resolvability, we propose the vision resolvability ellipsoid as a geometrical representation of the ability of different visual sensor configurations to resolve object positions and orientations. To show the ellipsoidal representation, the Jacobian mapping is decomposed into two mappings, one representing translational components and one representing rotational components. In all ellipsoid plots, the singular values of $\mathbf{J}$ and eigenvectors of $\mathbf{J}^T\mathbf{J}$ are given, where σ_i is the ith singular value and v_i is the ith eigenvector. The units of σ_i are pixel/meter for the translational case and are unitless for the rotational case.

In Figures 6.7 and 6.8, ellipsoids for a monocular system are shown in which the two examples have the same magnification (f/Z_C), but the object is located at different depths.

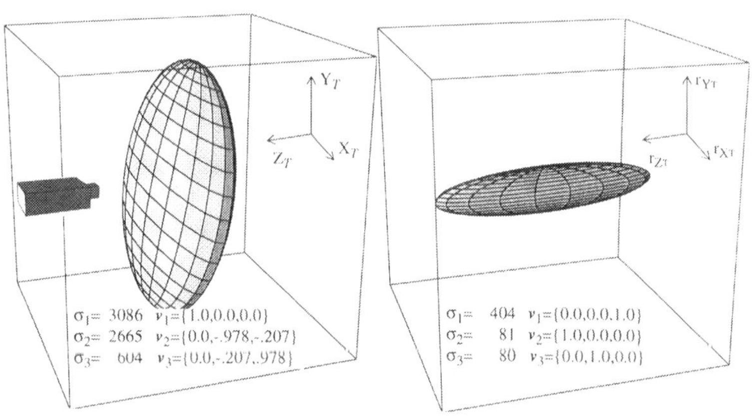

FIGURE 6.8
Resolvability ellipsoids: monocular system, $f = 12$ mm, depth = 0.5 m, two features located in the task frame at (0.1 m, 0.1 m, 0) and (−0.1 m, 0.1 m, 0).

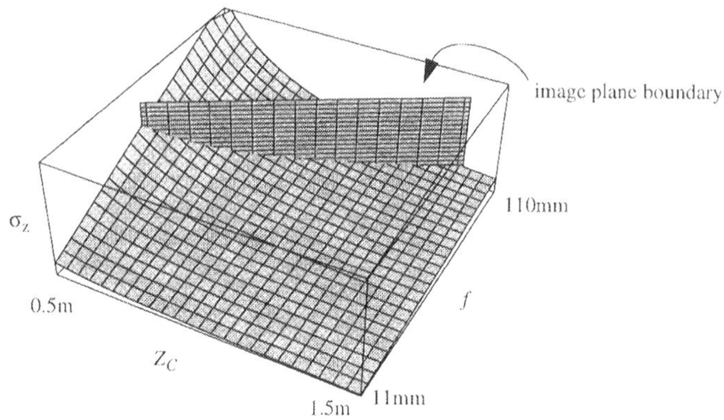

FIGURE 6.9
Resolvability of depth versus depth of object and focal length for two features located in the task frame at (0.05 m, 0, 0) and (− 0.05 m, 0, 0).

Figure 6.9 is a plot of resolvability in depth versus depth and focal length. From the plot one can observe that it is preferable to reduce the depth of an object rather than to increase focal length for a given magnification when attempting to maximize depth resolvability. In practice, depth becomes limited by the depth of field of the lens, and a trade-off must be made between focal length, depth, depth of field, and field of view [34]. Figure 6.10 shows the resolvability about the optical axes versus the position at which an object is observed on the image plane. The closer the object's projection to the boundary of the image plane, the greater the resolvability about the optical axis.

Figure 6.11 shows resolvability ellipsoids for a binocular system tracking a single feature. Depth can be resolved using a single feature, but not accurately relative to directions parallel to the image plane. Figure 6.12 shows a plot of resolvability in depth versus baseline and depth. This plot demonstrates that reducing depth is preferable to extending the baseline to improve resolvability in depth.

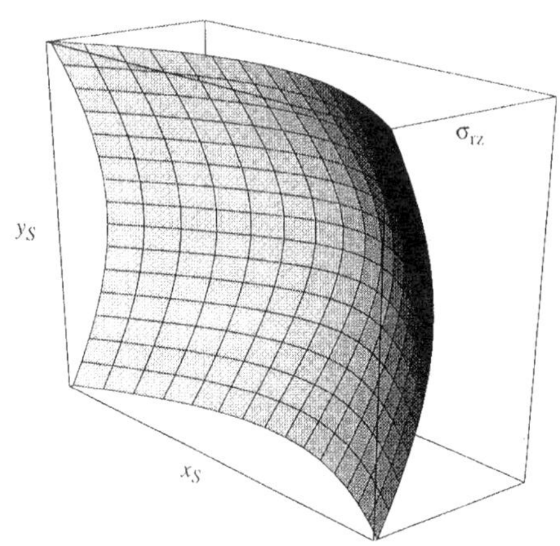

FIGURE 6.10
Resolvability in orientation about Z versus center of object projection onto the image plane.

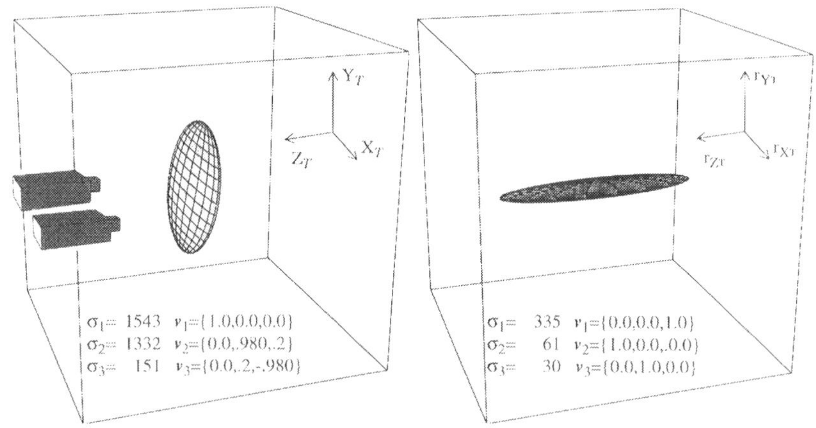

FIGURE 6.11
Resolvability ellipsoids: stereo pair — parallel optical axes, $f = 12$ mm, $b = 20$ cm, depth = 1.0 m, one feature located in the task frame at (0, 0.2 m, 0).

The resolvability ellipsoids for a binocular system with orthogonal optical axes are shown in Figure 6.13. The configuration provides a very well-conditioned Jacobian mapping from task space to sensor space, although resolvability about Y_T is still relatively low.

3.2 Force Resolvability

In order to assimilate the information provided by the disparate force and vision sensors, it is necessary to develop a model of the force sensor that allows a comparison of force and vision information. The concept of sensor resolvability is used for this comparison, in which the effect of infinitesimal task space displacements is viewed in sensor space. We desire an equation of the form

$$\delta \mathbf{x}_S = \mathbf{J}_f(\phi)\delta \mathbf{X}_T \tag{6.18}$$

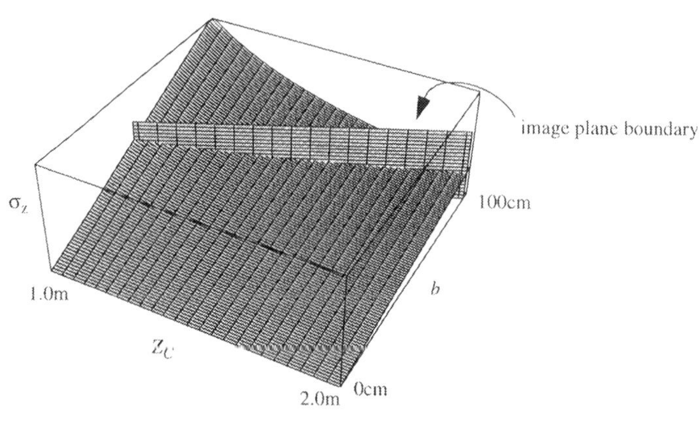

FIGURE 6.12
Resolvability in depth versus baseline length and depth of object for a stereo pair, parallel optical axes, $f = 12$ mm, and a single feature located at the origin of the task frame.

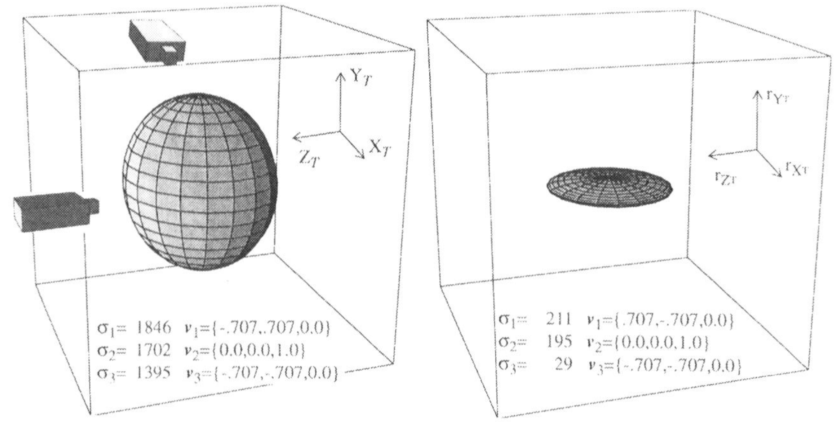

FIGURE 6.13
Resolvability ellipsoids: stereo pair—perpendicular optical axes, $f = 12$ mm, depth $= 1.0$ m, two features located in the task frame at $(-0.1\text{ m}, 0.1\text{ m}, 0)$, and $(0.1\text{ m}, -0.1\text{ m}, -0.1\text{ m})$.

where $\delta\mathbf{x}_S$ is an infinitesimal displacement vector in force sensor space; $\mathbf{J}_f(\phi)$ is the Jacobian mapping and may be time variable; and $\delta\mathbf{X}_T$ is an infinitesimal displacement vector in task space.

Figure 6.14 shows a typical wrist force sensor mounted at a manipulator end effector and the associated coordinate frame definitions. Force sensing is based on Hooke's law and is a

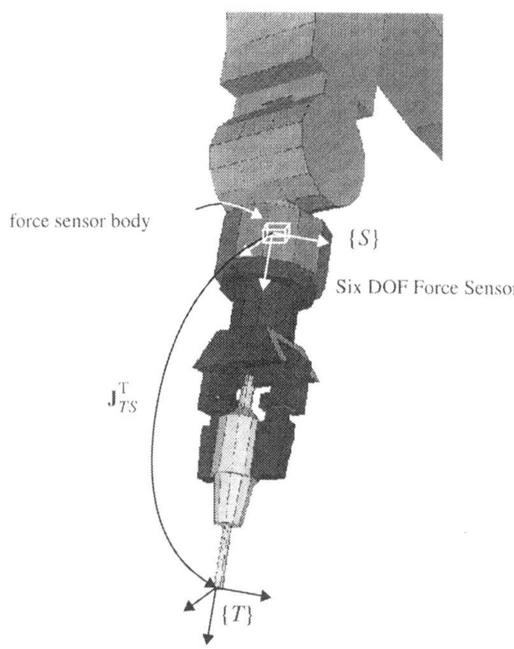

FIGURE 6.14
Coordinate frame definitions for a manipulation task that employs force sensing.

highly linear process, assuming induced strains remain within the elastic range of the material of the force sensor body. A measurement of strain $\delta\mathbf{x}_S$ taken from strain gauges mounted on the force sensor body is converted to a measurement of force $\mathbf{F}_S$ in the force sensor coordinate frame S through a force calibration matrix $\mathbf{C}_S$.

$$\delta\mathbf{x}_S = \mathbf{C}_S\mathbf{F}_S \tag{6.19}$$

$\mathbf{C}_S$ is a constant matrix that depends on the physical structure of the sensor body and the location of the strain gauges on the body. The number of strain gauges n is assumed to be greater than the number of desired force measurements m; therefore $\mathbf{C}_S^{-1}$ does not exist. Instead, the pseudoinverse of $\mathbf{C}_S$, $\mathbf{C}_S^+$, is used to obtain measured forces from measured strains. The solution to (6.19) is then

$$\mathbf{F}_S = \mathbf{C}_S^+\delta\mathbf{x}_S + (\mathbf{I} - \mathbf{C}_S^+\mathbf{C}_S)\mathbf{z} \tag{6.20}$$

where $\mathbf{z}$ is an arbitrary $m \times 1$ vector. Equation (6.20) minimizes the L_2 norm $\|\delta\mathbf{x}_S - \mathbf{C}_S\mathbf{F}_S\|$; however, the equation does not necessarily have a physical interpretation. By setting $\mathbf{z} = \mathbf{0}$, one can guarantee that $\mathbf{F}_S$ has a minimum magnitude. The result is a physically accurate solution for obtaining measured forces from measured strain gauge readings and is given by [46]

$$\mathbf{F}_S = \mathbf{C}_S^+\delta\mathbf{x}_S \tag{6.21}$$

The pseudoinverse of $\mathbf{C}_S$, $\mathbf{C}_S^+$, is

$$\mathbf{C}_S^+ = (\mathbf{C}_S^T\mathbf{C}_S)^{-1}\mathbf{C}_S^T \tag{6.22}$$

$\mathbf{F}_S$ is converted to a force $\mathbf{F}_T$ in task space T by the Jacobian mapping of the task frame with respect to the sensor frame

$$\mathbf{F}_T = \mathbf{J}_{TS}^T\mathbf{F}_S \tag{6.23}$$

Strain gauge measurements are then converted to forces in the task space via the equation

$$\mathbf{F}_T = \mathbf{J}_{TS}^T\mathbf{C}_S^+\delta\mathbf{x}_S \tag{6.24}$$

During contact stages of manipulation, particular components of $\mathbf{F}_T$ are the quantities to be controlled. However, when using force and vision feedback together, force measurements are meaningless in terms of visual feedback. Therefore, a system stiffness $\mathbf{K}$ is defined in order to arrive at a relationship between task displacement $\delta\mathbf{X}_T$ and task force $\mathbf{F}_T$.

$$\mathbf{F}_T = \mathbf{K}\delta\mathbf{X}_T \tag{6.25}$$

This formula applies to quasi-static cases only; therefore inertial and damping terms are ignored. This assumption is valid because we are concerned with the resolution of the sensor rather than its bandwidth properties.

In order to model the stiffness of the system $\mathbf{K}$, we must consider sources of compliance. We assume rigid objects are being manipulated, so no compliance exists in the objects. The sensor itself is obviously compliant, since it measures strain. Another important source of compliance is the manipulator itself. A rough stiffness analysis shows that the manipulator introduces the vast majority of task compliance; therefore we ignore sensor stiffness and concentrate on the compliance in the manipulator for the system stiffness model.

To analyze the relationship between end-effector stiffness and end-effector displacements, we use an augmented form of Kim's premultiplier diagram [52], shown in Figure 6.15. The premultiplier diagram describes the static relationships between manipulator forces and positions in both joint and end-effector coordinates for redundant and nonredundant manipulators. In addition to 6.25, the diagram also illustrates the following relationships

$$\delta\mathbf{X}_T = \mathbf{J}_M(\theta)\delta\theta \tag{6.26}$$

$$\tau = \mathbf{J}_M^T(\theta)\mathbf{F}_T \tag{6.27}$$

$$\tau = \mathbf{K}_\theta \delta\theta \tag{6.28}$$

where $\mathbf{J}_M(\theta)$ is the manipulator Jacobian matrix and varies with θ, the vector of joint positions; $\delta\theta$ is the infinitesimal displacement vector in joint space; τ is the vector of joint torques; and $\mathbf{K}_\theta$ is the joint stiffness matrix. Various vectors can be derived in terms of one another by traversing a path through the diagram and combining the proper mappings. As previously mentioned, we desire an expression for the system stiffness $\mathbf{K}$ in terms of known quantities for modeling purposes. By traversing the premultiplier diagram from $\mathbf{F}_T$ to $\delta\mathbf{X}_T$ via the joint variables τ and $\delta\theta$, we derive the expression

$$\mathbf{F}_T = \mathbf{J}_M^{-T}(\theta)\mathbf{K}_\theta\mathbf{J}_M^{-1}(\theta)\delta\mathbf{X}_T \tag{6.29}$$

$\mathbf{J}_M(\theta)$ is known because the kinematic structure of the manipulator is known. From the control law used to command joint torques, the joint stiffness $\mathbf{K}_\theta$ can be derived. For example, the most common strategy for controlling a manipulator is with inner loop PD (proportional derivative) position controllers at each joint. For a joint PD control scheme operating under quasi-static assumptions, the joint stiffness is simply the value of the proportional gain (neglecting joint friction). Therefore, the system stiffness can be expressed as

$$\mathbf{K} = \mathbf{J}_M^{-T}(\theta)\mathbf{K}_\theta\mathbf{J}_M^{-1}(\theta) \tag{6.30}$$

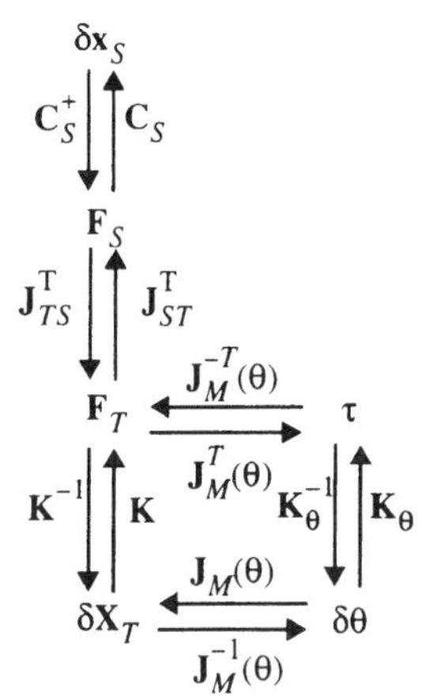

FIGURE 6.15
An augmented form of Kim's premultiplier diagram (Kim *et al.* 1992).

and depends on the configuration of the manipulator as well as the stiffness of the joint controllers.

From the augmented premultiplier diagram, the Jacobian mapping from task space to force sensor space is written as

$$\delta \mathbf{x}_S = \mathbf{C}_S \mathbf{J}_{ST}^T \mathbf{J}_M^{-T}(\theta) \mathbf{K}_\theta \mathbf{J}_M^{-1}(\theta) \delta \mathbf{X}_T \tag{6.31}$$

where

$$\mathbf{J}_f(k) = \mathbf{C}_S \mathbf{J}_{ST}^T \mathbf{J}_M^{-T}(\theta) \mathbf{K}_\theta \mathbf{J}_M^{-1}(\theta) \tag{6.32}$$

The principal components of this mapping can be used to determine the force resolvability of various sensor–manipulator–task configurations. These components are then compared with vision resolvability for assimilating information from the two disparate sensing modalities during task execution.

3.3 Fusing Vision and Force Feedback through Resolvability

In order to perform a comparison of the resolvability of force and vision feedback, the variance of sensor noise must be considered in terms of the resolvability of the sensor. For vision feedback, this variance is dependent on the tracking algorithm used, the size of the feature template, and the quality of the feature being tracked. For the experimental results to be presented in Section 6, the value is typically around 1.0 pixel. This variance is translated into the task space domain through the pseudoinverse of the image Jacobian used for vision resolvability

$$\sigma_T = \mathbf{J}_v^+(k) \sigma_S \tag{6.33}$$

where σ_T is the vector of positional variance in task space, $\mathbf{J}_v^+(k)$ is the pseudoinverse of the image Jacobian, and σ_S is the vector representing feature variance in sensor space. For the camera–lens configuration given in Figure 6.7, the task positional variance is on the order of 0.0003 m in a plane parallel to the image plane and 0.003 m along the optical axis.

To determine force sensitivity to task space displacements, we must invert the force resolvability matrix. This is written as

$$\sigma_T = \mathbf{J}_M(\theta) \mathbf{K}_\theta^{-1} \mathbf{J}_M^T \mathbf{J}_{TS}^T(\theta) \sigma_S \tag{6.34}$$

The force sensing system used to collect experimental results produces 12-bit strain gauge readings, which typically have a measured steady-state variance of 2.0 units. The stiffness of the manipulator is derived from the proportional gains on the joints of the manipulator used. These values are on the order of 1000–10000 Nm/rad for the first three Puma joints and 300–500 Nm/rad for the three wrist joints. For a typical configuration far from manipulator singularities, task space positional variances on the order of 10^{-6} m are calculated. Although the sensor is sensitive to displacements in the micron range, the noise introduced by inertial effects during manipulator motion on the strain gauge readings is significantly higher. This is discussed in more detail in Section 5.

As a task proceeds, the resolvability of the two sensors, force and vision, is continuously monitored. As a surface is approached, vision resolvability eventually becomes insufficient to provide meaningful control inputs to the manipulator Cartesian controller. This indicates that the force sensor can now provide valid feedback on the task even though contact has not actually occurred, and force sensor information should be considered as the primary sensing modality. This means that the task model must be capable of representing geometrical relationships among objects to be mated. This model could exist in 2-D image coordinates or in a 3-D world coordinate frame projected through the camera model, for example, Eqs. (6.4) and (6.5) for a monocular system.

It is important to note that at no time is the estimate of task displacement resolution used to control the task itself. The estimate is used only to compare the capabilities of each sensor. For vision resolvability, the variance of sensor noise is used as a threshold to determine when visual servoing is no longer relevant to the task. For force resolvability, the measure is used to determine the relative stiffness and force resolution of different manipulator–task configurations. The measure is also used to ensure that the resolution of the force sensor configuration provides more accurate positional feedback than the vision sensor, which is almost always the case.

From an analysis of force resolvability, it becomes evident that for a stiff manipulator very small displacements in the task frame result in relatively large measured strains in the force sensor space. If more compliant manipulator joint controllers are implemented, larger displacements in the task frame are needed in order to induce similar strains. In terms of resolvability, this means that stiff manipulators can more easily resolve task space displacements. Therefore, stiff manipulators can more accurately position objects based on force sensor readings. Of course, this is not the complete story, because stiff controllers are also much less stable in the face of modeling errors. As is the case with any control system, a trade-off must be made between performance, evaluated in this case with respect to positioning accuracy, and stability.

One should realize that the force control algorithm employed will, of course, have an effect on system stiffness. However, when evaluating force resolvability, the force control algorithm is not considered. This is because we determine when to switch to and from pure force control based on vision resolvability. Force resolvability is used only to ensure that the force sensor will provide more resolvable positional feedback than vision feedback will provide. Under visual servoing control a Cartesian velocity controller is used to drive the manipulator in Cartesian space; therefore, the system stiffness is a result of joint PD controllers driven by end-effector velocity commands. The quasi-static assumption used for the force resolvability derivation is valid, because as contact becomes imminent, the visual servoing controller reduces commanded velocities.

4 VISUAL SERVOING FORMULATION

4.1 Controller

The state equation for the visual servoing system is created by discretizing (6.13) and rewriting the discretized equation as

$$\mathbf{x}(k+1) = \mathbf{x}(k) + T\mathbf{J}_v(k)\mathbf{u}(k) \tag{6.35}$$

where $\mathbf{x}(k) \in \mathcal{R}^{2M}$ and represents the vector of feature states (i.e., feature coordinates in sensor space (M is the number of features being tracked)); T is the sampling period of the vision system; and $\mathbf{u}(k) = [\dot{X}_T \; \dot{Y}_T \; \dot{Z}_T \; \omega_{X_T} \; \omega_{Y_T} \; \omega_{Z_T}]^T$, the manipulator end-effector velocity. For the remainder of this chapter, $T\mathbf{J}_v(k)$ will be written as $\mathbf{J}_v(k)$ in order to simplify the formulas without loss of generality.

A control strategy can be derived using the controlled active vision paradigm [28]. The control objective of the visual tracking system is to control end-effector motion in order to place the image plane coordinates of features on the target at some desired position. The desired image plane coordinates could be constant or changing with time. The control strategy used to achieve the control objective is based on the minimization of an objective function at each time instant. The objective function places a cost on differences in feature positions from desired positions, as well as a cost on providing control input, and is of the form

$$F(k+1) = [\mathbf{x}(k+1) - \mathbf{x}_D(k+1)]^T \mathbf{Q}[\mathbf{x}(k+1) - \mathbf{x}_D(k+1)] + \mathbf{u}^T(k)\mathbf{L}\mathbf{u}(k) \tag{6.36}$$

where $\mathbf{x}_D(k+1)$ represents the desired feature state vector. This expression is minimized with respect to the current control input $\mathbf{u}(k)$. The end result yields the following expression for the control input:

$$\mathbf{u}(k) = -(\mathbf{J}_v^T(k)\mathbf{Q}\mathbf{J}_v(k) + \mathbf{L})^{-1}\mathbf{J}_v^T(k)\mathbf{Q}[\mathbf{x}(k) - \mathbf{x}_D(k+1)] \tag{6.37}$$

The weighting matrices $\mathbf{Q}$ and $\mathbf{L}$ allow the user to place more or less emphasis on the feature error and the control input. Their selection effects the stability and response of the tracking system. The $\mathbf{Q}$ matrix must be positive semidefinite, and $\mathbf{L}$ must be positive definite for a bounded response. Although no standard procedure exists for choosing the elements of $\mathbf{Q}$ and $\mathbf{L}$, general guidelines can be found in Papanikolopoulos *et al.* [53].

The system model and control derivation can be extended to account for system delays, modeling and control inaccuracies, and measurement noise. See Papanikolopoulos *et al.* [53] for a detailed explanation of how this can be accomplished.

4.2 Feature Tracking

The measurement of the motion of the feature on the image plane must be done continuously and quickly. The method used to measure this motion is based on optical flow techniques and is a modification of the method proposed in Anandan [54]. This technique is known as a sum-of-squares-differences (SSD) optical flow and is based on the assumption that the intensities around a feature point remain constant as that point moves across the image plane. The displacement of a point $\mathbf{p}_a = (x_S, y_S)$ at the next time increment to $\mathbf{p}_{a'} = (x_S + \delta x_S, y_S + \delta y_S)$, is determined by finding the displacement $\delta\mathbf{x}_S = (\delta x_S, \delta y_S)$ that minimizes the SSD measure

$$e(\mathbf{p}_a, \delta\mathbf{x}_S) = \sum_W [I_a(x_S + i, y_S + j) - I_{a'}(x_S + i + \delta x_S, y_S + j + \delta y_S)]^2 \tag{6.38}$$

where I_a and $I_{a'}$ are the intensity functions from two successive images and W is the window centered about the feature point that makes up the feature template. For the algorithm implemented, W is 16×16 pixels, and possible displacements of up to $x_S = y_S = 32$ pixels are considered. Features on the object that are to be tracked can be selected by the user, or a feature-selecting algorithm can be invoked. Features with strong intensity gradients in perpendicular directions, such as corners, are typically the best features to select.

In order to decrease the search space, a pyramidal search scheme, shown in Figure 6.16, has been implemented. The scheme first searches a coarse resolution of the image that has 1/16 the area of the original image, using a feature template in which a W that is originally

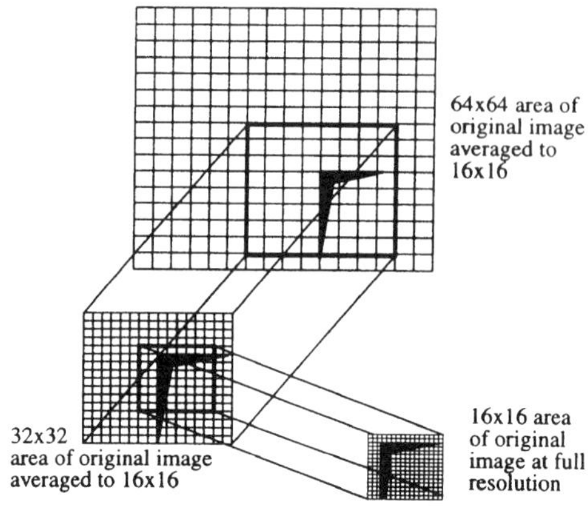

FIGURE 6.16
A pyramidal search scheme is used in the SSD optical flow algorithm in order to increase the overall sampling rate of the system.

32×32 is averaged to 8×8. After determining where the feature is in the coarse image, a finer resolution image that is 1/4 the original spatial resolution is searched with an original W of 16×16 that is averaged to 8×8 in an area centered about the location of the minimum SSD measure found in the coarse image. Finally, the full resolution image and the 16×16 feature template are used to pinpoint the location of the displaced feature.

The pyramidal scheme reduces the time required for the computation of the SSD algorithm by a factor of 5 for a single feature over the method of computing the feature locations at the full resolution alone. However, reliability can be sacrificed when the selected feature loses its tracking properties (strong perpendicular intensity gradients) at the coarser image resolutions. Since the search scheme first estimates where the feature is located based on the coarse image, it is critical that good features at coarse resolutions are tracked. When a user selects features, it is often not obvious that a particular feature may lose its tracking characteristics at coarse resolutions. Because of this, an automatic feature selector has been implemented based on Tomasi and Kanade [55] that accounts for the different levels of resolution in the pyramidal search scheme.

Depending on the tracking strategy chosen, the depth of the object from the camera may change in order to maximize the distance of the manipulator from singularities and joint limits. This slowly changes the size of the feature template based on the projection equations. In order to account for this change, the feature template can be periodically updated by using the matched feature window from a recent image as the new feature template.

5 VISION–FORCE SERVOING

In order to illustrate the advantages of assimilating disparate sensor feedback using our proposed method, we experimentally demonstrate the performance of the technique during contact transitions. To perform robotic manipulation tasks in uncertain environments quickly and efficiently, a robotic end effector must be able to approach successfully and contact objects rapidly using sensor feedback. In a rigid environment, this is difficult. With a stiff manipulator this becomes even more difficult because neither the surface nor the manipulator is able to dissipate excess energy easily upon contact. However, the most common form of force control uses a proportional derivative strategy with high proportional gains and low damping, which is an inherently stiff system. This type of force control is popular because it is simple to implement, choosing gains is easy, and it achieves a relatively high bandwidth once contact is successfully made. The problem with this strategy, however, is in achieving initial contact quickly and stably while maintaining low impact forces. Many researchers have studied this problem, and various impact strategies have been proposed, as discussed in Section 2.1. However, the fundamental problem of using force feedback alone to minimize impact forces while quickly achieving contact stably within imprecisely calibrated environments still exists. By combining vision feedback with force feedback using the concept of resolvability in a nonlinear control strategy, we demonstrate that fast stable contact transitions with a stiff manipulator in a rigid environment can be achieved.

The force control portion of our proposed visual–force servoing strategies is based on past work on hybrid force control. The implemented force control scheme is a combination of hybrid force–position control [11] and damping force control [56], resulting in a hybrid force–velocity control scheme. Because the dynamics, particularly friction, of the laboratory robot (a Puma 560) are difficult to accurately model, a simple Cartesian control scheme is used in which a manipulator Jacobian inverse converts Cartesian velocities to joint velocities, which are then integrated to joint reference positions. High servo rate (500 Hz) PD controllers are implemented for each joint in order to follow joint trajectories that achieve the desired Cartesian motion.

If simple force damping control is used to strike surfaces, a manipulator can easily become unstable unless force gains are tuned to extremely low values, resulting in unacceptably slow motion during the approach phase of the task. Because of this, most manipulation strategies use a guarded move to initiate contact with a surface. During a guarded move, surfaces are approached under position control while the force sensor is monitored. If the sensed force exceeds a threshold, motion is immediately stopped and a force control strategy can then be invoked. The main limitation of this strategy is that high contact forces can result unless the effective mass of the manipulator is low so that the end effector can be quickly stopped before contact forces increase significantly.

The proper use of visual feedback can overcome the problems exhibited by guarded move and pure force control strategies upon impact. Visual servoing improves manipulator performance during contact transitions by incorporating information regarding the proximity of the surface with which contact is to be made in the manipulator feedback loop. When the end effector is far from a surface, visual servoing commands fast end-effector motion. The speed of the approach decreases as the end effector comes closer to the surface. Contact can then be initiated stably through the use of low-gain force contollers. A generic control framework for visual–force servoing is shown in Figure 6.17.

A fundamental problem when sharing control between force and vision sensors is due to end-effector inertial effects. Because force sensors measure all forces (inertial, gravitational, and tactile), the inertial coupling of the end-effector mass beyond the sensor introduces inertial forces into force sensor readings. When the vision system commands motions, the resulting accelerations cause unstable excitations of the force control system. In order to compensate for the unstable excitations, it is necessary to develop robust strategies for avoiding the excitations. Thresholding of force readings is not feasible, because inertial effects can often be as large as desired contact forces. Figure 6.18 shows the magnitude of experimentally determined inertial forces and the associated measured Cartesian accelerations that cause these forces.

We have developed a robust vision–force control strategy based on the fact that large accelerations induce inertial forces. If visual servoing results in measurable end-effector accel-

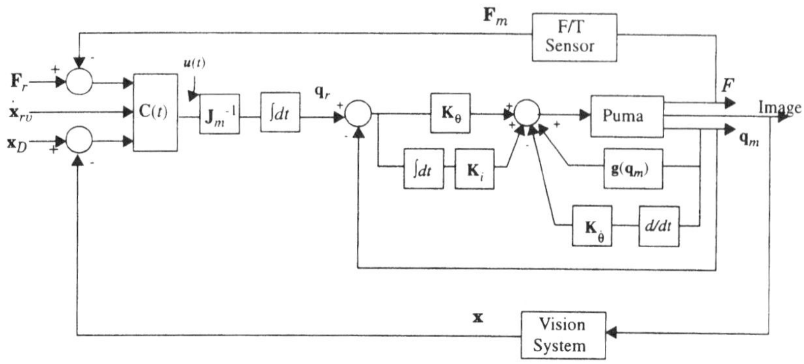

FIGURE 6.17
Force and vision in the feedback loop.

erations of sufficient magnitude, then force readings in directions opposite to these accelerations are induced. Because measured Cartesian accelerations are derived from joint encoder readings, thus requiring two differentiations of measured joint values and a transformation from joint space to Cartesian space, measured Cartesian accelerations are noisy. Therefore, we also consider the measured direction of end-effector motion. If measured Cartesian accelerations have been induced by visual servoing and if a measurable Cartesian velocity exists, then sensed forces must be due to inertial coupling, and force control commands should be ignored. This strategy can be written as

$$\dot{\mathbf{x}}_{ref_v} = -(\mathbf{J}_v^T(k)\mathbf{Q}\mathbf{J}_v(k) + \mathbf{L})^{-1}\mathbf{J}_v^T(k)\mathbf{Q}[\mathbf{x}(k) - \mathbf{x}_D(k+1)]$$

$$\dot{\mathbf{x}}_{ref_f} = \mathbf{S}_F\mathbf{G}_F(\mathbf{F}_r - \mathbf{F}_m(k))$$

for each axis, i {
  if$(((|\ddot{x}_{m_i}| > \varepsilon_a)$ and $(\dot{x}_{m_i} sign(F_{m_i}) < \varepsilon_v))$
    or $(\dot{x}_{ref_{vi}} F_{m_i} > 0.0)$ or $(|F_{m_i}| < F_T))$
   $\mathbf{S}_v[i,i] = 1.0 \quad \mathbf{S}_F[i,i] = 0.0$
  else
   $\mathbf{S}_v[i,i] = 0.0 \quad \mathbf{S}_F[i,i] = 1.0$
}

$$\mathbf{u}(k) = \mathbf{S}_v\dot{\mathbf{x}}_{ref_v} + \mathbf{S}_r\dot{\mathbf{x}}_r + \mathbf{S}_F\dot{\mathbf{x}}_{ref_f} \tag{6.39}$$

where $\mathbf{x}$ is the feature vector representing the object being servoed; $\mathbf{x}_D$ represents a state in feature space that will bring the object being servoed into contact with some surface; $\mathbf{S}_F$ is the matrix that selects axes along which force control will be applied; $\mathbf{G}_F$ is the matrix of force control gains; $\mathbf{F}_r$ and $\mathbf{F}_m$ represent reference and measured forces with respect to the task coordinate frame $\mathbf{T}$; $\dot{x}_{m_i}$ and $\ddot{x}_{m_i}$ represent measured Cartesian velocities and accelerations of the end effector in task space; $\dot{\mathbf{x}}_r$ is some desired reference end-effector velocity due, for example, to trackball input from a teleoperator; $\mathbf{S}_r$ is the matrix that selects axes along which this input will be applied; and ε_a, ε_v, and F_T threshold sensor noise.

For teleoperation tasks guided by visual servoing, compliant contact with the environment will occur if (6.39) is used alone, assuming the teleoperator adjusts the desired visual feature state to be at or below the surface to be touched. For autonomous manipulation, however, this strategy does not ensure that contact will occur if the actual location of the surface is

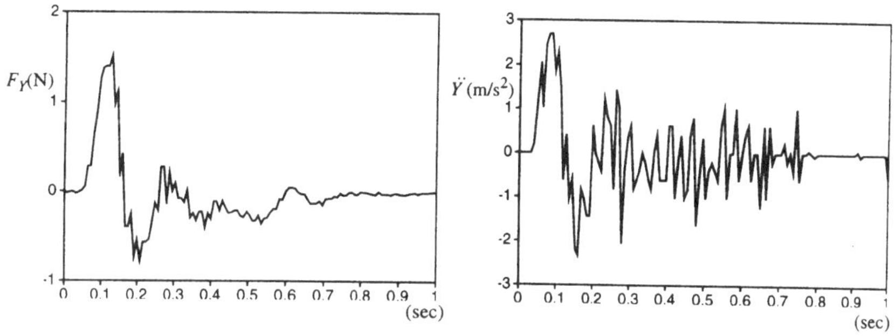

FIGURE 6.18
Inertial forces measured by the force sensor and the corresponding measured Cartesian accelerations that induced these forces.

beyond the visual estimate of the surface. During autonomous manipulation, the strategy given by (6.39) must be rewritten as

$$\mathbf{u}(k) = \begin{cases} (6.39), & \|\mathbf{x}_D(k) - \mathbf{x}(k)\| > \varepsilon \\ \mathbf{S}_F \mathbf{G}_F(\mathbf{F}_r - \mathbf{F}_m(k)), & \|\mathbf{x}_D(k) - \mathbf{x}(k)\| \leqslant \varepsilon \end{cases} \tag{6.40}$$

in order to ensure that contact will occur. Manipulator motion is first controlled by the strategy given in (6.39). The controller then switches to pure force control if the error between desired and measured visual feature states converges to within a threshold. This threshold is derived from the variance of the noise in the vision sensor,

$$\varepsilon = 2.0\|\sigma_S\| \tag{6.41}$$

where σ_S is the feature variance vector on the image plane and is determined experimentally. Stable impact with a surface can then be achieved, large contact forces can be minimized, and bounce can be avoided.

6 EXPERIMENTAL RESULTS

6.1 Hardware Setup

The vision–force servoing algorithms previously described have been implemented on a robotic assembly system consisting of three Puma 560s called the Troikabot. The Pumas are controlled from a VME bus with two Ironics IV-3230 (68030 CPU) processors, an IV-3220 (68020 CPU) processor that also communicates with a trackball, a Mercury floating point processor, and a Xycom parallel I/O board communicating with three Lord force sensors mounted on the Pumas' wrists. All processors on the controller VME run the Chimera 3.0 reconfigurable real-time operating system [57]. An adept robot is also used for providing accurate target motion. The entire system is shown in Figure 6.19.

A diagram of the hardware setup is shown in Figure 6.20. The vision system VME communicates with the controller VME using BIT3 VME-to-VME adapters. The Datacube Maxtower Vision System calculates the optical flow of the features using the SSD algorithm discussed in Section 4.2. A special high-performance floating-point processor on the

FIGURE 6.19
Laboratory setup used for performing vision–force servoing experiments.

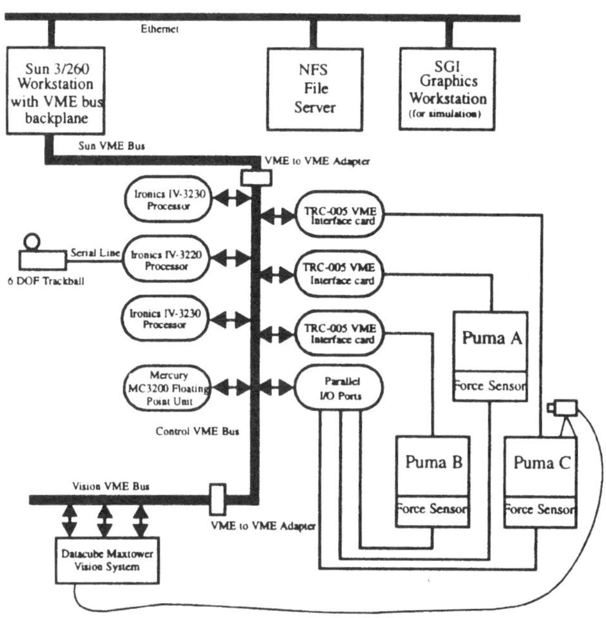

FIGURE 6.20
The Troikabot system architecture.

Datacube is used to calculate the optical flow of features, and a 68030 board, also on the vision system, computes the control input. An image can be grabbed and displacements for up to five 16 × 16 features in the scene can be determined at 30 Hz. A Lord model 15/50 force sensor provides force and torque values for each Cartesian axis at 100 Hz.

6.2 Results

Throughout this section, experimental results given will be referenced to the coordinate frames shown in Figure 6.21. For the initial set of experiments, the results of three trials are shown in which the desired goal position for the visual servoing strategy is purposely chosen to have differing magnitudes of error with respect to the true location of the surface. A final contact force of −2 N is desired. This allows us to evaluate the ability of our force–vision control strategy (6.40) to operate under conditions in which force information and vision information significantly disagree. Figure 6.22 shows the motion of the end effector on the image plane for the three trials. For trials 2 and 3 the desired image plane position of the end effector actually falls beneath the true surface. In trial 2 the error in surface position is 15 pixels, and in trial 3 the error is 45 pixels. For trial 1 the estimate of the surface and the true location are in close agreement, as would normally be the case.

In trials 2 and 3, the end effector strikes the surface after approximately 0.3 s, when motion of the end effector on the image plane abruptly stops. For trial 1, the surface is not touched until after approximately 0.5 s, because the manipulator purposely slows down before impact. The force plot in Figure 6.23 shows that this results in significantly reduced impact forces and a much quicker transition to the desired contact force of −2 N. When visual feedback incorrectly estimates the location of the surface, as in the case of the second and third trials, high contact forces occur. If the error in the estimate falls within the surface, as in trials 2 and 3, then the poorer the estimate of the surface, the higher the contact force because the

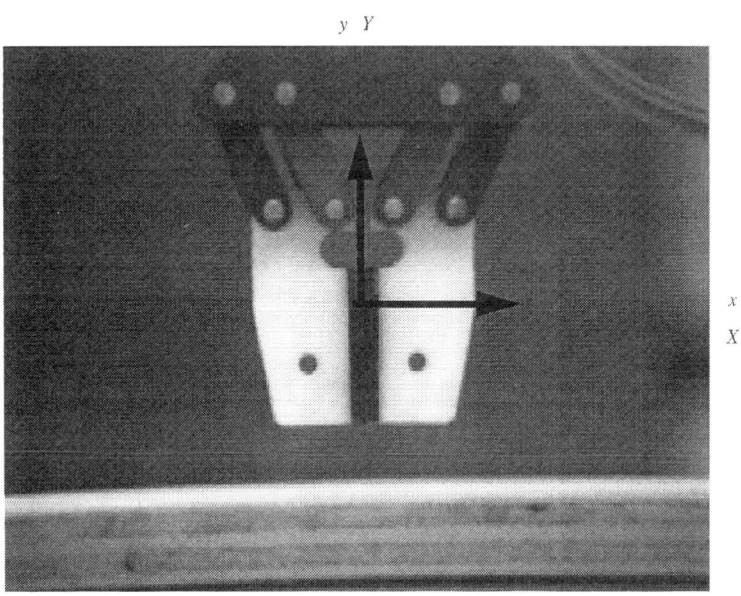

FIGURE 6.21
The camera view for visual servoing and coordinate axes on the image plane and in the task frame.

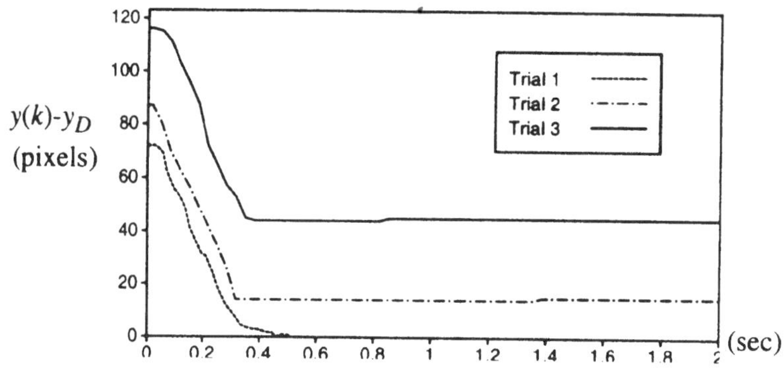

FIGURE 6.22

Vertical error between desired and measured end-effector position on the image plane for three different trials, each with a different error in the estimated location of the surface.

higher the commanded visual servoing velocity at impact. If the error in the surface location estimate is in the other direction, then the time it takes to initiate contact would increase directly with the magnitude of the error. The impact force, however, would be on the order of trial 1's impact force.

The commanded end-effector velocity for all three trials is shown in Figure 6.24. The solid lines correspond to (6.40), the dashed lines to the visual servoing velocity $\dot{\mathbf{x}}_{ref_v}$, and the dotted–dashed lines to the force servoing velocity $\dot{\mathbf{x}}_{ref_f}$. Visual servoing brings the end effector quickly toward the surface, and upon contact force servoing takes over. From the force plot in Figure 6.23, one can observe measurable inertial forces before contact actually occurs. These forces are of a magnitude greater than 1.5 N; however, our proposed control strategy (6.40) successfully rejects these observed forces because they are not the result of contact. From Figure 6.24, one can see that end-effector velocities have been clipped at 0.10 m/s. This is because the feature tracker can track only objects with a limited optical flow. Thus, the trial in which the surface location is in error of 45 pixels represents the worst-case impact force, because the manipulator is traveling at approximately 0.10 m/s at the time of impact. For these experimental results, force gains of 0.001 (m/s)/N were used, the diagonal elements of $\mathbf{Q}$ were chosen to be 2.0×10^{-6}, and the diagonal elements of $\mathbf{L}$ were chosen to be 10.0. Thresholds were experimentally chosen to be $\varepsilon_a = 0.01\ \text{m/s}^2$, $\varepsilon_v = 0.001$ m/s, and $F_T = 1.5$ N.

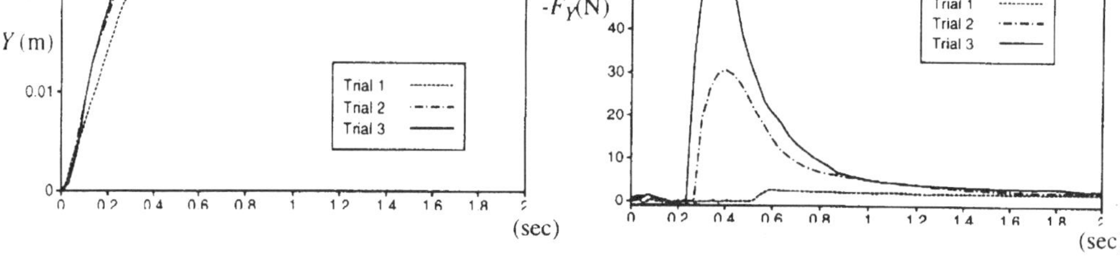

FIGURE 6.23

Vertical position of end effector in Cartesian space and force measured along the vertical direction versus time for all three trials.

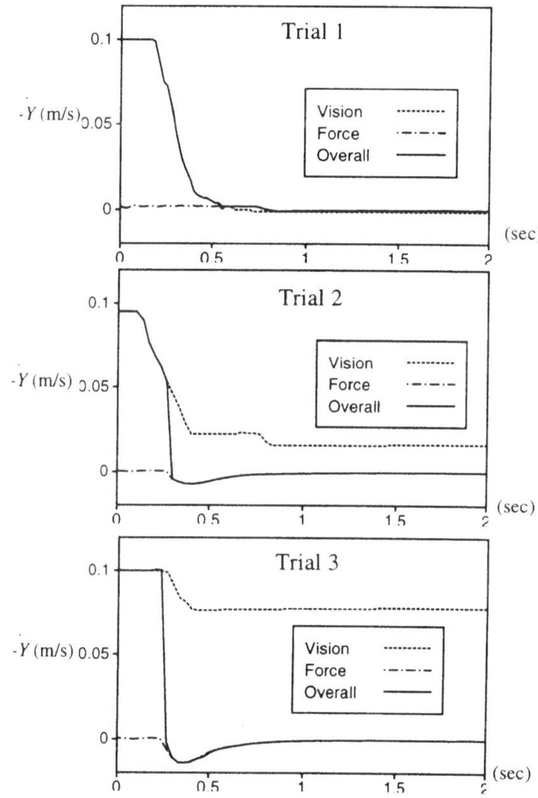

FIGURE 6.24

Commanded end-effector Cartesian velocity along $-Y$ for the three trials with varying error magnitudes in the estimated surface location. "Vision" corresponds to $\dot{\mathbf{x}}_{ref_v}$, "Force" corresponds to $\dot{\mathbf{x}}_{ref_f}$, and "Overall" corresponds to $u_y(t)$, in (6.39).

A second set of experimental results was collected in order to illustrate the advantages of our proposed force–vision strategy over two other common impact strategies, the guarded move and pure force control. Figure 6.25 shows results in which our proposed force–vision servoing algorithm (6.40) is used to servo the end effector to a surface 5.9 cm from the initial end-effector position. A force of -2 N between the end effector and the surface is maintained after contact. This strategy achieves contact after 1.43 s and achieves a stable -2 N contact force after approximately 4.5 s. With simple damping force control alone, the manipulator travels 5.9 cm in 3.1 s before reaching the surface. As soon as contact is made with the surface, the manipulator becomes unstable, as Figure 6.26 shows. The only way to achieve stable contact using damping control alone, given the force control implementation used, is to reduce the force gains to extremely low values, resulting in unacceptably slow motion. Figure 6.27 shows a force plot of a guarded move in which the force sensor is monitored at 100 Hz. High contact forces are created because of the finite amount of time required to stop the end effector after contact is sensed, illustrating the main limitation of a guarded move strategy.

Figure 6.28 shows a comparison of the motion and force time histories for the three impact strategies. The gains used in the force–vision control strategy are the same as the gains used in the previous set of experiments, including the force gain of 0.001 (m/s)/N. For force control

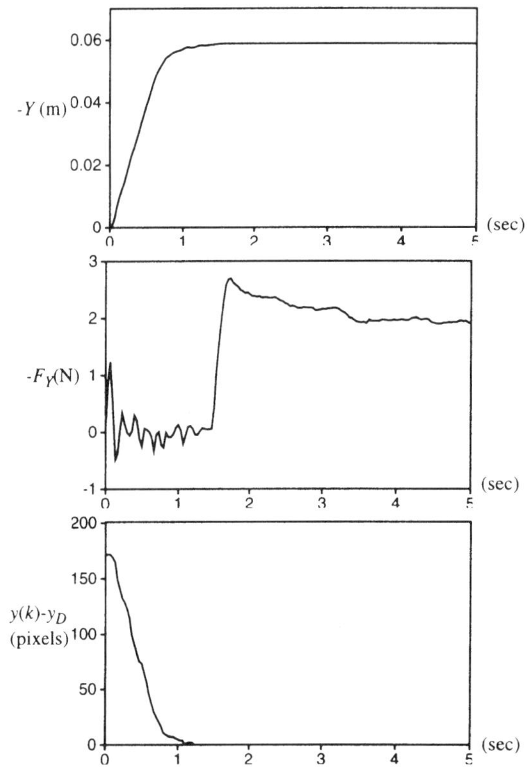

FIGURE 6.25

Vertical position of end effector in Cartesian space, force measured along the vertical direction, and pixel error versus time for force–vision control.

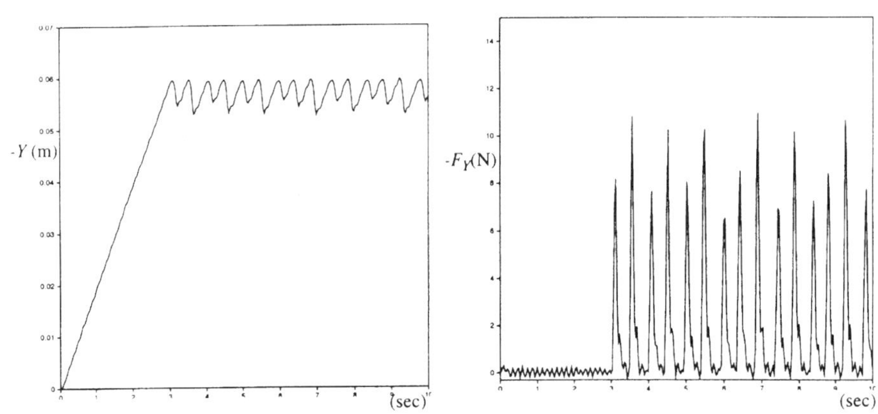

FIGURE 6.26

Vertical position of end effector in Cartesian space and force measured along the vertical direction versus time for simple damping force control upon impact with a surface.

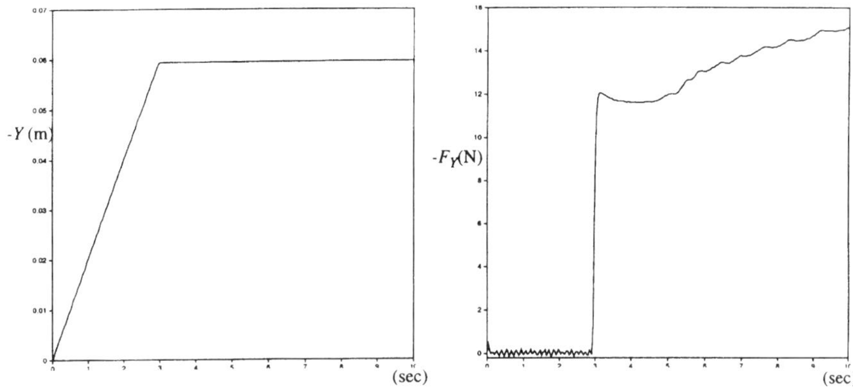

FIGURE 6.27
Vertical position of end effector in Cartesian space and force measured along the vertical direction versus time for a guarded move impact strategy that switches to position control.

alone, higher force gains (0.005 (m/s)/N had to be chosen in order to induce end-effector motion of a reasonable speed in free space, but this gain, while resulting in less than half the speed of visual servoing, still proved to be highly unstable. The guarded move strategy also allowed only moderate speeds (0.02 m/s) and still resulted in high impact forces. At higher speeds, extremely high impact forces would result, which could easily have damaged the manipulator. Using visual servoing to bring the manipulator near the surface provides a simple technique for slowing the end effector before contact is imminent. These results clearly show that visual servoing greatly simplifies the impact problem by providing low-level feedback on the proximity of the surface to the end effector. The result is a high approach velocity that generates low impact forces with no bounce.

7 CONCLUSION

Force and vision sensors provide complementary information, yet they are fundamentally different sensing modalities. This implies that traditional sensor integration techniques that require common data representations are not appropriate for combining the feedback from these two disparate sensors. In this chapter, vision and force sensor resolvabilities have been used to compare the ability of the two sensing modes to provide useful information during

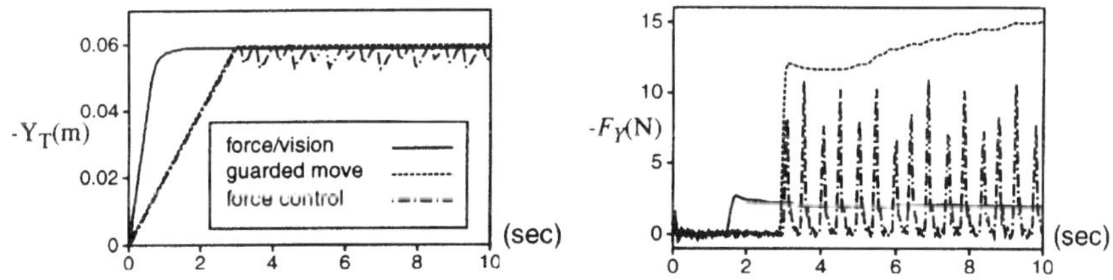

FIGURE 6.28
Combined plots of vertical position and measured force during impact for force–vision control, damping force control, and a guarded move.

robotic manipulation tasks. By monitoring the resolvability of the two sensing modes with respect to the task, the information provided by the disparate sensors can be seamlessly assimilated during task execution. A nonlinear force–vision servoing algorithm that uses force and vision resolvability to switch between sensing modes demonstrates the advantages of the assimilation technique. Contact transitions between a stiff manipulator and rigid environment, a system configuration that easily becomes unstable when force control alone is used, are robustly achieved. Experimental results show that the nonlinear controller is able to satisfy simultaneously the conflicting task requirements of fast approach velocities, maintaining stability, minimizing impact forces, and suppressing bounce between contact surfaces. The proper assimilation of force and vision feedback is the key to the success of this strategy.

REFERENCES

[1] R. C. Smith and P. Cheeseman. On the representation and estimation of spatial uncertainty. *Int. J. Robot. Res.* 5(4):56–68, 1986.

[2] H. F. Durrant-Whyte. *Integration, Coordination and Control of Multi-Sensor Robot Systems.* Kluwer Academic Publishers, Boston, 1988.

[3] J. M. Richardson and K. A. Marsh. Fusion of multisensor data. *Int. J. Robot. Res.* 7(6):78–96, 1988.

[4] G. D. Hager. *Task-Directed Sensor Fusion and Planning — A Computational Approach*, Kluwer Academic Publishers, 1990.

[5] R. Jain. Environment models and information assimilation. Tech. Rept. RJ 6866 (65692), IBM, Yorktown Heights, NY, 1989.

[6] B. Nelson and P. K. Khosla. Integrating sensor placement and visual tracking strategies. In *Experimental Robotics III: The Third International Symposium*, Kyoto, October 28–30, 1993, eds. T. Yoshikawa and F. Miyazaki. Springer-Verlag, London, 1994, pp. 169–181.

[7] B. J. Nelson and P. K. Khosla. The resolvability ellipsoid for visual servoing. *Proceedings of the 1994 Conference on Computer Vision and Pattern Recognition (CVPR94)*, 1994, pp. 829–832.

[8] B. J. Nelson and P. K. Khosla. Visually servoed manipulation using an active camera. *Proceedings Thirty-Third Annual Allerton Conference on Communication, Control, and Computing.* University of Illinois at Urbana-Champaign, Oct. 4–6, 1995.

[9] D. E. Whitney. Historical perspective and state of the art on robot force control. *Proceedings 1985 International Conference on Robotics and Automation.* IEEE, New York, 1985, pp. 262–268.

[10] N. Hogan. Impedance control: An approach to manipulation, Parts I–III. *ASME J. Dyn. Syst. Meas. Control* 107(1):1–24, 1985.

[11] M. H. Raibert and J. J. Craig. Hybrid position/force control of manipulators. *ASME J. Dyn. Syst. Meas. Control* 103(2):126–133, 1981.

[12] T. Yoshikawa. Dynamic hybrid position/force control of robot manipulators — description of hand constraints and calculation of joint driving force. *IEEE J. Robot. Automat.* 3(5):386–392, 1987.

[13] R. A. Volpe and P. K. Khosla. Experimental verification of a strategy for impact control. *Proceedings 1991 IEEE International Conference on Robotics and Automation.* IEEE, New York, 1991, pp. 1854–1860.

[14] C. An and J. Hollerbach. Dynamic stability issues in force control of manipulators. *Proceedings 1987 International Conference on Robotics and Automation.* IEEE, New York, 1987, pp. 860–896.

[15] S. Eppinger and W. Seering. Understanding bandwidth limitations on robot force control. *Proceedings 1987 International Conference on Robotics and Automation.* IEEE, New York, 1987, pp. 904–909.

[16] J. M. Hyde and M. R. Cutkosky. Contact transition: an experimental study. *Proceedings 1993 International Conference on Robotics and Automation.* IEEE, New York, 1993, pp. 363–368.

[17] N. Hogan. Stable execution of contact tasks using impedance control. *Proceedings 1987 International Conference on Robotics and Automation.* IEEE, New York, 1987, pp. 1047–1054.

[18] H. Kazerooni. Robust, non-linear impedance control for robot manipulators. *Proceedings 1987 International Conference on Robotics and Automation.* IEEE, New York, 1987, pp. 741–750.

[19] O. Khatib and J. Burdick. Motion and force control for robot manipulators. *Proceedings 1986 International Conference on Robotics and Automation.* IEEE, New York, 1986, pp. 1381–1386.

[20] H. P. Qian and J. De Schutter. Introducing active linear and nonlinear damping to enable stable high gain

force control in case of stiff contact. *Proceedings 1992 International Conference on Robotics and Automation*. IEEE, New York, 1992, pp. 1374–1379.

[21] Y. Xu, J. M. Hollerbach, and D. Ma. Force and contact transient control using nonlinear PD control. *Proceedings 1994 International Conference on Robotics and Automation*. IEEE, New York, 1994, pp. 924–930.

[22] S. Shirai and H. Inoue. Guiding a robot by visual feedback in assembling tasks. *Pattern Recogn.* 5:99–108, 1973.

[23] L. E. Weiss. Dynamic visual servo control of robots: An adaptive image-based approach. Ph.D. thesis CMU-RI-TR-84-16, Pittsburgh, Robotics Institute, Carnegie Mellon University, 1984.

[24] L. E. Weiss, A. C. Sanderson, and C. P. Neuman. Dynamic sensor-based control of robots with visual feedback. *IEEE J. Robot. Automat.* RA-3(5):404–417, 1987.

[25] P. K. Allen. Real-time motion tracking using spatio-temporal filters. *Proceedings Image Understanding Workshop*. Morgan Kaufmann, San Mateo, CA, 1989, pp. 695–701.

[26] P. I. Corke and R. P. Paul. Video-rate visual servoing for robots. *Lecture Notes in Control and Information Science*, eds. V. Hayward and O. Khatib. Springer-Verlag, London, 1989, pp. 429–451.

[27] J. T. Feddema and C. S. G. Lee. Adaptive image feature prediction and control for visual tracking with a hand-eye coordinated camera. *IEEE Trans. Syst. Man Cybernet.* 20:1172–1183, 1940.

[28] N. P. Papanikolopoulos, P. K. Khosla, and T. Kanade. Adaptive robotic visual tracking. *Proceedings of the American Control Conference*. Evanston, IL, American Autom. Control Council, 1991, pp. 962–967.

[29] B. Espiau, F. Chaumette, and P. Rives. A new approach to visual servoing in robotics. *IEEE Trans. Robot. Automat.* 8(3):313–326, 1992.

[30] B. K. Ghosh, M. Jankovic, and Y. T. Wu. Some problems in perspective system theory and its application to machine vision. *Proceedings 1992 IEEE/RSJ International Conference on Intelligent Robots and Systems* (IROS92). IEEE, New York, 1992, pp. 139–146.

[31] K. Hashimoto and H. Kimura. LQ optimal and nonlinear approaches to visual servoing. In *Visual Servoing: Real-Time Control of Robot Manipulators Based on Visual Sensory Feedback*, ed. K. Hashimoto. World Scientific, London, 1993, pp. 165–198.

[32] W. J. Wilson. Visual servo control of robots using Kalman filter estimates of robot pose relative to work-pieces. In *Visual Servoing: Real-Time Control of Robot Manipulators Based on Visual Sensory Feedback*, ed. K. Hashimoto. World Scientific, London, 1993, pp. 71–104.

[33] A. J. Koivo and N. Houshang. Real-time vision feedback for servoing of a robotic manipulator with self-tuning controller. *IEEE Trans. Syst. Man Cybernet.* 21(1):134–142, 1991.

[34] B. Nelson, N. P. Papanikolopoulos, and P. K. Khosla. Visual servoing for robotic assembly. In *Visual Servoing Real-Time Control of Robot Manipulators Based on Visual Sensory Feedback*, ed. K. Hashimoto. World Scientific, River Edge, NJ, 1993, pp. 139–164.

[35] A. Castaño and S. Hutchinson. Visual compliance: Task-directed visual servo control. *IEEE Trans. Robot. Automat.* 10:334–342, 1994.

[36] G. D. Hager. Robot feedback control based on stereo vision: towards calibration-free hand-eye coordination. *Proceedings 1994 International Conference on Robotics and Automation*. IEEE, New York, 1994, pp. 2850–2856.

[37] N. Maru, H. Kase, S. Yamada, A. Nishikawa, and F. Miyazaki. Manipulator control by visual servoing with the stereo vision. *Proceedings of the 1993 IEEE/RSJ International Conference on Intelligent Robots and System (IROS-93)*, 1993, pp. 1866–1870.

[38] K. Hosoda and M. Asada. Versatile visual servoing without knowledge of true Jacobian. *Proceedings 1994 IEEE/RSJ International Conference on Intelligent Robots and Systems (IROS94)*. IEEE, New York, 1994, pp. 186–193.

[39] C. K. Cowan and P. D. Kovesi. Automatic sensor placement from vision task requirements. *IEEE Trans. Pattern Anal. Machine Intell.* 12:407–416, 1988.

[40] K. Tarabanis, R. Y. Tsai, and P. K. Allen. Satisfying the resolution constraint in the "MVP" machine vision planning system. *Proceedings of the 1990 Darpa Image Understanding Workshop*, 1990, pp. 850–860.

[41] S. Yi, R. M. Haralick, and L. G. Shapiro. Automatic sensor and light positioning for machine vision. *Proceedings of the 10th International Conference on Pattern Recognition*, 1990, pp. 55–59.

[42] S. Das and N. Ahuga. A comparative study of stereo, vergence, and focus as depth cues for active vision. *Proceedings of the 1992 Computer Vision and Pattern Recognition Conference* (CVPR-92), 1993, pp. 194–199.

[43a] R. Sharma and S. Hutchinson. On the observability of robot motion under active camera control. *Proceedings 1994 International Conference on Robotics and Automation*. IEEE, New York, 1994, pp. 162–167.

[43b] R. Sharma and S. Hutchinson. Optimizing hand-eye configuration for visual-servo systems. *Proceedings 1995 International Conference on Robotics and Automation*. IEEE, New York, 1995, pp. 172–177.

[44] T. Yoshikawa. Manipulability of robotic mechanisms. In *Robotic Research 2*, eds. H. Hanafusa and H. Inoue, MIT Press, Cambridge, MA, 1995, pp. 439–446.

[45] Y. Nakamura, T. Yoshikawa, and I. Futamata. Design and signal processing of six-axis force sensors. *Robot. Res. 4*, eds. R. Bolles and B. Roth, MIT Press, Cambridge, MA, 1988, pp. 75–81.

[46] M. Uchiyama, E. Bayo, and E. Palma-Villalon. A systematic design procedure to minimize a performance index for robot force sensors. *J. Dyn. Syst. Meas. Control* 113(9):388–394, 1991.

[47] J. Ishikawa, K. Kosuge, and K. Furuta. Intelligent control of assembling robot using vision sensor. *Proceedings 1985 International Conference on Robotics and Automation*. IEEE, New York, 1990, pp. 1904–1909.

[48] H. F. Durrant-Whyte. Consistent integration and propagation of disparate sensor observation. *Int. J. Robot. Res.* 6(3):3–24, 1987.

[49] P. K. Allen. Integrating vision and touch for object recognition tasks. *Int. J. Robot. Res.* 7(6):15–33, 1988.

[50] S. A. Stansfield. A robotic perceptual system utilizing passive vision and active touch. *Int. J. Robot. Res.* 7(6):138–161, 1988.

[51] V. C. Klema and A. J. Laub. The singular value decomposition: Its computation and some applications. *IEEE Trans. Automat. Control* 25(2):164–176, 1980.

[52] J. O. Kim, P. K. Khosla, and W. K. Chung. Static modeling and control of redundant manipulators. *Robot. Comput. Integrated Manuf.* 9(2):145–157, 1992.

[53] N. P. Papanikolopoulos, B. Nelson, and P. K. Khosla. Full 3-d tracking using the controlled active vision paradigm. *Proceedings 1992 IEEE International Symposium on Intelligent Control (ISIC-92)*. IEEE, New York, 1992, pp. 267–274.

[54] P. Anandan. Measuring visual motion from image sequences. Tech. Rept. COINS-TR-87-21, University of Massachusetts COINS Department, Amherst, 1987.

[55] C. Tomasi and T. Kanade. Detection and tracking of point features. Tech. Rept. CMU-CS-91-132. Pittsburgh, Carnegie Mellon University School of Computer Science, 1991.

[56] D. E. Whitney. Force feedback control of manipulator fine motions. *ASME J. Dyn. Syst. Meas. Control* June: 91–97, 1977.

[57] D. B. Stewart, D. E. Schmitz, and P. K. Khosla. The Chimera II real-time operating system for advanced sensor-based control systems. *IEEE Trans. Syst. Man Cybernetics* 22:1282–1295, 1992.

CHAPTER 7

Sensor Referenced Impact Control in Robotics

RACHEL YUNYING WU

Computer Sciences Corporation, Fairview Heights, Illinois

TZYH-JONG TARN

Department of Systems Science and Mathematics, Washington University, St. Louis, Missouri

NING XI

Department of Electrical & Computer Engineering, Michigan State University, East Lansing, Michigan

ALBERTO ISIDORI

Department of Systems Science and Mathematics, Washington University, St. Louis, Missouri

1 INTRODUCTION

This chapter addresses the issues of force regulating and contact transition stabilizing control or so-called impact control. Impact control is required in most assembly tasks involving the interaction of the manipulator and the workpiece. For workpieces with low tolerance, force feedback is necessary to reduce the effect of the uncertain location and unknown stiffness of the environment.

In a general contact operation, a certain approach velocity is needed for high productivity, resulting in a collision with the contact surface. Collision is considered as a dangerous phenomenon compromising the safety of the equipment.

In this chapter, we study a robust force control design via positive acceleration feedback combined with a switching control strategy for the impact control and force regulation. The transient force response during impact is controlled to limit the peak impact force. The system can be stabilized after a finite number of switches. The output forces in the contact directions and output positions in the moving directions can be regulated simultaneously after contact is established. Accidental loss of contact can be corrected automatically. A great amount of effort will be spent on the stability analysis, leading to a theoretical foundation in the impact control area.

2 HISTORY AND BACKGROUND

The evolution of robotics is closely tied to the development of the technology of controls, sensors, and computers. The industrial robots of the early 1980s were mostly used in relatively simple operations such as machine tending, material transfer, painting, and welding. Since the late 1980s, other applications have also become important, especially in assembly, which is now dominating robot applications. Most assembly tasks in industry require the manipulator end effector to come into contact with the workpiece. Force sensing and feedback can be used to guarantee positive contact between two mating parts and can be used to monitor the forces of interaction to ensure that they do not exceed a safe limit. The increasing demand for advanced contact control has resulted in a growth of interest in force regulation and impact control.

Historically, the representative categories of robot control have been independent joint control, computed torque control, resolved motion control, nonlinear feedback control, force control, hybrid force–position control, and impact control. These evolved from joint-level control to task-level control and from non–model-based control to model-based control. The advantage of model-based task-level control is that dynamic control can be realized. Most industrial robots still use independent joint control, driven only by servo error. Given the improvements in computation and sensing technology, task-level dynamic control and force control will eventually replace classical joint-level kinematic control in order to increase productivity by increasing operational speed and system flexibility.

The manipulator and the environment are two separate systems before contact. Reaction forces are created afer the end effector and the environment are pushed together. The two systems are then coupled by the reaction forces, introducing an uncertain nonlinearity in the dynamics of the coupled system. The coupled systems can become separated again if the contact between the two systems is broken.

The contact tasks can be divided into free motion mode, impact mode, and contact motion mode. In the free motion mode, the manipulator will move from free space to the contact surface. Impact mode will be triggered by initial contact between the manipulator and the surface. The system switches to the contact motion mode after the contact is established and the manipulator can move along a specified trajectory on the surface while maintaining a desired contact force.

Previous research studies dealt mostly with the force control in contact motion mode without considering that the manipulator may lose contact with the environment during operation. Assuming that the manipulator maintains contact during operation is equivalent to assuming stability without proof. Another difficulty encountered here is the unknown and nonlinear relationship between reaction force and deformation (position). The force responses and position responses are coupled and the transient responses cannot be accurately controlled.

Because most assembly tasks require the end effector of the manipulator to come into contact with a workpiece at a certain speed, an impact is inevitable. Transition control from free motion to constrained motion or from constrained motion to free motion takes on the same importance as classical force control. In addition to all the difficulties of force control, impact control introduces its own challenges. Impact phenomena exist during transitions, resulting in dangerous collisions. The manipulator may bounce back and forth on the contact surface after the transition, and there is no clear distinction between the impact mode and contact motion mode. Thus, the ideal controller should be able to control the transient force response during impact, keep the same control structure from impact mode to contact motion mode, and remain robust in an uncertain environment.

The area of robot force control actually evolved from remote manipulator and artificial arm control in the 1950s and 1960s. These allowed users to have control by utilizing hands or muscles in a natural way. In the late 1960s and in the 1970s, the first computer controls using force feedback were proposed to replace the human operators. It was then that the stability problem was encountered. The main approaches to force control have been the following [1]:

Logic branching feedback method (Ernst 1961): The essence of this method is a set of discrete moves terminated by a discrete event such as contact. IF-THEN logic is used in the control strategy.

Continuous feedback method (Groome 1972): This was applied to assembly and edge-following tasks to maintain continuous contact. It was an early effort in continuous force feedback control.

Damping method (Whitney 1977; Paul and Shimano 1976): An integrating controller is used in which sensed forces give rise to velocity modification.

Active compliance (Salisbury 1980) [2]: A desired force is calculated based on the difference between the desired and actual hand position.

Passive compliance (Watson 1976): The end effector is equipped with a passive mechanical device composed of springs and dampers. It is capable of quick responses and is relatively inexpensive. The application of such devices is limited to very specific tasks.

Impedance control (Hogan 1980) [3]: An attempt is made to control the dynamic interaction between the manipulator and the environment. The dynamic relationship between force and position is controlled instead of pure force or position. The control structure is simple and the performance is robust. The drawback is the limitation of the dynamic performance. The output force cannot be regulated unless an accurate environmental model can be obtained.

Explicit forces control (Nevins and Whitney 1973) This employs a desired force input rather than a position or velocity input.

Implicit force control (Borrel 1979): No sensor is used. A particular stiffness matrix can be obtained by adjusting the joint servo gains.

Hybrid force–position control (Raibert and Craig 1981; Mason 1981) [4, 5]: This is one of the most popular force control schemes today. A selection matrix is used to determine the force-controlled direction and position-controlled position. In the force-controlled direction, various force control methods can be utilized. The one most frequently used is explicit force control [6–8].

In recent years, force control has been intensively studied. Various control methods have been proposed including the operational space approach [6], dynamic hybrid position–force control [9], compliant motion control [10, 11], object impedance control [12], hybrid control considering motor dynamics [13], and adaptive force control [14–16]. The inherent stability and other problems involved in force control were also discussed by many researchers. Problems addressed included kinematic and dynamic stability issues [17, 18], friction and stiction in force control [2], asymptotic stability of hybrid position–force control [19, 20], stability of impedance control [21], the stability problem in Cartesian compliance [22], dynamic problems [23], important considerations in implementation [24], and the bandwidth limitation [25]. Research work and investigations based on the aforementioned formulations have been presented [26–30].

Control of the phase transition has also been studied by many researchers and various schemes have been proposed. The initial work was mainly focused on the modeling of a robot in collision with its environment [31–33]. Models of the quantitative relation between the impulsive forces and torques of the constraint and the collision were derived. In [6], impact control was briefly discussed and treated as a transient with the same force controller structure. Maximal damping was employed during the impact phase. Impedance control was considered as an effective impact control strategy because of its unified and stable control structure.

Other approaches to impact control include generalized dynamical systems [34, 35], discontinuous control [7, 8, 36, 37], adaptive force control [35], jump impact control [38], and event-based impact control [39].

Basically, the existing schemes can be divided into two categories. The first category is impedance control. It provides a stable and unified control structure for the three operation modes. The output force cannot be regulated after contact has been established unless the exact environmental model is known and integrated into the motion plan. The second category includes hybrid control schemes. The output force can be regulated with the assumption that contact is maintained after impact. Stability cannot be guaranteed and peak impact force cannot be limited.

Hybrid control and impedance control are well known for their simple and unified control structures. However, they show good performance only under some restrictive conditions or assumptions. The assumption most frequently made is that the environmental model is a linear spring. The problem of lack of a rigorous stability analysis exists in all the previous results on impact control. Local linearization is a common technique in the stability analysis.

3 IMPACT DYNAMICS

In general manipulator operations, high productivity can be achieved only by minimizing the operation time. This implies that the speed of operation has to be increased. Because in an assembly operation, a certain percentage of time will be spent on the contact transition, increasing the speed of transition becomes a priority task in order to satisfy the requirement of the overall performance of the system. Hence, the contact velocity has to be larger than a minimum acceptable velocity. However, for relatively high impact velocities, large impact forces can be generated, exceeding the tolerance of the equipment. The end effector may rebound from the contact surface and cause instability. Therefore there is a trade-off between productivity and safety. It should be noted that impact velocity may also be caused by inaccurate planning due to environmental uncertainties.

As we know from classical physics, the dynamics of impact are very different from the dynamics of a rigid body. Impact dynamics are quite complicated. The behavior that occurs at the interface is highly dependent on material properties such as the coefficients of restitution and the mass of the end effector. There are two ways to describe impact. One way is to calculate the velocity change after impact and neglect the behavior during the collision. Another way is to model the local dynamics of the object or the relation between the deformation and the impact force.

In the literature there are many different deformation models, such as the linear elastic model, linear plastic model, Hertz model, and nonlinear oscillator model. In order to avoid damaging the workpieces or the robot, we have to ensure that no permanent deformation occurs after impact. This implies that only elastic collisions are allowed. Compared with other existing impact models, the Hertz impact theorem [40] provides a good elastic model of the collision of two spheres or the impact of a sphere against a thick plate, provided the materials are relatively hard and the initial velocity is relatively low.

3.1 Hertz Impact Model

Consider the nature of the collision between the manipulator and the contact surface. It is reasonable to assume that the deformation is elastic, the contact force is monotone increasing with the deformation, and the contact surface is static.

In the Hertz model, the normal force, F, is related to the relative displacement, α, by

$$F = k\alpha^{3/2}$$

where

$$k = (4\sqrt{\bar{r}}/3\pi)(k_1 + k_2)$$
$$k_i = \{(1 - v_i^2)/\pi\}E_i, \quad i = 1, 2$$

v_i and E_i are Poisson's ratio and Young's modulus for the end effector and the environment, and $\bar{r} = \dfrac{r_1 r_2}{r_1 + r_2}$, where r_1, r_2 are the radii of the contact masses. Since for a thick plate, r_2 goes to infinity, $\bar{r} = r_1$ is the contact radius of the end effector. The relative velocity at any time during contact is

$$\dot{\alpha}^2 = \dot{\alpha}_1^2 - \frac{2}{m}\int_0^{\alpha} k\alpha^{3/2}d\alpha = \dot{\alpha}_1^2 - \frac{4k\alpha^{5/2}}{5m}$$

The maximum displacement, α_{max}, can be obtained by setting $\dot{\alpha}$ to zero:

$$\alpha_{max} = (5mv_0^2/4k)^{2/5}$$

where $v_0 = \dot{\alpha}_1$ is the impact velocity and m is the mass of the end effector. Hence, the maximum impact force is

$$F_{max} = k\alpha_{max}^{3/2} \tag{7.1}$$

The duration of impact, T_c, is

$$T_c = 3.214(m/k)^{2/5}v_0^{-1/5} \tag{7.2}$$

In this way, the relationship between the impact velocity and the maximum impact force is obtained. For a free mass collision, the peak impact force is proportional to the impact velocity and the mass of the end effector. For a controlled collision such as the contact transition of a manipulator, the peak impact force should be made as small as possible. Thus, it must be smaller than the peak impact force for a free mass collision with the same impact velocity. Therefore, if we require that the peak impact force in Eq. (7.1) be less than the maximum allowable impact force, the resulting impact velocity derived from (7.1) will be safe. From Eq. (7.1), we have

$$v_0 < 0.89m^{-1/2}k^{-1/3}F_{safe}^{5/6}$$

For digital implementation, the sampling rate needs to be taken into consideration. The force controller will be activated on the detection of the impact forces. If the sampling rate of the force measurement is too slow, the impact force may exceed the limit before the control can switch over to force control and the end effector may rebound from the contact surface. If the sampling time can be reduced to less than one half of the impact duration (i.e., $2T_s < T_c$), the chances of rebound will drop to zero. From Eq. (7.2), the upper bound of impact velocity is

$$v_0 < 10.72m^2/k^2T_s^5$$

The limitation of the magnitude of the control inputs is another factor that needs to be taken into consideration. From Eq. (7.1), we have

$$v_0 < 0.89\rho^{-1}m^{-1/2}k^{-1/3}F_{\text{control}}^{5/6}$$

where ρ is a weighting factor less than 1 and depends on the sampling rate.

To meet the preceding two constraints, the upper bound of the impact velocity that can be applied should be

$$v_{\max} = \min\{10.7m^2/(k^2T_s^5),\ 0.9m^{-1/2}k^{-1/3}F_{\text{limit}}^{5/6}\}$$

where $F_{\text{limit}} = \max\{F_{\text{safe}}, F_{\text{control}}\}$. The theoretical analysis provides us with a basic idea of the limitations of the impact velocity. It is consistent with our intuition that the higher the sampling rate or the control limit, the higher the impact velocity safety bound. Of course, the real world is more complicated. Even though it is hard to get an accurate safety range of the impact velocity, at least we can make the velocity far below the calculated upper bound to protect the manipulator and the equipment.

4 ROBUST IMPACT CONTROL

The dynamic equation for a manipulator with n degrees of freedom is given by

$$\begin{cases} D(q)\ddot{q} + C(q, \dot{q}) + G(q) + J(q)^T f = \tau \\ y = h(q) \end{cases}$$

where $D(q)$, $C(q, \dot{q})$, $G(q)$ are inertia, centripetal and Coriolis forces, and gravities; $q \in R^n$ is the vector of joint angles; $\dot{q}$, $\ddot{q}$ are the joint space velcity and acceleration, respectively; $h(q)$ is the forward kinematics; $J(q)$ is the $n \times n$ Jacobian matrix relating joint space velocity to task space velocity (i.e., $\dot{y} = J(q)\dot{q}$); τ is the $n \times 1$ vector of joint driving torques; y is the output position and orientation in task space; and f is the output force in task space.

4.1 Feedback Linearization and Decoupling

Introducing the well-known nonlinear feedback linearization control law

$$\tau = D(q)J(q)^{-1}(v - \dot{J}(q)\dot{q}) + C(q, \dot{q}) + G(q) + J(q)^T f$$

the dynamics of the robot system can be linearized and decoupled as

$$\ddot{y} = v \tag{7.3}$$

where v is the vector of auxiliary inputs (commanded acceleration).

Letting subscript u denote a quantity in the unconstrained directions and subscript c denote a quantity in the constrained directions, Eq. (7.3) can be written as $\ddot{y}_u = v_u$, $\ddot{y}_c = v_c$. The controllers are decoupled. Using the traditional proportional derivative (PD) position control before impact, we have

$$v = \ddot{y}^d + k_v(\dot{y}^d - \dot{y}) + k_p(y^d - y), \quad t \geqslant 0$$

After impact, we keep using PD position control in the motion directions,

$$v_u = \ddot{y}_u^d + k_{vu}(\dot{y}_u^d - \dot{y}_u) + k_{pu}(y_u^d - y_u), \quad t \geqslant t_{sw}$$

where t_{sw} is the time instant of the detected impact. The preceding closed-loop systems are asymptotically stable.

4.2 Positive Acceleration Feedback

To establish safe and smooth contact with the environment, the transient force response during impact should be controlled with minimum overshoot and oscillation. The contact force introduces a complicated nonlinearity in the closed-loop force feedback system.

The closed-loop force dynamics in the contact directions after impact are

$$\ddot{y}_c + k_d\dot{y}_c - k_f(f_c^d - f_c) = 0, \quad t \geqslant t_{sw}$$

where f_c is the reaction force. In state space, we have

$$\begin{cases} \dot{x}_1 = x_2 \\ \dot{x}_2 = -k_d x_2 + k_f(f_c^d - f_c) \end{cases}$$

where $x_1 = y_c$, $x_2 = \dot{y}_c$. Force control is equivalent to high-gain position feedback control, and the transient response could be very oscillatory. Consider a linear environmental model $f_c = k_e x_1$, where k_e is the stiffness,

$$\begin{cases} \dot{x}_1 = x_2 \\ \varepsilon\dot{x}_2 = -k_1 k_e x_1 - k_2 x_2 \end{cases}$$

where ε is a positive value close to zero. It is clear that this singular perturbed linear system is asymptotically stable. The approximate dynamics of x_1 are reduced to a first-order system,

$$\dot{x}_1 = -k_1 k_e x_1 / k_2 + O(\varepsilon)$$

which is more robust in an uncertain environment.

For an unknown nonlinear environment, consider force feedback control with positive acceleration for $t \geqslant t_{sw}$. Let

$$v_c = \alpha \ddot{y}_c - k_d \dot{y}_c + k_f e_f + k_I \int e_f \, dt$$

where α is close to but less than 1, $e_f = f_c^d - f_c$. Assume the reaction force, f_c, is a nonlinear monotone increasing function of the deformation, that is, $f_c = g(y_c)$. Let y_c^d denote the corresponding position to the desired force f_c^d. Thus $f_c^d = g(y_c^d)$. The closed-loop dynamics in the constrained direction, after the control is switched, become

$$\varepsilon \ddot{y}_c + k_d \dot{y}_c + k_f (g(y_c) - f_c^d) + k_I \int_0^t (g(y_c) - f_c^d) \, dt = 0 \tag{7.4}$$

Here $\varepsilon = 1 - \alpha > 0$. Equation (7.4) is a singularly perturbed nonlinear system. Denote

$$w = -k_d \dot{y}_c + k_f (f_c^d - g(y_c)) + k_I \int (f_c^d - g(y_c)) \, dt$$

In state space, we have

$$\begin{cases} \dot{x}_1 = x_2 \\ \varepsilon \dot{x}_2 = w \end{cases}$$

where $x_1 = y_c$, $x_2 = \dot{y}_c$. The approximate dynamics of x_1 are

$$\dot{x}_1 = \frac{k_f}{k_d} (f_c^d - g(y_c)) + \frac{k_I}{k_d} \int (f_c^d - g(y_c)) \, dt + O(\varepsilon)$$

The closed-loop dynamics are reduced to $w = 0$ if $\varepsilon = 0$. The dominant dynamics of f_c become $w = 0$, for small integral gain, and the transient response of f_c is close to the response of a first-order system and will be robust for uncertain environments.

Current technology does not allow us to get an accurate measurement of the acceleration. Instead, we choose to use the previous commanded acceleration as the feedback. Together with the robot model, we have

$$\tau = \hat{D}(q) \hat{J}(q)^{-1} (v - \hat{\dot{J}}(q) \dot{q}) + \hat{C}(q, \dot{q}) + \hat{G}(q) + \hat{J}(q)^T f$$

where $\hat{D}$, $\hat{J}$, $\hat{C}$, $\hat{G}$ are the estimates of the parameters of the robot model. The closed-loop system actually includes a modeling error term $\ddot{y}_c = v + \eta$, where η is the modeling error and

$$\eta = JD^{-1}((\hat{D}\hat{J}^{-1} - DJ^{-1})v + DJ^{-1}\dot{J}\dot{q} - \hat{D}\hat{J}^{-1}\hat{\dot{J}}\dot{q} + \hat{C} - C + \hat{G} - G + \hat{J}^T f - J^T f)$$

Let $v = \alpha v_{\text{previous}} + w$, where v_{previous} is the commanded acceleration to the previous sampling time instant. Because $v \approx v_{\text{previous}}$, $\eta \approx \eta_{\text{previous}}$, and $\ddot{y}_c \approx \ddot{y}_{c_{\text{previous}}}$ for a high sampling rate, we have

$$\varepsilon \ddot{y}_c - w + \varepsilon \eta = 0$$

The dominant dynamics of x_1 are reduced to $w = 0$, which is close to a first-order system. In addition, the modeling error term in the closed-loop system is reduced to $\varepsilon\eta$. It turns out that the control design is robust to bounded modeling uncertainties and environmental uncertainties.

4.3 Stability

From the error dynamics in the constrained directions after impact, together with Liapunov's direct method, it can be shown that the equilibrium point is asymptotically stable for the closed-loop system if the trajectory remains in the constrained space.

Let $f_c = g(y_c)$ and y_c^d denote the corresponding position to the desired force f^d. Hence $f_c^d = g(y_c^d)$. Due to the fact that the bigger the deformation, the larger the reaction force, we can assume $g(y_c)$ is a monotone increasing function of the deformation. This leads to

$$(g(y_c) - g(y_c^d))(y_c - y_c^d) > 0 \quad \text{if } y_c - y_c^d \neq 0$$
$$(g(y_c) - g(y_c^d))(y_c - y_c^d) = 0 \quad \text{iff } y_c = y_c^d$$

Theorem 1 *Consider the closed-loop system* (7.4), *and define*

$$Z = \begin{cases} z_1 = y_c - y_c^d \\ z_2 = \dot{y}_c \\ z_3 = \displaystyle\int_0^t (f_c - f_c^d)\, dt \end{cases}$$

as the state of the system. Assume that (1) *the trajectory remains in the constrained space,* (2) *all the gains are positive, and* (3) *the actual force f is an unknown monotone increasing function of the deformation. Then a sufficient condition for the system to be asymptotically stable is*

$$k_d k_f - k_I \varepsilon > 0 \tag{7.5}$$

In the new coordinate system, the closed-loop system (7.4) becomes

$$\varepsilon \dot{z}_2 + k_d z_2 + k_f \dot{z}_3 + k_I z_3 = 0 \tag{7.6}$$

Introducing the state space representation:

$$\begin{cases} \dot{z}_1 = z_2 \\ \dot{z}_2 = \{-k_d z_2 - k_f\{g(z_1 + y_c^d) - f_c^d\} - k_I z_3\}/\varepsilon \\ \dot{z}_3 = g(z_1 + y_c^d) - f_c^d \end{cases}$$

recall that $z_1 + y_c^d = y_c$. Then, $Z = (z_1, z_2, z_3)^T = 0$ is the equilibrium point. Now, take

$$\begin{aligned} V &= b \int_0^{z_1} (g(\xi + y_c^d) - f_c^d)\, d\xi + z_2^2/2 + c z_2 z_3 + a z_3^2/2 \\ &= b \int_0^{z_1} h(\xi)\, d\xi + z_2^2/2 + c z_2 z_3 + a z_3^2/2 \end{aligned} \tag{7.7}$$

as the Liapunov function candidate, where

$$\begin{cases} a = k_I k_f / k_d \varepsilon \\ b = (k_f/\varepsilon - k_I/k_d) \\ c = k_I/k_d \\ h(\xi) = g(\xi + y_c^d) - f^d = g(\xi + y_c^d) - g(y_c^d) \end{cases}$$

If the sufficient condition (7.5) is satisfied, then $a - c^2 > 0$, which implies $z_2^2/2 + cz_2z_3 + az_3^2/2$ is a positive definite function. Since $b\int_0^{z_1} h(\xi)\, d\xi$ is positive if $z_1 \neq 0$, V is a positive definite function. Now consider

$$\dot{V} = -(k_d z_2 + k_I z_3)^2 / k_d \varepsilon$$

Therefore, $\dot{V} \leqslant 0$. If $\dot{V} = 0$, then $k_d z_2 + k_I z_3 = 0$. Take the derivative to obtain $k_d \dot{z}_2 + k_I \dot{z}_3 = 0$. According to Eq. (7.6), we have $\varepsilon \dot{z}_2 + k_f \dot{z}_3 = 0$. Since $k_d k_f - k_I \varepsilon > 0$, this implies $\dot{z}_3 = 0$, $\dot{z}_2 = 0$. By the definition of z_3, we have $\dot{z}_3 = f_c - f_c^d$, and $\dot{z}_3 = 0$ means $f_c = f_c^d$, which is equivalent to $y_c = y_c^d$. Consequently, $\dot{y}_c = \dot{y}_c^d = 0$, which implies $z_2 = 0$, and this leads to $z_3 = 0$. Therefore $\dot{V} = 0$ contains only the trivial trajectory. According to LaSalle's theorem, the system is asymptotically stable.

5 SWITCHING CONTROL

A controller with one time switch cannot guarantee that the manipulator will never lose contact. Using force control only after impact may be dangerous if the manipulator leaves the surface; the manipulator will be dragged back to the surface with very high speed by the force controller and instability may be excited. Instead, position control should be used to drag the manipulator back to the surface when necessary. To make sure that the manipulator maintains contact with the surface, a switching control law is designed to eliminate unexpected rebounds and reestablish contact.

5.1 Control Design

The minimum detectable force f_{sw}, dependent on the sensitivity of the force–torque sensor and sampling rate, is used as the switching condition. The controller is designed to switch between a position controller for the free motion and a force controller for the constrained motion. The position $y_{c,sw}$ corresponding to f_{sw} is unknown but can be recorded on line to serve as the desired position for position control. When the measured force is greater than or equal to f_{sw}, the controller is switched to force control; otherwise the controller will be switched to position control to reestablish contact.

The sampling rate of the measurement should be considerably higher than the bouncing frequency for the implementation. It is reasonable to assume that impact occurs in a very short period, which implies that $y_{c,sw}$ can be considered as a constant. Thus, the desired velocity and desired acceleration corresponding to $y_{c,sw}$ are zero for a soft landing. According to the preceding analyis, in the constrained directions, we design the rigorous switching logic

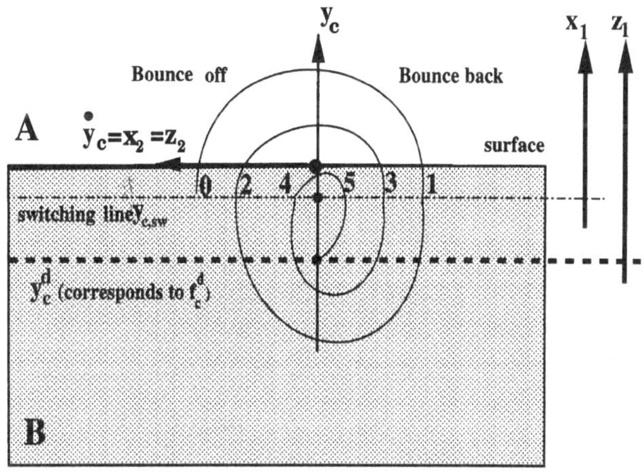

FIGURE 7.1
Phase portrait of the switching system.

as (see Figure 7.1)

$$\begin{cases} A\colon\ (i)\ f_c < f_{sw} \quad or \quad (ii)\ f_c = f_{sw}\ \&\ \dot{y}_c(t) > 0 \\ B\colon\ (i)\ f_c > f_{sw} \quad or \quad (ii)\ f_c = f_{sw}\ \&\ \dot{y}_c(t) \leqslant 0 \end{cases} \tag{7.8}$$

Clearly, at any moment in time either A holds or B holds. It is also apparent by inspection that A and B are disjoint. Apparently, the control will switch to the force controller if the state of the system is in region B and to the position controller if the state of the system is in region A, that is,

If A is true, then position control
If B is true, then force control

The controller is designed as

$$v_c = \begin{cases} -k_v\dot{y}_c - k_p(y_c - y_{c,sw}) & (A) \\ a\ddot{y}_c - k_d\dot{y}_c + k_f e_f + k_I \displaystyle\int_{t_{sw}}^{t} e_f\, dt & (B) \end{cases}$$

The closed-loop system is

$$\begin{cases} \ddot{y}_c + k_v\dot{y}_c + k_p(y_c - y_{c,sw}) = 0 & (A) \\ \varepsilon\ddot{y}_c + k_d\dot{y}_c - k_f e_f - k_I \displaystyle\int_{t_{sw}}^{t} e_f\, dt = 0 & (B) \end{cases} \tag{7.9}$$

It can be shown that if bouncing occurs after impact, the manipulator will establish contact with the environment after a finite number of switches and the desired force can be achieved simultaneously.

We prove that if the parameters k_v and k_p of the closed-loop system for the unconstrained motion,

$$\ddot{y}_c + k_v \dot{y}_c + k_p(y_c - y_{c,sw}) = 0$$

are appropriately chosen, then the switching system (7.7) with the proposed switching logic (7.6) will asymptotically converge to the equilibrium. To this purpose, set $k_v = 2\zeta\omega_n$, $k_p = \omega_n^2$, and $\zeta < 1$, and we have:

Theorem 2 *If condition*

$$k_a k_f (1 - e^{-2\zeta\pi/\beta}) - k_I \varepsilon > 0 \tag{7.10}$$

is satisfied, where $\beta = \sqrt{1-\zeta^2}$, then the switching system (7.9) *with the proposed switching logic* (7.8) *will asymptotically converge to the equilibrium $Z = 0$ after finite switches.*

The complete proof can be found in the next section. The outline of the proof is as follows: It is seen that condition (7.10) implies (7.5). To prove the theorem, it can be shown that if the number of switchings in a trajectory is not finite, then the trajectory itself will asymptotically approach a set that is entirely contained in the set B of Eq. (7.8). On the contrary, Theorem 1 shows that every trajectory that remains entirely in B converges to the equilibrium of (7.4) as $t \to \infty$. Thus, in B there is no other invariant set but the equilibrium of (7.4), and this contradicts the existence of an infinite number of switchings. Thus, the number of switchings is finite and the last switching always corresponds to the transition from free space to constrained space after which the system converges to $Z = 0$ as $t \to \infty$ by Theorem 1.

5.2 Stability Proof

Lemma 1 *Consider the closed-loop system in the unconstrained direction,*

$$\ddot{y}_c + k_v \dot{y}_c + k_p(y_c - y_{c,sw}) = 0$$

which is an asymptotically stable system. Define

$$X = \begin{cases} x_1 = y_c - y_{c,sw} \\ x_2 = \dot{y}_c \end{cases}$$

as the state of the system. Let $k_v = 2\zeta\omega_n$, $k_p = \omega_n^2$, and $\zeta < 1$. Suppose at time $t = 0$ there is a transition from constrained space to free space (bouncing off). Let $x_1(0) = 0$ and $x_2(0) = v_0 > 0$. Then there is a time $T > 0$ at which a transition from free space to constrained space occurs (bouncing back) and $x_1(T) = 0$, $x_2(T) = -v_0 e^{-\zeta\pi/\beta}$ (where $\beta = \sqrt{1-\zeta^2}$ and $e^{-\zeta\pi/\beta} < 1$).

Proof

$$V_1 = k_p(y_c - y_{c,sw})^2/2 + \dot{y}_c^2/2 = k_p x_1^2/2 + x_2^2/2$$

is the Liapunov function of the system. Let $\bar{y} = x_1 = y_c - y_{c,sw}$. Since $y_{c,sw}$ is a constant

$\dot{\bar{y}} = \dot{y}_c, \ddot{\bar{y}} = \ddot{y}_c$. The system will become

$$\ddot{\bar{y}} + k_v \dot{\bar{y}} + k_p \bar{y} = 0$$

Taking the Laplace transform, we have

$$(s^2 + 2\zeta\omega_n s + \omega_n^2)\bar{Y}(s) = s\bar{y}(0) + \dot{\bar{y}}(0) + 2\zeta\omega_n \bar{y}(0)$$

Since the initial conditions are $\bar{y}(0) = x_1(0) = 0$, $\dot{\bar{y}}(0) = x_2(0) = v_0$,

$$(s^2 + 2\zeta\omega_n s + \omega_n^2)\bar{Y}(s) = \dot{\bar{y}}(0) = v_0$$

The corresponding time domain response is

$$\bar{y}(t) = \left(\frac{v_0}{\omega_n \beta}\right) e^{-\zeta\omega_n t} \sin \omega_n \beta t$$

Since $\beta = \sqrt{1 - \zeta^2}$, we can define $\sin \phi = \beta$, $\cos \phi = \zeta$. Then

$$\dot{\bar{y}}(t) = -\left(\frac{v_0}{\beta}\right) e^{-\zeta\omega_n t} \sin(\beta\omega_n t - \phi)$$

At the time when $\bar{y}(T) = x_1(T) = 0$, $\sin \omega_n \beta T = 0$. Choose $\omega_n \beta T = \pi$; then,

$$\dot{\bar{y}}(T) = x_2(T) = -v_0 e^{-\zeta\omega_n T} \sin(\pi - \phi)/\beta = -v_0 e^{-\zeta\omega_n T} = -e^{-\zeta\pi/\beta} \tag{7.11}$$

Lemma 2 *Let*

$$\mu = \max\left\{\frac{a\gamma + a\rho}{a\gamma - c^2}, \frac{a}{a - c^2}\right\} \tag{7.12}$$

where $\gamma = 1 - e^{-2\zeta\pi/\beta}$, ρ is an arbitrary positive number. If condition (7.10) is satisfied and $\alpha > \mu$, then we have $\alpha > 1$ and $a - c^2 > 0$.

Proof Since $\gamma < 1$ and

$$a\gamma - c^2 = (k_I k_f / k_d \varepsilon)\gamma - k_I^2/k_d^2 = \frac{k_I}{k_d^2 \varepsilon}(k_d k_f \gamma - k_I \varepsilon) > 0$$

we have $a - c^2 > a\gamma - c^2 > 0$. Also, it is easy to verify that $\mu > 1$, which implies $\alpha > 1$.

Let the switching time sequence denoted by

$$t_0, t_1, t_2, \ldots, t_{2i}, t_{2i+1}, t_{2i+2}, \ldots$$

where t_0 is the time when the manipulator starts to bounce off from the constrained space (Figure 7.2). Therefore, $Z_0 = (z_1(t_0), z_2(t_0), z_3(t_0))$ is the last state of the trajectory in the constrained space. It corresponds to the initial state $X_0 = (0, z_2(t_0))$ in the free space at time

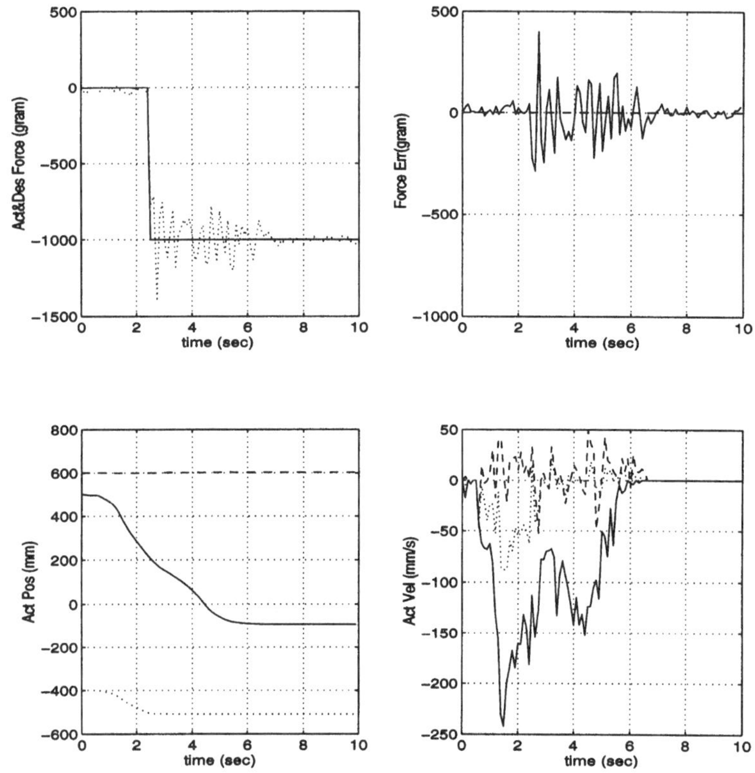

FIGURE 7.2
Impact control on a rigid flat table made of compound material.

t_0. Note also that $z_1(t_0) = \bar{z}_1 = y_{c,sw} - y_c^d$. Similarly, at switching time t_{2i}, we have $Z_{2i} = (\bar{z}_1, z_2(t_{2i}), z_3(t_{2i}))$ and $X_{2i} = (0, z_2(t_{2i}))$.

Lemma 3 *If condition* (7.10) *is satisfied, then there exist a Liapunov function* V_2 *for the closed-loop system in the constrained space* (*B in* (7.9)) *and a positive number* ρ *such that at even switching indices* V_2 *is strictly decreasing and*

$$V_2(Z_{2i}) - V_2(Z_{2i+2}) > \rho z_2^2(t_{2i})/2 \tag{7.13}$$

Proof Choose ρ as an arbitrary positive number and $\alpha = \mu + 1$, where μ is defined in Eq. (7.12). Let

$$V_1(X) = k_p(y_c - y_{c,sw})^2/2 + \dot{y}_c^2/2$$

$$V_2(Z) = \alpha b \int_0^{z_1} h(\xi)\, d\xi + \alpha a z_3^2/2 + \alpha z_2^2/2 + \alpha c z_2 z_3$$

be the Liapunov functions for the closed-loop system (7.9) in free space (A) and constrained space (B), respectively.

By Lemma 1,

$$V_1(X_{2i}) > V_1(X_{2i+1}) \tag{7.14}$$

and by Theorem 1,

$$V_2(Z_{2i+1}) > V_2(Z_{2i+2}) \tag{7.15}$$

where $2i$ indicates the $2i$th switching instant. At the $2i$th switching instant,

$$V_2(Z_{2i}) - V_1(X_{2i}) = \left\{\alpha b \int_0^{z_1} h(\xi)\, d\xi + \alpha a z_3^2/2 + (\alpha - 1)z_2^2/2 + \alpha c z_2 z_3\right\}_{2i}$$

By Lemma 2, $\alpha > 1$ and $a - c^2 > 0$, together with (7.12), we have:

$$\begin{vmatrix} \alpha a & \alpha c \\ \alpha c & \alpha - 1 \end{vmatrix} = \alpha[(a - c^2)\alpha - a] > 0$$

This implies that $V_2(Z_{2i}) - V_1(X_{2i})$ is positive.

Also, at the $(2i + 1)$th switching instant,

$$V_2(Z_{2i+1}) - V_1(X_{2i+1}) = \left\{\alpha b \int_0^{z_1} h(\xi)\, d\xi + (\alpha - 1)z_2^2/2\right\}_{2i+1}$$

Therefore:

$$\begin{aligned}[V_2(Z_{2i}) - V_1(X_{2i})] - [V_2(Z_{2i+1}) - V_1(X_{2i+1})] &= \{\alpha a z_3^2/2 + (\alpha - 1)z_2^2/2 + \alpha c z_2 z_3\}_{2i} \\ &\quad - \{(\alpha - 1)z_2^2/2\}_{2i+1} = I\end{aligned}$$

Claim

$$I > \rho z_2^2(t_{2i})/2 \tag{7.16}$$

By Lemma 1,

$$\frac{\dot{y}_c(t_{2i+1})}{\dot{y}_c(t_{2i})} = \frac{z_2(t_{2i+1})}{z_2(t_{2i})} = -e^{-\zeta\pi/\beta}$$

which implies

$$I - \rho z_2^2(t_{2i})/2 = \{\alpha a z_3^2/2 + ((\alpha - 1)\gamma - \rho)z_2^2/2 + \alpha c z_2 z_3\}_{2i}$$

where $\gamma = 1 - e^{-2\zeta\pi/\beta}$ as in Lemma 2.

By Lemma 2, together with (7.12), we have $\mu > \dfrac{\alpha\gamma + a\rho}{\alpha\gamma - c^2}$, $\alpha > \mu$. Thus, $\alpha > \dfrac{a\gamma + a\rho}{a\gamma - c^2}$, which implies $\alpha(a\gamma - c^2) > (a\gamma + a\rho)$. Therefore,

$$\begin{vmatrix} \alpha a & \alpha c \\ \alpha c & (\alpha - 1)\gamma - \rho \end{vmatrix} = \alpha[(a\gamma - c^2)\alpha - (a\gamma + a\rho)] > 0$$

Hence $I - \rho z_2^2(t_{2i})/2$ is positive. Therefore,

$$V_2(Z_{2i}) - V_1(X_{2i}) > V_2(Z_{2i+1}) - V_1(X_{2i+1}) + \rho z_2^2(t_{2i})/2$$

Based on Theorem 1, Lemma 1, and Eqs. (7.14) and (7.15), we have

$$\begin{aligned} V_2(Z_{2i}) - V_2(Z_{2i+2}) &= V_2(Z_{2i}) - V_1(X_{2i}) + V_1(X_{2i}) - V_2(Z_{2i+2}) \\ &\geqslant V_2(Z_{2i+1}) - V_1(X_{2i+1}) + \rho z_2^2(t_{2i})/2 + V_1(X_{2i}) - V_2(Z_{2i+2}) \\ &= V_2(Z_{2i+1}) - V_2(Z_{2i+2}) + V_1(X_{2i}) - V_1(X_{2i+1}) + \rho z_2^2(t_{2i})/2 \\ &> \rho z_2^2(t_{2i})/2 \end{aligned}$$

Proof of Theorem 2 Suppose the system has infinite switchings. Since $V_2(Z_{2i})$ is strictly decreasing and bounded below by zero, $\lim_{i\to\infty} V_2(Z_{2i})$ exists. Let $\lim_{i\to\infty} V_2(Z_{2i}) = L$. By Lemma 3,

$$V_2(Z_{2i}) - V_2(Z_{2i+2}) \geqslant \rho z_2^2(t_{2i})/2$$

Taking the limits on both sides,

$$\lim_{i\to\infty} V_2(Z_{2i}) - \lim_{i\to\infty} V_2(Z_{2i+2}) \geqslant \lim_{i\to\infty} \rho z_2^2(t_{2i})/2 \geqslant 0$$

Hence, we have $0 = L - L \geqslant \lim_{i\to\infty} \rho z_2^2(t_{2i})/2 \geqslant 0$, which implies $\lim_{i\to\infty} \rho z_2^2(t_{2i})/2 = 0$, which implies $\lim_{i\to\infty} z_2(t_{2i}) = 0$. By Lemma 1, $|z_2(t_{2i+1})| < |z_2(t_{2i})|$, which implies $\lim_{i\to\infty} z_2(t_{2i+1}) = 0$, which implies $\lim_{i\to\infty} z_2(t_i) = \dot{y}_c(t_i) = 0$.

It can also be shown that $\lim_{i\to\infty} z_3(t_{2i+2}) = 0$. In fact, (7.15) implies

$$V_2(Z_{2i+1}) = \alpha b \int_0^{\bar{z}_1} h(\xi)\, d\xi + \alpha \frac{z_2^2(t_{2i+1})}{2}$$

$$V_2(Z_{2i+2}) = \alpha b \int_0^{\bar{z}_1} h(\xi)\, d\xi + \alpha \frac{z_2^2(t_{2i+2})}{2} + \frac{\alpha a z_3^2(t_{2i+2})}{2} + \alpha c z_2(t_{2i+2}) z_3(t_{2i+2})$$

$$\begin{aligned} 0 < V_2(Z_{2i+1}) - V_2(Z_{2i+2}) &= \alpha z_2^2(t_{2i+1})/2 - \alpha z_2^2(t_{2i+2})/2 \\ &\quad - z_3(t_{2i+2})[\alpha a z_3(t_{2i+2})/2 + \alpha c z_2(t_{2i+2})] \end{aligned}$$

from which, since $\lim_{i\to\infty} z_2(t_i) = 0$, the result follows.

Recall that

$$V_2(Z_{2i+1}) = D + \alpha \frac{(z_2(t_{2i+1}))^2}{2}$$

$$V_2(Z_{2i+2}) = D + \alpha \frac{(z_2(t_{2i+2}))^2}{2} + z_3(t_{2i+2}) \left[\alpha a \frac{z_3(t_{2i+2})}{2} + \alpha c z_2(t_{2i+2}) \right]$$

where $D = V_2(\bar{z}_1, 0, 0)$ and $\bar{z}_1 = y_{c,sw} - y_c^d$. Since

$$\lim_{i \to \infty} z_2(t_{2i+1}) = 0$$

$$\lim_{i \to \infty} z_2(t_{2i+2}) = 0$$

$$\lim_{i \to \infty} z_3(t_{2i+2}) = 0$$

we obtain that, for any $\varepsilon > 0$, there is i^* such that for $i > i^*$,

$$|V_2(Z_{2i+1}) - D| < \varepsilon$$

$$|V_2(Z_{2i+2}) - D| < \varepsilon$$

Therefore, since $V_2(Z(t))$ is decreasing for $t_{2i+1} < t < t_{2i+2}$, we have proved that for any $\varepsilon > 0$, there exists i^* such that, for all $i > i^*$ and all $t \in [t_{2i+1}, t_{2i+2}]$,

$$D + \varepsilon > V_2(Z(t)) > D - \varepsilon$$

In other words, for all $i > i^*$ and all $t \in [t_{2i+1}, t_{2i+2}]$, the trajectory $Z(t)$ is in an ε-neighborhood of the level set

$$\mathscr{A} = \{Z \in R^n : V_2(Z) = V_2(\bar{z}_1, 0, 0)\}$$

This, together with the observation that the trajectory in the free space is itself in an ε-neighborhood of the point $\{z_1 = \bar{z}_1, z_2 = 0\}$, completes the proof that for any $\varepsilon > 0$, there is T such that for all $t > T$ the trajectory is contained in an ε-neighborhood of $\mathscr{A}$.

Since $\mathscr{A}$ is entirely contained in the set B, by Theorem 1, $\mathscr{A}$ cannot contain any nontrivial invariant set. This proves that there cannot be infinitely many switchings. In view of Lemma 1, the last switching is from the free space to the constrained space.

6 EXPERIMENTS

The intensive simulation and experimental studies were conducted at the Center for Robotics and Automation in order to verify the theoretical results. Experimental comparisons with other typical contact control schemes were also studied. The trajectory tracked consists of a straight line in free space and a straight line on the constrained surface.

A 6-DOF dual-arm PUMA 560 manipulator [41] manufactured by Unimation Inc. was used for the experiments. The manipulator is hardware interfaced to a Motion Tek universal motion controller [42], which is an electronic system used to control a number of motors for the robot system. It is highly modular in configuration and is capable of up to 1000 Hz servo rate on every axis. Each controller stack consists of five modules. The power supply module provides the logic and servo power for the entire control stack. The joint processor module contains a 10-MHz, 32-bit NS 32016 microprocessor that runs both the configuration set up and the servo code generator; up to 16 axes can be controlled. Two joint interface modules contain the hardware to drive the motors and provide raw feedback from each joint; each one can drive four motors. All position and velocity information is obtained from a single incremental encoder on the motor. The user processor module provides the capability to run user-developed software; its 128 K RAM memory can be used as shared memory or communication between the UMC and other high-level computing devices. The expansion module contains the interface board to connect the user process multibus with the VME bus of high-level computers. The motors are driven by a 30-V power tap from the power supply module.

The high-level computing device is an SGI 4D/340 VGX graphics workstation [43]. It has four symmetric R 3000/3010 RISC processors with a 33-MHz rate and can deliver up to 117 MIPS or 36 MFLOPS. The SGI workstation is interfaced with the UMC controller via shared memory mapping and provides sufficiently fast communication.

The robot arms were equipped with California Cybernetics FSA-3254 six-axis wrist-mounted force–torque sensors [44]. The sensors are interfaced with the controller.

Our control algorithm was implemented on one of the robot arms equipped with the force–torque sensor. The force–torque was measured at a rate of 1000 Hz, and the filtered force–torque was computed at a rate of 500 Hz. The sampling rate for joint position–velocity measurement was 1000 Hz. The feedback was also computed at a rate of 1000 Hz. Since the time constant of the joint motor of the PUMA 560 is 3 ms, the sampling and feedback rates are high. A high sampling rate will increase the control accuracy and reduce the effect of uncertainties [45], especially for force control [46]. In the experiments, the positions in the x, y, and z directions and orientation were commanded for free motion. After the detection of contact, only force was commanded in the z direction. The positions in the x and y directions together with orientation were commanded for constrained motion. Three kinds of surface material were tested: a rigid flat table made of a compound material, a flat plastic foam, and a steel surface with a "camel-back" shape. The location, shape, and stiffness of the contact surface are uncertain.

The secondary input is

$$v = \alpha v_{\text{previous}} - k_d \dot{y}_c + h_f(f_c^d - f_c) + k_I \int (f_c^d - f_c)\,dt$$

where α, k_f, k_I, and k_d were chosen to be 0.85, 0.06, 0.8, and 15.0 for every test case. The output y was calculated from measured joint angles. Since the kinematic modeling error can be reduced by calibration and the PUMA 560 arm is rigid, the performance will not be significantly affected by the kinematic modeling error and the joint flexibility. The path considered consists of two segments, one a straight line from free space to the constrained surface, the other a straight line from the contact point to a final point on the constrained surface. The impact velocity is chosen to be 0.1 m/s. The first experiment, shown in Figure 7.2, tested the impact control scheme on a flat rigid table made of a compound material. It showed the force response for a desired force value of -1.0 kg. The force error goes to zero smoothly and quickly. The second experiment, shown in Figure 7.3, tested the switching control scheme on the same table; the system is stabilized after a few switches. The third experiment, shown in Figure 7.4, tested the impact control scheme on a flat plastic foam; a similar stable transition response can be observed. The fourth experiment, shown in Figure 7.5, tested the switching control scheme on the same plastic foam, resulting in a robust and stable contact transition. The fifth experiment, shown in Figure 7.6, tested the impact control

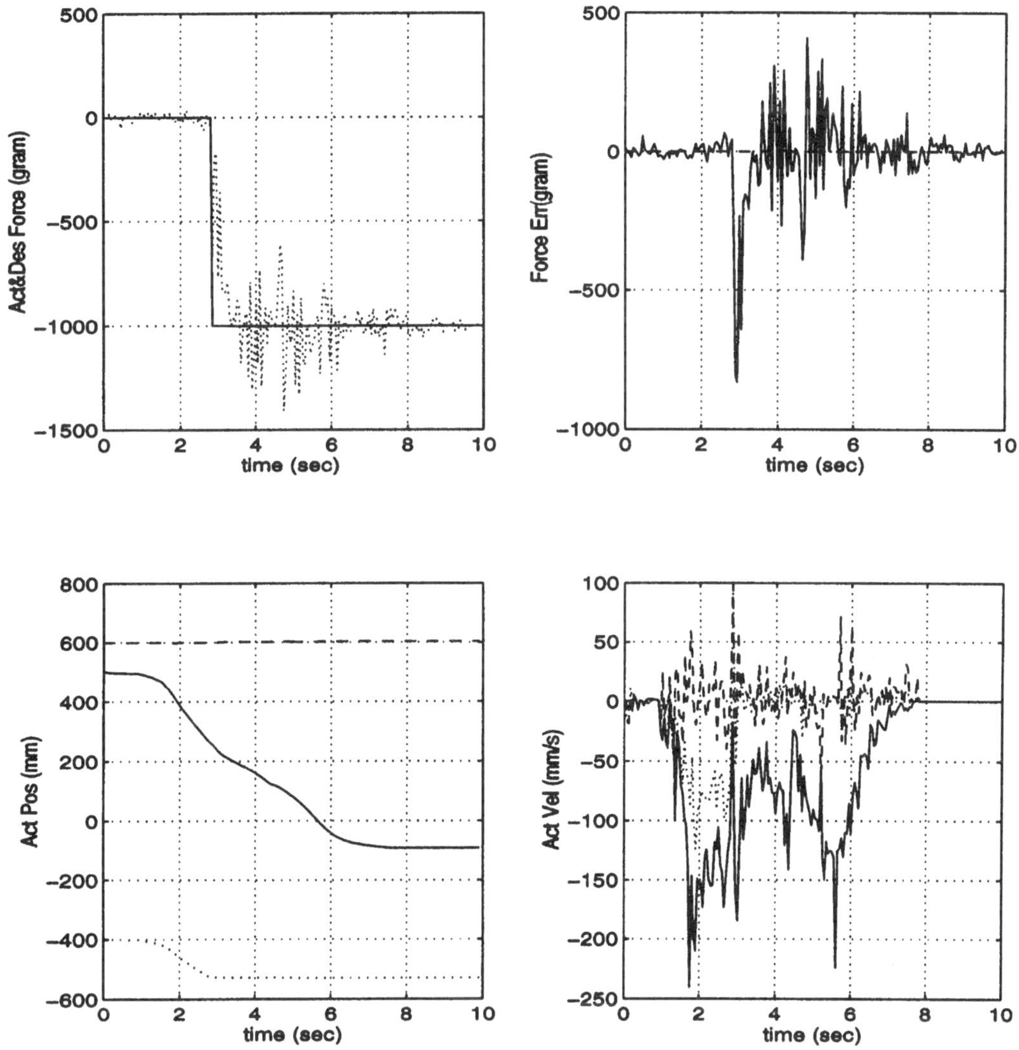

FIGURE 7.3
Impact control with switching on a rigid flat table made of compound material.

on a camel back–shaped steel surface of 0.1 inch thickness. A smooth and stable transition without loss of contact can be observed. The sixth experiment, shown in Figure 7.7, tested the switching control scheme on the camel back–shaped surface. Again, a robust and stable response is observed.

7 SUMMARY

This chapter discussed the remaining problems in the area of impact control and force regulation. The unknown and nonlinear relation between force and position is no longer a

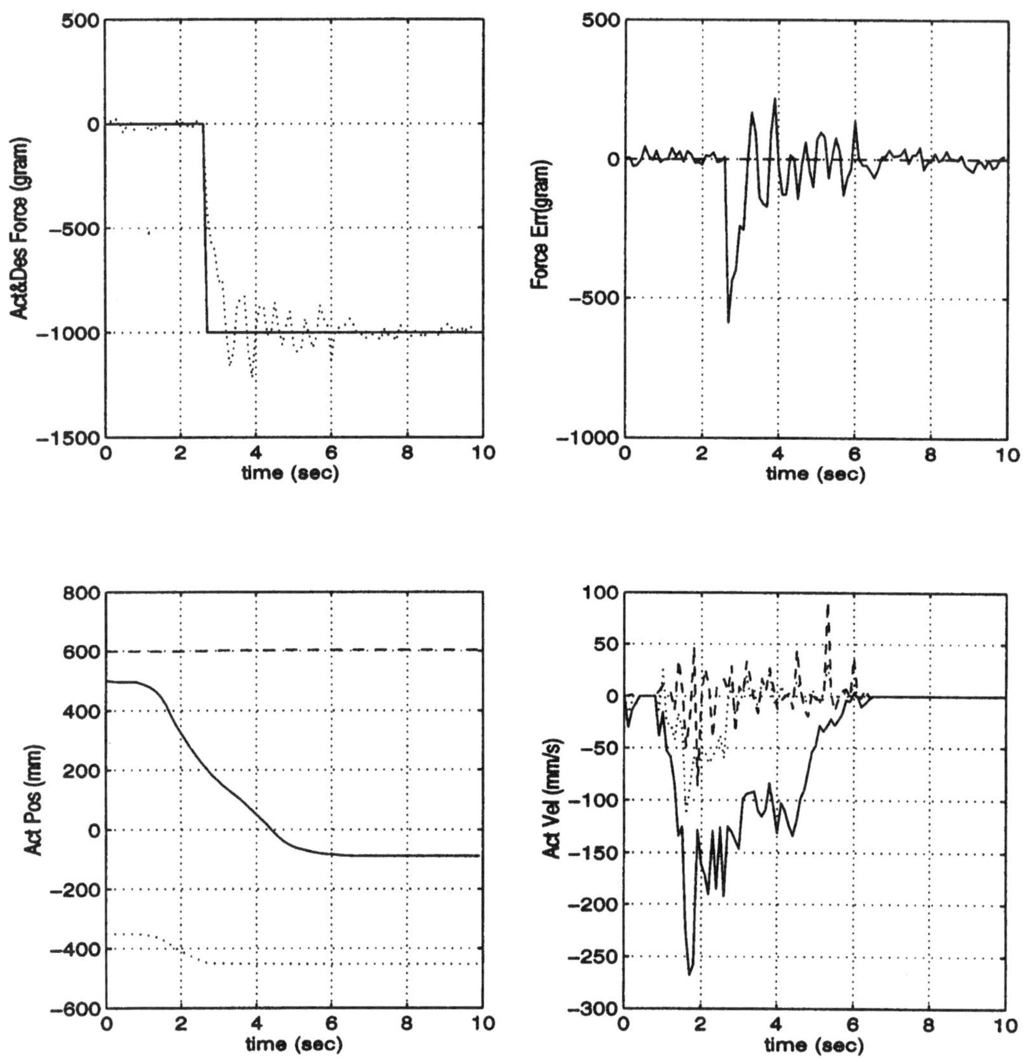

FIGURE 7.4
Impact control on a flat plastic foam.

difficulty in control design and stability proof. A robust and stable impact control strategy was found. The controller can regulate both position and force simultaneously. The transient force response during impact can be successfully controlled. The controller switches between position control and force control on a detectable force threshold to eliminate accidental loss of contact and reestablish contact softly.

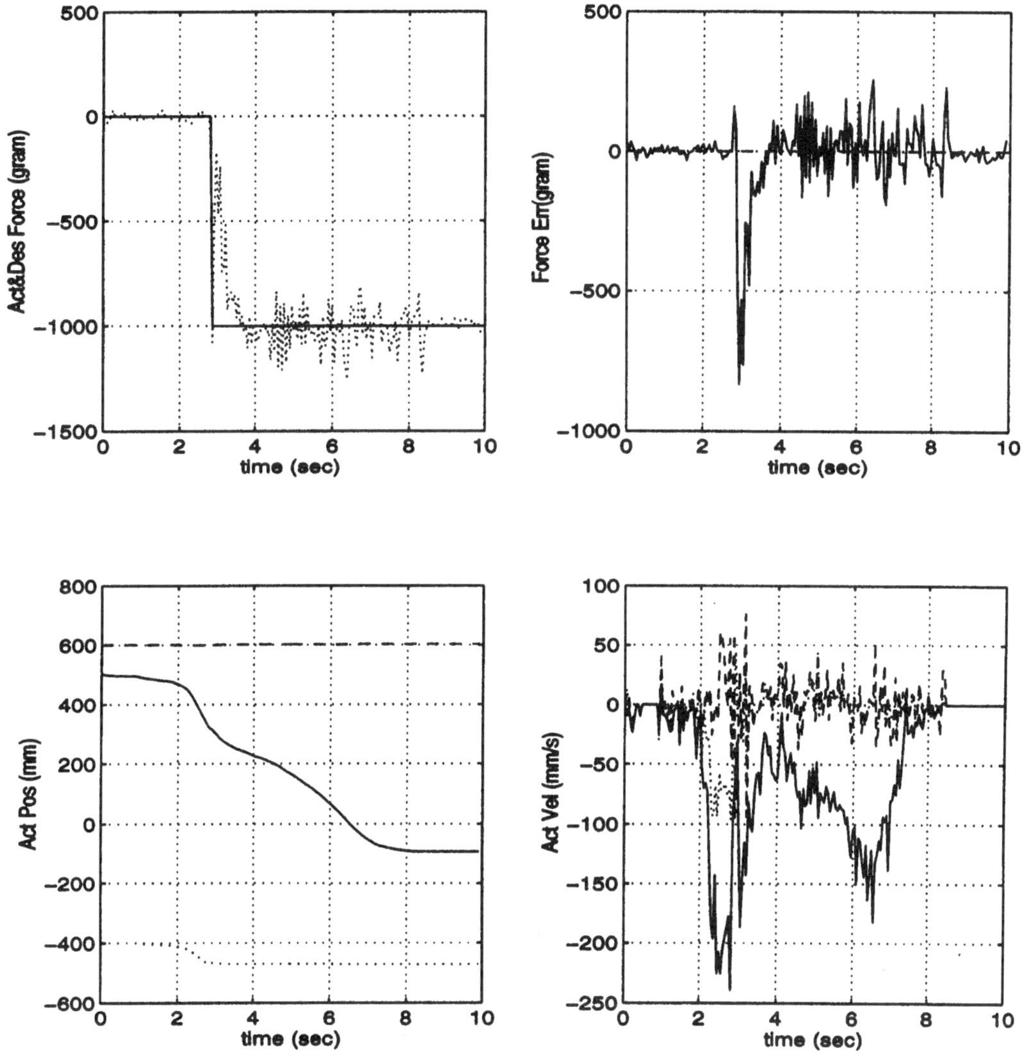

FIGURE 7.5
Impact control with switching on a flat plastic foam.

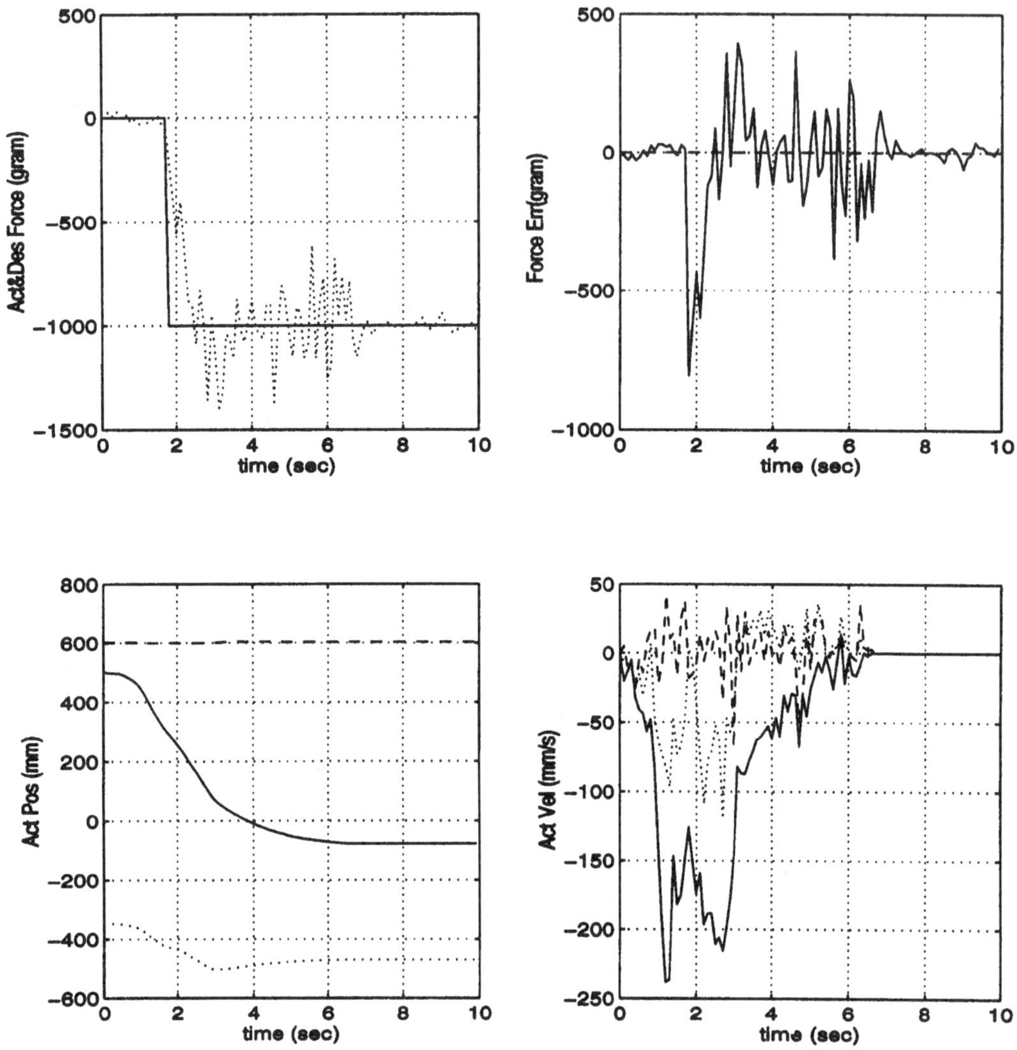

FIGURE 7.6
Impact control on a camel back–shaped steel surface.

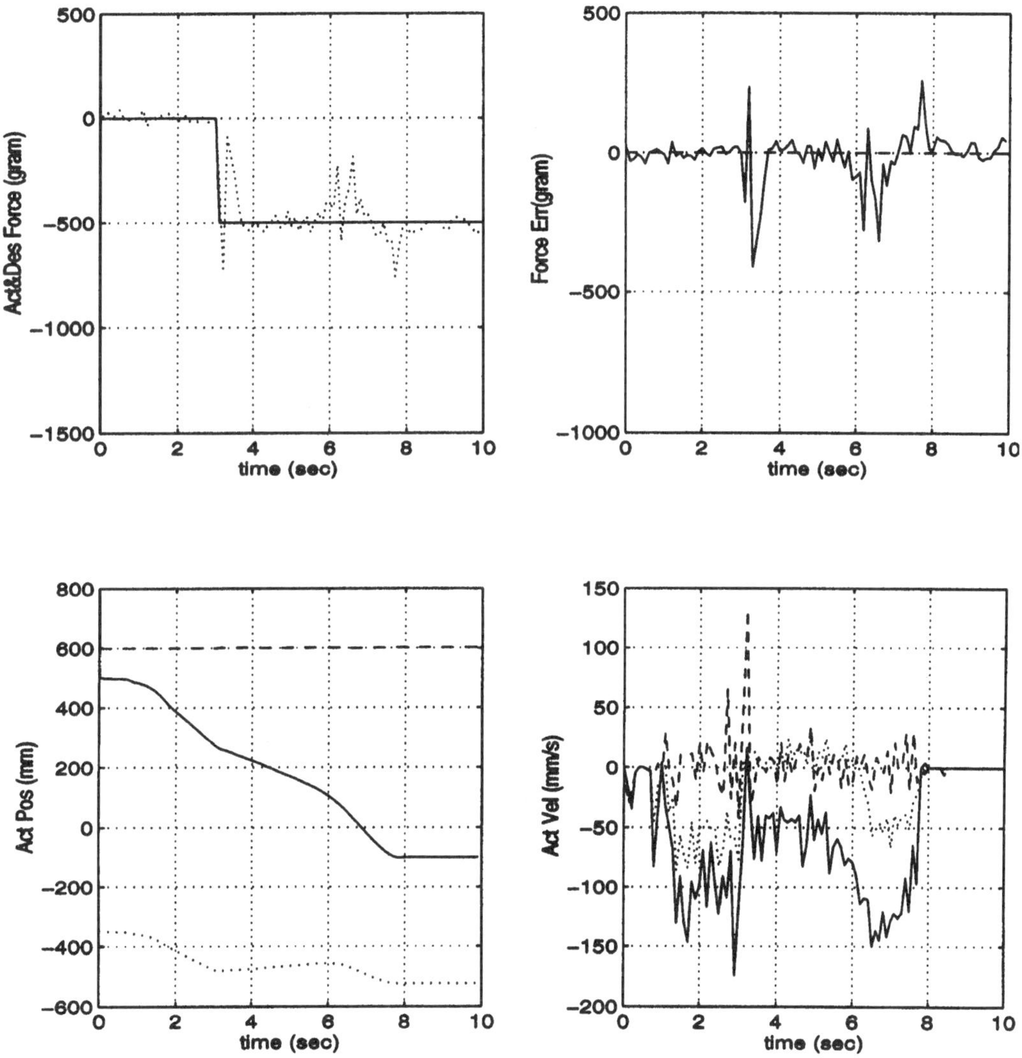

FIGURE 7.7
Impact control with switching on a camel back–shaped steel surface.

REFERENCES

[1] D. E. Whitney. Historical perspective and the state of the art in robot force control. *Int. J. Robot. Res.* 6(1):3–14, Spring 1987.

[2] W. T. Townsend and J. K. Salisbury. The effect of coulomb friction and stiction on force control. In *Proceedings of the IEEE International Conference on Robotics and Automation*, 1987, pp. 883–889.

[3] N. Hogan. Impedance control: An approach to manipulation: Part i — theory; part ii — implementation; part iii — applications. *J. Dynam. Syst. Meas. Control* 107:1–24, March 1985.

[4] M. H. Raibert and J. J. Craig. Hybrid position/force control of manipulators. *Trans. ASME* 102:126–133, June 1981.

[5] M. T. Mason. Compliance and force control for computer controlled manipulators. *IEEE Trans. Syst. Man Cybernet.* SMC-11(6):418–432, June 1981.

[6] O. Khatib. A unified approach for motion and force control of robot manipulations. The operational space formulation. *IEEE J. Trans. Robot. Automat.* RA-3(1):43–53, Feb. 1987.

[7] R. Volpe and P. Khosla. A theoretical and experimental investigation of impact control for manipulators. *Int. J. Robot. Res.* 12(4):352–365, Aug. 1993.

[8] G. T. Marth, T. J. Tarn, and A. K. Bejczy. Stable phase transition control for robot arm motion. In *Proceedings of the IEEE International Conference on Robotics and Automation*, 1993, pp. 355–362.

[9] T. Yoshikawa. Dynamic hybrid position/force control of robot manipulators — description of hand constraints and calculation of joint driving force. *IEEE Trans. Robot. Automat.* RA-3(5):386–392, Oct. 1987.

[10] H. Kazerooni, B. Waibel, and S. Kim. On the stability of robot compliant motion control: Theory and experiments. *Trans. ASME J. Dynam. Syst. Meas. Control* 112:417–425, 1990.

[11] N. H. McClamroch and D. Wang. Feedback stabilization and tracking of constrained robots. *IEEE Trans. Robot. Automat.* 33(5):419–426, 1988.

[12] S. A. Schneider and R. H. Cannon, Jr. Object impedance control for cooperative manipulation: Theory and experimental results. *IEEE Trans. Robot. Automat.* 8(3):383–395, June 1992.

[13] T. J. Tarn, A. K. Bejczy, and Ning, Xi. Hybrid position/force control of redundant robot with consideration of motor dynamics. In *13th Annual Allerton Conference on Communication, Control, and Computing*, Monticello, IL, 1992, pp. 702–711.

[14] R. R. Y. Zhen and A. A. Goldenberg. An adaptive approach to constrained robot motion control. In *Proceedings of the IEEE International Conference on Robotics and Automation*, vol. 2, 1995, pp. 1833–1838.

[15] B. Yao and M. Tomizuka. Robust adaptive constrained motion and force control of manipulator with guaranteed transient performance. In *Proceedings of the IEEE International Conference on Robotics and Automation*, vol. 1, 1995, pp. 893–898.

[16] L. Whitcomb, S. Arimoto, and F. Ozakki. Experiments in adaptive model-based force control. In *Proceedings of the IEEE International Conference on Robotics and Automation*, 1995, pp. 1846–1853.

[17] C. H. An and J. M. Hollerbach. Kinematic stability issues in force control of manipulators. In *Proceedings of the IEEE International Conference on Robotics and Automation*, Raleigh, NC, 1987, pp. 897–903.

[18] C. H. An and J. M. Hollerbach. Dynamical stability issues in force control of manipulators. In *Proceedings of the IEEE International Conference on Robotics and Automation*, Raleigh, NC, 1987, pp. 890–896.

[19] T. Yabuta, A. J. Chona, and G. Beni. On the asymptotic stability of the hybrid position/force control scheme for robot manipulators. In *Proceedings of the IEEE International Conference on Robotics and Automation*, Philadelphia, 1988, pp. 338–343.

[20] T. Yabuta, A. J. Chona, and G. Beni. On the asymptotic stability of the hybrid position/force control scheme for robot manipulators. In *Proceedings of the IEEE International Conference on Robotics and Automation*, 1988, pp. 338–343.

[21] N. Hogan. On the stability of manipulators performing contact tasks. *IEEE Trans. Robot. Automat.* 4(6):677–686, Dec. 1988.

[22] Tae-Sang Chung. An inherent stability problem in cartesian compliance and an alternative structure of compliance control. *IEEE Trans. Robot. Automat.* 7(1):21–30, Feb. 1991.

[23] S. D. Eppinger and W. P. Seering. Three dynamic problems in robot force control. *IEEE Trans. Robot. Automat.* 8(6):751–758, Dec. 1992.

[24] E. Paljug, V. Kumar, T. Sugar, and X. Yun. Some important considerations in force control implementation. In *Proceedings of the IEEE International Conference on Robotics and Automation*, 1992, pp. 1270–1275.

[25] S. D. Eppinger and W. P. Seering. Understanding bandwidth limitations in robot force control. In *Proceedings of the IEEE International Conference on Robotics and Automation*, Raleigh, NC, 1987, pp. 904–909.

[26] B. Bona and M. Indri. Exact decoupling of the force-position control using the operational space formulation. In *Proceedings of the IEEE International Conference on Robotics and Automation*, 1992, pp. 1394–1398.

[27] W. D. Fisher and M. S. Mujtaba. Hybrid position/force control: A correct formulation. *The International Journal of Robot. Res.* 11(4):299–311, Aug. 1992.
[28] W. McCormick nd H. M. Schwartz. An investigation of impedance control for robot manipulators. *Int. J. Robot. Res.* 12(5):473–489, Oct. 1993.
[29] R. Volpe and P. Khosla. A theoretical and experimental investigation of explicit force control strategies for manipulators. *IEEE Trans. Automat. Control* 38(11):1634–1650, Nov. 1993.
[30] D. W. Meer and S. M. Rock. Coupled-system stability of flexible-object impedance control. In *Proceedings of the IEEE International Conference on Robotics and Automation*, 1995, pp. 1839–1845.
[31] Y. F. Zheng and H. Hemami. Mathematical modeling of a robot collision with its environment. *J. Robot. Syst.* 2(3):289–230, 1985.
[32] Y. Wang and M. Mason. Modeling impact dynamics for robotic operations. In *Proceedings of the IEEE International Conference on Robotics and Automation*, Raleigh, NC, April 1987, pp. 678–685.
[33] K. Youcef-Toumi and D. A. Gutz. Impact and force control. In *Proceedings of the IEEE International Conference on Robotics and Automation*, Scottsdale, AZ, 1989, pp. 410–416.
[34] D. M. Lokhorst and J. K. Mills. Implementation of a discontinuous control law on a robot during collision with a stiff environment. In *Proceedings of the IEEE International Conference on Robotics and Automation*, Cincinnati, 1990, pp. 56–61.
[35] P. Akella, V. Parra-Vega, S. Arimoto, and K. Tanie. Discontinuous model-based adaptive control for robots executing free and constrained tasks. In *Proceedings of the IEEE International Conference on Robotics and Automation*, vol. 4, 1994, pp. 3000–3007.
[36] J. Hyde and M. Cutkosky. Contact transition control: An experimental study. In *Proceedings of the IEEE International Conference on Robotics and Automation*, 1993, pp. 363–368.
[37] B. Brogliato, P. Orhant, and R. Lozano. On the transition phase in robotics—part 1: Impact models and dynamics, part 2: Control. In *Proceedings of the IFAC Symposium on Robot Control*, vol. 2, Capri, Italy, 1994, pp. 562–576.
[38] D. Chiu and S. Lee. Robust jump impact controller for manipulators. In *Proceedings of the 1995 IEEE/RSJ International Conference on Intelligent Robots and Systems*, vol. 1, 1995, pp. 299–304.
[39] G. T. Marth. *Sensor Referenced Event Based Control for Stable Phase Transitions to Constrained Manipulator Motion.* Ph.D. thesis, Washingon University, 1993.
[40] R. M. Brach. *Mechanical Impact Dynamics.* John Wiley & Sons, 1991.
[41] Unimation 500 series equipment. Manual.
[42] Motion Tek manual for the universal motion controller.
[43] Technical report of the Silicon Graphics power series.
[44] User's manual for fsa-3254-f and fsa 3254-p force/torque sensors. California Cybernetics, 1992.
[45] T. J. Tarn, A. K. Bejczy, G. T. Marth, and A. K. Ramadorai. Performance comparison of four manipulator servo schemes. *IEEE Control. Syst. Mag.* 13(1):22–29, Feb. 1993.
[46] T. J. Tarn, N. Xi, and A. K. Bejczy. Motion planning in phase space for intelligent robot arm control. In *Proceedings IEEE/RSJ International Conference on Robots and Systems*, Raleigh, NC, 1992, pp. 1507–1514.

SUGGESTED READINGS

F. Berlin, W. M. Grimm, and P. M. Frank. On robot motion control with acceleration feedback. In *IFAC Symposium on Robot Control '88*, Karlsruhe, 1988, pp. 60.1–60.6.
F. Berlin, W. M. Grimm, and P. M. Frank. On robot motion control with acceleration feedback. In *International Federation of Automatic ontrol, Symposium on Robot Control*, Karlsruhe, Oct. 1988, pp. 60.1–60.6.
John J. Craig. *Introduction to Robotics, Mechanics and Control.* Addison-Wesley, 1989.
M. R. Cutkosky, and I. Kao. Computing and controlling the compliance of a robotic hand. *IEEE Trans. Robot. Automat.* RA-5(2):151–165, April 1989.
A. De Luca and C. Manes. On the modeling of robots in contact with a dynamic environment. In *Proceedings of 5th International Conference on Advanced Robotics, ICAR '91*, June 1991.
A. De Luca and C. Manes. Hybrid force–position control for robots in contact with dynamic environment. In *Preprints of 3rd IFAC Symposium on Robot Control, SYROCO '91*, Vienna, 1991, pp. 377–382.
S. D. Eppinger and W. P. Seering. On dynamic models of robot force control. In *Proceedings of the IEEE International Conference on Robotics and Automation*, San Francisco, 1986, pp. 29–34.
K. S. Fu, R. C. Gonzalez, and C. S. G. Lee. *Robotics: Control, Sensing, Vision, and Intelligence.* McGraw-Hill, New York, 1987.
W. Hahn. *Stability of Motion.* Springer-Verlag, New York, 1967.

A. Isidori. *Nonlinear Control Systems — An Introduction*, 2nd ed. Springer-Verlag, 1989.

O. Khatib and J. Burdick. Motion and force control of robot manipulator. In *Proceedings of the IEEE International Conference on Robotics and Automation*, 1986, pp. 1381–1386.

P. V. Kokotouic, H. K. Khalil, and J. O'Reilly. *Singular Perturbation Methods in Control: Analysis and Design*. Academic Press, 1986.

M. B. Leahy, Jr. Model-based control of industrial manipulators: An experimental analysis. *J. Robot. Syst.* 7(5):741–758, 1990.

J. Y. S. Luh. An anatomy of industrial robots and their control. In R. C. Gonzalez, C. S. G. Lee, and K. S. Fu, eds. *Tutorial on Robotics*. IEEE Computer Society Press, 1983, pp. 5–25.

J. K. Mills. Manipulator transition to and from contact tasks: A discontinuous control approach. In *Proceedings of the IEEE International Conference on Robotics and Automation*, Cincinnati, OH, 1990, pp. 56–61.

J. K. Mills. Stability of robotic manipulators during transition to and from compliant motion. *Automatica* 26(5):861–874, 1990.

J. K. Mills and D. M. Lokhorst. Experimental results in manipulator contact task control. In *Proceedings of the IEEE International Conference on Robotics and Automation*, Sacramento, CA, April 1991, pp. 1645–1651.

J. K. Mills and C. V. Nguyen. Robotic manipulator collisions: Modeling and simulation. *Trans. ASME J. Dynam. Syst. Meas. Control* 114:650–658, Dec. 1992.

J. K. Mills and D. M. Lokhorst. Control of robotic manipulators during general task execution: A discontinuous control approach. *Int. J. Robot. Res.* 12(2):861–874, April 1993.

R. P. Paul and B. Shimano. Compliance and control. In *Proceedings Joint Automatic Control Conference*, 1976, pp. 694–699.

M. C. Readman. *Flexible Joint Robots*. CRC Press, 1994.

J. K. Salisbury. Active stiffness control of a manipulator in cartesian coordinates. In *IEEE Conference on Decision and Control*, Dec. 1980, pp. 95–100.

M. W. Spong and M. Vidyasagar. *Robot Dynamics and Control*. John Wiley & Sons, 1989.

T. J. Tarn, A. K. Bejczy, A. K. Isidori, and Y. L. Chen. Nonlinear feedback in robot arm control. In *Proceedings of 23rd IEEE Conference on Decision and Control*, Las Vegas, 1984.

T. J. Tarn, A. K. Bejczy, and X. Yun. Dynamic equations for Puma 560 robot arm. Lab Report SSM-RL-85-02, Dept. of SSM, Washington University, St. Louis, 1985.

T. J. Tarn, A. K. Bejczy, and X. Yun. Robot arm force control through system linearization by nonlinear feedback. In *Proceedings of the IEEE International Conference on Robotics and Automation*, Philadelphia, 1988, pp. 1618–1625.

T. J. Tarn, A. K. Bejczy, Z. Li, and S. Ganguly. Dynamic equations for the Puma 560 robot arm (revised). Robotics Laboratory Report SSM-RL-85-13, Dept. of Systems Science and Mathematics, Washington University, St. Louis, Oct. 1989.

T. J. Tarn, G. T. Marth, A. K., Bejczy, and A. K. Ramadorai. Kinematic characterization of the Puma 560 manipulator. Lab Report SSM-RL-91-15, Dept. of SSM, Washington University, St. Louis, Dec. 1991.

T. J. Tarn, Y. Wu, N. Xi, A. Isidori. Force regulation and contact transition control. *IEEE Control Systems Magazine*, vol. 161, Feb. 1996, pp. 32–40.

M. Vidyasagar. *Nonlinear Systems Analysis*. Prentice-Hall International, Englewood Cliffs, NJ, 1993.

C. H. Wu and R. P. Paul. Resolved motion force control of robot manipulator. *IEEE Trans. Syst. Man Cybernet.* SMC-12(3):266–275, June 1982.

Y. Wu, T. J. Tarn and N. Xi. Force and transition control with environmental uncertainties. *Proceedings of the IEEE International Conference on Robotics and Automation*, Vol. 1, 1995, pp. 899–904.

Y. Wu, T. J. Tarn, N. Xi, and A. Isidori. On robust impact control via positive acceleration feedback for robot manipulators. *Proceedings of the IEEE International Conference on Robotics and Automation*, Vol. 2, 1996, pp. 1891–1896.

T. Yoshikawa, T. Sugie, and M. Tanaka. Dynamic hybrid position/force control of robot manipulators — controller design and experiment. *IEEE J. Robot. Automat.* 4(6):699–705, Dec. 1988.

R. R. Y. Zhen and A. A. Goldenberg. Robust position and force control of robots using sliding mode. In *Proceedings of the IEEE International Conference on Robotics and Automation*, San Diego, 1994, pp. 623–628.

Y. F. Zheng and F. R. Sias, Jr. Two robot arms in assembly. In *Proceedings of the IEEE International Conference on Robotics and Automation*, San Francisco, 1986, pp. 1230–1235.

SECTION IV

System Integration, Modeling, and Controller Design

CHAPTER 8

A Modular Approach to Sensor Integration

R. J. ANDERSON
Intelligent Systems and Robotics Center, Sandia National Laboratories, Albuquerque, New Mexico

1 INTRODUCTION

In this chapter we will demonstrate a method whereby sensor information can be integrated into a robot–telerobot system in a modular fashion. The focus will be on constraint information imposed by proximity sensors such as capacitive and ultrasonic sensors, but the method can be extended to vision or force sensor feedback.

The framework is a nonlinear control system called SMART (Sandia's Modular Architecture for Robotics and Teleoperation) that utilizes passivity and network concepts [1–3]. Each module in the SMART system represents an input device, a robot, a constraint, a kinematic mapping, or a sensor input and can be represented by a network equivalent. Systems are derived by combining modules in different telerobotic behavioral modes. SMART has been described in various other publications [4–7], and has been applied to problems as diverse as multirobot control, redundant robot control, and waste storage tank operations. In this chapter, however, we will focus on sensor integration issues. In so doing we will delve into the theoretical specifics of SMART, demonstrating how modularity is achieved in a discrete domain and how nonlinear mappings can be integrated with sensor data to enforce constraints. The system we will develop will be modular, both theoretically and in actual implementations. Any module in a SMART system can be run on any processing unit in a multiprocessor environment.

Modularity is especially important when integrating multiple proximity sensors in telerobotic systems, because there are various potential sensors with complementary capabilities, which can be configured in numerous different ways. For instance, capacitive sensors such as the Sandia WHAP (Whole Arm Proximity) sensor are good for short-range (4 to 12 inches) detection of metals and other materials with a distinct dielectric constant. They can see these objects instantaneously without reflections and independently of surface properties. Ultrasonic sensors, on the other hand, operate independently of object materials but are sensitive to surface textures and have an inherent lag in reading distances. Infrared sensors are instantaneous and are independent of surface materials, but suffer problems in dusty, hazy conditions.

A robot working in a hazardous environment may include a complement of sensors whose primary aim is to prevent the robot from running into unmapped obstacles. In this case the sensors must fulfill a number of requirements. First, they must not interfere with the normal operation of the arm. Second, they should never cause the arm to become unstable, even when multiple sensors from different sources are providing conflicting information. Third, when an object is detected, the sensor must provide information to guide the manipulator away from the object. Finally, the system must be flexible enough to allow the addition or deletion of sensors.

Combining feedback from multiple sensors in a modular environment is problematic. In order to accommodate sensor feedback in a modular fashion, we must first overcome the problem of algebraic instability. This problem is first described and then overcome by the use of scattering theory. A simple example will be used to describe the problem and then motivate the solution.

Next, constraints are fundamentally nonlinear and thus have the potential for all kinds of unpredictable behavior if imposed carelessly. We must introduce constraint behavior in a fashion that will guarantee the existence, uniqueness, and stability of smooth trajectories. This will be addressed in the subsequent sections.

Finally, a number of examples using the SMART architecture are given showing how sensors, telerobotic input devices, and robots can be coordinated in a modular fashion. We then demonstrate how nonlinear boundary function behavior can be implemented in a stable fashion consistent with this architecture, utilizing various proximity sensors in various different modes of operation.

2 TERMINOLOGY

In this chapter networks will follow the effort flow analogy [8], which relates force (effort) to voltage and velocity (flow) to current. We will use lowercase letters to represent signals in the system and uppercase letters to represent operators. Subscripts will denote the space in which these signals are based, which for robot systems is typically $\Re^n$.

Operators that map signals from a space into itself will have a single subscript denoting the space. Thus, if input and output signals are both in $\Re^n$, the operator is an $n \times n$ square matrix with a single subscript. If the operator maps from one space to a different space, such as the Jacobian operator, then two subscripts separated by a backslash will be used denoting the two spaces. If the operator maps the Cartesian product of two spaces, an $i \times j$ subscript is used. The operators derived from this analogy are summarized in Tables 8.1 and 8.2.

Also, a number of unfamiliar terms (at least to a robotics audience) will be introduced here [9]. These are summarized in Table 8.3 for completeness.

3 THE PROBLEM OF ALGEBRAIC LOOPS

In a network representation of a system we are accustomed to thinking in terms of standard state variables, such as effort, flow, force, velocity, voltage, and current. Unfortunately, when the system is discretized, either for simulation or for control, these variables are generally the wrong variables, as their usage can introduce nonpassivity and instability into a system whose ideal continuous representation would be passive and robustly stable. Overcoming

Table 8.1 Network Variables

Symbol	Description
s	Laplace frequency
d	Unit delay ($d = z^{-1}$)
v_i	Velocity, angular velocity
f_i, fd_i	Force or torque, desired force or torque
x_i, xd_i	Position, desired position
q_i, R_i	Orientation (quaternion, rotation matrix)

Table 8.2 Network Operators

Symbol	Name/analogy	Equation	Symbol
M_i	Inertial/inductance	$f_i = \frac{d}{dt}(M_i v_i)$	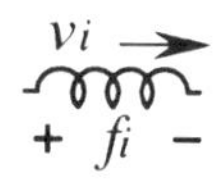
B_i	Damping/resistance	$f_i = B_i v_i$	
K_i	Stiffness/capacitance	$f_i = \int K_i dx_i$	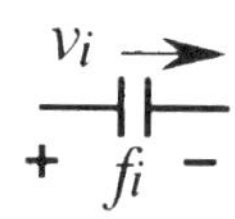
$J_{j/i}$	Jacobian/transformer	$v_j = J_{j/i} v_i$ $f_i = J_{j/i}^T f_j$	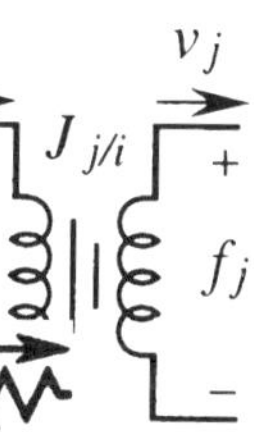
$Z(s)$	Impedance	$f_i = Z_i(s) v_i$	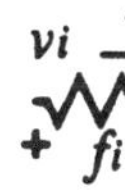

Table 8.3 Scattering Operator Terminology

Symbol	Name	Equation
$Z_0, Z0_p$	Characteristic impedance	$Z0_p = Z_0 I_p$ (scaled identity)
a_i	Input wave	$a_i = f_i + Z_0 v_i$
b_i	Output wave	$b_i = f_i - Z_0 v_i$
$S_i, S_{i \times j}$	Scattering operator	$b_i = S_i a_i$ $\begin{bmatrix} b_i \\ b_j \end{bmatrix} = S_{i \times j} \begin{bmatrix} a_i \\ a_j \end{bmatrix}$

this problem is the key to designing SMART modules. Before developing these techniques for our prototype telerobot system, we shall start with a much simpler example.

3.1 Modularizing a Velocity Divider Network

Consider Figure 8.1.

It consists of an ideal velocity source, $v_d(t)$, and two damping terms, B_1 and B_2. The constituent equations for the network elements are

$$f_1 = -B_1 v_1 \tag{8.1}$$

$$f_2 = B_2 v_2 \tag{8.2}$$

and the connection equations are given by

$$v_2 = v_1 + v_d \tag{8.3}$$

$$f_1 = f_2 \tag{8.4}$$

which, when combined, result in the current (velocity) divider equations:

$$v_1 = -(B_1 + B_2)^{-1} B_2 v_d \qquad v_2 = (B_1 + B_2)^{-1} B_1 v_d \tag{8.5}$$

Suppose, however, that these equations are to be implemented in a modular fashion, and the constituent equations and the connection equations for the network are computed separately.

In this case a sample delay is incurred, because some computations depend on results of other computations. If the connection equations can be computed without sample delay, that is, for each time instant k,

$$v_2(k) = v_1(k) + v_d(k) \tag{8.6}$$

$$f_1(k) = f_2(k) \tag{8.7}$$

then the constituent equations must contain some sampling delay, as shown below:

$$v_1(k) = -B_1^{-1} f_1(k) \tag{8.8}$$

$$f_2(k) = B_2 v_2(k-1) \tag{8.9}$$

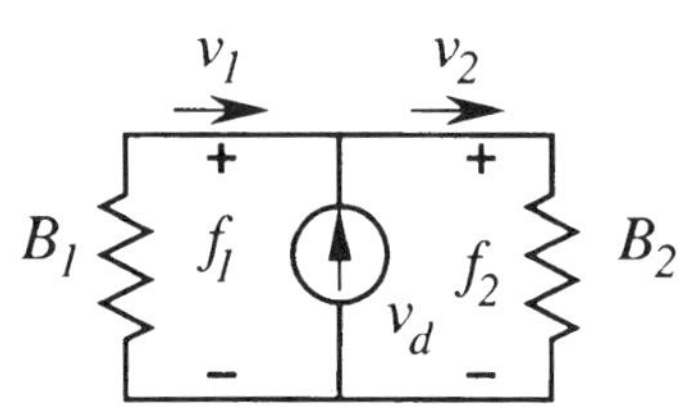

FIGURE 8.1
Velocity divider network.

Combining these equations yields

$$v_2(k) = v_1(k) + v_d(k) = -B_1^{-1}f_1(k) + v_d(k) = -B_1^{-1}B_2v_1(k-1) + v_d(k) \tag{8.10}$$

Letting $d = z^{-1}$ represent a unit sample delay, Eq. (8.10) can be solved to get the characteristic equation

$$0 = z + B_1^{-1}B_2 \tag{8.11}$$

For a scalar system this will be stable if and only if $|B_2| < |B_1|$. For a matrix-valued system with diagonal damping operators B_1 and B_2, each element of B_2 must be less than the corresponding diagonal element of B_1, and for matrix-valued damping the induced *two-norm*[1] condition

$$\|B_1^{-1}B_2\|_2 < 1 \tag{8.12}$$

must be satisfied.

Sampling in a different order may reverse the stability requirements, requiring that B_1 be less than B_2, but it will not eliminate the problem. As long as effort and flow variables are sampled, instability may arise due to sampling. This problem is called *algebraic instability*, where the ideal continuous system contains elements of the same order, which, when separated into modular components, result in additional discrete-time states. Thus, in our example, a zero-order damping system becomes a first-order system in the discrete domain. Likewise, two first-order coupled systems might result in a third-order discretized system. It is important to note that this problem is *independent of the sampling rate*. Thus the common belief that "sampling fast enough" will cause our discrete system to exhibit the behavior of the ideal continuous system is in general invalid for modular systems.

Now if the only goal was to design linear, time-invariant robotic control systems for single, rigid-body robots with known *a priori* interactions, then present methods would be adequate. The entire system could be combined into one large equation before sampling (as in Eq. (8.5)) or the stringent coupled stability requirements of Eq. (8.12) could be met. However, if the goal is to build flexible, expandable, control systems such as SMART, then discretization is mandatory and modularity is required, not because it is conceptually and computationally pleasing but because it is the only way to handle complex systems.

Luckily, there is a body of theory that can be applied, where stability is always maintained after discretization, where components can be designed in a completely modular fashion, and where the "sampling fast enough" adage will be true, even in the presence of additional discrete time states. The theory is called *scattering theory*.

4 SCATTERING THEORY

Scattering theory was developed to handle problems in transmission line analysis. Ideal lossless transmission lines transmit effort and flow (voltage and current in this case) across distances without losing energy and without changing the steady-state behavior of the signal. However, depending on the line impedance, Z_0, and the impedances terminating either end of the line, the transient behavior of the signal will be affected.

[1]The two-norm for a matrix, A is equal to the maximum singular values, that is, the maximum eigenvalue of the matrix A^TA.

4.1 Wave Variables

Scattering theory involves a change of basis. Instead of using effort–flow variables to describe dynamic time-delay phenomena at port connections, *wave variables* are used. Wave variables represent a bilinear map of the standard effort–flow variables. The input wave, a, is defined as the effort plus the scaled flow entering the port,

$$a = f + Z_0 v \tag{8.13}$$

where f is the effort (force) at the port terminals, v is the flow (velocity) entering the port at its terminals, and Z_0 is a constant representing the characteristic impedance. The output wave, b, is defined as the effort minus the scaled flow,

$$b = f - Z_0 v \tag{8.14}$$

For a two-port (Figure 8.2) with flow v_p entering the port from the previous port, with flow v_n exiting the port to enter the next port, and with efforts f_p, and f_n at the port terminals, the corresponding wave variables are defined as

$$a_p = f_p + Z_0 v_p \tag{8.15}$$

$$a_n = f_n - Z_0 v_n \tag{8.16}$$

$$b_p = f_p - Z_0 v_p \tag{8.17}$$

$$b_n = f_n + Z_0 v_n \tag{8.18}$$

If the underlying reference frame is matrix valued, then the same equations still apply, except that in this case Z_0 represents the scaled identity matrix.

4.2 Scattering Operators

Using this mapping of effort and flow variables into wave variables, we can determine how impedance, admittance, and hybrid operators can be mapped into *scattering operators*. A scattering operator, S, is the mapping between input waves variables, a, and output wave variables, b.

$$b = Sa \tag{8.19}$$

and can be readily derived from the constituent equations for an element.

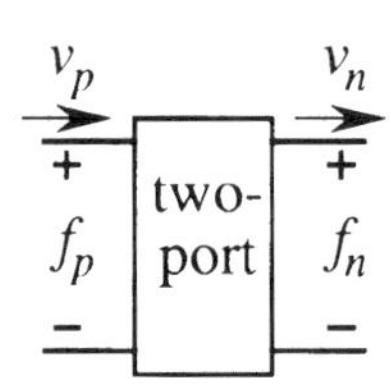

FIGURE 8.2
Two-port network.

Suppose a one-port is represented by its equivalent impedance, $Z(s)$,

$$f(s) = Z(s)v(s) \tag{8.20}$$

Then its input wave is defined by

$$a(s) = f(s) + Z_0 v(s) = (Z(s) + Z_0)v(s) \tag{8.21}$$

and its output wave is defined by

$$b(s) = f(s) - Z_0 v(s) = (Z(s) - Z_0)v(s) = (Z(s) - Z_0)(Z(s) + Z_0)^{-1}a(s) \tag{8.22}$$

and thus the scattering operator is defined by

$$S(s) = (Z(s) - Z_0)(Z(s) + Z_0)^{-1} \tag{8.23}$$

The inverse, $(Z(s) + Z_0)^{-1}$, whether scalar or matrix valued, is always well defined, since $Z(s)$ is positive real and Z_0 is strictly positive.

The beauty of the scattering approach is that it transforms passive impedances, which have Nyquist frequency plots in the right half-plane, to passive scattering operators, which have Nyquist frequency plots in the unit circle.

Suppose we have a linear time-invariant (LTI) one-port consisting solely of passive elements (e.g., masses, springs, dampers, and Jacobians) with driving point impedance $Z(s)$. It is a well-known result of network theory [1] that the Nyquist plot of the impedance will lie entirely in the right half-plane. Likewise, the singular values for a passive two-port impedance operator or admittance operator will also lie entirely in the closed right half-plane.

Thus, for an LTI system all of the following are equivalent:

- An n-port is passive.
- An n-port can be represented by a network of passive elements.
- An n-port has an impedance operator with a Nyquist plot contained entirely in the right half-plane.
- An n-port has a scattering operator with frequency response in the unit circle.
- An n-port has a scattering operator with norm less than or equal to one.

The real beauty of the scattering operator approach for robotic systems is that the scattering operator representation is as valid for nonlinear systems as it is for linear systems. Thus, although we cannot talk about Nyquist plots for nonlinear systems, we can still maintain the following equivalence:

- An n-port is passive.
- An n-port can be represented by a network of passive elements.
- An n-port has a scattering operator with norm less than or equal to one.

4.3 Revisiting the Velocity Divider Problem

Let us now apply the scattering approach to the velocity divider problem. In terms of wave variables, the connection equations are given by

$$a_1(k) = f_1(k) - Z_0 v_1(k) = f_2(k) - Z_0 v_2(k) + Z_0 v_d(k) = b_2(k) + Z_0 v_d(k) \tag{8.24}$$

$$a_2(k) = f_2(k) + Z_0 v_2(k) = f_1(k) + Z_0 v_1(k) + Z_0 v_d(k) = b_1(k) + Z_0 v_d(k) \tag{8.25}$$

and the constituent equations are given by

$$b_1(k) = S_1 a_1(k) \tag{8.26}$$

$$b_2(k) = S_2 a_2(k-1) \tag{8.27}$$

where $S_1 = (B_1 - Z_0)(B_1 + Z_0)^{-1}$ and $S_2 = (B_2 - Z_0)(B_2 + Z_0)^{-1}$.

Combining these equations then yields

$$\begin{aligned} a_2(k) = b_1(k) + Z_0 v_d(k) &= S_1 a_1(k) + Z_0 v_d(k) = S_1(b_2(k) + Z_0 v_d(k)) + Z_0 v_d(k) \\ &= S_1 S_2 a_2(k-1) + (S_1 Z_0 + Z_0) v_d(k) \end{aligned} \tag{8.28}$$

which has the characteristic equation

$$0 = z - S_1 S_2 \tag{8.29}$$

Equation (8.29) has all of its poles in the unit circle, since $\|S_1\| < 1$ and $\|S_2\| < 1$ for any positive damping B_1 and B_2. Furthermore, this will be true no matter what sampling strategy is chosen. This will also be true even if B is a nonlinear function of position and/or velocity, as long as it remains positive definite at all times.

Thus, when a system represented by effort–flow variables is modularized and sampled directly (as in Section 3.1), it is likely to become unstable. If, on the other hand, its impedance is represented in terms of scattering operators, wave variables rather than effort–flow variables are sampled. Thus, when the system is then modularized into individual ports, a discrete time modular system is derived that is as stable and robust as the original compound continuous system. The reason for this is summarized by the following theorem.

Theorem 1 *Any delayed passive scattering operator, $e^{-sT}S(s)$, remains passive.*

Proof

$$\|e^{-sT}S(s)\| \leqslant \|e^{-sT}\| \, \|S(s)\| \leqslant \|S(s)\| \leqslant 1 \quad \square \tag{8.30}$$

Thus, because a pure phase shift cannot take the frequency response of the scattering operator outside the unit circle, it cannot affect the passivity of the scattering operator. Because all elements in the system remain passive, even after sampling, the stability properties are unchanged. As long as we can represent modules in the system in terms of scattering operators, it is possible to convert between discrete domain and continuous domains without introducing instability mechanisms.

5 APPLYING SCATTERING THEORY TO ROBOT MODULES

To illustrate how scattering operator techniques can be applied to robot control modules, we will develop scattering operator maps for a number of simple ports. We will first derive

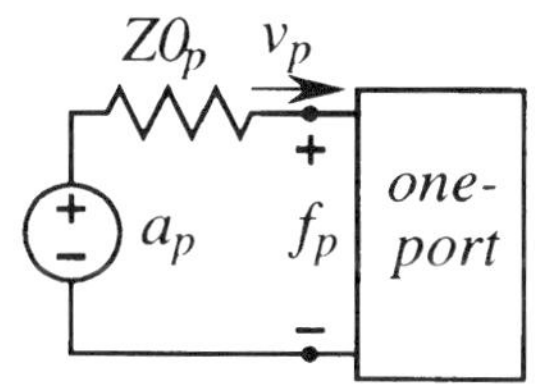

FIGURE 8.3
One-port network with driving force a_p.

the scattering operator equations for two linear time-invariant ports and then tackle operator inputs and kinematic mappings.

5.1 One-Port Scattering Operators

The simplest way to compute the scattering operator for a network is to apply a wave effort source in series with the characteristic impedance at each port and then apply superposition of the inputs. For example, for a one-port, as shown in Figure 8.3, we can apply a driving signal, a_p, and a series impedance, Z_0, which results in the standard wave equation,

$$a_p = f_p + Z_0 v_p \tag{8.31}$$

Then, using standard network analysis techniques we can derive

$$b_p = f_p - Z_0 v_p \tag{8.32}$$

as a function of the input wave, a_p.

Example 1 KB1 Port Scattering Operator. Consider the stiffness-damping one-port (KB1 port) shown in Figure 8.4 with driving wave attached.

Applying Kirchhoff's laws, we get

$$a_n = -(K/s + B + Z_0)v_n \tag{8.33}$$

Therefore we get for the wave equation

$$b_n = a_n + 2Z_0 v_n = \left(I - 2Z_0\left(K\left(\frac{1}{s}\right) + B + Z_0\right)^{-1}\right)a_n$$

$$= (K + Bs - Z_0 s)(K + Bs + Z_0 s)^{-1} a_n = S_n a_n \tag{8.34}$$

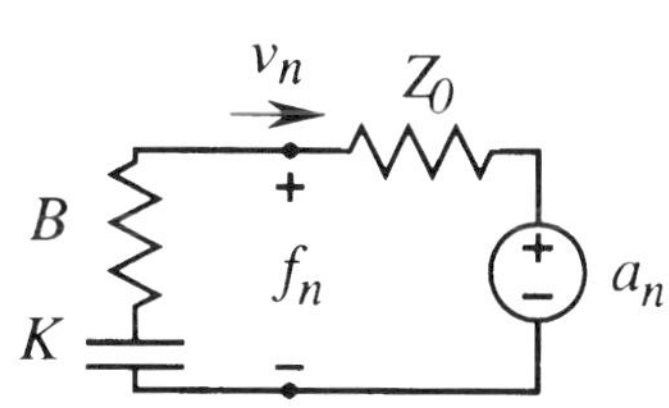

FIGURE 8.4
KB1 port with driving wave.

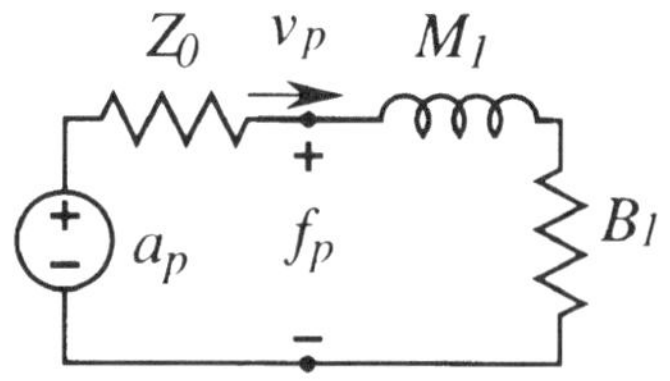

FIGURE 8.5
MB1 port with driving force a_p.

where the scattering operator, S_n, is given by

$$S_n = (K + Bs - Z_0 s)(K + Bs + Z_0 s)^{-1} \tag{8.35}$$

Example 2 MB1 Port Scattering Operator. Using the same technique for the MB1 port (Figure 8.5), we get

$$a_p = (M_1 s + B_1 + Z_0) v_p \tag{8.36}$$

Therefore we can compute the wave equation

$$\begin{aligned} b_p = a_p - 2Z_0 v_p &= (I - 2Z_0(M_1 s + B_1 + Z_0)^{-1}) a_p \\ &= (M_1 s + B_1 - Z_0)(M_1 s + B_1 + Z_0)^{-1} a_p = S a_p \end{aligned} \tag{8.37}$$

where the scattering operator for the MB1 port is given by

$$S = (M_1 s + B_1 - Z_0)(M_1 s + B_1 + Z_0)^{-1} \tag{8.38}$$

5.2 Two-Port Scattering Operators

We can apply the same technique to two-ports, by connecting a input driving wave at both port connections, as shown in Figure 8.6.

Example 3 Spaceball Port. Consider the spaceball two-port in Figure 8.7 with both driving waves attached. The spaceball adds a desired velocity command signal, v_d, to the input from the previous port to get the velocity input for the next port. It also adds a small amount of symmetrically placed damping, B, to help reduce wave reflections in the module.

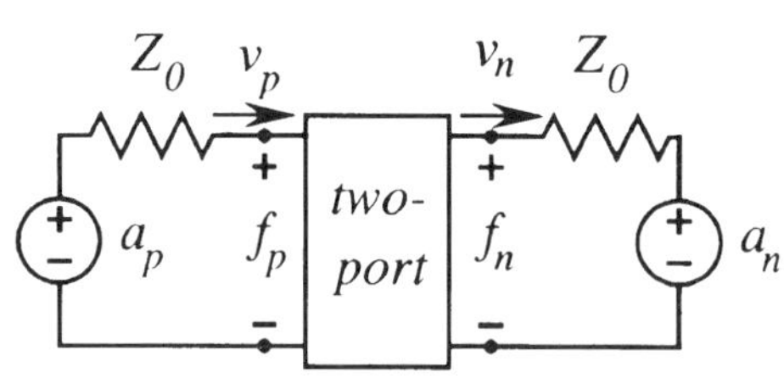

FIGURE 8.6
Generic two-port with driving waves.

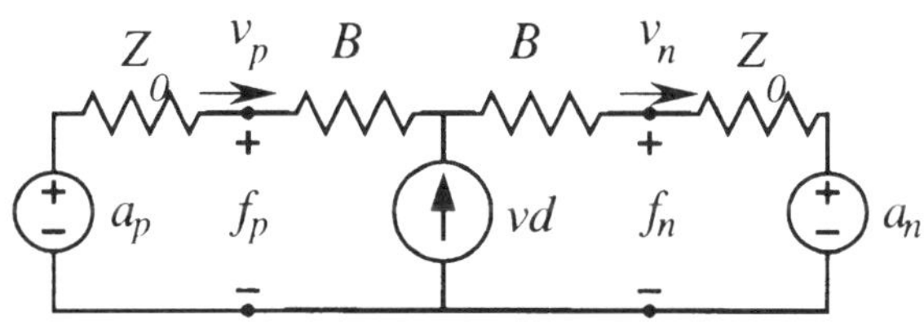

FIGURE 8.7
Spaceball port with driving waves.

Applying Kirchhoff's laws and utilizing symmetry,

$$v_p = -v_n = \tfrac{1}{2}(Z_0 + B)^{-1}(a_p - a_n) + \tfrac{1}{2}v_d \tag{8.39}$$

and thus for the wave equations[2]:

$$\begin{bmatrix} b_p \\ b_n \end{bmatrix} = \begin{bmatrix} a_p \\ a_n \end{bmatrix} - 2Z_0\begin{bmatrix} v_p \\ -v_n \end{bmatrix} = \begin{bmatrix} a_p \\ a_n \end{bmatrix} - Z_0(Z_0 + B)^{-1}\begin{bmatrix} I & -I \\ -I & I \end{bmatrix}\begin{bmatrix} a_p \\ a_n \end{bmatrix} - Z_0\begin{bmatrix} vd \\ vd \end{bmatrix}$$

$$= (Z_0 + B)^{-1}\begin{bmatrix} B & Z_0 \\ Z_0 & B \end{bmatrix}\begin{bmatrix} a_p \\ a_n \end{bmatrix} - Z_0\begin{bmatrix} vd \\ vd \end{bmatrix} = S\begin{bmatrix} a_p \\ a_n \end{bmatrix} - Z_0\begin{bmatrix} vd \\ vd \end{bmatrix} \tag{8.40}$$

6 COMPUTING THE JACOBIAN SCATTERING OPERATOR

A Jacobian operator, J, can be used to define any mapping between reference frames that satisfies the principle of virtual work. In robotics it typically refers to the velocity mapping between a robot's joint space and its world space motions. In SMART the Jacobian is used for a wide variety of variable mappings. It is used to map between different frames on a manipulator, it is used to implement nonlinear boundary functions, and it is used to map between spaces with different degrees of freedom. More than any other element, it enables the versatility and generality of the vector network approach.

Unlike the dynamic elements in the previous section, the Jacobian is memoryless; that is, no energy is stored in the element and no states are needed to describe its behavior. Thus the discrete-time implementation should be identical to the continuous-time implementation. For this reason it is a good location for representing any nonlinear behavior in the system. Nevertheless, when a system is implemented in a modular form, the use of Jacobian mappings can lead to algebraic loops and discrete instability. Furthermore, because the Jacobian can be highly nonlinear, proving stability for the sampled system may be difficult using standard discretization techniques. For this reason, Jacobian mappings will always be implemented in a scattering operator form. In this section the scattering operator representation for the Jacobian element will be derived.

Consider the Jacobian element shown in Figure 8.8.

The Jacobian, $J_{n/p}$, maps the velocities from space p to space n, according to the equation $v_n = J_{n/p}v_p$, and thus has the force mapping, $f_p = J^T_{n/p}f_n$. The goal is to determine the

[2]This scattering operator will work directly only for the three linear degrees of freedom (DOFs). The angular velocity DOFs can be computed in a similar fashion only by premultiplying and postmultiplying the wave variables by the appropriate rotation matrix. This technique is beyond the scope of this chapter.

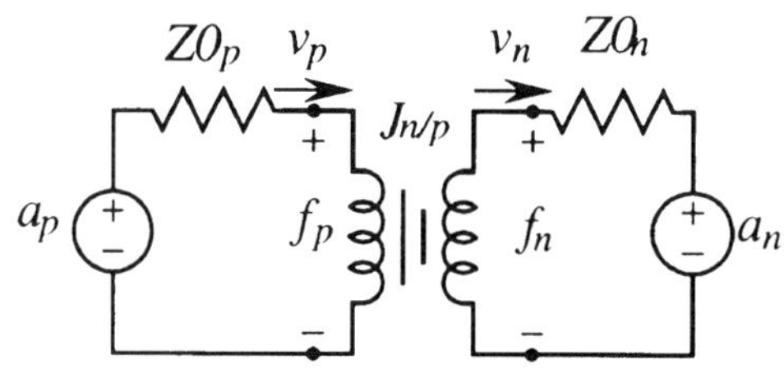

FIGURE 8.8
Jacobian element driven by wave inputs.

mapping between the input waves,

$$\begin{bmatrix} a_p \\ a_n \end{bmatrix} = \begin{bmatrix} f_p + Z0_p v_p \\ f_n - Z0_n v_n \end{bmatrix} \tag{8.41}$$

and the output waves,

$$\begin{bmatrix} b_p \\ b_n \end{bmatrix} = \begin{bmatrix} f_p - Z0_p v_p \\ f_n + Z0_n v_n \end{bmatrix} \tag{8.42}$$

Applying the loop equation to Figure 8.8 gives

$$a_p - Z0_p v_p = J_{n/p}^T a_n + J_{n/p}^T Z0_n v_n \tag{8.43}$$

since the impedance operators are just scaled identity operators, $Z0_p = Z_0 I_p$ and $Z0_n = Z_0 I_n$, it follows that

$$a_p - J_{n/p}^T a_n = (I_p + J_{n/p}^T J_{n/p}) Z0_p v_p \tag{8.44}$$

By defining the matrix

$$D_p = 2(I_p + J_{n/p}^T J_{n/p})^{-1} \tag{8.45}$$

the scaled output velocity is given by

$$2Z0_p v_p = D_p(a_p - J_{n/p}^T a_n) \tag{8.46}$$

and thus the output waves can be written as

$$\begin{bmatrix} b_p \\ b_n \end{bmatrix} = \begin{bmatrix} f_p - Z0_p v_p \\ f_n + Z0_n v_n \end{bmatrix} = \begin{bmatrix} a_p - 2Z0_p v_p \\ a_n + 2Z0_n J_{n/p} v_p \end{bmatrix} = \begin{bmatrix} I_p - D_p & D_p J_{n/p}^T \\ J_{n/p} D_p & I_n - J_{n/p} D_p J_{n/p}^T \end{bmatrix} \begin{bmatrix} a_p \\ a_n \end{bmatrix} \tag{8.47}$$

which implies that the scattering operator for the two-port Jacobian, $J_{n/p}$, is given by

$$S_{p \times n} = \begin{bmatrix} I_p - D_p & D_p J_{n/p}^T \\ J_{n/p} D_p & I_n - J_{n/p} D_p J_{n/p}^T \end{bmatrix} \tag{8.48}$$

It can also be shown that an equivalent representation for the scattering operator for the Jacobian two-port is

$$S_{p\times n} = \begin{bmatrix} J_{n/p}^T D_n J_{n/p} - I_p & J_{n/p}^T D_n \\ D_n J_{n/p} & D_n - I_n \end{bmatrix} \tag{8.49}$$

where

$$D_n = 2(I_n + J_{n/p} J_{n/p}^T)^{-1} \tag{8.50}$$

Clearly, the choice of implementation for $S_{p\times n}$ depends on which space is of lower rank, n or p, since the time to compute the matrix inversion (Eq. 8.45 or 8.50) is highly dependent on the rank of the matrix and is the dominant computation factor in computing the scattering operator.

In any case, the matrix inversion is always easily computable because the desired inverses are always positive definite symmetric. This is an important point; even if the Jacobian is singular or nonsquare, the scattering operator derived from the Jacobian is always well defined.

6.1 Computing the Norm of the Jacobian Scattering Operator

The imposed two-norm on a matrix-valued system is defined as the maximum singular value of the matrix. To compute the singular values for $S_{p\times n}$, we need to compute the eigenvalues of

$$\begin{aligned} S_{p\times n}^T S_{p\times n} &= \begin{bmatrix} I_p - D_p D_p J_{n/p}^T \\ J_{n/p} D_p D_n - I_n \end{bmatrix} \begin{bmatrix} I_p - D_p D_p J_{n/p}^T \\ J_{n/p} D_p D_n - I_n \end{bmatrix} \\ &= \begin{bmatrix} I - 2D_p + D_p^2 + D_p J^T J D_p & D_p J^T D_n - D_p^2 J^T \\ D_n J D_p - J D_p^2 & J D_p^2 J^T + D_n^2 - 2D_n + I \end{bmatrix} \end{aligned} \tag{8.51}$$

but $D_p J_{n/p}^T = J_{n/p}^T D_n$, and $D_p J^T J D_p + D_p^2 = 2D_p$, and $D_n J J^T D_n + D_n^2 = 2D_n$. Therefore,

$$S_{p\times n}^T S_{p\times n} = \begin{bmatrix} I_p & 0_{p/n} \\ 0_{n/p} & I_n \end{bmatrix} \tag{8.52}$$

and consequently all of the eigenvalues are identity, and thus

$$\|S_{p\times n}\|_2 = 1 \tag{8.53}$$

for any choice of $J_{n/p}$.

So, despite the fact that the Jacobian can vary widely as a function of position, the resulting scattering operator is always orthonormal with a norm of one. This somewhat surprising mathematical result is not surprising at all when the physics of the system are analyzed. Because the Jacobian represents a lossless transformation and the norm of the scattering operator represents the power gain for the element, it follows that the norm should always be unity, as no power is ever gained or dissipated in a lossless element.

By using wave variables rather than effort–flow variables for the Jacobian element, all of the problems associated with reduced rank Jacobians is kinematics computations are eliminated, since J^{-1} and J^{-T} are never directly computed.

7 DISCRETIZING DYNAMIC NETWORKS

Once the continuous scattering operator for a subnetwork has been determined, a method for discretizing the scattering operator while maintaining passivity must be found. In this section a method for discretizing linear time-invariant ports is presented which ensures that the resulting discrete scattering operator satisfies the same norm conditions as the original continuous-time system. Assuming that any nonlinearities of the system have been segmented out of the circuit into separate memoryless one-ports, we then have a comprehensive method for developing coupled nonlinear control laws for nonlinear systems.

Let $S_c(s)$ represent the continuous scattering operator for a linear time-invariant passive network port, where the argument, s, represents the Laplace frequency, $s = \sigma + j\omega$. Because the network is passive, it follows that the 2-norm of $S_c(s)$ is less than or equal to one. We need to find a discretization map, $T; S_c(s) \Rightarrow S_d(s)$, that maintains passivity, namely

$$\|S_c(j\omega)\| \leqslant 1 \forall \omega \in [0, \infty) \Rightarrow \|S_d(d)\| = \|S_d(e^{j\theta})\| \leqslant 1 \forall \theta \in [0, 2\pi] \tag{8.54}$$

where $d = e^{-sT}$ is the delay operator and S_d is the discrete-time version of the continuous scattering operator.

7.1 Tustin's Method: A Passivity Preserving Discretization Technique

Now let us consider Tustin's method. Tustin's method (also called the bilinear approximation) is one of the simplest discretization methods, requiring only a straight substitution. We simply replace every occurrence of the Laplace variable, s, with the function

$$s = \frac{2}{T}\left(\frac{1-d}{1+d}\right) \tag{8.55}$$

where $d = e^{-sT}$ is the delay operator for fixed sampling period T.

Theorem 2 (Tustin's Passivity Preservation). *If Tustin's method is applied to a passive scattering operator, $S_c(s)$, then the resulting discrete scattering operator, $S_d(s)$ is also passive.*

Proof A scattering operator is passive if and only if its norm is less than or equal to one. Computing the norm for the discretized scattering operator,

$$\|S_d(d)\| = \max_\theta \|S_d(e^{j\theta})\| \quad 0 \in [-\pi, \pi] = \max_\theta \left\|S_c\left(\frac{2}{T}\left(\frac{1-e^{j\theta}}{1+e^{j\theta}}\right)\right)\right\| \quad \theta \in [-\pi, \pi] \tag{8.56}$$

but

$$\left(\frac{1 - e^{j\theta}}{1 + e^{j\theta}}\right) = j\tan(\theta) \tag{8.57}$$

Making the variable substitution $\phi = \tan(\theta)$, we get

$$\|S_d(d)\| = \max_{\phi} \|S_c(j\phi)\| \phi \in [-\infty, \infty] = \|S_c(s)\| \tag{8.58}$$

Thus, if $\|S_c(s)\| \leqslant 1$ it follows that $\|S_d(s)\| \leqslant 1$ and passivity is preserved.

Example 4 Discretizing the KB1 Port. For example, we can take the continuous time equation for the KB1 port (Eq. (8.34))

$$S_c(s) = (K + Bs - Z_0 s)(K + Bs + Z_0 s)^{-1} \tag{8.59}$$

If each of the matrices is diagonal, then all of the degrees of freedom are uncoupled, and we compute the scattering operator for each degree of freedom. Plugging in for s, we have

$$\begin{aligned} S(d) &= \frac{K + (B - Z_0)\frac{2}{T}\left(\frac{1-d}{1+d}\right)}{K + (B + Z_0)\frac{2}{T}\left(\frac{1-d}{1+d}\right)} = \frac{K(1+d)T - (B - Z_0)2(1-d)}{K(1+d)T + (B + Z_0)2(1-d)} \\ &= \frac{\left(\frac{KT - 2B + 2Z_0}{KT + 2B + 2Z_0}\right) + d\left(\frac{KT + 2B - 2Z_0}{KT + 2B + 2Z_0}\right)}{1 - d\left(\frac{KT - 2B - 2Z_0}{KT + 2B + 2Z_0}\right)} \end{aligned} \tag{8.60}$$

This filter would then be implemented in a run-time system by first precomputing the filter constants

$$\alpha_0 = \left(\frac{KT - 2B + 2Z_0}{KT + 2B + 2Z_0}\right) \qquad \alpha_1 = \left(\frac{KT + 2B - 2Z_0}{KT + 2B + 2Z_0}\right) \qquad \beta_0 = \left(\frac{KT - 2B - 2Z_0}{KT + 2B + 2Z_0}\right) \tag{8.61}$$

and then computing the wave equation

$$b_n(k) = \beta_0 b_n(k-1) + \alpha_0 a_n(k) + \alpha_1 a_n(k-1) \tag{8.62}$$

in real time in the feedback loop. If the system is n-dimensional, then this would be applied using precomputed filter constants for each DOF.

Example 5 Discretizing the MB1 Port. The MB1 port is almost identical to the KB1 port. Suppose we take the MB1 port of Example 2 in Section 5.1. Here the scattering operator is given by

$$S_c(s) = \frac{Ms + B - Z_0}{Ms + B + Z_0} \tag{8.63}$$

Applying Tustin's method,

$$S(d) = \frac{M\frac{2}{T}\left(\frac{1-d}{1+d}\right) + B - Z_0}{M\frac{2}{T}\left(\frac{1-d}{1+d}\right) + B + Z_0} = \frac{\left(\frac{2M + BT - Z_0T}{2M + BT + Z_0T}\right) - d\left(\frac{2M - BT + Z_0T}{2M + BT + Z_0T}\right)}{1 - d\left(\frac{2M - BT - Z_0T}{2M + BT + Z_0T}\right)} \tag{8.64}$$

As with the KB1 module, this filter would then be implemented in a run-time system by first precomputing the filter constants

$$\alpha_0 = \frac{2M + BT - Z_0T}{2M + BT + Z_0T} \qquad \alpha_1 = \frac{2M - BT + Z_0T}{2M + BT + Z_0T} \qquad \beta_0 = \frac{2M - BT - Z_0T}{2M + BT + Z_0T} \tag{8.65}$$

and then computing the wave equation

$$b(k) = \beta_0 b(k-1) + \alpha_0 a(k) + \alpha_1 a(k-1) \tag{8.66}$$

in real time in the feedback loop.

8 IMPOSING NONLINEAR CONSTRAINTS USING SENSOR FEEDBACK

Standard robot control approaches tend to consider the manipulator only in linear regimes and thus tend to apply linear control laws. Joint motion is conducted between joint travel limits and at speeds below the joint velocity limits. Obstacles are avoided via path planning. But how do we design a system to respond to constraint information derived from sensors? How can we avoid obstacles and prevent collisions even during teleoperation? How can we utilize multiple sensors without inducing instability? To solve these problems we need to introduce nonlinear constraint functions into our modular control architecture.

By using the network approach presented in this chapter, we can implement nonlinear boundary functions in a bilateral environment and prevent the robot from entering undesirable regimes when either teleoperated or controlled remotely. The key to the approach is to extend the concept of a Jacobian. Rather than mapping sensor data directly into the system and possible introducing instabilities, we will introduce sensor data indirectly, by changing the functional mapping of a Jacobian.

Consider the simple massless lever system shown in Figure 8.9.

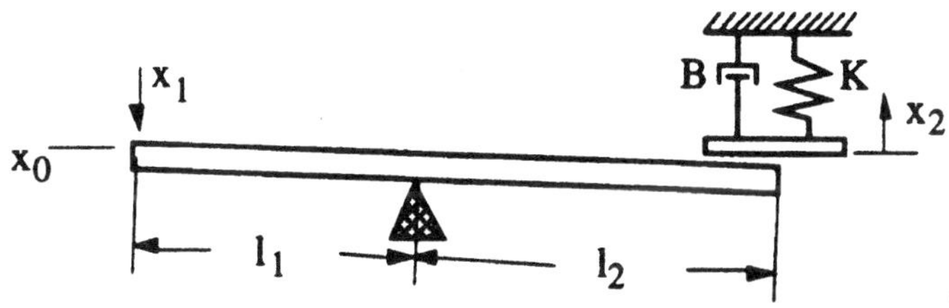

FIGURE 8.9
Simple lever system.

The lever is driven at one end of the fulcrum with input velocity v_1 and contacts a linear-spring damper system at the other end of the fulcrum. The force at the spring–damper interface, f_2, is then related to the spring–damper velocity, v_2, by the linear equation

$$f_2 = Bv_2 + Kx_2 = \left(B + \frac{K}{s}\right) v_2(s) \tag{8.67}$$

Although the force mapping is linear, the mapping between the input lever velocity and the velocity of the spring–damper surface is nonlinear, because the surface will not move until the lever makes contact with it. This type of nonlinearity is called a *switching* nonlinearity. It is encountered whenever a manipulator moves through different regimes, such as from a noncontact state to a contact state. It may also occur when a proximity sensor determines that contact is imminent.

The velocity relationship for the lever's switching function is defined as

$$v_2 = \begin{cases} \dfrac{l_2}{l_1} v_1 & x_1 > x_0 \\ 0 & \text{otherwise} \end{cases} \tag{8.68}$$

Likewise, the force is also a nonlinear relationship defined by

$$f_1 = \begin{cases} \dfrac{l_2}{l_1} f_2 & x_1 > x_0 \\ 0 & \text{otherwise} \end{cases} \tag{8.69}$$

Equations (8.68) and (8.69) represent the force–velocity map between two different velocity frames and thus define a Jacobian mapping. In particular, the switching Jacobian for the two frames is defined by

$$J_{2/1}(x_1) = \begin{cases} \dfrac{l_2}{l_1} & x_1 > x_0 \\ 0 & \text{otherwise} \end{cases} \tag{8.70}$$

By using this Jacobian, the linear dynamic portion of the system (the spring–damper system) has been separated from the nonlinear memoryless portion of the system (the switching condition). Both elements of the system can then be individually computed and discretized using the wave variable techniques presented so far. This is done in the following.

The wave equation for the spring–damper system is given by

$$b_2 = \frac{B + \dfrac{K}{s} - Z_0}{B + \dfrac{K}{s} + Z_0} a_2 = \frac{(B - Z_0)(1 - d) + \dfrac{KT}{2}(1 + d)}{(B + Z_0)(1 - d) + \dfrac{KT}{2}(1 + d} a_2 \tag{8.71}$$

and the wave equation describing the Jacobian two-port is

$$\begin{bmatrix} b_1 \\ a_2 \end{bmatrix} = \begin{bmatrix} I - D & J_{2/1}D \\ DJ_{2/1}^T & I - J_{2/1}DJ_{2/1}^T \end{bmatrix} \begin{bmatrix} a_1 \\ b_2 \end{bmatrix} \tag{8.72}$$

where

$$D = 2(I + J_{2/1}^T J_{2/1})^{-1} = \begin{cases} \dfrac{2}{1 + (l_2/l_1)^2} & x_1 > x_0 \\ 2 & \text{otherwise} \end{cases} \tag{8.73}$$

which leads to the scattering operator for the constraint two-port,

$$S_{2/1} = \begin{cases} \begin{bmatrix} -\dfrac{(l_2/l_1)^2 - 1}{1 + (l_2/l_1)^2} & \dfrac{l_2/l_1}{1 + (l_2/l_1)^2} \\ \dfrac{l_2/l_1}{1 + (l_2/l_1)^2} & \dfrac{1 - (l_2/l_1)^2}{1 + (l_2/l_1)^2} \end{bmatrix} & x_1 > x_0 \\ \begin{bmatrix} -1 & 0 \\ 0 & 1 \end{bmatrix} & \text{otherwise} \end{cases} \tag{8.74}$$

Thus we have a stable Jacobian-based method to compute the switching nonlinearity for the simple lever system. A network representation of this system is shown in Figure 8.10.

So far, the fulcrum position has been held constant, and the ratio l_2/l_1 is therefore fixed. What would happen, however, if the fulcrum position were slid back and forth as a function of the tip position, constantly changing the ratio of l_2 to l_1? Surprisingly, the equations describing the system (Eqs. (8.71) and (8.72)) would remain the same. The linear spring damper would remain a spring damper and the scattering operator derived from the Jacobian

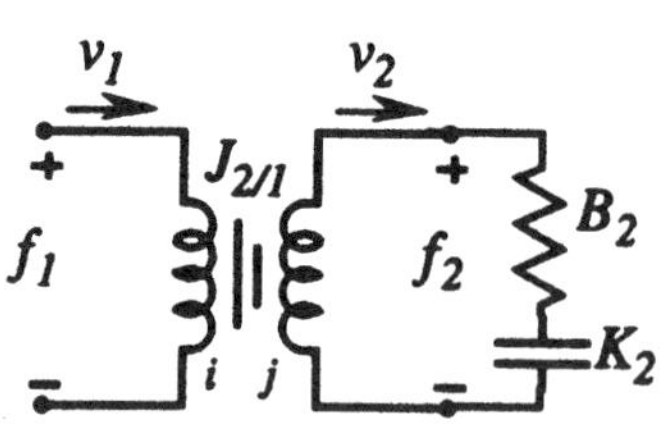

FIGURE 8.10
Network representation of simple lever.

relationship (Eq. (8.70)) would still be valid, except that now the ratio l_2/l_1 would have to be recomputed for every position. The equivalent stiffness as seen from the inputs, however, would now vary continuously, rather than just switching from zero to a positive constant.

As an example, suppose the fulcrum was adjusted according to the following function:

$$\frac{l_2}{l_1} = \begin{cases} 0 & x_1 < x_0 \\ \dfrac{x_1}{\Delta x_1} & x_0 < x_1 < x_0 + \Delta x_1 \\ 1 & x_0 + \Delta x_1 < x_1 \end{cases} \tag{8.75}$$

This would result in the equivalent stiffness varying smoothly as a function of position, until a maximum effective value of K was achieved, as shown in Figure 8.11.

The contact force can be computed directly from the constituent equations as shown here:

$$f_1 = J_{2/1}^T \int K_2 J_{2/1} v_1 \, dt = \begin{cases} 0 & x_1 < x_0 \\ \dfrac{K_2}{2\Delta x_1^2}(x_1 - x_0)^3 & x_0 < x_1 < x_0 + \Delta x_1 \\ \dfrac{K_2 \Delta x_1}{2}(x_1 - x_0 - \Delta x_1) & x_0 + \Delta x_1 < x_1 \end{cases} \tag{8.76}$$

By using a large stiffness, K, and damper, B, and a small threshold region, Δx, a large penalty force can be exerted in a very small region. Unlike the pure switching nonlinearity, however this force will be applied smoothly to the manipulator. If, on the other hand, the stiffness and damping are not ramped up from zero, then as soon as the surface was first contacted a force of $f_1 = B(l_2/l_1)^2 v_1$ would be reflected to the system, jolting the manipulator.

It is important to note that the fulcrum could have been moved to the point $l_2/l_1 = \infty$, which would have resulted in an infinite stiffness and an impenetrable barrier function.

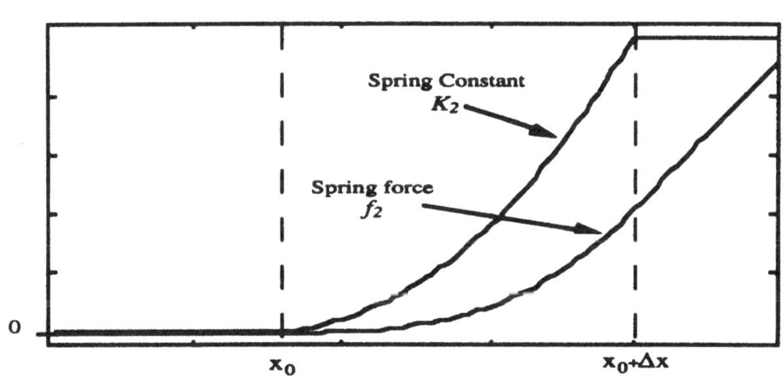

FIGURE 8.11
Equivalent lever stiffness and stiffness force at contact for moving fulcrum.

Plugging this into the scattering operator gives the simple-to-implement equation

$$S = \begin{bmatrix} -\dfrac{\infty^2 - 1}{1 + \infty^2} & \dfrac{\infty}{1 + \infty^2} \\ \dfrac{\infty}{1 + \infty^2} & \dfrac{1 - \infty^2}{1 + \infty^2} \end{bmatrix} = \begin{bmatrix} 1 & 0 \\ 0 & -1 \end{bmatrix} \tag{8.77}$$

The problem with this is that an infinite stiffness, although theoretically appealing, is impossible to achieve in a sampled data system. When this equation is implemented the system suffers from large wave reflections due to the infinite termination impedance and still has an effective compliance of T/Z_0, where T is the sample delay and Z_0 is the line impedance. Thus, even though the continuous spring could be sent to infinity, the effective discrete spring constant would still be finite.

By mapping proximity sensor data into the fulcrum position, we can indirectly impose constraints on any motion of the manipulator. Furthermore, we an superimpose the effects of multiple sensors without affecting the stability of the resulting system.

9 IMPLEMENTATION

Our approach to building a modular telerobot control system consists of the following steps. First represent the robot, any input devices, any sensors, and the controller as n-ports in a robot control system. Second, for each n-port module, break the system into linear time-invariant sections and nonlinear Jacobian sections. Third, apply Tustin's passivity-preserving discretization method to all linear time-invariant components of the system. Fourth, apply scattering theory to all network components to get input–output mappings between wave variables for every module, and precompute all filter constants. Fifth, distribute the modules across the processors in the system and connect the output waves of each module to the input waves of the next module using pointers in dual-ported memory. This approach is the basis for the SMART (Sandia's Modular Architecture for Robotics and Teleoperation) architecture.

If all of the modules in the system are designed using the techniques described here, they will have the same stability and performance independent of the order that they are sampled. If sampling is increased, the performance will more closely approximate the continuous system upon which they are based. The system will remain stable even if modules are sampled at different rates, miss sample updates, or stop updating all together.

10 EXAMPLES

In this section we give numerous examples of how sensors, robots, and input devices can be integrated in a consistent fashion using the SMART architecture.

The first example (Figure 8.12) shows an autonomous controller for a PUMA 760 manipulator that has been outfitted with a WHAP sensor and is driven by an autonomous trajectory generator in joint space. The TRAJECTORY module generates trajectories that flow to the PUMA_JOINTS module, which drives the robot. Constraints are implemented in identical fashion for the LIMITS module and the sensor module. The LIMITS module

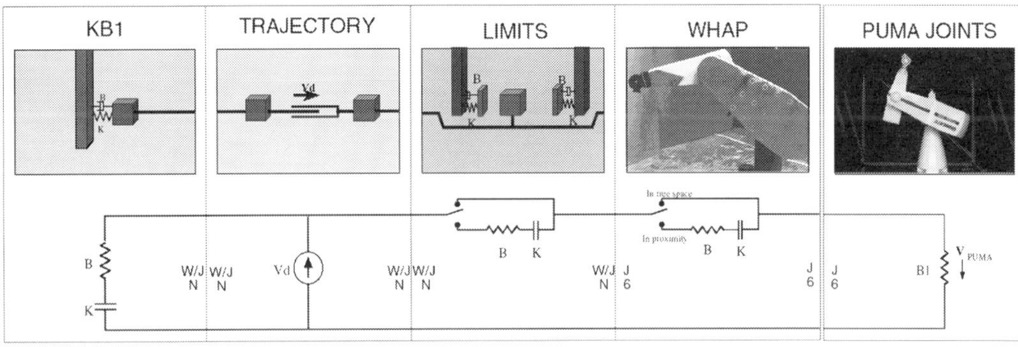

FIGURE 8.12
PUMA manipulator with WHAP sensor.

shunts the flow generated from the trajectory module into a spring damper as described in Section 8, whenever the manipulator approaches a joint limit. The spring damper generates forces that are detected across the ports of the TRAJECTORY module. This will halt the motion generation once a predetermined force threshold is reached. Furthermore, because of the bilateral nature of the SMART architecture, the system can recover from the limit constraint by subsequently moving in any direction that reduces the stored energy in the system, that is, such that

$$f_n^T v_n < 0 \tag{8.78}$$

at the ports of the TRAJECTORY module.

The WHAP sensor module behaves in the same way as the LIMITS module. No force occurs across the port unless the sensors detect proximity to an object. When proximity is detected, the manipulator's flow is shunted into a spring damper, generating enough force to turn off the flow source.

The second example uses a teleoperator input device (the CIS Dimension 6 force–torque ball) to command motion, and rather than using a capacitive proximity sensor, it utilizes an ultrasonic sensor head attached to the gripper. Because the sensors are attached to the tip of the robot, rather than being distributed along the arm, the sensor forces are mapped into tool space. Thus the system also includes a PUMA_KIN module, which takes the force and

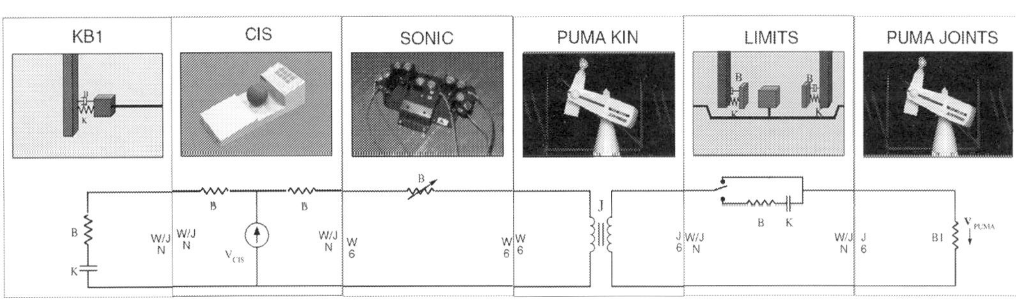

FIGURE 8.13
PUMA manipultor with ultrasonic sensor.

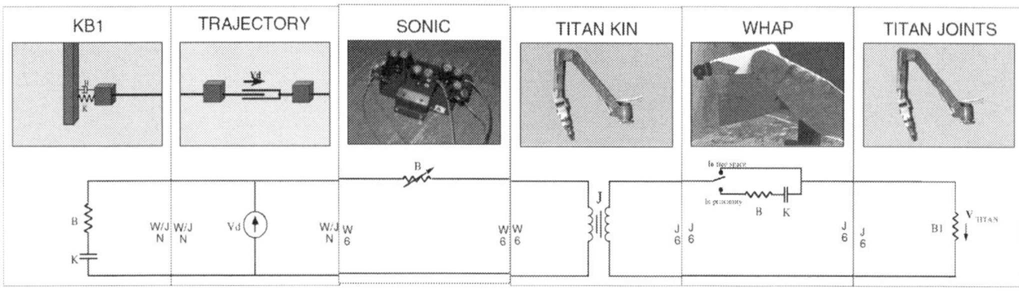

FIGURE 8.14
Titan manipulator with WHAP and ultrasonic sensor.

velocities in world space and maps them into joint space. In joint space the LIMITS module and PUMA_JOINTS module act as before.

The third example combines both the ultrasonic sensor operating in world space and the capacitive WHAP sensor operating in joint space. Instead of being attached to a PUMA robot, however, these are used to provide motion constraints for a Schilling Titan2 manipulator. Any force generated in joint space due to the WHAP sensor is mapped into world space by the TITAN_KIN module and then summed with the forces generated from the SONIC module. The resulting force is then seen across the ports of the TRAJECTORY module consistent with Kirchhoff's laws. Because neither sensor module generates any energy, they cannot destabilize the robot. If the sensors are wired incorrectly or suffer a failure, the worst that can happen is that the arm halts at inappropriate times or fails to halt when an object is detected. It cannot become unstable, however, for any sensor reading under any condition.

The final example shows a bilateral teleoperation system operating in joint space. The system utilizes both a WHAP proximity sensor and a JR3 force–torque senor. The inclusion of the force sensor allows the operator to feel contact forces directly while driving the manipulator. The forces from the force sensor are then superimposed on forces from the WHAP module and fed back to the teleoperation. Notice, however, that the module utilizes an effort source to insert the force sensor measurements directly. Because the forces measured across the force sensor are dependent on the position of the sensor, the effort source is actually a dependent source and instability can arise in the system.

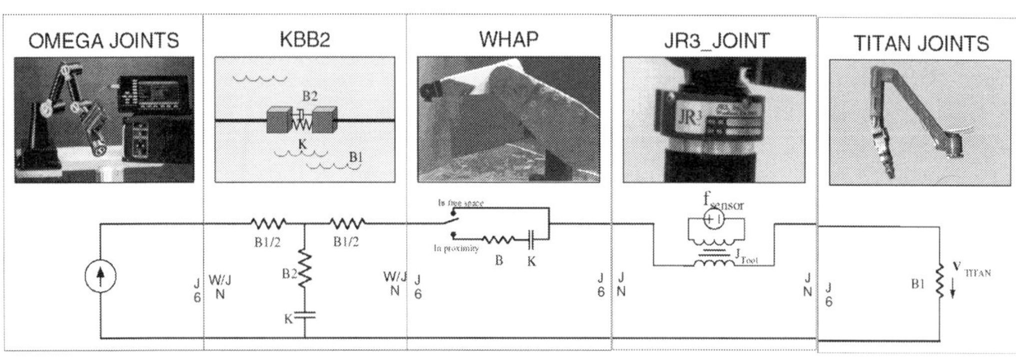

FIGURE 8.15
Titan manipulator with WHAP sensor and JR3 force–torque sensor.

In an ideal system with a perfect rigid robot response and unfiltered continuous force sensor readings, the forces measured across the force sensor would always be within 90 degrees of phase of the velocity and the module would always be passive. Because of any number of potential problems (robot latency, sensor filter, noncollocation), this will not be true at high frequencies, and the system may become unstable unless proper filtering and gain reduction are applied to the force sensor feedback.

11 CONCLUSIONS

We have described techniques for incorporating sensor data in a modular fashion without affecting the stability of the telerobot control system. The sensor integration approach is compatible with various teleoperator and control modalities and is robust to delays in receiving sensor data. By using the inherent stability of passive systems, a plug-and-play control architecture whereby individual modules can be placed arbitrarily in the system without affecting the overall system stability has been developed.

Because the constraint modules implement boundary functions via scattering operators, the system is very robust. Additional sensor modules can be added without adversely affecting the system. As long as no objects are detected, each module lies dormant and does not affect the behavior of the system. The first module to detect an object will typically provide the overriding signal that determines the behavior of the system. If multiple sensors are active at once, the system may respond a bit faster but will not be destabilized. Furthermore, the chattering phenomena that can occur when two different sensors with opposing values are applied sequentially cannot occur with the scattering implementation shown here.

The network-based, scattering operator approach described here has been and continues to be the basis for the SMART modular telerobotic control system [10]. To date, sensor modules for Sandia's and ORNL's capacitive sensor system and for a Sandia-developed ultrasonic system have been developed. In addition, modules for both JR3 and ATI force sensors have been developed and have been integrated into the SMART architecture as force sources.

These sensor modules complement a number of modules that have been written incorporating world model information (such as the LIMITS module) both in Cartesian space [5] and in configuration space and modules that implement different behaviors when forces are felt along different vectors.

REFERENCES

[1] B. D. O. Anderson. *Network Analysis*. Prentice Hall, Englewood Cliffs, NJ, 1973.

[2] N. Hogan. Impedance control: An approach to manipulation. *ASME J. Dynam. Syst, Meas. Control* 107:1–7, March 1985.

[3] G. Raju, G. C. Verghese, and T. B. Sheridan. Design issues in 2 port network models of bilateral remote manipulation. *IEEE International Conference on Robotics and Automation*, Scottsdale, AZ, 1989, pp. 1316–1321.

[4] R. J. Anderson. SMART: A modular control architecture for telerobotics. *IEEE Robot. Automat. Soc. Mag.*, Sept. 1995, pp. 10–18.

[5] R. J. Anderson. Teleoperation with virtual force feedback. *Proceedings of the '93 SPIE International Symposium on Optical Tools for Manufacturing and Advanced Automation*, Boston, Sept. 1993.

[6] R. J. Anderson and K. Lilly. A modular approach to multi-robot control. *Proceedings of the 34th Conference on Decision & Control*, New Orleans, Dec. 1995, pp. 1021–1027.
[7] R. J. Anderson and B. Davies. Using virtual objects to aid underground storage tank teleoperation. *Proceedings of '94 IEEE International Conference on Robotics and Automation*, San Diego, May 1994, pp. 1421–1426.
[8] R. Rosenberg, and D. Karnop. *Introduction to Physical System Dynamics*. McGraw-Hill, New York, 1983.
[9] R. Chipman. *Transmission Lines*. New York: McGraw-Hill, 1968.
[10] R. J. Anderson. Modular Architecture for Robotics and Teleoperation. U.S. Patent 5,581,666, December 3, 1996.

CHAPTER 9

A Circuit-Theoretic Analysis of Robot Dynamics and Control

SUGURU ARIMOTO
Department of Robotics, Ritsumeikan University, Kusatsu, Shiga, Japan

ABSTRACT

This chapter is aimed at solving the problem of making robot arms and mechanical hands acquire the ability to manipulate things smoothly and dexterously without using a complicated mathematical model of nonlinear dynamics pertaining to robot tasks. Such abilities as investigated in this chapter are related to adaptive control, iterative learning, and impedance matching when robots manipulate soft and deformable objects or touch a rigid object with their soft fingers. It is shown that these abilities can be realized in robots by devising corresponding control schemes on the basis of physical properties such as passivity and dissipativity inherent in the dynamics of robot tasks. The essential idea is to observe that a large class of robot dynamics can be expressed by a nonlinear position-dependent circuit and the impedance concept inherent in linear electric circuits can be extended to cope with such a nonlinear circuit via the passivity and dissipativity. As a by-product of this analysis, it is shown that iterative learning based on such physical properties can be interpreted as steady progress of impedance matching. In the process of this interpretation, the concept of impedance matching is generalized by means of passivity to cope with nonlinear dynamics in the case in which a mechanical hand whose finger ends are covered by soft material grasps rigid objects. It is finally claimed that robot control must be advanced in a direction from using full-model-based control (e.g., the computed torque method) to simple control schemes without using models of dynamics or using at most approximate models, as seen in human motion when tasks are executed dexterously.

1 INTRODUCTION

In the early stage of robotics research before 1980, it was thought that the design of robot controllers is a difficult problem because Lagrange's equation of motion of a robot with rotational joints (anthropoid robots) is nonlinear and has strong couplings between joints. In fact, its dynamics become more complicated when it comes in contact with an object in an environment, and, furthermore, the dynamics of a setup of multiple fingers with multiple joints grasping an object are much more complicated. To cope with the difficulty caused by this complexity of robot dynamics, the so-called computed torque method, which is based on real-time computation of the left-hand side of Lagrange's equation of motion, was proposed independently by Vukobratovich [1], Bejczy [2], and Takase [3], who predicted the rapid development of microprocessor technology from 1970 to 1980.

In 1980 two rapid methods for computing torque were found independently by Hollerbach [4] and Luh *et al.* [5]. Their findings drew much attention from research workers in robotics, and subsequently various methods for implementation of the algorithms in very large scale integration (VLSI) chips were proposed in the literature. However, these VLSI chips are not used in present industrial robots. The most essential reason is that physical parameters such as inertia moments of links, link masses, and friction coefficients cannot be evaluated with sufficient precision, and hence calculating the control input by using these parameters results in a discrepancy between the actual robot's motion and the desired motion large as time elapses. Then in 1987, a novel idea was proposed by Slotine and Li [6], who, by using regressors and estimators for uncertain dynamic parameters appearing linearly in Lagrange's equation of motion, showed that the discrepancy between the robot's motion and the desired one approaches zero with increasing time. Subsequently, Sadegh and Horowitz [7] showed that there is no need to compute regressors in real time. This finding is quite important, because the complexity of the computation is almost the same as that of fast computed torque methods.

A fairly simple control scheme for set point position control of robot manipulators of anthropoid type was first proposed by Takegaki and Arimoto [8], which is now called proportional derivative (PD) control with gravity compensation. This scheme requires only knowledge of the gravity term in Lagrange's equation of motion at the desired position. The PD feedback at each joint can be determined locally and independently without referring to that at other joints. Subsequently, in 1983, a linear proportional integral and derivative (PID) feedback scheme of distributed type (independent at each joint) was proposed by Arimoto and Miyazaki [9] that establishes asymptotic stability of set-point position control without compensating for the gravity term. However, this asymptotic stability cannot be established in a global sense.

Since then, the problem of finding the simplest feedback scheme that attains global asymptotic stability without using any knowledge of robot dynamics remained unsolved until 1994. Then the global asymptotic stability of set point position control was established by one of the authors [10, 11] by using a distributed saturated proportional (position), integral and derivative) (SP-ID) feedback scheme. The essence was to find that if a positive linear combination of the angular velocity vector and a saturated position angle error vector is taken as an output, then the pair of this output and the torque input satisfies dissipativity concerning the robot dynamics with PD feedback. Thus, a negative feedback connection of the robot dynamics with an appropriate feedback of the SP and D output and a linear feedback through integration (its transfer function matrix is C_1/s, where C_1 is an appropriate positive diagonal matrix and s is a complex variable) establishes global asymptotic stability by virtue of Popov's hyperstability theorem. Once the global asymptotic stability of such an SP-ID feedback scheme is established, it is possible to show that the structure of robot dynamics with SP-ID feedback plays an essential role in iterative learning, model-based adaptive control, and H-infinity control in the sense of disturbance attenuation [12]. At the same time, the discovery that such a physical structure with SP-ID feedback satisfies passivity or dissipativity suggested the idea of describing the structure similarly to electric circuits [12, 13].

In Section 5 the idea of making robots free from friction or gravity forces is presented on the basis of compensating the terms of friction and gravity forces by constructing a circuit of control blocks to be connected with the original circuit of robot dynamics. In Section 6 the problem of impedance control is treated from the circuit-theoretic viewpoint and the concept of impedance matching is generalized via passivity and dissipativity. As a by-product of this generalization, the structure of iterative learning is reinterpreted as a monotone increase of the grade of matching between the internal dynamics consisting of a learning update law and the external (load) dynamics. The concluding section summarizes the advance of robot control from model-based control methods to control schemes without referring to a robot model or with use of only an approximate model. At the same time, answers to the question of why simple control schemes for robot tasks work well regardless of complexity of their dynamics are discussed.

2 PASSIVITY OF ROBOT DYNAMICS AND NONLINEAR POSITION-DEPENDENT CIRCUITS

Dynamics of a robotic arm with all rotational joints are expressed in the following form:

$$\left\{H(q)\frac{d}{dt} + \frac{1}{2}\dot{H}(q)\right\}\dot{q} + S(q, \dot{q})\dot{q} + r(\dot{q}) + g(q) = u \tag{9.1}$$

where $q = (q_1, \ldots, q_n)^T$ denotes the joint angle vector, $H(q)$ the inertia matrix, $S(q, \dot{q})$ a skew-symmetric matrix, $r(\dot{q})$ the damping term including frictional forces, $g(q)$ the gravity term, and u the control input vector whose components are torques generated by joint actuators. It should be noted that $H(q)$ is symmetric and positive definite. Moreover, because each entry of $H(q)$ is constant or a sinusoidal or cosine function of components of q, $S(q, \dot{q})$ is linear and homogeneous in $\dot{q}$ with property $S(q, 0) = 0$ and each nondiagonal entry of $S(q, \dot{q})$ is a sinusoidal or cosine function. Further, it is natural to assume that $\dot{q}^T r(\dot{q}) \geqslant 0$ and there exists a potential function $U(q)$ such that $\partial U/\partial q = g(q)^T$. As $U(q)$ is a linear combination of sinusoidal and cosine functions of q, it is possible to take the constant term of $U(q)$ so that $\min_q U(q) = 0$.

As to the robot dynamics of Eq. (9.1), the pair of input u and output $\dot{q}$ satisfies passivity, that is,

$$\int_0^t \dot{q}^T(\tau)u(\tau)\, d\tau \geqslant E(t) - E(0) + \int_0^t \dot{q}^T(\tau)r(\dot{q}(\tau))\, d\tau \geqslant -E(0) = -\gamma_0^2 \tag{9.2}$$

where E denotes the total energy expressed as $E = \frac{1}{2}\dot{q}^T H(q)\dot{q} + U(q)$. This passivity relation plays a crucial role in building a bridge between the energy conservation law in physics and the operational input–output characterization in system theory. In light of the passivity of robot dynamics, the following two types of servo controller for set point position control were introduced by Takegaki and Arimoto [8] and Arimoto and Miyazaki [9], respectively:

$$u(t) = -B_1\dot{q}(t) - A_1\Delta q(t) + g(q_d) \tag{9.3}$$

$$u(t) = -B_1\dot{q}(t) - A_1\Delta q(t) - C_1\int_0^t \Delta q(\tau)\,d\tau \tag{9.4}$$

where q_d denotes a given desired position, $\Delta q = q - q_d$, and A_1, B_1, and C_1 are constant diagonal matrices with positive diagonal elements. When the damping term $r(\dot{q})$ does not include static and Coulomb frictions (this case will be treated in Section 4), it is possible to show that substituting Eq. (9.3) into Eq. (9.2) and differentiating the resultant equation in time t yield

$$\frac{d}{dt}V(q, \dot{q}) = -\dot{q}^T B_1 \dot{q} - \dot{q}^T r(\dot{q}) \tag{9.5}$$

where

$$V(q, \dot{q}) = \tfrac{1}{2}\dot{q}^T H(q)\dot{q} + \{U(q) - U(q_d) - \Delta q^T g(q_d) + \tfrac{1}{2}\Delta q^T A_1 \Delta q\} \tag{9.6}$$

Because $U(q)$ is a combination of sinusoidal and cosine functions of components of q, it is possible to choose $A_1 > 0$ so that the content of brackets $\{\}$ on the right-hand side of Eq. (9.6) is positive definite in Δq because the maximum of the maximum eigenvalue of the Hessian matrix of $U(q)$ for all q is bounded. This means that the linear feedback $-A_1\Delta q$ with gravity compensation $g(q_d)$ makes the shape of the resultant potential positive definite in Δq. Therefore, the feedback scheme of Eq. (9.3) is now called the "energy shaping with damping injection" method.

The robot dynamics of Eq. (9.1) can be expressed by a nonlinear position-dependent circuit depicted in Figure 9.1. This circuit expresses the energy storage and flow. In fact, the block indicated by an inductor-like symbol expresses a function of storage of the kinetic

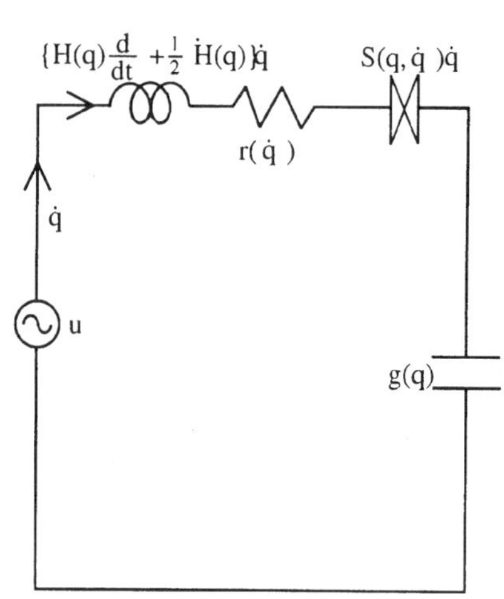

FIGURE 9.1
A nonlinear position-dependent circuit expresses dynamics of Eq. (9.1).

energy, since the inner product between the torque (corresponding to voltage drop) across the block and the angular velocity (current) reproduces the time rate of the kinetic energy, that is, $\dot{q}^T\{H(q)d/dt + \frac{1}{2}\dot{H}(q)\}\dot{q} = d\{\frac{1}{2}\dot{q}^T H(q)\dot{q}\}/dt$. In the circuit, the resistance block with torque (voltage drop) $r(\dot{q})$ means a nonlinear resistance, because $\dot{q}^T\dot{r}(\dot{q}) \geqslant 0$ means the time rate of energy consumption. Another block of torque $S(q, \dot{q})\dot{q}$ neither stores nor dissipates energy, because the inner product between $\dot{q}$ and $S(q, \dot{q})\dot{q}$ vanishes due to the skew symmetry of $S(q, \dot{q})$. Hence this block can be called a nonlinear "gyrator." Finally, the block depicted by a capacitor-like symbol stocks the potential energy $U(q)$ because the inner product between $\dot{q}$ and $g(q)$ leads to the time rate of the potential, that is, $\dot{q}^T g(q) = dU(q)/dt$. Note that, in contrast to conventional electric circuits, whether linear or nonlinear, nonlinear circuits expressing nonlinear dynamics of mechanical systems must be position dependent. For example, nonlinearity of an inductor-like block must be dependent on position q (which corresponds to electric charge since $\dot{q}$ corresponds to current). In electric circuits, there do not exist any charge-dependent inductors.

It is now quite easy to see that the closed-loop system when u of Eq. (9.3) is substituted into Eq. (9.1) can be expressed by the circuit in Figure 9.2. Then it is intuitively possible to understand that the overall block of capacitor-like elements in Figure 9.2 satisfies the following: (1) the solution to $A\Delta q + g(q_d + \Delta q) - g(q_d) = 0$ is unique, that is, $q = q_d$ is only one solution to the preceding equation, and (2) the inner product $\dot{q}^T\{A\Delta q + g(q_d + \Delta q) - g(q_d)\} = d\{\frac{1}{2}\Delta q^T A \Delta q + U(q) - U(q_d) - \Delta q^T g(q_d)\}/dt$ induces the time rate of the positive definite potential. If these two conditions are satisfied, then the global asymptotic stability of position control is established.

Robot dynamics when the tool endpoint is holonomically constrained on an environmental surface can also be expressed by a nonlinear position-dependent circuit. Further, dynamics of robots with joint flexibility and/or with coupling with internal dynamics of actuators together with electric circuits of armatures in motors can be analyzed by their expressions in terms of nonlinear position-dependent circuits. More detailed discussions can be found in the literature (see [12, 13]).

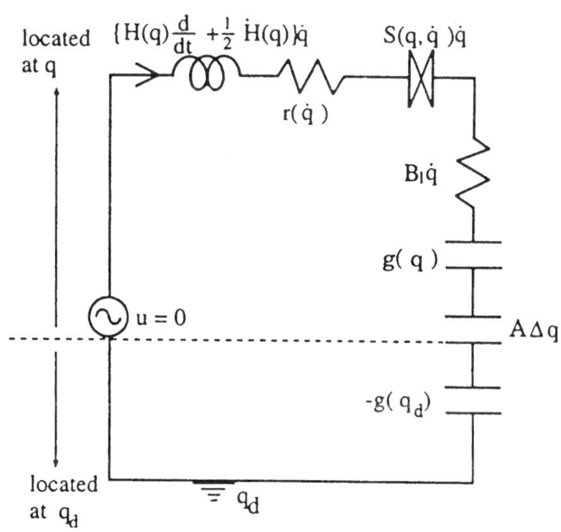

FIGURE 9.2
A circuit expresses dynamics of the closed-loop system of Eq. (9.1) when a PD control defined by Eq. (9.3) is employed.

3 SP-ID CONTROL

The PD feedback with gravity compensation defined by Eq. (9.3) assumes knowledge of the gravity term $g(q)$. However, pure PD feedback without compensating for the gravity term gives rise to a steady-state position error called "offset." To avoid incurring such an offset in positioning, a distributed PID feedback scheme defined by Eq. (9.4) was introduced. However, this PID control scheme ensures asymptotic stability in positioning only in a local sense.

The global asymptotic stability of set point position control was first established by one of the authors [10, 11] for a kind of PID control of the form

$$u(t) = -B_0\dot{q} - C_1\Delta q - \alpha C_0 s(\Delta q) - \alpha C_1 \int_0^t s(\Delta q)\, d\tau \tag{9.7}$$

or

$$u(t) = -B_1\dot{q} - A\Delta q - \left(C_0 + C_1 \int_0^t d\tau\right) y \tag{9.8}$$

where

$$y = \dot{q} + \alpha s(\Delta q) \tag{9.9}$$

and $\alpha > 0$, $s = (s_1, \ldots, s_n)^T$, and each $s_i(\Delta q_i)$ is a saturated function with the profile exhibited in Figure 9.3. The key to this result is the fact that the pair of output y and input v for the closed-loop system

$$\left\{H(q)\frac{d}{dt} + \frac{1}{2}\dot{H}(q)\right\}\dot{q} + S(q, \dot{q})\dot{q} + r(\dot{q}) + B_1\dot{q} + A\Delta q + g(q) - g(q_d) = v \tag{9.10}$$

satisfies both passivity and dissipativity by selecting diagonal matrices A and B_1 and a constant α appropriately (the details may be found in Arimoto [13]). Then, according to Popov's theorem concerning negative feedback connection of one passive block with another dissipative block, the closed-loop system with the structure depicted in Figure 9.4 becomes globally asymptotically stable. Note that both feedback schemes (9.7) and (9.8) are distributed and independent at each joint in the sense that each joint servo loop does not need to

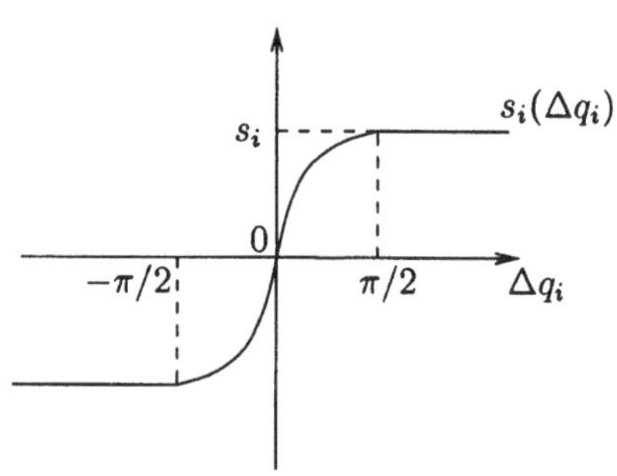

FIGURE 9.3
$s_i(\Delta q_i)$ is a saturated nonlinear function.

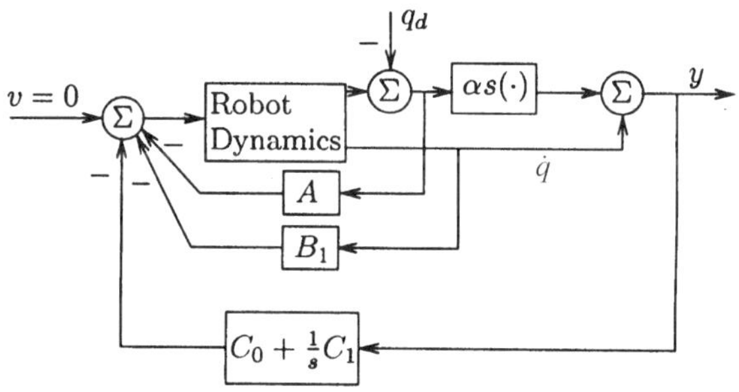

FIGURE 9.4
Negative feedback connection of two passive systems.

know the measurement data for position and velocity at other joints, since all gain matrices in Eqs. (9.7) and (9.8) can be selected as positive diagonal matrices. Further note that Eq. (9.7) can be reduced to the form of Eq. (9.8) except for the constant term if $B_0 = B_1 + C_0$ and $A = 0$. Both schemes of Eqs. (9.7) and (9.8) are therefore termed SP-ID feedback control. It should be further noted that a type of sliding mode control can be obtained by letting the shape of the nonlinear function $s_i(\Delta q_i)$ tend to the signature function $\mathrm{sgn}(\Delta q_i)$. In that case, C_0 in Eqs. (9.7) and (9.8) should be set to zero to avoid the occurrence of chattering [14].

4 ADAPTABILITY AND LEARNABILITY

Next consider the problem of trajectory tracking control for a given joint trajectory $q_d(t)$ defined for some finite interval $[0, T]$. It is reasonable to assume the existence of a desired control input $u_d(t)$ realizing the joint trajectory $q_d(t)$, which can be written in the form:

$$u_d = \left\{H(q_d)\frac{d}{dt} + \frac{1}{2}\dot{H}(q_d)\right\}\dot{q}_d + S(q_d, \dot{q}_d)\dot{q}_d + B\dot{q}_d + g(q_d) \tag{9.11}$$

where we assume $r(\dot{q}) = B\dot{q}$ in Eq. (9.1), that is, the damping term contains only viscous frictions (after compensating static and Coulomb's frictions as discussed in the next section). Then it is possible to write down the equation of Δq by subtracting Eq. (9.11) from Eq. (9.1) with an additional position feedback $-A\Delta q - D\Delta y$ in the following way:

$$\left\{H(q)\frac{d}{dt} + \frac{1}{2}\dot{H}(q)\right\}\Delta\dot{q} + S(q, \dot{q})\Delta\dot{q} + B\Delta\dot{q} + A\Delta q + D\Delta y + h(\Delta q, \Delta\dot{q}) = \Delta u \tag{9.12}$$

where

$$\Delta y = \Delta\dot{q} + \alpha s(\Delta q) \tag{9.13}$$

and

$$h(\Delta q, \Delta\dot{q}) = \{H(q_d + \Delta q) - H(q_d)\}\ddot{q}_d$$
$$+ \{\tfrac{1}{2}\dot{H}(q_d + \Delta q) + S(q_d + \Delta q, \dot{q}_d + \Delta\dot{q}) - \tfrac{1}{2}\dot{H}(q_d) - S(q_d, \dot{q}_d)\}\dot{q}_d + g(q_d + \Delta q) - g(q_d) \tag{9.14}$$

Note that the nonlinear term $h(\Delta q, \Delta\dot{q})$ is linear in $\Delta\dot{q}$ and small, of the order of Δq. By virtue of this property, it is possible to show that the pair of the input Δu and the output Δy concerning Eq. (9.12) satisfies dissipativity; that is, there exist a storage function $V(\Delta q, \Delta\dot{q})$ positive definite in Δq and $\Delta\dot{q}$ and a scalar-valued dissipation function $\xi(y) > 0$ with $\xi(0) = 0$ such that

$$\int_0^t \Delta y^T \Delta u d\tau \geqslant V(\Delta q(t), \Delta\dot{q}(t)) - V(\Delta q(0), \Delta\dot{q}(0)) + \int_0^t \xi(y(\tau))\, d\tau \tag{9.15}$$

Further, it is possible to adjust the velocity gain B and the feedback gain D by damping injection so that $\xi(\Delta y)$ is of the order of a quadratic function of Δy,

$$\xi(\Delta y) \geqslant \beta\|\Delta y\|^2 \tag{9.16}$$

with some $\beta > 0$. The most important result concerning the robot's learnability is that robots can acquire the desired motion through repeating practice when the control input is updated iteratively by the law (see Figure 9.5)

$$u_{k+1}(t) = u_k(t) - \Phi\Delta y_k(t) \tag{9.17}$$

where Φ is an appropriate control gain matrix. It has been shown (see Arimoto [5]) that if $\Phi < 2\beta I$ then $\Delta y_k \to 0$ as $k \to \infty$ in the sense of $L^2(0, T)$. Since $\Delta\dot{q}_k + \alpha s(\Delta q) \to 0$ as $k \to \infty$, $\Delta q_k(t) \to 0$ as $k \to \infty$.

The most noteworthy merit of using the output Δy in the learning update law instead of $\Delta\dot{q}$ is that the initial setting at the beginning of every trial is unnecessary if the desired output q_d satisfies $q_d(0) = q_d(T)$ and $\dot{q}_d(0) = \dot{q}_d(T)$. Figure 9.6 shows the experimental data for comparison of the use of $\Delta\dot{q}$ in Figure 9.6(a) with the use of Δy in Figure 9.6(b), where a DD arm with three degrees of freedom was used. The details of the experiment were reported in a dissertation at Yamaguchi University (see Naniwa [15]). Another noteworthy advantage of using an SP-D output Δy as defined in Eq. (9.14) is that a model-based adaptive controller can be designed in such a manner as

$$u(t) = -B_1\Delta\dot{q} - A\Delta q - D\Delta y + \hat{u}_d \tag{9.18}$$

where $\hat{u}_d$ is an estimate of the desired input u_d for a desired joint trajectory q_d as defined by Eq. (9.11). Referring to the well-known fact that the left-hand side of Eq. (9.1) can be characterized by a set of dynamic parameters $\Theta = (\theta_1, \ldots, \theta_m)^T$ appearing linearly and a

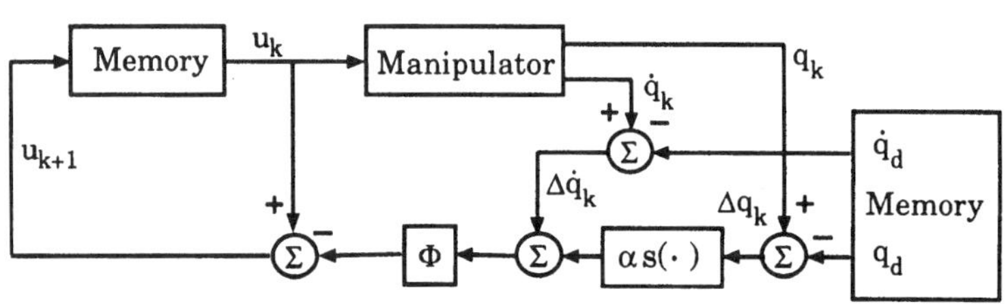

FIGURE 9.5
Iterative learning based on the SP-D–type update law.

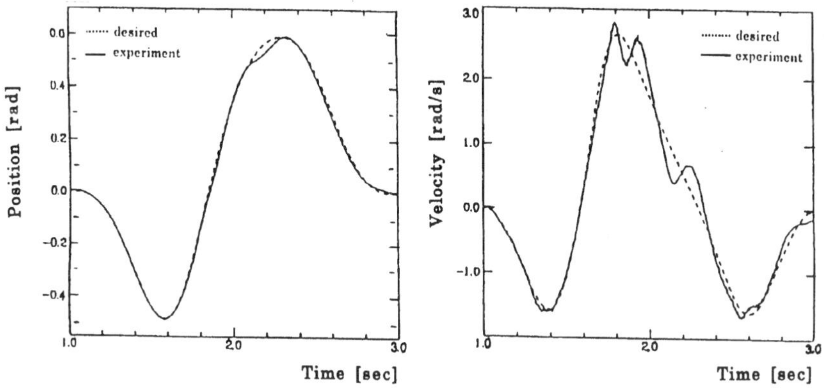

FIGURE 9.6 (a)
Joint angle position and velocity trajectories of link 2 after 30 trials.

regressor $Y(q, \dot{q}, \ddot{q})$ (see Slotine and Li [6]), that is,

$$H(q)\ddot{q} + \{\tfrac{1}{2}\dot{H}(q) + S(q, \dot{q}) + B\}\dot{q} + g(q) = Y(q, \dot{q}, \ddot{q})\Theta \tag{9.19}$$

it is convenient to define the estimate $\hat{u}_d$ in the form

$$\hat{u}_d = Y(q_d, \dot{q}_d, \ddot{q}_d)\hat{\Theta} \tag{9.20}$$

Further, note that the estimate for the unknown parameter can be calculated in a causal way:

$$\hat{\Theta}(t) = \hat{\Theta}(0) - \int_0^t \Gamma^{-1} Y^T(q_d(\tau), \dot{q}_d(\tau), \ddot{q}_d(\tau))\Delta y(\tau)\, d\tau \tag{9.21}$$

where Γ is an $m \times m$ constant positive diagonal matrix. Then, substituting Eq. (9.18) into Eq.

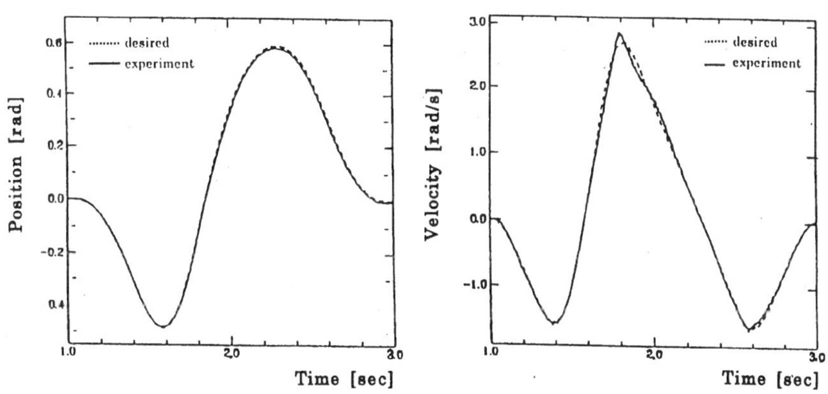

FIGURE 9.6 (b)
Joint angle position and velocity trajectories of link 2 after 30 trials using Δy.

(9.1) yields

$$\left\{H(q)\frac{dt}{dt}+\frac{1}{2}\dot{H}(q)\right\}\Delta\dot{q}+\{S(q,\dot{q})+B+B_1\}\Delta\dot{q}+A\Delta q+D\Delta y+h(\Delta q,\Delta\dot{q})$$

$$=\hat{u}_d-u_d=\Delta u=Y(q_d,\dot{q}_d,\ddot{q}_d)\Delta\Theta \tag{9.22}$$

where $\Delta\Theta=\hat{\Theta}-\Theta$ and h is a nonlinear function of Δq and $\Delta\dot{q}$ defined as in Eq. (9.14).

It is now important to see that taking an inner product between Δy and Eq. (9.21) yields

$$\frac{d}{dt}\left\{V(\Delta q,\Delta\dot{q})+\frac{1}{2}\Delta\Theta^T\Gamma\Delta\Theta\right\}\leqslant-\xi(\Delta y)\leqslant-\beta\|\Delta y\|^2 \tag{9.23}$$

where V and ξ can be determined as they appeared in Eq. (9.15). Thus, it is possible to show that $\Delta q(t)\to 0$ and $\Delta\dot{q}(t)\to 0$ as $t\to\infty$; that is, the joint trajectory tracking is established in an asymptotic manner.

A similar argument can be applied for the ability of iterative learning and adaptive trajectory tracking in such two cases that a robot tool endpoint is holonomically constrained and a set of multiple robots manipulates a rigid object cooperatively. Detailed discussions with experimental and computer simulation results may be found in Whitcomb *et al.* [15] and Naniwa and Arimoto [17, 18].

5 REALIZATION OF FRICTION/GRAVITY-FREE ROBOTS

Motion of actual mechanical robots is subject to unknown parasitic dynamics mainly caused by friction in motors themselves and transmission mechanisms. Therefore, the damping term $r(\dot{q})$ in Eq. (9.1) cannot be modeled explicitly. However, principal contributions to the term $r(\dot{q})$ must be static, Coulomb, and viscous frictions, all of which can be compensated by updating the estimates for those friction coefficients on the basis of measured regressors. In fact, suppose that

$$r(\dot{q})=B\dot{q}+\bar{C}\,\mathrm{sgn}(\dot{q}) \tag{9.24}$$

where B and $\bar{C}$ signify diagonal matrices whose diagonal components express coefficients of viscous and Coulomb frictions, respectively, and $\mathrm{sgn}(\dot{q})=(\mathrm{sgn}(\dot{q}_1),\ldots,\mathrm{sgn}(\dot{q}_n))^T$. Static friction is defined as the maximum torque level $\bar{c}_i+\delta_i$ with $\delta_i>0$ that can start to rotate the corresponding joint from the rest position (see Figure 9.7). In the case of DD (direct drive) robots, it can be assumed that the level of static friction at the jth joint is almost coincident with the coefficients of Coulomb friction $\bar{c}_i$ of the DD motor itself (i.e., $\delta_i\approx 0$). In this case, compensation for both friction and gravity forces can be realized by constructing regressors for the friction term $r(\dot{q})$ and the gravity term $g(q)$ in such a manner as

$$u=-A\Delta q-B_1\dot{q}+Z(q,\dot{q})\hat{\Theta} \tag{9.25}$$

where $\hat{\Theta}$ denotes the estimate of unknown parameters $\Theta=(\theta_1,\ldots,\theta_m)^T$ consisting of diagonal components of B and $\bar{C}$ and link masses appearing linearly in $g(q)$, and $Z(q,\dot{q})$

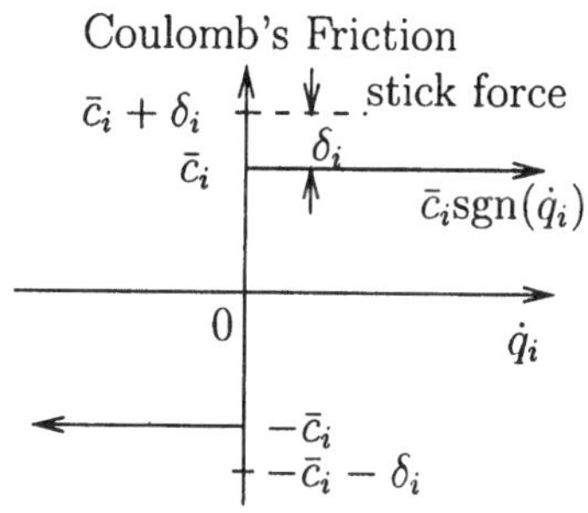

FIGURE 9.7
Characteristics of Coulomb's friction.

signifies the regressor for $r(\dot{q})$ and $g(q)$, that is,

$$Z(q, \dot{q})\Theta = B\dot{q} + \bar{C}\,\mathrm{sgn}(\dot{q}) + g(q) \tag{9.26}$$

The estimator $\hat{\Theta}$ should be updated in the following way:

$$\hat{\Theta}(t) = \hat{\Theta}(0) - \int_0^t \Gamma^{-1} Z^T(q(\tau), \dot{q}(\tau)) y(\tau)\, d\tau \tag{9.27}$$

where y is defined as in Eq. (9.9).

An experimental study of the realization of friction/gravity-free robots was carried out by using a DD robot with three degrees of freedom (DOF) at Yamaguchi University. Experimental data are available in a master-course dissertation [19], in which compensation for static friction was also fulfilled (see [20] for compensation for static).

6 GENERALIZATION OF IMPEDANCE MATCHING TO NONLINEAR DYNAMICS

The concept of impedance is inherent in linear dynamical systems such as lumped-parameter electric circuits. However, the concept can be generalized for nonlinear mechanical systems in terms of input–output relations called "passivity" and "dissipativity." Preliminary discussions of this problem were published by one of the authors [12, 13] in relation to coping with design problems of controllers for nonlinear mechanical systems such as anthropoid robots and mechanical hands. Another direction of extension of the impedance concept is to generalizing the framework of impedance matching to nonlinear dynamics. Consider the simplest problem of impedance control depicted in Figure 9.8, where a single DOF tool whose end is covered with soft material must press a rigid object at the desired force f_d. In ordinary situations the mass M of the tool is uncertain and the nonlinear characteristics $f(\Delta x)$ of reproducing force with respect to displacement Δx are unknown (see Figure 9.9). The dynamics of the system can be described by

$$M\ddot{x} + c\dot{x} = -f + u \tag{9.28}$$

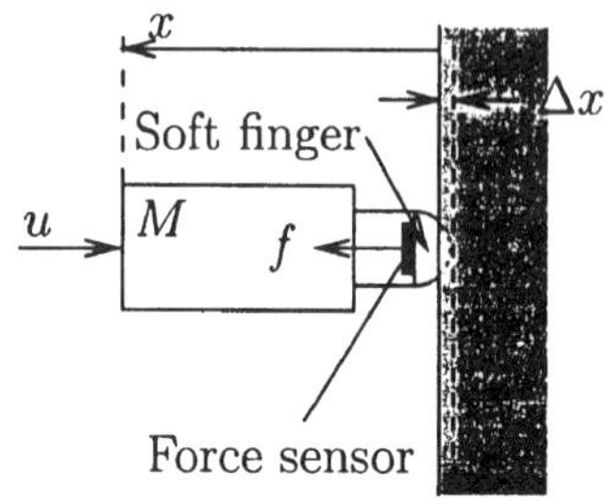

FIGURE 9.8
Impedance control for a single DOF system.

where u denotes the control input. A reasonable way to design u is as

$$u = f_d + \hat{M}\dot{r} + \hat{c}r + v \tag{9.29}$$

where $\hat{M}$ and $\hat{c}$ stand for estimates for M and c, r is an appropriate signal defined later, and v is an extra input. Substituting Eq. (9.29) into Eq. (9.28) yields

$$M(\ddot{x} - \dot{r}) + c(\dot{x} - r) + \Delta M\dot{r} + \Delta cr = -\Delta f + v \tag{9.30}$$

where $\Delta M = M - \hat{M}$ and $\Delta c = c - \hat{c}$. By denoting $y = \dot{x} - r$ and setting

$$\hat{M}(t) = \hat{M}(0) - \int_0^t \gamma_M^{-1}\dot{r}(\tau)y(\tau)\, d\tau \tag{9.31}$$

$$\hat{c}(t) = \hat{c}(0) - \int_0^t \gamma_c^{-1}r(\tau)y(\tau)\, d\tau \tag{9.32}$$

it is possible to see that an inner product between y and Eq. (9.30) yields

$$\frac{d}{dt}\frac{1}{2}\{My^2 + \gamma_M\Delta M^2 + \gamma_c\Delta c^2\} + cy^2 = -y(\Delta f + v) \tag{9.33}$$

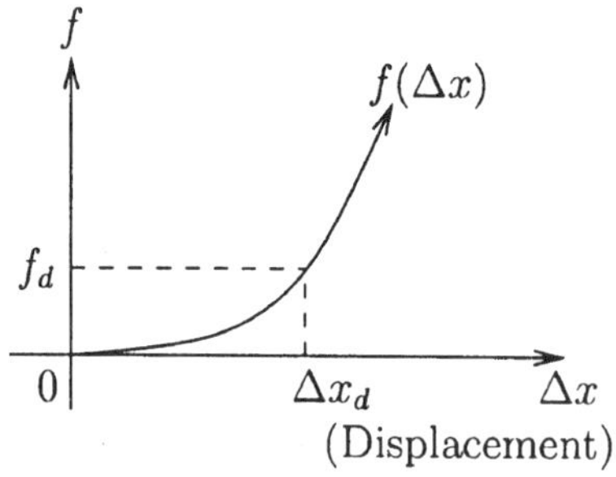

FIGURE 9.9
Nonlinear characteristics of reproducing force.

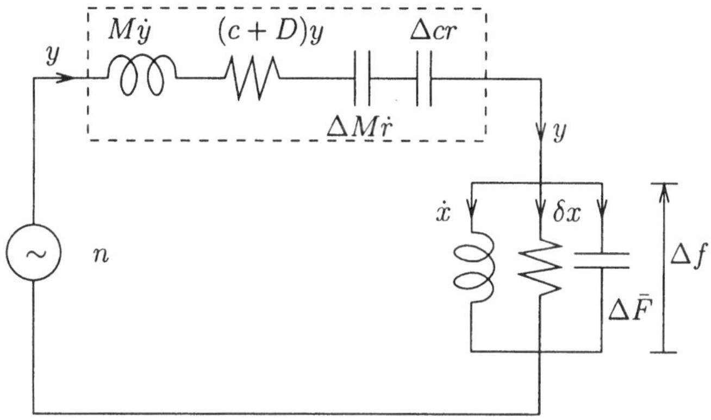

FIGURE 9.10
A circuit-theoretic expression of impedance control.

This form suggests the best design for signal y through the design of signal r. In other words, if the pair $\{y, \Delta f\}$ satisfies passivity or dissipativity, then the input v and the output y of the overall system satisfy passivity or dissipativity. This observation leads to the definition

$$r = -\alpha\Delta x - \beta\Delta F, \qquad \Delta F = \int_0^t \Delta f \, d\tau \tag{9.34}$$

where α and β are appropriate positive constants. Then,

$$\begin{aligned} y &= \dot{x} + \alpha\Delta x + \beta\Delta F \\ &= \dot{x} + \alpha\delta x + \beta\Delta\bar{F} \end{aligned} \tag{9.35}$$

where $\delta x = \Delta x - \Delta x_d$, $f(\Delta x_d) = f_d$, and $\Delta\bar{F} = \Delta F + \alpha\Delta x_d/\beta$. If v is considered to be an extra damping injection Dy plus an original disturbance n, then Eq. (9.35) can be written in the form

$$M\dot{y} + (c + D)y + (\Delta M\dot{r} + \Delta cr) = -\Delta f + n \tag{9.36}$$

The pair of Eqs. (9.35) and (9.36) can be expressed in a circuit as shown in Figure 9.10. Note that the circuit of Figure 9.10 is just a kind of nonlinear version of an electric circuit depicted in Figure 9.11. A further analysis of this system in relation to the theorem of maximum power supply as well as preliminary experimental results may be found in [21, 22].

7 LEARNING AS MAKING PROGRESS TOWARD IMPEDANCE MATCHING

Iterative learning can be considered to be a process of acquiring a desired control $u_d(t)$ realizing the given desired motion $q_d(t)$ for $t \in [0, T]$ by making progress toward impedance matching through repeated practice. To see this, it is necessary to simplify the argument and gain physical insight into the problem by treating the simplest case in which the objective system is linear and time invariant, and satisfies dissipativity. First note that the circuit in

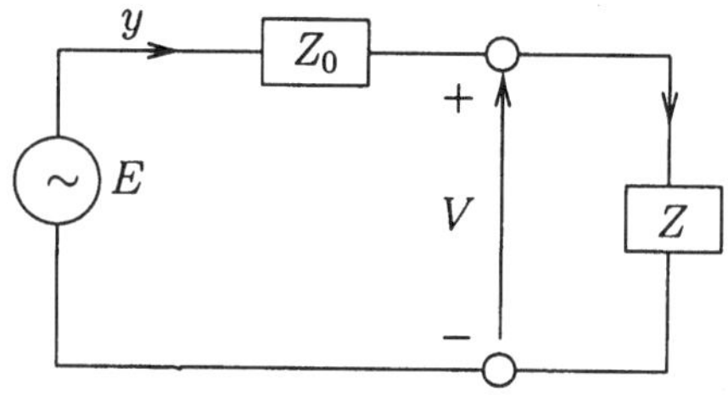

FIGURE 9.11
Impedance matching is realized when $Z = Z_0^*$. Z_0^* is the complex conjugate of Z_0.

Figure 9.11 can be rewritten in the form of a negative feedback control system as shown in Figure 9.12. Suppose that, given a desired periodic output $y_d(t)$ (corresponding to E in Figure 9.12) with period T, that is, $y_d(t) = y_d(t + T)$, the problem is to find the desired u_d (corresponding to V in Figure 9.12) that realizes $Zu_d = y_d$. It should be noted that the impedance function of the objective system is strictly positive real and the internal impedance Z_0 must be very close to zero. Then, as discussed in Section 3, the iterative learning control system can be depicted as in Figure 9.13, where the learning update law is described by Eq. (9.17). Because the strict positive realness of Z implies the existence of a positive constant $\gamma > 0$ such that

$$\int_0^t \Delta u_k \Delta y_k \, d\tau \geqslant V_k(t) - V_k(0) + \int_0^t \gamma \|\Delta y_k\|^2 \, d\tau \tag{9.37}$$

where V expresses a storage function that is nonnegative (see [23]), it is possible to obtain the following inequality (subtract u_d from both sides of Eq. (9.17) and take an inner product of both sides of the resultant equation through Φ^{-1}):

$$\Phi^{-1}\|\Delta u_{k+1}\|^2 + V_{k+1} \leqslant \Phi^{-1}\|\Delta u_k\|^2 + V_k + (\Phi - 2\gamma)\|\Delta y_k\|^2 \tag{9.38}$$

where $y_k(t) = y(t + kT)$, $u_k(t) = u(t + kT)$, $V_k(t) = V(t + kT)$, and $\|\Delta u\|$ denotes the norm of Δu in $L^2[0, T]$. Hence, if Φ is chosen so that $0 < \Phi < 2\gamma$, then Eq. (9.38) means that $y_k(t) \to y_d(t)$ in $L^2[0, T]$ as $k \to \infty$. This indicates that the forward path in the system of Figure 9.13 tends to realize the zero impedance (the infinitely large admittance), because $\Delta y_k \to 0$ and $u_k \to u_d$ as $k \to \infty$.

Another type of learning with nonzero real part of the internal impedance (i.e., $\mathrm{Re}(Z_0) \neq 0$) can be discussed in a similar way. Some results concerning this problem together with its extension to nonlinear dynamics will be presented at the ILC (Iterative Learning Control) Workshop preceding the IEEE CDC '98.

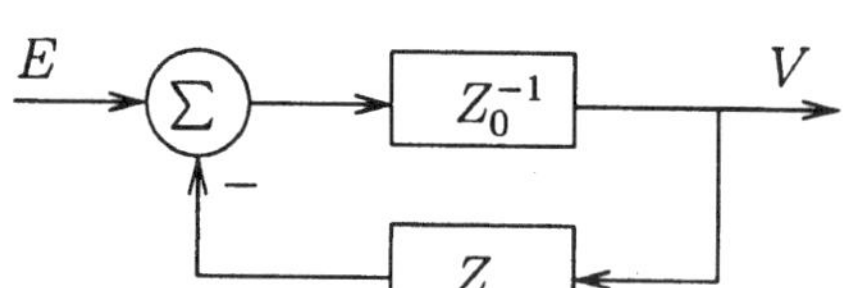

FIGURE 9.12
This negative feedback structure is equivalent to the circuit depicted in Figure 9.11.

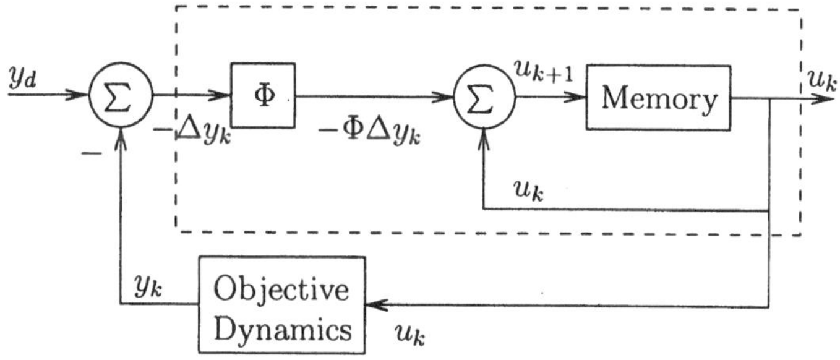

FIGURE 9.13
Iterative learning as impedance matching with internal zero-impedance.

8 CONCLUSION

This chapter attempted to unveil some secrets of the complicated dynamics of a robot that can fulfill prescribed tasks with a relatively simple control scheme without referring to the full knowledge of its physical parameters, regardless of the fact that Lagrange's equation of its motion is nonlinear and has strong couplings between joints. The most important result is the observation that a generalized concept of impedance matching to nonlinear mechanical systems is essential for a robot executing given tasks with sufficient smoothness and with the use of less knowledge of its dynamics.

In parallel with this research, we are also attempting to show that exact calculation of Jacobian matrices of the camera coordinates and the task coordinates with respect to the joint coordinates is unnecessary in the cases of visual feedback and hybrid position–force control under constraint (see [24]).

REFERENCES

[1] M. Vukobratovic. *Scientific Fundamentals of Robotics*, vols. 1 and 2. Springer-Verlag, Berlin, 1982.

[2] A. K. Bejczy. Robot arms dynamics and control. TM: 33-69, Jet Propulsion Laboratory, 1974.

[3] K. Takase. General decomposition and control of a motions of a manipulator. *Trans. SICE* (Society of Instrumentation and Control Engineers) 12:300–306, 1976 (in Japanese).

[4] J. Hollerbach. A recursive Lagrangian formulation of manipulator dynamics and a comparative study of dynamics formulation complexity. *IEEE Trans. Syst. Man Cybernet.* 10:730–736, 1980.

[5] J. Y. S. Luh, M. W. Walker, and R. P. C. Pul. On-line computational scheme for mechanical manipulators. *Trans. ASME J. Dynamic Syst. Meas. Control* 102:69–76, 1980.

[6] J. J. Slotine and W. Li. On the adaptive control of robot manipulators. *Int. J. Robot. Res.* 6:49–59, 1987.

[7] N. Sadegh and R. Holowitz. Stability and robustness of adaptive controllers for robotic manipulators. *Int. J. Robot. Res.* 9:74–92, 1990.

[8] M. Takegaki and S. Arimoto. A new feedback method for dynamic control of manipulators. *ASME J. Dynam. Syst. Meas. Control* 103:119–125, 1981.

[9] S. Arimoto and F. Miyazaki. Stability and robustness of PID feedback control for robot manipulators of sensory capability, in Robotics Research: First Int. Symp. (ed. M. Brady and R. P. Paul), MIT Press, Cambridge, MA, pp. 783–799.

[10] S. Arimoto. A class of quasi-natural potential and hyper-stable PID servoloops for nonlinear robotic systems. *Trans. SICE* (Society of Instrumentation and Control Engineers) 30:1005–1012, 1994.

[11] S. Arimoto. Fundamental problems of robot control: Part I. Innovation in the realm of robot servo-loops. Part II. A nonlinear circuit theory towards an understanding of dexterous motions. *Robotica* 13:19–27, 111–122, 1995.

[12] S. Arimoto and T. Nakayama. Another language for describing motions of mechatronics systems. *IEEE/ASME Trans. Mechatronics* 1:168–180, 1996.

[13] S. Arimoto. *Control Theory of Nonlinear Mechanical Systems: A Passivity-Based and Circuit-Theoretic Approach.* Oxford University Press, Oxford, 1996.

[14] V. Parra-Vega and S. Arimoto. An exponentially convergent adaptive sliding mode control of robot manipulators. *Int. J. Syst. Sci.* 26:2263–2276, 1995.

[15] T. Naniwa. Trajectory control of a DD robot by learning: A dissertation conducted by H. Muraoka at Yamaguchi University, 1996 (in Japanese).

[16] L. L. Whitcomb, S. Arimoto, T. Naniwa, and F. Ozaki. Adaptive model-based hybrid control of geometrically constrained robot arms. *IEEE Trans. Robot. Automat.* 13:105–116, 1997.

[17] T. Naniwa and S. Arimoto. Learning control for robot tasks under geometric endpoint constraints. *IEEE Trans. Robot Automation*, vol. 11, pp. 432–441, 1995.

[18] T. Naniwa, S. Arimoto, and K. Wada. Learning and adaptive controls for coordination of multiple manipulators holding a geometrically constrained object. *Proceedings of IROS'97*, Grenoble, France, 1997, pp. 484–490.

[19] T. Naniwa. Learning control using saturated position errors with fiction compensation: A master-course dissertation conducted by H. Tsutsui at Yamaguchi University, 1997 (in Japanese).

[20] S. Arimoto, H. Koga, and T. Naniwa. Proposal of the law-of-inertia (friction/gravity-free) robots. *Proc. of the 1997 IEEE ICRA*, Albuquerque, NM, 1997, pp. 2232–2239.

[21] S. Arimoto. A generalization of impedance matching for evaluation of robot skill, *Proceedings of the 3rd ECPD Conference on Advanced Robotics, Intelligent Automation and Active Systems*, Bremen, Germany, 1997, pp. 62–67.

[22] S. Arimoto, H.-Y. Han, C. C. Cheah, and S. Kawamura. Generalization of impedance matching to nonlinear dynamics of robot tasks, Preprints of the 4th IFAC NOLCOS'98, Enschede, The Netherlands, 1998, pp. 539–544.

[23] S. Arimoto, S. Kawamura, and H.-Y. Han. System structure rendering iterative learning convergent. *Proceedings of the IEEE CDC'98*, Tampa, FL, in press.

[24] C. C. Cheah, S. Kawamura, and S. Arimoto. Hybrid position and force control for robot manipulator with a class of constraint uncertainty, *Proceedings of Japan-USA Symposium on Flexible Automation*, Ohtsu, Shiga, Japan, 1998, pp. 501–507.

CHAPTER 10

Sensor-Based Planning and Control in Telerobotics

WILLIAM R. HAMEL
The University of Tennessee, Knoxville, Tennessee
and
Oak Ridge National Laboratory, Oak Ridge, Tennessee

1 INTRODUCTION

The notion of telerobotics is relatively new. Although considerable research has addressed telerobotics since the 1970s, actual accomplishments of real-world working systems have been limited. Most work has been at the laboratory scale and has addressed elements of telerobotics-related concepts rather than integrated systems working under realistic conditions. Integrated telerobotic systems that can have impact must provide robust operations under the unabridged constraints of the application domain being addressed. Sensing, planning, and control issues are most usefully considered in the context of the realities of real-world applications. One of the most challenging application domains for telerobotics is nuclear remote operations, where everyday workers attempt to perform complex maintenance and manufacturing operations through remote control. Activities in the nuclear domain are broadly applicable to other domains such as space and undersea operations. To provide context, this chapter will use details of nuclear applications to steer the discussion concerning planning, sensing, and control issues as they relate to robust telerobotics. The nuclear context is very specific, yet it also provides general considerations that are widely relevant in other applications.

Various forms of remote handling systems have been in use since humans have dealt with hazardous environments. Around 1940, research in atomic physics led to a new era in remote handling as scientists sought to explore the nature of the atom safely in the presence of ionizing radiation. As related experiments and developments became more complex, mechanical manipulator systems were created that allowed operators to perform increasingly complex tasks safely behind thick biological shielding. These mechanical systems then evolved into electrical systems that allowed larger work volumes to be covered, such as the large production plants that were being built to produce nuclear weapons. Incredible engineering achievements occurred in a brief period of 15 years, primarily within the Remote

Control Division of the Argonne National Laboratory. Even though this era represented tremendous technical achievement, it went further in illustrating the instrinsic complexity of remote operations. The best work performance achieved with sophisticated teleoperations-based remote systems is quite poor in comparison with the work efficiency that human workers can achieve with direct contact operations and common tools. Typically, this form of teleoperation (i.e., manual control over a physical distance or barrier) is ten to hundreds of times slower than contact operations! Remote operations are extremely expensive and time consuming because of this effect and have been the continual target of engineering improvements [1].

Many research and development efforts have focused on different avenues for improving the work efficiency of remote operations. These efforts have included the development of better manipulators, control stations, control algorithms, and so forth, all intended to enhance reliability and maintainability. In the late 1960s and early 1970s, as digital electronics became more cost effective, interest began to emerge in the integration of automation with teleoperations as a scheme for effectively increasing the work efficiency of remote operations. It was around this time that industrial robot concepts were also introduced. Combining selective automation of specific subtasks with traditional teleoperations offers the potential to reduce labor requirements and to improve the quality of repetitive task executions.

This integration of automation with teleoperation became the foundation of what is now termed "telerobotics." From the 1970s until today, telerobotics has been an active area of research and development in many different domains, including nuclear, space, and military applications. Unlike manufacturing automation, remote operations in hazardous and unstructured work task environments necessitate human-in-the-loop control, or teleoperations, as a backstop to ensure safe operations. Human-in-the-loop control greatly enhances the likelihood of successful recovery from automated task faults and failures.

This chapter is intended to provide the reader with an introduction to the concepts and issues that pertain to the realization of practical, useful, and cost-effective telerobotic systems. Sensing, planning, and control functions are the fundamental ingredients necessary to implement robust automation within the hybrid character of a telerobot. This discussion is intended to be an introduction to the concepts and technical challenges that telerobotics involves. The first section provides the reader with background on the history of teleoperations and remote handling, which ultimately defines the baseline of performance against which practical telerobotic systems will be evaluated. The next section discusses the definition of a telerobotic system from the different prevailing perspectives and presents recent examples of actual systems that have been built. As with any engineering systems, practical telerobot concepts are totally driven by application requirements. The next section discusses a nuclear application domain and the inherent problems and constraints included therein. With this requirements-driven foundation, the technical features of a robust telerobot are presented in their idealized form. The chapter concludes with discussions of current research and remaining challenges pertaining to the realization of the robust telerobot.

2 HISTORY OF TELEOPERATIONS AND REMOTE HANDLING

As will be discussed in more detail in Section 3, telerobotics, by definition, involves the combination, or integration, of teleoperations with automation. Much can be gained toward analyzing the synergy between manual and automatic control in telerobotics by first understanding the foundations of manual control as represented in teleoperations. The standards of established remote operations design and performance are the reference states against which telerobotics concepts will ultimately be judged. It is useful to review the history of teleoperations and remote handling to acquire a perspective on these important reference levels of performance.

The creation of special tools for remote handling has been an integral part of humans' natural adaptation to their environment throughout history. This process intensified greatly as scientists explored atomic physics and natural radioactivity at the turn of the century. The engineering implications of production-scale nuclear weapon manufacture resulted in a new generation of remote handling technology. During this era, remote engineering projects occurred throughout the U.S. Atomic Energy Commission (AEC) complex, in Europe, and in the former Soviet Union. Early remote handling systems were surprisingly simple. Walls were built from suitable shielding materials, which were usually lead bricks. Periscopes and mirror arrangments were configured to see over the shielding walls into the work task areas. Imaginative forms of long-handled tools were fabricated to allow workers to manipulate objects from behind and over the walls as necessary to accomplish experiments and tests involving radiation.

As the science and experiments became more complicated, it became increasingly difficult to accomplish the research effectively with unnatural long-handled tools and mirrors. It was determined that it would be desirable to have transparent shielding that one could look straight through to observe the work area and to have some type of mechanical arms in the radiation area that could do a much better job than long-handled tools in emulating human motions. These ideas represented schemes whereby human capabilities could be more completely "projected" into the remote work environment. The tacit assumption was that human projection would naturally result in improved remote work efficiencies. The Remote Control Division of the Argonne National Laboratory was created to explore this type of R&D in the early 1940s for the AEC. This extraordinary group was led by Ray Goertz, and over a period of approximately 15 years the division achieved incredible engineering accomplishments in manipulators, teleoperations, and other aspects of remote handling.

One of the first challenges that the Argonne group tackled was the development of mechanical master–slave manipulator systems that would allow operators to perform more complex tasks in remote areas. The basic concept was to create a mechanism that would have a master controller side where an operator could provide position and orientation commands to a slave-side mechanism or linkage system that would "replicate" motions and forces in the remote work area. From the beginning, Goertz felt that force reflection to and from the master and slave systems was essential for the operator's sensory awareness of the task execution. Human factors experimentation has repeatedly verified the significance of both kinesthetic (muscular) and tactical (touch) feedback in performing more complicated tasks [2]. The first test unit of the mechanical master–slave manipulator idea is shown in Figure 10.1.

This initial engineering development hardware ultimately led to the development of modern mechanical master–slave manipulators (MSMs) such as the one depicted in Figure 10.2. Today, thousands of MSMs are in use around the world in nuclear, biological, and other types of hazardous remote experimentation and operations. MSMs have been refined over the years in terms of their payloads, kinematics, friction properties, and reflected inertia properties. Metal tape drives and pulley systems that minimize friction and inertia are normally used to transmit master motion and torque to the slave side with incredible fidelity. For example, MSMs are routinely used in remote hot cells to perform chemistry experiments constructed from standard laboratory glassware. The remote work efficiency of a dual-arm MSM with shield-window viewing is about 5 to 10 times slower than that of equivalent

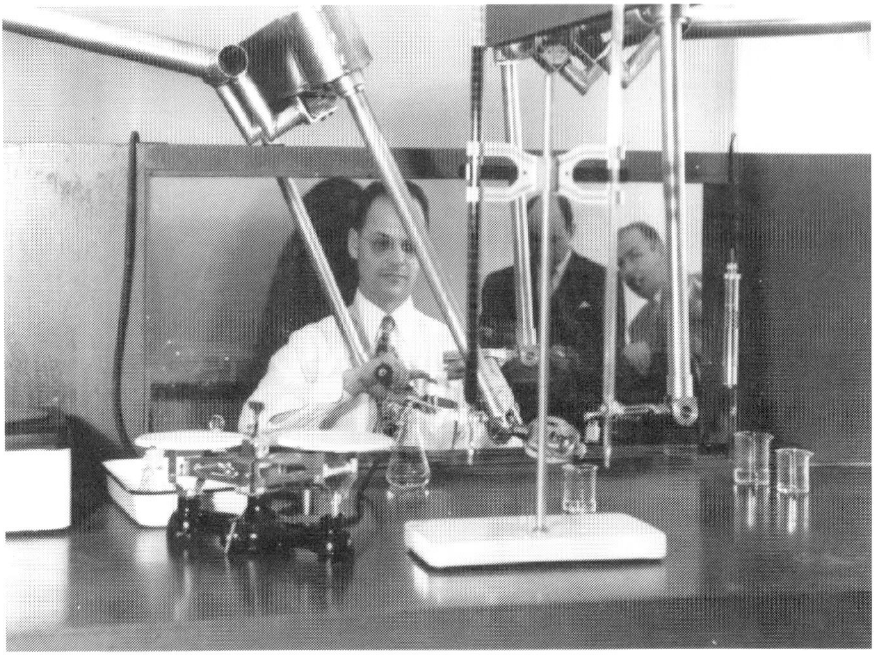

FIGURE 10.1
First mechanical master–slave manipulator. (Courtesy of Oak Ridge National Laboratory.)

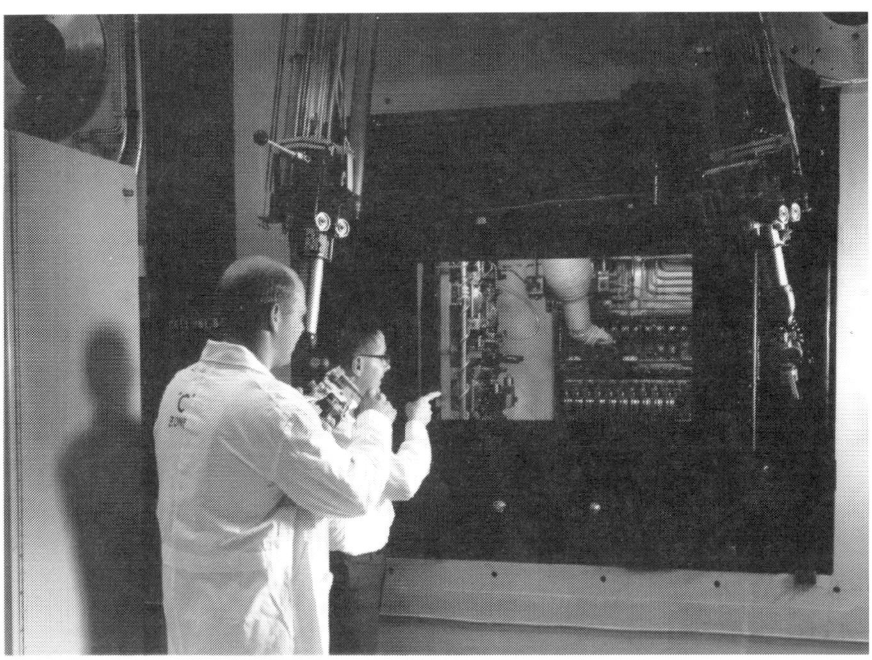

FIGURE 10.2
Modern mechanical master–slave manipulator. (Courtesy of Oak Ridge National Laboratory.)

contact operations. Most remote handling technologists consider the remote work efficiency of MSMs as a standard of comparison for the performance evaluation of alternative systems. The reader should save this fact for later reference during the discussion of telerobotics and the basic concept of selective automation as a scheme to enhance remote work efficiency.

MSMs provide noteworthy teleoperations capabilities; however, they have a very severe limitation because of their purely mechanical construction. Because the master and slave sections are mechanically coupled through the metal-tape drive transmission, the physical separation that can exist between the safe operating area and the hazardous remote work area is limited to a maximum of approximately 10 m. As shown in Figure 10.2, the through the shielding wall physical arrangement of the MSMs results in an elbows-up kinematic configuration that is best suited for tabletop operations rather than reaching into work areas. Because of these characteristics, MSMs are quite restrictive in many applications and often have overly constrained the physical design of remote cells. Goertz and others recognized that it would be much better to have the equivalent of a fly-by-wire MSM in which the physical separation of the master and slave would be essentially unconstrained. This need led to the development of electrical master–slave manipulators (EMSs), which are commonly called electrical servomanipulators [3, 4]. R&D on the EMS began in the late 1940s and continued into the early 1950s.

The Argonne group was at that time limited by the available electrical control technology. Nonetheless, they made noteworthy progress toward integrated systems as depicted in Figure 10.3. The system shown is a dual-arm anthropomorphic system with head-aiming remote television viewing and bilateral force reflection. This system provided (at least at the research level) an unprecedented degree of human projection in remote operations. One has the sense that the Argonne group had all of the correct fundamental concepts but lacked the

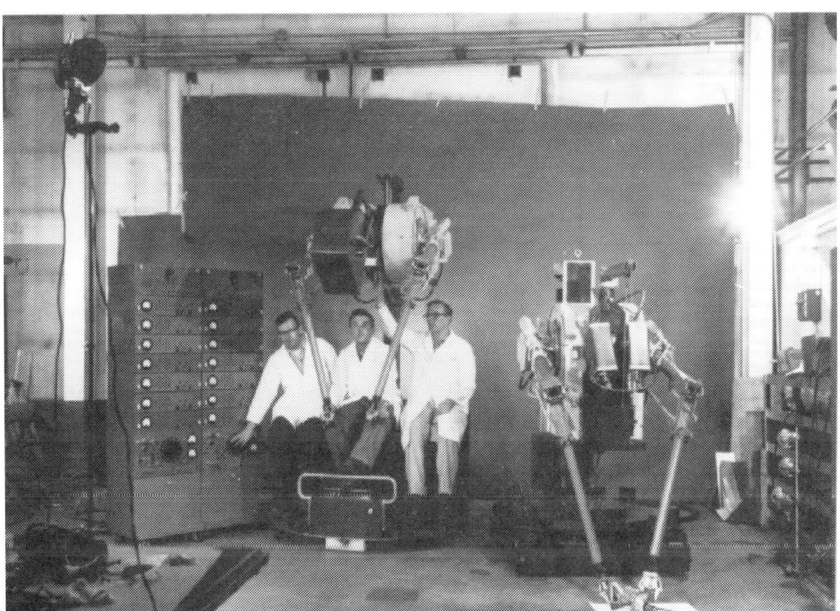

FIGURE 10.3
An integrated electrical master–slave manipulator system. (Courtesy of Oak Ridge National Laboratory.)

supporting technology necessary to implement practical and cost-effective systems. In the late 1950s, the Argonne team began to disband because of declining support from the AEC, who had concluded that remote handling technology had reached an apex constrained by supporting technologies and that further R&D would yield limited returns.

The Argonne group performed outstanding R&D in a wide range of remote handling technologies, especially manipulator systems. Their fundamental research on remote manipulators is still quite valid today. Worldwide use of MSMs really stems from the pioneering work done at Argonne. In terms of modern telerobotics, the history of teleoperations and remote handling offers some important lessons. Good force-reflecting teleoperators with effective remote viewing can be used to perform tasks with the complexity typical of contact operations but with slowdown factors of 5 to 10. Prior experience clearly defines the high sensitivity of remote operations to remote sensing and/or viewing and the anthropomorphism of the manipulation system. It seems clear that new telerobotic systems must exceed the established performance standards set by early teleoperations, or they will be avoided because of the increased costs and risk factors associated with their perceived complexity. New ideas survive only if they pay off.

The early work at Argonne led to the invention of robot manipulators and made great strides toward effective teleoperators for remote operations. After the Remote Control Division was disbanded, limited R&D occurred until the 1970s, when commercial nuclear power growth was driving a number of research programs in the United States, West Germany, France, and Japan. During this time, the programs in the United States and France were by far the most aggressive; several generations of electrical servomanipulator systems were developed, most of which incorporated emerging microprocessor technology [5, 6]. The Central Research Laboratories model M2 system, shown in Figure 10.4, was jointly developed with the Oak Ridge National Laboratory and was the first force-reflecting servomanipulator system to use distributed digital electronics to implement position–position force reflection with multiplexed serial communications between the master and slave. This system also incorporated a menu-driven alphanumeric operator interface that greatly improved operator efficiency and reduced training time [5]. The model M2 system was used over the years to perform a wide range of complex demonstration tasks for military, space, and nuclear applications. In addition to mechanical assembly and disassembly tasks, difficult tasks such as manual electric welding have been performed through force-reflecting teleoperations. The M2 was used to replicate the assembly of the NASA ACCESS space truss assembly, and excellent results were obtained regarding robot replacement of astronaut tasks for many operations. Refer to Figure 10.5. This system played a major role in demonstrating the potential of electrical servomanipulator systems for efficient teleoperations in highly unstructured task environments.

The development of the advanced servomanipulator (ASM) followed the M2 in an effort to improve the remote maintainability of the remote manipulators themselves by making them mechanically modular so that one robot system could be used to repair another. The motivation for this work was to reduce maintenance technician radiation exposure and to increase the overall availability of the remote maintenance system. The ASM was designed from the beginning to provide a foundation for telerobotics in addition to effective teleoperations [7]. See Figure 10.6. Its control system, which was advanced for its day, was a modern distributed digital system with a complete "glass cockpit" approach to the controls and displays. The ASM was considered a successful demonstration of remote manipulator maintainability. In fact, the M2 manipulator was used to dissassemble and successfully reassemble one of the ASM manipulator arms in a total of $8\frac{1}{2}$ hours of remote operating time. Trajectory teach–playback and automated tool-changing functions were also demonstrated.

FIGURE 10.4
CRL model M2 servomanipulator maintainance system. (Courtesy of Oak Ridge National Laboratory.)

Unfortunately, the Department of Energy (DOE) program supporting the work was canceled along with the nation's breeder nuclear reactor program before telerobotic automation functions could be incorporated and evaluated, although simple teach and playback trajectory control using the master controller as the teaching pendant was implemented.

At the time of the M2 and ASM developments, Jean Vertut and his colleagues in the Commission d'Énergie Atomique (CEA) focused their research on the development of telerobotic functionality for their MA-23 electrical servomanipulator systems as shown in Figure 10.7. This work may very well be one of the earliest experimental demonstrations of telerobotic functions [6].They called their concept computer-assisted teleoperations, and it included both operator assists and robotic teach–playback functions. Operator assists included software jigs and fixtures designed for the improvement of the remote operation of tools such as saws and drills. For example, they demonstrated how an imaginary software plane could be defined to constrain the motion of a rotating disc cutter, much like a carpenter's miter box. Similarly, drilling along an arbitrarily oriented centerline was demonstrated. In both of these cases, the operator would "feel" excursions from the constraint through the system force reflection. They also implemented surface or object-tracking assists using end-effector proximity sensing [6]. These telerobotic assists were studied in laboratory-scale experiments and demonstrated significantly improved (with respect to both task time and work quality) remote disc cutting and drilling.

FIGURE 10.5
CRL model M2 manipulator system performing space truss assembly. (Courtesy of Oak Ridge National Laboratory.)

In the 1980s and 1990s, nuclear remote operations technology migrated into other areas such as space and the military as nuclear power activities began to decline. Nuclear-style teleoperations influenced the Space Shuttle remote manipulator system and the short-lived Flight Telerobotic Servicer program. Lesser but nonetheless important influences occurred in undersea remote technology as well. As we look forward to robust telerobotics, it is important that we understand what has been accomplished and learned to date. Hazardous environments are highly unstructured in comparison with repetitive manufacturing environments. Task "unstructuredness" ultimately results in high task uncertainty. Task uncertainty drives the use of systems based on effective teleoperations, because human-in-the-loop operations ensure the availability of human cognition, creativity, and innovation to counteract the certain occurrence of unexpected events. Experience has shown that minor perturbations in any planned scenario can disrupt automated task execution. There is a noteworthy track record of effective teleoperations in highly complex environments. This track record is the baseline, or reference state, against which any telerobotics schemes will be judged. For ideas to go beyond the R&D laboratory, their performance in the field must exceed these established standards.

This history discussion touches the surface of teleoperations developments and experience. We can draw a few generalizations from the baseline that should be applied to the pursuit

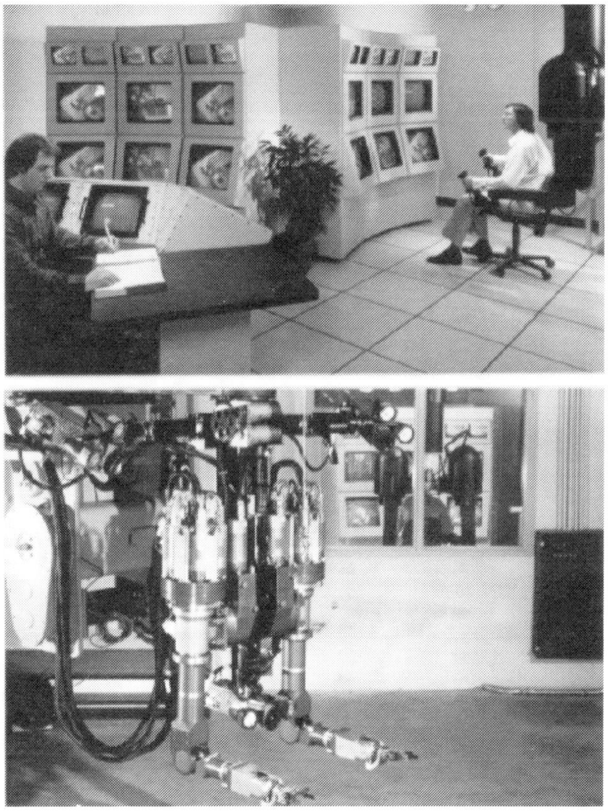

FIGURE 10.6
Advanced servomanipulator system. (Courtesy of Oak Ridge National Laboratory.)

of more robust telerobotics. First, the mind-set of people who perform actual remote operations is very focused on the task at hand. There is zero tolerance for systems and equipment that do not perform their intended functions correctly and reliably. This mind-set is also a conservative one toward new ideas, and rightly so because there are many instances in which hardware failures have resulted in total failure of the overall mission. Only systems that perform reliably out of the box are used consistently. Second, the functionality and performance capabilities of the state of the art in modern teleoperations set the standards of expectation. New telerobotic system concepts must do more than equal this performance level; they must exceed it so that users are motivated to try something new. The teleoperations baseline (see [6] for more details of other systems than those discussed here) that provides remote work efficiencies of 5 to 10 relative to contact operation can be characterized as follows:

- Dual-arm manipulator configurations for anthropomorphism and the ability to handle tools and object simultaneously and in a coordinated manner.
- Manipulators with human-compatible operating characteristics:
  (1) No load tip speeds between 36 and 48 in/s
  (2) Link lengths near human upper and lower arm sizes
  (3) High degree of anthropomorphism with respect to size and kinematic arrangement

FIGURE 10.7
Computer-assisted teleoperation experiment.

(4) Systems that feed back forces (i.e., provide force reflection) from the remote task environment to the operator; modern teleoperators have achieved 0.5–1.0% of maximum payload.

- Systems that provide high-fidelity sensory feedback from the remote task environment to the operator:
  (1) Stereo and multiview remote television that is easily steered and zoomed
  (2) Audio remote feedback
  (3) Kinesthesia and tactility through the master controller and the manipulator force reflection.
- Humanly sensible controls and displays:
  (1) Manipulator master controllers with intuitive correspondence to the slave manipulators
  (2) Intuitive kinematics in all system articulations including the manipulators
  (3) Ergonomically designed controls and displays.
- Systems that were designed for and have the innate ability to handle and operate standard and special tooling efficiently.

The complex synergy of these characteristics (at least these; there are probably many others) establishes the remote work efficiency of an integrated system. There are impressive and promising new hardware and software technologies that vastly enhance our ability to implement effective telerobotics including improved teleoperator functions. New ideas built around these technologies must provide a clear benefit/cost ratio relative to the existing teleoperations baseline.

3 THE NOTION OF TELEROBOTICS

Given the history of teleroperations, let us now step back to discuss the fundamental structures associated with telerobotic systems as a foundation for later consideration of related sensing, planning, and control issues. Several groups and individuals have claimed the creation of the concept of a telerobot. Although it is not known who really deserves the full honor for inventing the new terminology, it is clear what a reasonable definition is: A telerobot is a system that beneficially combines human interaction and automation in a single robot system. This distinction can be quite general, but for the purposes of this discussion, the concept is applied to the general problem of remote operations in hazardous environments. A telerobot in remote operations is best thought of as a system that provides a continuum of remote capabilities from fully manual operations through fully autonomous task execution where the mixture of operations is chosen to improve overall system performance [8].

Fully manual operations in the sense of remote manipulation are basically classical teleoperations in which a human operator uses some form of manual controller (master) to direct continuously the operation of the remote manipulation system (slave). The system architecture of telerobots as they were envisioned in the late 1970s is shown in Figure 10.8. The nature of this control interconnection can be quite sophisticated and variable in the

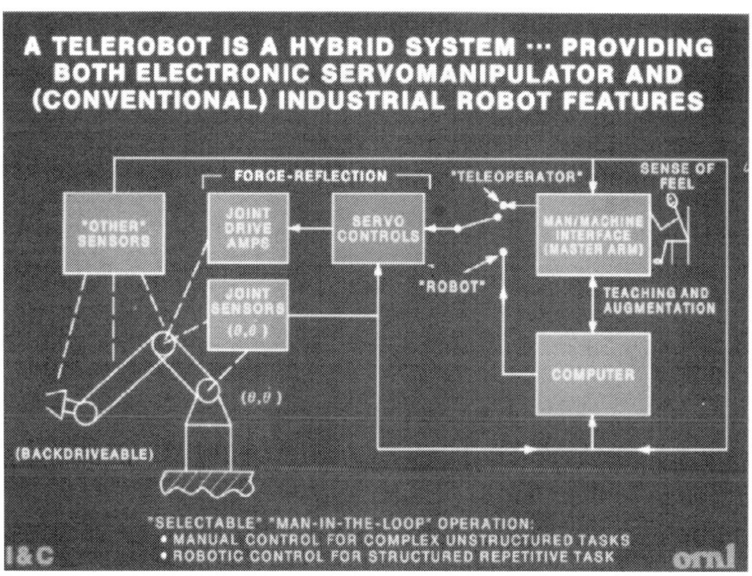

FIGURE 10.8
Telerobot architecture circa 1978.

overall work efficiency it provides the human operator. The most elaborate systems are designed to provide the operator with kinesthetic feedback, humanly compatible dynamic performance, and humanly compatible controls and displays, but the best of such systems are quite inefficient compared with equivalent direct human task execution. The key issue with regard to teleoperations, regardless of the remote work efficiency, is that remote work is performed in one-to-one correspondence with a human operator. That is to say, there is always a dedicated operator with each system. Consequently, the aggregate work efficiency can be no better than that of a single human operator. This fundamental limitation of teleoperations has been one of the factors that has motivated the concept of telerobotics.

The fundamental idea embodied in telerobotics is to enhance performance, especially task efficiency, by integrating automated functions that can be selected and used on demand by the operator. In this sense, the telerobot can be viewed as a combined teleoperator and robot. Many factors are critical to the value and utility of the concept. The telerobot system must preserve effective teleoperational features. The automatic functions must be readily usable under extremely unstructured task conditions, which places high demands on the *in situ* programmability. The perspective assumed in telerobotics is very important. One can consider the telerobot as a modified robot or as a modified teleoperator. These two perspectives provide fundamentally different results. Symbolically, this can be summarized as

$$Tr = Teleoperator \cup Robot \tag{10.1}$$

$$tR = Robot \cup Teleoperator \tag{10.2}$$

$$Tr \neq tR \tag{10.3}$$

These simple expressions mask the complexity of the underlying differences. A Tr system is one that is foremost a teleoperator but that can be "programmed" to perform specific subtasks automatically. A tR system is a programmable robot that has been augmented with control features that allow it to be operated manually in a teleoperator mode. Robot manipulators, which are used in tR systems, are commercially available and are designed to fundamentally different criteria than manipulators typically used in remote operations [9]. Manufacturing-driven robot manipulators place high value on position repeatability, cost, and other factors such as factory floor environmental requirements. These emphases result in manipulator designs with extreme structural rigidity and corresponding bulk (i.e., weight and physical size). They also generally involve digital controllers that are specialized, proprietary, and with narrow functional capabilities. In comparison with modern force-reflecting telemanipulators developed for nuclear applications, efforts to convert robots into teleoperators (e.g., space balls and PUMAs) have resulted in less effective systems in terms of remote dexterity and work efficiency.

Remote force-reflecting manipulators are usually designed to optimize the ability to "reflect" the force and moment conditions at the remote slave end effector "back" to the human operator. This critical sensory feedback effectively allows the operator to "feel" what is happening. Kinesthetic, tactile, and visual feedbacks are extremely important in executing the complex types of tasks that are typical of hazardous work environments [2]. This class of robot manipulators places very high value on the minimization of inertia, friction, and physical cross sections in the interest of obtaining sensitive force reflection and minimal arm cross section for maneuverability. Although these design objectives enhance teleoperation, they result in (compared with industrial robots) compliant structural characteristics that make precise programmed trajectory execution difficult. One can say that industrial robots do not make very good teleoperators (tRs) and teleoperators do not make very good robots

(Trs). One would ask, "What is the best approach for telerobots that must be able to perform both basic functions?"

Because a telerobot in the sense of this discussion (i.e., a Tr) is an extension of teleoperations intended to improve (in some sense) remote operations, it is imperative that the system provide effective teleoperability first. In the applications discussed in Section 4, nuclear radiation levels are often sufficiently high that remote operations without any amount of human intervention are necessary. Under these conditions, the baseline for comparison of operational alternatives is classical remote operations. Classical remote operations using tools such as force-reflecting teleoperators represent established and acceptable methods. Any alternative, such as telerobotics, must provide clear benefits in terms of performance, safety, or work quality.

The fundamental telerobot concept is very simple and obvious. It is not so obvious how one should balance the trade-offs already discussed or how one can actually realize a robust system with the task versatility that real-world hazardous environments require. It has been about 20 years since the concept was first discussed, and to this date a comprehensive telerobotic system has not actually been used in a real-world remote application. Clearly, much work remains to achieve the ideals established for telerobots. Let us look to an example of an actual application domain to study further the challenges implicit in robust telerobotics.

4 TYPICAL APPLICATION DOMAIN

There are many application domains for telerobots in the hazardous operations associated with undersea, space, nuclear, and other dangerous environments. Using the author's direct experience, the focus of this discussion will be on applications of telerobotics to the decontamination and decommissioning (D&D) of defunct nuclear facilities located around the world, but with emphasis on U.S. facilities. Nuclear D&D examples are quite useful because they present a range of problems and challenges due to the wide variety of facilities being considered. This range and variety provide excellent examples that typify unstructured and hazardous work environments in general.

D&D is essentially the task of demolishing and cleaning up old radioactive facilities, buildings, processing equipment, and sites such that buildings and property can be rendered to a safe condition or possibly reused. As shown in Figure 10.9, D&D is an iterative process of characterizing, decontaminating, and dismantling process systems and buildings until specific criteria for radiation release and other factors are achieved. Characterization is the process of making measurements to determine the concentration and location of contaminants. Decontamination is the process of removing contaminants. Dismantlement is the process of tearing down equipment and facilities. Within these fundamental operations, there are numerous possibilities for the beneficial integration of new technologies such as telerobotics. Telerobotic systems can contribute in several general respects: (1) reduction of human exposure to radiation and hazardous materials through remote operations; (2) increased productivity and quality of operations through selective automation of specific tasks; (3) reduction of secondary waste generation during D&D; and (4) facilitation of *in situ* survey, decontamination, and dismantling operations.

Within governement facilities and in areas such as the former Soviet Union, there is a wide range of D&D requirements that stem directly from the numerous types of research and production facilities which face demolition in the future. Facilities range from production-scale nuclear reactors and processing plants to small hot-cell and glove box research facilities. Physical requirements for load capacity and reach range from kilograms to metric tons and

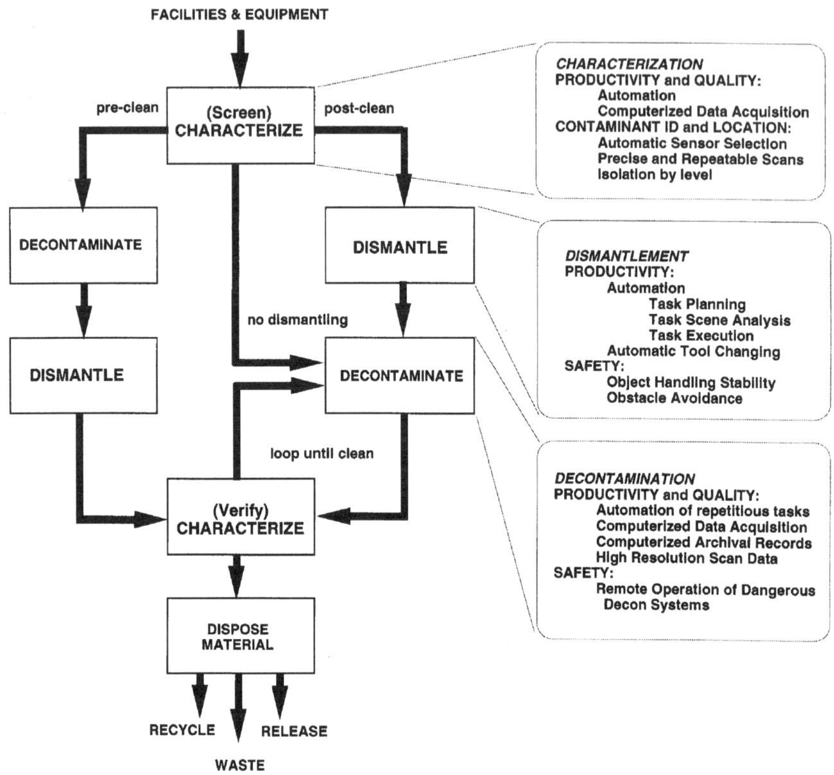

FIGURE 10.9
D&D operational flow diagram.

from centimeters to tens of meters [10]. These facilities include a correspondingly wide range of mobility and access requirements, from overhead crane-type configurations to very demanding floor terrains involving climbing and stepping over obstacles. Nuclear radiation levels range from benign, but legally significant, conditions to lethal exposures.

From a general telerobotics perspective, one can characterize these task environments as geometrically complex, visually confusing, occluded, and environmentally harsh (i.e., radiation, chemicals, humidity, and temperature). Figure 10.10 depicts a typical task scene from a distance and Figure 10.11 shows a close view. Notice the complex equipment and piping arrangements, which involve occlusions, poor contrast, nonspecular reflective properties of surfaces, and high vertical reach requirements. Imagine a telerobotic system that approaches this scene. The goal will be to dismantle the equipment and structure systematically for removal and further materials processing. Obviously, dismantling the entire structure will require multiple stages of operation. The equipment module is a layered structure when considered from a D&D perspective. Inner layers cannot be seen or sensed until outer layers are removed. The dismantlement sequence must be based on a strategy that takes into account the structural design characteristics of the equipment module. Care must be taken to ensure that the cutaway strategy will sustain structural integrity to ensure safety of the telerobot and other equipment in the area. A telerobotic system that could function effectively in this domain requires high mobility, comprehensive manipulation and tools, and extremely robust sensory perception and task-planning capabilities. Subtask automation will require

FIGURE 10.10
Typical D&D far-field task scene. (Courtesy of Oak Ridge National Laboratory.)

that the immediate outer equipment layers be at least geometrically characterized (e.g., *in-situ* construction of a mathematical representation of the relative location of the components). This will require steerable and robust range imaging sensors. The sensors will have to work with poor lighting and nonideal surface properties. Work and task planning must be very smart to be able to include position and dexterity constraints associated with the mobile telerobot, to be able to account for equipment-defined task sequence constraints, and to be able to include multiple tooling systems. Tooling, mobility, and manipulation controls that provide robust automatic subtask execution under these realistic conditions are difficult to realize.

The established approach to D&D operations of the type shown in Figure 10.11 would be remote operations based on teleoperations in which the human operators perform all of the dismantlement functions directly and sequentially. Work planning would be formalized but would be performed off line by the work team. *In-situ* sensing would include remote television viewing with audio monitoring. Human involvement with man-in-the-loop control of virtually every operation would provide a level of robustness that can be quite high in well-trained teams. Fault detection and recovery would be accomplished by the team as necessary. This mode of operation has been used numerous times successfully around the world to the level of proficiency represented in a 5 to 100 slowdown factor (relative to contact

FIGURE 10.11
Typical D&D near-field task scene.

operations). This level of teleoperations proficiency represents a norm, or baseline, that must be exceeded by a sufficient margin to justify the additional complexity and cost of any telerobot. Ideas and concepts for telerobots that can possibly meet this challenge abound. The far more uncertain issue is the achievement of sufficient operational robustness. The next section discusses what is meant by robustness and what types of system structures will provide such robustness.

5 A ROBUST TELEROBOTIC CONCEPT

In terms of this overall discussion, a "robust" telerobot is defined as one that (1) provides effective teleoperability, (2) permits efficient *in situ* subtask automation, (3) is able to monitor and detect fault conditions reliably, (4) is able to facilitate smooth or bumpless transfer between manual and automatic control modes, and (5) performs these functions reliably under realistic hazardous task conditions.

Robust telerobotics requires that the system be capable of reliably identifying fault conditions and that the system provide operator-interactive features that allow the operator to commingle with system operation, to assist in planning operations, and to assist in the recovery from fault conditions. This level of human interaction is essential, given the state of intelligent systems technology in comparison with the extreme multidimensional complexity of typical hazardous work environments such as those described in the previous section. A robust telerobot must emphasize human interaction not only for effective teleoperability but also for the aspects of robotic execution. Specifically, the system must provide the ability to program robotic functions quickly and efficiently so that when the operator chooses to automate a particular subtask it can be done faster than simply executing the task under manual control. Also, it is extremely important that the system provides the operator with "seamless transfer" between teleoperation and robotic execution. This intrinsic robustness is essential if the system is to function effectively under the highly unstructured task conditions that are typical of hazardous environments. The operator must be able to "maneuver" the system in and out of automated task execution to ensure smooth, continuous, and time-efficient overall operation.

The robust telerobotic system must be able to detect tooling malfunctions of all types (e.g., when a nut runner exceeds reasonable torque levels while attempting to loosen corroded fasteners). Task planners must be able to deal with a range of tooling systems, some of which are functionally redundant. Tooling and remote handling system monitors must be able to detect the effects of unexpected task hardware responses such as load shifting or structural faults. The system must deal with realistic image sensing and analysis requirements. It cannot afford the time that might be involved in being computationally confused by scene complexity or obscured sensors. There are also a host of functions associated with telerobotic system infrastructure, such as data communications networks that move data and command signals back and forth from the operator to the remote systems. Many of these requirements are relevant to any remote system, whether it is automated or not. Today's teleoperator systems are quite complex. The situation is substantially more complex when automation is integrated into the basic system functionality. The automation functions are essentially autonomous subfunctions. Autonomous subfunctions that are slow or "get lost" as a result of task complexity effects will not be successful in real-world remote operations environments such as nuclear D&D. Telerobotics comes down to the implementation of autonomous subfunction behaviors within the context of a human-controlled and supervised remote systems. These autonomous subfunctions are, in principle, not different from any autonomous system abilities that have been and continue to be the subject of innumerable research efforts. Perhaps these practical applications of autonomy to things such as nuclear D&D will provide the proverbial acid test for much of the basic research embodied within today's intelligent systems R&D.

The robotic functionality of the telerobot requires that the system have all of the necessary features (e.g., sensing, planning, controls) to characterize and program a specific subtask *in situ*. As a minimum, the telerobot must have the ability to construct geometric models adequate for trajectory and tooling sequence planning. Almost always, the geometry of the task environment is highly unstructured and uncertain. Likewise, the precision and accuracy of the requisite geometric knowledge vary from task to task, as does the extent of the task space itself. Also, a significant fraction of the tasks to be performed are complex by any standard. These factors put full automation of many large tasks (e.g., removing an entire equipment module) beyond the reach of current technology. However, there are certain subtasks that are amenable to automatic planning and execution; interjection of telerobotic subtasks into the overall sequences is the most attainable exploitation of automation benefits in the foreseeable future.

Automation of a task requires complete quantitative data about the task and/or subtasks to be performed, the manipulation systems, and the tooling systems to be used. Task space scene analysis (*TSSA*) [11] refers to the phase in which the telerobotic work system gathers

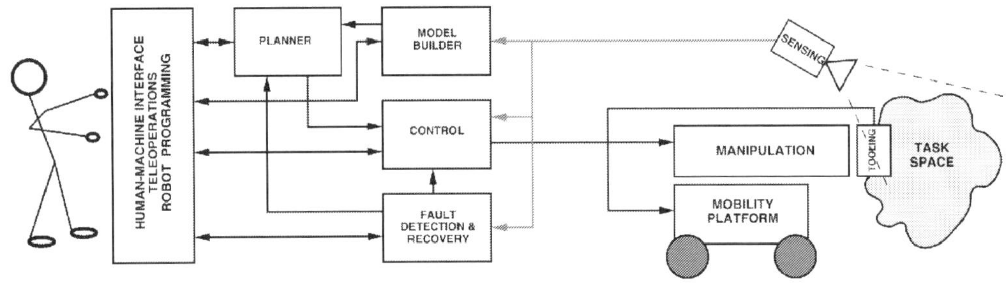

FIGURE 10.12
Functional architecture of a robust telerobot.

geometrical and other types of information that are necessary to characterize, analyze, and plan the automated subtask execution. For example, in a dismantlement scenario, the subtask may be to remove a segment of process piping using remote manipulators and cutting tools. If such a task is to be automated, then it is necessary to describe the location and orientation of each piping element to be removed with respect to the telerobot system. This data representation, or model, must be complete and accurate to the extent dictated by the specific tool to being used; positioning a shear cutter demands less accuracy than achieving the proper standoff distance for a plasma torch. Once a sufficient model is available, planning the manipulator motions is relatively straightforward. The TSSA process is in essence a model builder of the near field of view of the telerobotic work system. Current research [11, 12] has addressed both human-interactive and fully automated TSSA features. The functional architecture of a robust telerobot that includes TSSA is shown in Figure 10.12. We see that a robust telerobot is a hybrid machine that must allow sequences of manual and automated subtasks to be interleaved as shown in Figure 10.13. The telerobotic subtask sequence will consist of the modeling, planning, execution, and verification steps also shown in Figure 10.13. This diagram describes the nominal operational case in which subtask execution evolves as planned. As discussed earlier, there are numerous events that can disrupt the planned execution in a realistic unstructured task environment. The operational flow symbolized in Figure 10.14 depicts the occurrence of off-nominal faults that halt or defeat the planned task execution. In the simplest case, the telerobot would include sufficient intelligence to analyze the fault and replan an alternative execution. In the worst case, it is necessary for the human operator to take over control to rectify the situation if possible. Seamless transfer between manual and autonomous operation across all of the basic telerobot functions is critically important in both of these cases. At this point, it is constructive to look at relevant examples of current research as discussed in the next section.

6 CURRENT RESEARCH IN INTEGRATED D&D TELEROBOTICS

A next-generation D&D telerobot [13] is being developed in the U.S. Department of Energy Robotics Technology Development Program to provide a platform for the exploration of modern telerobotic principles and how they might be used to improve D&D remote operations. The system can be thought of as a "tool kit" consisting of a number of modular and reconfigurable subsystems for mobility, manipulation, tooling, and other support functions. Figure 10.15 shows graphical descriptions of the three modes of mobilty deployment (i.e., overhead, crane, and floor mounted) that are envisioned. Another goal of this

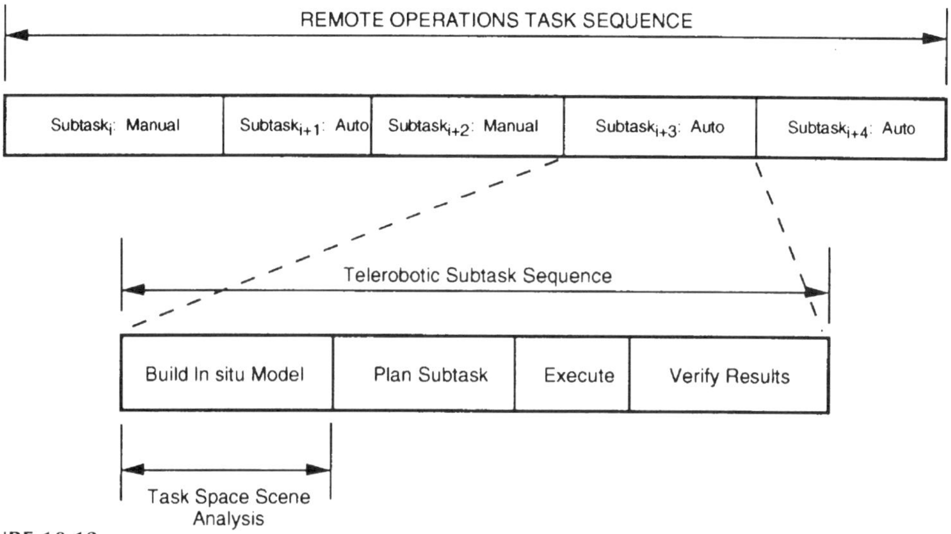

FIGURE 10.13
Telerobot operational sequence diagram.

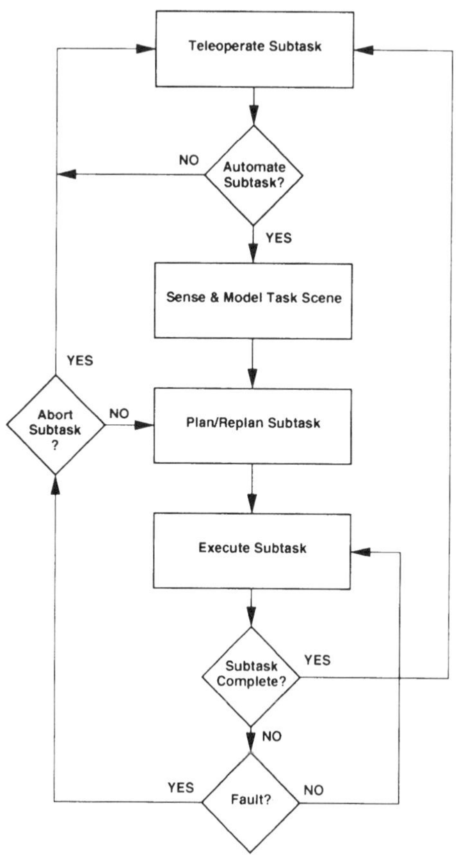

FIGURE 10.14
Telerobot operational flow diagram.

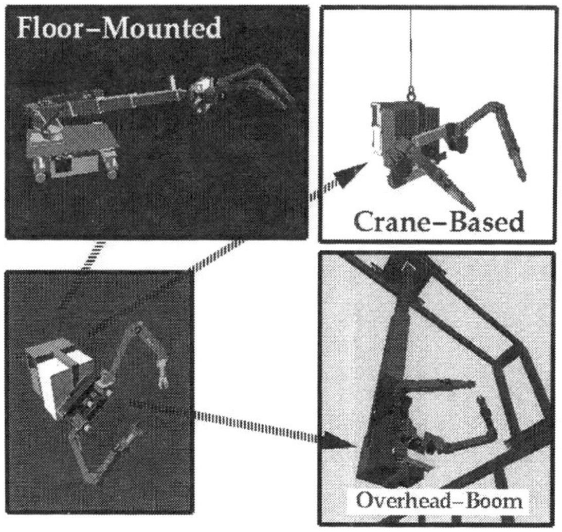

FIGURE 10.15
Selective equipment removal system concept.

design approach is to facilitate scaling of the system module sizes to accommodate a wide range of application requirements with minimal engineering development and modifications. In this way, a majority of the hardware and software R&D costs can be amoritized across a large number of D&D applications. This system has been named the Selective Equipment Removal System (SERS). The modifier "selective" was coined to describe the SERS ability to be used as a highly mobile and versatile system to dismantle problematic (i.e., unusually high radiation) equipment in otherwise benign areas, as well as projects requiring comprehensive remote operations. The system will be used in quantitative tests and evaluations of advanced teleoperation and telerobotic functions under full-scale D&D demonstrations in the Robotics Technology Assessment Facility at the Oak Ridge National Laboratory (ORNL). The dual-arm work module (DAWM) is a principal element of SERS. The DAWM, as shown in Figure 10.16, incorporates dual Schilling Titan II hydraulic manipulators with six degrees of freedom (DOF) mounted on a torso mechanism. The torso mechanism provides an addition five degrees of freedom that include left and right shoulder separation, chest rotation, and manipulator base rotations. The Titan II hydraulic manipulators are used to provide high payloads of approximately 200 lb. The torso mechanism is designed to allow the dual arms to be configured and positioned to handle large and cumbersome objects, which are typical in dismantlement tasks. The total 17 DOF of the DAWM provide task flexibility but lead to significant control challenges. The operator control station for the DAWM is a modified version of the control station for the Advanced Integrated Maintenance System shown in Figure 10.6. Note that the manipulator master controllers [i.e., elbows down, P-R-P-(P,Y,R)] are an entirely different kinematic configuration than the Titan II slaves (i.e., Y-P-P-P-R-P). The initial control implementation [14] of the DAWM was designed to provide effective force-reflecting teleoperation with dissimilar master and slave kinematics. To accomplish this goal, the slave is controlled in a Cartesian space associated with the end effector. Cartesian space control of the slave is straightforward in the automatic trajectory control mode. In manual control, commands from the kinematically dissimilar master controllers are transformed by treating the controller grip inputs as equivalent

FIGURE 10.16
Dual-arm work module. (Courtesy of Oak Ridge National Laboratory.)

Cartesian commands for the Cartesian positioning of the slave end effector. This scheme works reasonably well, although it is somewhat cumbersome in certain manipulator configurations. It involves human factors attributes that are not intuitive and have not been fully evaluated.

The most interesting mobility module is the ROSIE vehicle [15], which was developed by RedZone Robotics. ROSIE is an omnidirectional wheeled floor transporter with an integral three-DOF heavy positioning manipulator that has a 28-ft reach and a payload capacity of approximately 2500 lb. See Figure 10.17. Consequently, ROSIE has the dexterity and payload to position the DAWM in a very large work volume relative to the floor. The ROSIE/DAWM SERS configuration would be capable of dismantling the large structure shown in Figure 10.10, from the floor. Although ROSIE provides maneuverability of a high-capacity work module, it is large and heavy (it weighs approximately 12,000 lb). The system requires a 60-hp onboard hydraulic power supply. Many nuclear applications restrict the use of flammable fuels such as gasoline in remote areas. In this initial configuration, ROSIE is powered and controlled through a special tether and tether management system. Although the tether is a highly reliable connection to the remote work system, its mass and size restrict mobility. A human-interactive stereo range analysis TSSA system [11] is an integral part of the SERS. This system shows real promise as a technique for the *in situ* development of near-field-of-view geometric models. Because the approach is human inter-

FIGURE 10.17
Robot mobile transporter, ROSIE. (Photography by Matt Bulvony and courtesy of RedZone Robotics, Inc.)

active, it is useful for extremely complex task scenes. The operator uses point and click tools with the remote video views to define objects of interest in the right and left camera views. The stereo disparity calculations become straightforward in complex scenes when the points of correspondence are specified by the operator. The prototype stereo TSSA at ORNL has achieved accuracies of about 0.5 in at a standoff and concurrence viewing distance of about 15 feet in a pipe modeling experiment. Carnegie Mellon University [12] is working on an automated TSSA system called Artisan. This work uses surface reconstruction and object recognition schemes to analyze the near field of range camera scenes to produce model primitives of objects in the scene.

Several full-scale mock-ups of actual DOE facility equipment systems that are scheduled for D&D have been constructed for nonradioactive testing in the next few years. These tests will address many important issues such as remote procedure development, tooling evaluations, and remote sensing evaluations. In addition, basic telerobotic functions will be quantitatively evaluated against baseline teleoperation execution. Specifically, a human-interactive scene analysis subsystem called the task space scene analyzer will be used to construct a geometric model of a piping and support structure to be removed from the near field of view with respect to the SERS. These results will be used to define task tooling to be used, operational sequences to be followed, and end-effector trajectories necessary to accomplish the pipe removal automatically. The SERS telerobotic control system will be used to transform all of these planning results into executable forms. The same tasks will be performed with direct teleoperation. Parameters such as total task execution time, manipulation errors, and work quality measures will be carefully recorded for the comparison of the two approaches.

7 KEY REMAINING CHALLENGES AND SUMMARY

When one considers the implementation of practical telerobotic systems such as the SERS in work task environments such as D&D environments, a long stream of research needs begin to surface; however, there are several key research areas that involve fundamental challenges. These fundamental challenges are associated with essential system functions that are at the very heart of robust telerobotics.

As has been mentioned, a high degree of operational robustness will be essential to achieve the confidence levels that operators routinely expect. Many of the ingredients required for this level of robustness exist. However, there are several critical and fundamental areas in which the concepts are well defined, but yet practical and working implementations are lacking. It is believed that basic and applied research is needed to bridge this gap with results that are ready for integrated systems. These critical areas represent the principal remaining challenges:

Fault Detection, Isolation, and Interpretation. A system must be able to realize when it cannot achieve automated task objectives. It must be able to alert the human operator to likely causes for malfunction, and it must be able to assist the operator in developing work around strategies and alternatives. This will require on-line diagnostic capabilities that go far beyond existing system diagnostic concepts.

Recovery Interaction. The workable telerobotic system must provide operator utilities that facilitate effective and fast replanning, including detailed safety and consequence assessments (i.e., operational readiness restrictions in operations involving nuclear materials can be extremely demanding and stringent). The nature and complexity of hazardous environment operations will require that humans be an integral part of, and have final approval for, recovery plans. R&D is needed to develop such interactive systems that give proper consideration to the human engineering issues.

Efficient *in Situ* Programmability. The programming cycle time for selected subtasks must be short enough to make telerobotic execution advantageous. Refer back to Figure 10.13. Programming must be accomplished on station as needs and opportunities arise. Sensors and visualization schemes that facilitate task parameterization must be developed. Efficient interactive task planners that can handle the complexity must be developed.

Seamless Transfer. Seamless transfer is the process whereby a telerobot can make a transition between manual operation and automatic control effectively. For the telerobotic concept to be viable in practical applications such as D&D, dynamic seamless transfer is needed to give operators the freedoms they will need to drive these systems through their complex tasks efficiently and effectively. The realization of systems that can literally jump from manual and automated machine states reliably and confidently is a major challenge. R&D is needed to develop the fundamental real-time software concepts and structures to implement these systems.

Human–Machine Interfaces. It has been argued that human interaction is the glue that will hold telerobot systems together in the sense of ensuring flexibility and robustness. Human–machine interfaces (HMIs) cross-cut virtually every system function that has been discussed, including the key remaining challenges. Unfortunately, modern intelligent systems research too often tacitly treats HMI issues as secondary matters. HMIs are often whatever workstation window structures programmers define in the course of building their systems. Experiences in aerospace and nuclear applications have shown that the efficacy of the HMI is always a first-order factor in terms of overall system functionality and efficiency. This is certainly the case in robust telerobotics, and much more research investment must be devoted to the human factors issues of human interaction at all levels.

In summary, we have discussed the straightforward concepts associated with the notion of telerobotics. We have found that, in practice, robust telerobotic systems are virtually nonexistent for a host of reasons, some of which are technical and many of which are nontechnical. Remote systems applications in hazardous environments provide valuable experience in the teleoperations dimension of telerobotics and richly define the requirements facing robust telerobotics systems. New research into the human interaction functions of these systems is needed in key areas associated with fault handling, fault recovery, *in situ* programming, and system control. Robust telerobotics should be a central focus of human-centered robotics research.

REFERENCES

[1] W. R. Hamel and M. J. Feldman. The advancement of remote technology: Past perspectives and future plans. In *Proceedings of the 1984 National Topical Meeting on Robotics and Remote Handling in Hostile Environments*, American Nuclear Society, Gatlinburg, TN, 1984.

[2] J. V. Draper *et al.* Effects of force reflection on servomanipulator performance. In *Proceedings of the 1987 International Topical Meeting on Robotics and Remote Handling in Hostile Environments*, American Nuclear Society, Pasco, WA, 1987.

[3] R. Goertz *et al.* Manipulator systems development at ANL. In *Proceedings of the 12th Conference on Remote Systems Technology*, American Nuclear Society, 1962.

[4] R. Goertz. Some work on manipulator systems at ANL, past, present, and a look at the future. In *Proceedings of the 1964 Seminars on Remotely Operated Special Equipment*, vol. 1, pp. 27–69, Atomic Energy Commission, Germantown, MD, May 1964.

[5] J. N. Herndon *et al.* The state of the art model m2 maintenance system. In *Proceedings of the 1984 National Topical Meeting on Robotics and Remote Handling in Hostile Environments*, American Nuclear Society, Gatlinburgh, TN, May 1984.

[6] J. Vertut and P. Coiffet. *Teleoperation and Robotics, Evolution and Development, Robot Technology*, vol. 3A, pp. 302–307. Hermes Publishing, 1985.

[7] D. Kuban and H. L. Martin. An advanced remotely maintainable servomanipulator concept. In *Proceedings of the 1984 National Topical Meeting on Robotics and Remote Handling in Hostile Environments*, American Nuclear Society, Gatlinburgh, TN, April 1984.

[8] H. L. Martin and W. R. Hamel. Joining teleoperations with robotics for remote manipulation in hazardous environments. In *Proceedings of Robots 8*, Robotics International, Detroit, 1984.

[9] W. R. Hamel and H. L. Martin. Robotics-related technology in the nuclear industry. In *Proceedings of the SPIE*, vol. 442, Robotics and Robot Sensing Systems, San Diego, Aug. 1983.

[10] W. R. Hamel and D. C. Haley. Advanced robotics for decontamination and dismantlement. In *Fifth International Symposium on Robotics and Manufacturing, ISRAM '94*, Maui, HI, Aug. 1984.

[11] S. Thayer *et al.* Three-dimensional sensing, graphics, and interactive control in a human–machine interface for decontamination and decommissioning applications. In *Proceedings of the Sensor Fusion V Conference*, SPIE, pp. 74–85, Boston, Nov. 1992.

[12] A. E. Johnson *et al.* 3-d object modeling and recognition or telerobotic manipulation. In *Proceedings of the IEEE Conference on Intelligent Robots and Systems*, vol. I, Aug. 1995.

[13] W. R. Hamel *et al.* Dual-arm manipulation module for use in decontamination and decommissioning operations. In *1994 International Symposium on Decontamination and Decommissioning*, Knoxville, TN, April 1994.

[14] J. F. Jansen and R. L. Kress. Hydraulically powered dissimilar teleoperated system controller design. In *Proceedings of the 1996 IEEE International Conference on Robotics and Automation*, Minneapolis, April 1996.

[15] L. Conley, W. R. Hamel, and B. R. Thompson. Rosie: A mobile worksystem for decontamination and dismantlement operations. In *Proceedings of the ANS 6th Topical Meeting on Robotics and Remote Systems*, Monterey, CA, Feb. 1995.

SECTION V

Application

CHAPTER 11

Automated Integration of Multiple Sensors

J. E. BAKER
Oak Ridge National Laboratory, Oak Ridge, Tennessee

1 INTRODUCTION

Multisensor integration (MSI) is the combining of data and information from more than one source in order to generate a more reliable and consistent representation of the environment [1]. This chapter focuses on specific MSI algorithms developed for recent Department of Energy (DOE) applications. These applications included autonomous robotic guidance for deployment in hazardous environments, scene mapping and environmental characterization, buried waste localization and identification, sensor self-calibration through automated scene quality assessment, and waste stream object identification.

To be useful in many real-world applications, such as autonomous or teleoperated robotics, real-time feedback is critical [2, 3]. Unfortunately, many MSI–image processing algorithms require significant processing time [4–7]. This is especially true of feature extraction, object isolation, and object recognition algorithms because of their typical reliance on global or large neighborhood information. The techniques presented in this chapter attempt to exploit the speed currently available in state-of-the-art digitizers and highly parallel processing systems by developing MSI algorithms based on pixel-level rather than global-level features. Hence, the basic direction of these approaches to conceptual enhancement is the potentially faster and more robust formation of cluster from pixels rather than the slower process of segmenting images into clusters. (Note that although this process and the resulting representation are generally called "segmentation" in the literature [8–11], I will use the term "clustering" to reinforce this distinction in the basic direction of the approaches.)

The "data enhancement" algorithm presented is a specific sensor anomaly detection and remediation technique. This technique is invoked to remedy an anomaly that can reduce the applicability of the sensor data and is given as an example of a traditional MSI approach. Addressing such anomalies is necessary if real sensor data are to be targeted. It is a great temptation to develop and evaluate image processing (IP) algorithms on simulated data with the rationalization that the results can be easily adapted to real images. Although this may well be true theoretically, often the problems and difficulties inherent in processing real sensor data are such that simulation-based techniques resolve few of the real problems. That is, the distortions, sensor anomalies, noise, and so on inherent in real images so dominate the processing problem and either the simulation-based technique is solving such a trivial component or the anticipated mapping function from real images to simulation quality images is at least as complex as the original IP problem. It is our view, therefore, that real sensor data must be the focus from day one for all but function encoding verification, graphical explanation, or preliminary concept generation. The techniques presented in this chapter are developed and evaluated on actual multimodal sensor data consistent with real-world robotic applications, that is, a laser range camera used for robotic navigation and ground conductivity data used for buried waste assessment.

The following section describes the traditional approach and provides a comparative overview of our adaptive learning approach to MSI. This approach is detailed in Section 3. Section 4 describes the sensory target domains, including a description of the physical nature of each sensor system (laser range camera, LRC, and ground conductivity, GC, sensor) and their targeting environments. Section 5 presents a specific sensor anomaly common to LRCs and our MSI approach to resolving its distortions, complete with empirical results. The sixth section presents an empirical evalution of the performance of our automated MSI system applied to LRC images. Section 6 validates these results empirically in the learned (LRC) and unlearned (GC) domains. A summary section follows.

2 BACKGROUND

Although MSI has a generally accepted overall purpose, that is, to generate a more reliable and consistent representation, the *specific* goals of each MSI implementation vary from one application to another [4, 5, 7]. MSI techniques combine multimodal sensor images (e.g., sonar and visual [12], distance and reflectance [1, 13]), multiple single modal sensor images (e.g., multisampling, a single sensor's images displaced in time or in space [14]), or combinations of these.

MSI is necessary because of basic ambiguities inherent in our current sensor imaging technologies. Sensor ambiguities derive from two basic causes: limitations of the physical attribute being measured by the sensor (e.g., visible light's inability to permeate opaque surfaces, sonar edge effects) and the sensor's inaccuracies in making measurements (e.g., noise, resolution). Ambiguity exists as long as the mapping from reality to image is not one to one. That is, if different "realities" lead to identical images, a single image cannot reveal the particular reality that was the truth; for example, a two-dimensional (2-D) visual image of an opaque object cannot reveal its interior or its hidden surfaces, hence an infinite number of "realities" would result in the same sensor image.

MSI techniques attempt to resolve some of these ambiguities by appropriately coupling complementary images to eliminate possible inverse mappings. What constitutes the most advantageous MSI technique is dependent on the given application domain, available sensors, and task requirements. Multisampling is perhaps the simplest MSI technique and is used primarily to reduce noise. Merging multiple single modal images [14] displaced in space can improve three-dimensional information, reveal otherwise occluded areas, and reduce orientation-induced artifacts, such as glare or sonar positional anomalities. Merging multiple single modal images displaced in time can be used to reduce temporal effects or to isolate them. Employing multimodal sensors permits exploitation of each sensor's strengths without suffering their intrinsic weaknesses; for example, using both sonar and vision can provide both accurate distance (sonar) and edge detection (vision) [12]. Clearly, the choice of sensors and MSI techniques is critical to achieving performance gains from MSI; MSI provides nothing if the images to be merged are not complementary.

The goal of MSI is to improve the representation's reliability and consistency, and the means of achieving this goal can be divided into three categories based on the information content of the original images relative to that of the desired representation.

In the first case, "detail enhancement," the relative information content of the original images is less rich than that of the desired representation. This case occurs whenever the original images are integrations of the target (e.g., raw nuclear magnetic resonance, X-ray photography) or undergo a distorting function (e.g., unfocused lenses, interference patterns). These images can be translated into a more accurate and detailed representation if the original integration or distortion function is sufficiently known to permit an inverse mapping function to be well approximated [10].

In the second case, "data enhancement," the MSI techniques are concerned with improving the accuracy of the data rather than either increasing or decreasing the level of detail. Techniques in this category inclode noise reduction, resolution enhancement, and reduction of sensor artifacts. One such technique is discussed in Section 5.

In the third case, "conceptual enhancement," the image contains more detail than is desired, making it difficult to recognize objects of interest easily. In these images one can group together pixels corresponding to the same conceptual object and thereby reduce the level of extraneous detail. This task may require significant amounts of global knowledge and processing time. For example, if one wishes to distinguish "walls" from "floors," one could separate objects on planar edges; however, this would subdivide complex structures into their many facets. Collecting these facets into single multiplanar objects without combining separate, adjacent objects requires detailed knowledge of each complex structure of interest (object definition) and recognition of that structure within the image (object recognition). This problem extends to all sensor modalities as well (e.g., color, texture, composition, shape). Either one accepts the shortcomings inherent in making clustering decisions independent of global knowledge or one accepts the time and computational complexity associated with object definition and recognition code. This chapter concerns real-time applications and, hence, restricts its focus to optimizing the performance ability of rapidly executable approaches.

Rapid execution is not only useful but also indispensable for many applications involving autonomous or teleoperated robotics. Unfortunately, many image processing algorithms require significant processing time. This is especially true of feature extraction, object isolation, and object recognition algorithms because of their typical reliance on global or large neighborhood information. For example, many object isolation algorithms are based on expected object templates, feature extraction, and "hypothetical" feature extension (i.e., extension of edges to possible intersection points in order to form closed polygons from only partial edge line segments). These methods have the advantage of global knowledge, large neighborhood features, *a priori* expectations, and so on but are, for the same reasons, rather slow and domain or target specific.

State-of-the-art image digitizers are capable of performing complex functions at the pixel level at full image acquisition rates [15]. This permits one to obtain pixel-level features with little processing delay time and may permit MSI images to be produced at near-original image acquisition rates, leading to virtual "multimodal sensors." The methods described herein exploit this speed capability by developing MSI algorithms based on pixel-level features. Hence, the basic direction of this approach to conceptual enhancement is the potentially faster and more robust formation of clusters from pixels rather than the slower process of segmenting images into clusters.

This pixel-level approach to conceptual enhancement assumes that the sensors have sufficient resolution, relative to the objects, to ensure that adjacent pixels corresponding to the same object are similar, that is, in zero, first, or second order derivatives, in at least one of the sensor modalities and that the objects of interest span several pixels. If these conditions are not met, either the objects are out of range or the sensors' modalities are inappropriate for the task.

The following two subsections provide a basic overview of the traditional approach and our adaptive learning approach to multisensor integration.

2.1 Traditional Approach

Current state-of-the-art in MSI application involves a series of sequential, time-consuming steps that often only highly qualified and experienced researchers in sensors and image processing can conduct [6, 7]. As the knowledge and experience gained in one step are often critical to performing the next successfully, a single researcher or team must generally perform most of the work. This substantially constrains the critical path of the schedule; that is, in general, because the steps cannot be performed in parallel, the length of the overall schedule will not be shortened by the addition of researchers to the development team but may be seriously lengthened by their departure. These basic steps are as follows:

1. *Understand and characterize the application domain.* The conditions within the application domain that must be considered include ambient conditions (e.g., temperature, background radiation, lighting, humidity, air quality), targeting conditions (e.g., specific characteristics of the target, false targets, obstructions, background objects), and anomalistic effects (e.g., glare, shadows, changing air conditions). Any conditions in the operating environment that will affect the choice of sensors, their performance, and the image interpretation techniques must be sufficiently understood to direct selection and exploitation of the sensor suite. It is important to note that specific conditions affecting sensor operations are a function of the sensors themselves. Therefore, this step cannot be executed without a detailed familiarity with the candidate sensors, their physical operation, and corresponding interpretation techniques. This step interdependence and required sensor and image processing expertise necessitate the high qualifications required of the MSI researchers.
2. *Determine and define the appropriate sensor suite.* This step requires an in-depth knowledge of the various available sensors (i.e., their advantages, disadvantages, and complementary behaviors with respect to the application domain and task requirements) and modifications that potential manufacturers could implement. Sensor capabilities must be weighed against both the operating domain and available potential image processing techniques. Furthermore, secondary considerations (e.g., weight, kinetic shock survivability, radiation hardening, signal-to-noise ratios, long-term product support, power requirements, field-of-view) must also be understood and considered when determining the best sensor suite for a given application.
3. *Analyze, characterize, and calibrate the sensor suite.* An empirical evaluation of the sensor(s) must be performed within the designated operating domain and must include realistic targets, obstacles, and all expected ambient conditions. The purposes of this step are to validate the sensor selection, drive image interpretation software development, and understand any sensor anomalous behavior. It includes sensor accuracy as

a function of measurements discontinuity (e.g. color quality at boundaries), impact of theoretically unrelated modalities (e.g., range accuracy as a function of surface reflectance), scene dynamics (e.g., moving targets and/or moving obstacles), active sensor "cross talk" (e.g., acceptable proximity and frame rate of sonar ranging sensors), and ambient noise (e.g., presence of an ambient light source that corresponds to a laser camera's operating frequency). If a sensor proves to be unacceptable, one may need to return to step 1 for both the single erroneous sensor and potentially the entire sensor suite. The entire suite may be affected, because in some cases individual sensors are selected for their behavioral compatibility and/or complementary modalities. Hence, the unexpected replacement of an unacceptable sensor may affect the utility of the others.

4. *Recognize and accommodate sensor interactions.* This step involves two distinct types of sensor interactions. In the first type, the individual accuracy of the data is affected (e.g., "cross talk" interactions affecting the interpretation of data correspondences). As mentioned in step 3, cross talk can occur whenever two or more active sensors confuse each other's signals for their own. This is a common potential problem for sonar range sensors and is usually redressed by controlled firing. Clearly, this changes the effective data stream rate. Other interactions occur because of incompatibilities between the physical measurement properties of the sensors. For example, the performance of a magnetic transducer–based positional system or magnetic conductivity sensor can be adversely affected by the near proximity of a highly conductive object, such as another sensor system [16]. Further, since some sensors are based on physically unreliable phenomena, multiple physical characteristics of each point must be determined in order to represent the scene accurately. For example, in some range cameras, the accuracy of the range data is a function of the reflective quality of the surface being measured. Hence, by coupling a range sensor with a reflectance sensor, one can obtain the information to correct the range image.

   The second type of sensor interaction that must be understood and addressed is the accurate interpretation of data correspondences. Determining which pixels in multiple distinct images of the same or similar scenes correspond to each other is a very difficult problem and has yet to be fully solved [17–19]. The problem, called image or data registration, is complicated by the possibilities of occlusions, dynamic changes in the scene, variations in sensor quality as a function of position or time, and so forth, which may prevent the existence of *any* correctly corresponding pixel. The image registration problem is not directly a subject of this chapter but must be addressed if spatially or temporally (for dynamic scenes) disjoint images are to be merged. Generally, one concentrates on registering images exhibiting only trivial changes or one determines the precise relative spatial positions of the sensors so that simple geometry can be used. However, even if these conditions are met, the actual integration can be extremely complicated by the association and resolution problems [17–19]. That is, given two distinctly different pixels that are geometrically determined to lie in very close four-dimensional proximity (space and time), (1) should they be merged (and by what function)? (2) is one anomalous or less reliable because of its sensor's point-of-view (e.g., due to lighting, object sheen, timing)? or (3) do they represent distinct, albeit spatially proximate, objects? In most cases, the problem is even worse because four-dimensional placement is unknown (e.g., CCD images effectively yield vector data, not point data).

5. *Develop and optimize robust merging code.* Based on the results obtained in the preceding four steps and the MSI goal (e.g., improved human display, rapid evaluation

of vast sensor data, target detection, object characterization, target tracking, self-location), specific merging software must be designed, implemented, tested, and empirically tuned. Because of the variety of potential MSI goals, the resulting merging code varies tremendously from one application to another. Developing this code involves issues similar to those for image registration described earlier. If two perfectly registered and corresponding pixels have different values for the same physical characteristic, (1) is one erroneous and to be ignored? (2) are both erroneous and can any information be garnered from them? (3) if they are to be merged, by what function—simple average, weighted average, the minimum, the maximum, or some complex weighted function of secondary variables (e.g., temperature, air quality, timing, lens focal length)? The number of merging possibilities grows substantially if the two pixels represent different physical properties, such as distance and reflectance.

Once completed, such carefully orchestrated and expertly crafted MSI systems are expensive, fragile, and suboptimal. Any change (e.g., to the sensor suite, environmental domain, task specification, target description) may require a complete reanalysis. This problem is exacerbated by the typically multiple human-years necessary to complete the cycle for real-world applications, as replacing a researcher lost to natural attrition can often necessitate a significantly time-consuming learning curve. Similarly, because most state-of-the-art MSI implementations are performed on proof-of-principle demonstration test beds, the entire development cycle will need to be reperformed even if the initial development was successful (i.e., prototyping the hardware system will probably result in sensor system changes due to obsolete models, new weight or power constraints, etc.).

Another critical deficiency in this approach is that because of the development costs of even a single MSI "solution," the only empirical quality comparison generally available is against the original, non-MSI system. Hence, traditionally developed MSI systems can undergo little more than a minor tuning from their original design parameters. As it is financially and timewise infeasible to develop fully even two MSI systems to compare competing sensor suites, initial design decisions must be accepted on faith unless they prove unworkable; optimizing for them is cost prohibitive.

2.2 Adaptive Learning Approach

Much of the MSI development process described in the previous subsection could benefit from adaptive learning [1]. That is, an Automated MSI Solution Generator (AMSG) could be designed that would take actual sensor data and the desired merged results as input during a training stage in order to develop or determine adaptively an effective method of merging the sensor data (see Figure 11.1). An AMSG would significantly reduce the time required to apply MSI to a given problem, while increasing the quality of the final result and providing both an objective analysis and comparison of competing MSI techniques and sensor suites.

The AMSG that we developed consists of four basic components (see Figure 11.2):

1. *Search space.* This is not a "component" in the strictest sense, but the specification of the search space, more than anything else, defines the potential quality of the AMSG. The search space must be composed of algorithms sufficiently specialized to provide high-quality MSI performance and sufficiently robust to permit working latitude in the fielded system. Furthermore, the search space must be sufficiently broad to include solutions applicable to a wide variety of domains and sufficiently small to permit reasonable search (preferably no more than perhaps 10^{20} candidate algorithms).

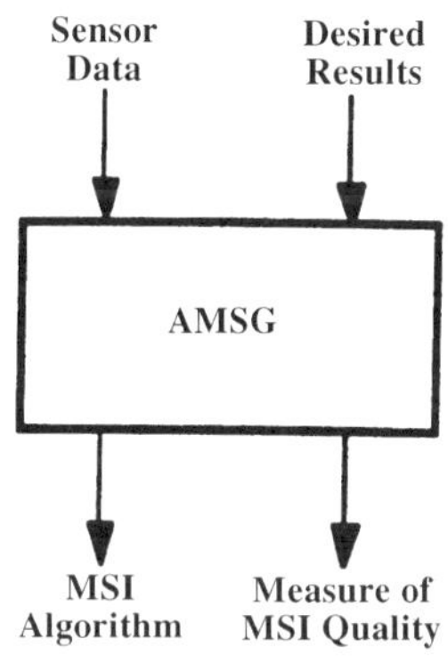

FIGURE 11.1
Data interface with AMSG.

2. *MSI implementation code.* This component is responsible for applying an MSI algorithm to the incoming sensor data at near-real-time rates. Once an appropriate MSI algorithm is generated for a given application domain, this component will constitute the entire fielded system; see Figure 11.3.
3. *Learning strategy.* The learning strategy must select an MSI algorithm from the candidate solutions within the search space according to the quantified performances of prior selections. This learning strategy must efficiently exploit the feedback so that a sufficiently high-performing solution can be found in a reasonable amount of time.
4. *Evaluation function.* The evaluation function compares the MSI results of the candidate algorithm with the user-provided desired results and quantifies the algorithm's quality. This quantification can be used as an objective comparison of MSI algorithms. To guide the AMSG efficiently, this function should have high resolution and monotonically encourage MSI quality; that is, it should be such that even small improvements in the MSI algorithm will be reflected in the evaluation measure and have a positive impact on the learning process.

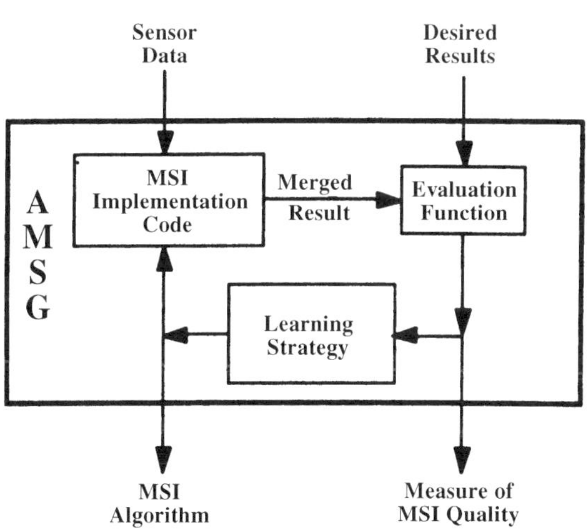

FIGURE 11.2
Basic schematic of AMSG data flow.

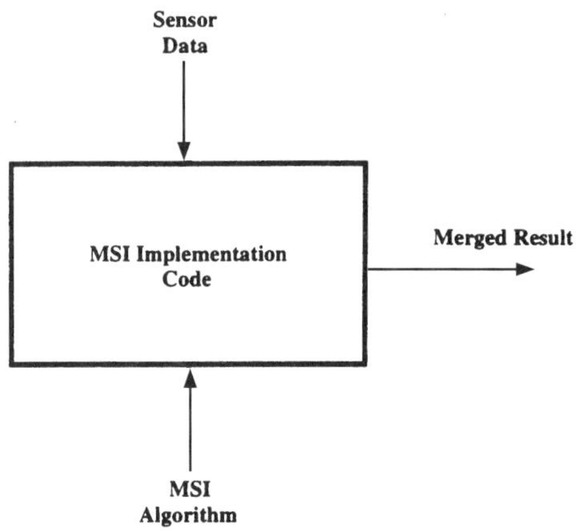

FIGURE 11.3
Fielded MSI system (single component from AMSG).

During the learning stage, the AMSG is provided with a set of sensor data and the corresponding desired MSI results. The system then progresses through a basic feedback learning cycle (see Figure 11.2): (1) apply candidate MSI algorithm to sensor data, (2) compare results with desired results and quantify performance quality, and (3) use the quality measure to determine the next MSI algorithm to evaluate. This cycle proceeds until some user-specified stopping criterion is met (e.g., minimum performance quality, given number of iterations). When the user is satisfied with the performance quality of the MSI algorithm generated, that MSI algorithm can be directly applied to incoming sensor data; see Figure 11.3.

Each of these components will be detailed in Section 3 and the implemented system empirically analyzed in Sections 6 and 7.

3 AUTOMATED MSI SYSTEM

Although an AMSG could be developed for a variety of MSI tasks, we will restrict our consideration to conceptual enhancement, wherein the original image contains more detail than is desired, making it difficult to recognize objects of interest easily. This research focuses on merging pixels into "homogeneous" clusters and leaving the unification of conceptually related clusters for some more abstract, postprocessing phase.

Our approach to conceptual enhancement is based on the premise that given appropriate sensors (i.e., those capable of distinguishing the significant conceptual regions, or *facets*) and sufficient resolution (i.e., providing multiple pixels of each such facet), clusters of pixels can be formed that correspond to the facets by accurately answering the fundamental question for each pair of adjacent pixels: "Do these pixels belong to the same facet?" Given an accurate response for each pair of adjacent pixels, forming accurate clusters is trivial. This question applies to cluster formation regardless of the sensor or application domain. The answer, in the form of a confidence value, provides a universal and uniform interface—applicable to any sensor system and MSI technique to be integrated.

This fundamental question suggests simply comparing the adjacent values across each sensor modality. But when posing the same question in the negative, "Is there a surface edge lying between these two pixels?" one is drawn to compare the pixel neighborhoods on each side of the questioned interface, for example, comparing the linear extrapolation of pixels on the first side of the interface with the second pixel's value, comparing the average adjacent pixel value variation on each side of the interface, and so forth. The AMSG is a system that searches for a function that best answers this question for each sensor in the suite simultaneously and merges their results, and optimizes those functions in accordance with the user-defined, desired results.

As outlined in the previous section, the AMSG consists of four primary components: the search space, the implementation code, the learning strategy, and the evaluation function. This section will detail each of these components.

3.1 Search Space

There are competing interests driving the search space design. First, the space must be searchable; that is, there must be sufficient regularity for an automated global search technique to be applicable. Second, grossly dissimilar sensors must be accommodated. Third, all sensor inputs must be transformed into a single format for merging. Fourth, the MSI algorithm must be very robust and general purpose to permit maximum domain application. Fifth, the MSI algorithm must be highly specialized to ensure quality solutions. And sixth, the search space design must be extendable to as yet unforeseen sensor types and suites.

To these ends, we have chosen a hierarchical organization of highly parameterized functions. Each sensor's input is transformed into an edge confidence map for subsequent confidence combination merger. The edge confidence map was chosen as a universal–fundamental interface that is based on perhaps the most reliable and basic property of sensors—the ability to detect discontinuities in some physical phenomenon. (That is, although absolute measurements may be erroneous and the detectors may not even be measuring the physical property intended by their designers, significant changes in output can be reasonably inferred to correspond to some significant change in some physical property in the scene. Hence, at the very least, an edge confidence map can be generated.) Edge confidence maps have the added advantage of being sufficiently universal that almost any source (e.g., CAD model, intelligence reports) can be translated into one and thereby merged by this sytem.

In our approach, each sensor to be merged is analyzed separately, according to its "affinity function." This real-valued, parameterized function provides a "degree of match" for each pair of adjacent pixels within that sensor's image. This operation translates the image into an edge confidence map. These maps are then merged across the sensor suite by combining the confidences for each pixel interface. This merging operation results in a single edge confidence map for the entire sensor suite; see Figure 11.4. A fully segmented image can then be produced by threshold-based region growing. Hence, a *single MSI algorithm* is defined by

1. The pixel interface statistics used by each sensor's affinity function.
2. The affinity function and corresponding parameter values for each sensor in the suite.
3. The merge function and corresponding parameter values that define how to combine confidences across the sensor suite.
4. The region growing algorithm and corresponding parameter values that indicate cohesion or disjunction for a given pixel interface.

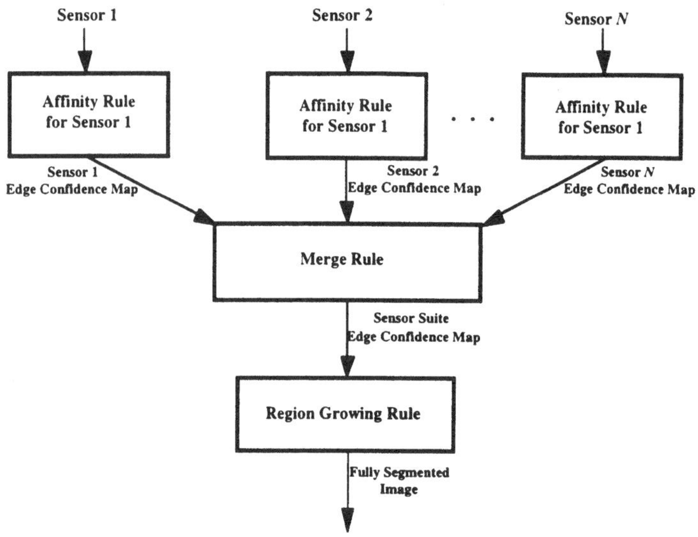

FIGURE 11.4
Schematic of sensory data flow through execution of an AMSG's MSI algorithm.

Hence, the *search space* consists of

1. *Statistical definitions:* A set of low-level pixel-interface statistics (e.g., change in pixel value, change in slope of adjacent pixel values).
2. *Affinity function definitions:* A set of affinity functions, each of which is a parameterized function of various pixel-interface statistics.
3. *Merge function definitions:* A set of merge functions, each of which is a parameterized function of affinity values.
4. *Region growing algorithm definitions:* A set of region growing algorithms, each of which is a parameterized function of merge function values.
5. *Parameter/threshold values:* The possible ranges and precisions of parameter values for each of the preceding functions or algorithms.

To be included in the statistics and function sets, a definition need only be considered potentially useful in distinguishing facet interface boundaries in at least *some* sensor modality. This search space design has several significant advantages:

1. It is applicable to most sensor systems, sensor modalities, and application domains, by simply augmenting or modifying the definition sets or parameter spaces.
2. Its design is applicable to spatial or temporal images; a pixel interface need not be considered a uniform spatial quantity.
3. Its hierarchy permits arbitrarily complex definitions at each stage; for example, an image edge finding algorithm or a misregistered, inaccurate world map could supply affinity values. The AMSG's architecture permits the results from diverse systems to be readily merged using standard confidence combination approaches.
4. Its reliance on bottom-up image processing; that is, forming clusters from pixels rather than segmenting images into clusters permits extremely robust statistic and function definitions, applicable to a wide variety of sensor systems and modalities.

5. Changing the sensor suite has little impact on the AMSG, requiring at most augmenting, not changing, the definition sets.
6. State-of-the-art image digitizers [15] are capable of performing complex functions at the pixel level at full image acquisition rates. This may permit one to execute a given MSI algorithm with little processing delay time, leading to virtual "multimodal sensors."
7. It transforms the MSI development problem into a global search optimization problem. This permits the system to leverage existing technology in the area of global optimization.
8. By enabling an automated global search mechanism to "find" an appropriate MSI algorithm, new application and sensor domains can be targeted with minimal development costs and increased speed and quality. This, in turn, should permit a better and more objective analysis of the synergistic capabilities of various sensor suites while reducing the need for the developer to recognize and understand the "best" sensor interactions to be exploited for a given task.

Statistical Definitions

The statistical definitions are determined by the needs of the affinity function definitions. The statistical values can either be precalculated and stored or determined at runtime for each affinity function invocation. This decision is an implementation detail, and the relative trade-off analysis of memory space versus learning stage runtime depends largely on the available hardware system.

Affinity Functions

In this chapter we define four distinct affinity functions for comparison and evaluation. These differ in the size and use of pixel neighborhoods. For the following definitions, refer to Figure 11.5 and let

$E(x, y)^z$ = the edge confidence for the interface between pixels x and y using affinity function z, for any sensor

$L(x, y, z)_\alpha$ = the linear extrapolation at point α, using a least squares fit to the points x, y, and z

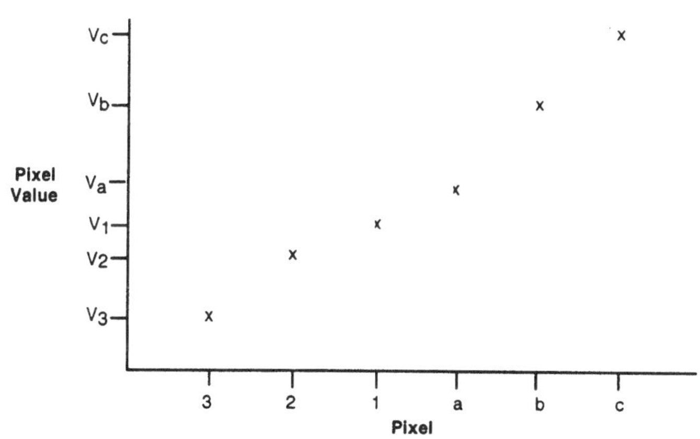

FIGURE 11.5
Hypothetical plot of pixel values, used solely for explanatory purposes.

1. "2-pt"—uses only the two pixels adjacent to the interface, that is,

$$E(1, a)^{\text{2-pt}} = |V_1 - V_a|$$

   This function is a simple delta value across the interface. It is likely to be too simplistic for real images because of its required assumptions of uniform pixel quality across the detector's sensitivity range and sensor's image. That is, a single edge-indicating threshold parameter probably does not exist that would be appropriate everywhere, even in a single sensor image.
2. "4-pt General"—uses the best linear fit to any four adjacent pixels spanning the interface, that is,

$$\begin{aligned} E(1, a)^{\text{4-pt General}} = \min\{&\max\{|L(3, 2, 1)_a - V_a|, |L(3, 2, a)_1 - V_1|\}, \\ &\max\{|L(2, 1, b)_a - V_a|, |L(2, a, b)_1 - V_1|\}, \\ &\max\{|L(1, b, c)_a - V_a|, |L(a, b, c)_1 - V_1|\}\} \end{aligned}$$

   This affinity function assumes that the facet is at least 4 pixels in width. Hence, each contiguous 4 pixels, which span the interface, are considered separately. If both pixels are common to one linear facet, then every linear fit involving those pixels should yield a low error. Hence, for each 4-pixel set, we take the worst (max) linear fit. However, since only *one* of the three 4-pixel sets needs to match, we take the best (min) of these three matches.
3. "4-pt Predictive"—uses the three adjacent pixels on each side of the interface to predict the pixel on the other, that is,

$$E(1, a)^{\text{4-pt Predictive}} = \min\{|L(3, 2, 1)_a - V_a|, |L(a, b, c)_1 - V_1|\}$$

   This affinity function is a subset of 4-pt general. It still assumes a 4-pixel-wide facet but also assumes that at least 3 pixels are on one side. This assumption significantly increases the implementation speed but may undesirably reduce the effectiveness.
4. "6-pt"—uses the three adjacent pixels on each side of the interface to measure the linear discontinuity at the interface, that is,

$$E(1, a)^{\text{6-pt}} = \max\{|L(3, 2, 1)_a - V_a|, |L(a, b, c)_1 - V_1|\}$$

   This, the most restrictive affinity function tested, assumes a 6-pixel-wide facet and requires a low 4-point linear fit error across the interface from *both* directions.

From these definitions, it is clear that many other potentially useful affinity functions exist. These four affinity functions provide a sample which differ in their minimum size requirements of the facets that can be detected. These facet size assumptions are detailed in Table 11.1.

Merge Functions

Four merge functions are defined, two basic merge functions and two relative merge functions. The two basic merge functions differ in the assumed independence of the sensors

Table 11.1. Comparison of affinity modes by the strictness of the size requirements of interfaced facets (i.e., how "wide" do two facets need to be for their interface to be detectable?)

Method	Min. Size of Interfaced Facets	Therefore...
2-pt	1 and 1	One cluster $\geqslant$ 1 pixel
4-pt General	2 and 2 or 3 and 1	One cluster $\geqslant$ 2 pixels
4-pt Predictive	3 and 1	One cluster $\geqslant$ 3 pixels
6-pt	3 and 3	Both clusters $\geqslant$ 3 pixels

in the suite, that is, the expectation for multiple sensory conformation. For the following definitions let

$E(x, y)_S$ = the edge confidence for the interface between pixels x and y using sensor S, for any affinity function

$BM(x, y)^z$ = the basic merged edge confidence for the interface between pixels x and y using merge function z, across the sensor suite

W_S = affinity function weighting factor for sensor S

1. "OR"—assumes the pixels belong to the same facet if *any* of the sensors confirm it. This function is more appropriate if the sensors being merged have dissimilar modalities, because it assumes facet continuity if any of the sensors are able to confirm it. Hence, the basic merged edge confidence is equal to the minimum edge confidence across the sensor suite for that pixel interface, that is,

$$BM(x, y)^{\mathrm{OR}} = \min\{W_S E(x, y)_S \mid \forall S \in \text{Sensor Suite}\}$$

2. "AND"—assumes that the pixels belong to the same facet only if *all* of the sensors confirm it. This function is more appropriate if the sensors being merged have common, or similar, modalities, as it assumes that all of the sensors should be able to confirm the facet continuity if it is appropriate. Hence, the basic merged edge confidence is equal to the maximum edge confidence across the sensor suite for that pixel interface, that is,

$$BM(x, y)^{\mathrm{AND}} = \max\{W_S E(x, y)_S \mid \forall S \in \text{Sensor Suite}\}$$

Unfortunately, these two functions may be too absolute. Just as "2-pt" affinity is likely to be insufficiently robust across an entire image because of variations in sensor measurement quality, precision, and so on, these two basic merge functions are assuming a uniform edge confidence quality and precision. By examining the consistency of the basic merged edge confidence map in the pixel interface neighborhood, a relative indication of discontinuity is readily available. This technique provides an adaptive measure of local pixel consistency. By examining the ratio of the current interface's edge confidence to that of its neighbors, this measure becomes more strict in smooth regions (e.g., corresponding to an artificial surface or a region of low noise) and more forgiving in rough regions (e.g., corresponding to natural surfaces or to regions of high signal-to-noise ratios). In this way, relative merge functions vary their clustering requirements over the image based on local measuring precision or quality.

The two relative merge functions provided in this chapter use the basic merged edge confidences obtained by either "OR" or "AND" as a meta-level description of the pixel interfaces. For the following definitions, let

$n(x, y)_S^z$ = the edge confidence for the interface between pixels x and y normalized with the edge confidence for the interface between pixels y and z, where z is the pixel adjacent to y on the opposite side from x, using sensor S, for any affinity function

$N(x, y)_S$ = the combined edge confidence for the interface between pixels x and y based on the interface's two adjacent normalized confidences, using sensor S, for any affinity function

$RM(x, y)^z$ = the relative merged edge confidence for the interface between pixels x and y using merge function z, across the sensor suite

W_S' = merge function weighting factor for sensor S

Specifically, referring to Figure 11.5, the two normalized confidences for the interface between pixels 1 and a are

$$n(a, 1)_S^2 = (|BM(1, a)^\alpha| + 1)/(|BM(2, 1)^\alpha| + 1) \ni \alpha \in \text{merge functions}$$
$$n(1, a)_S^b = (|BM(1, a)^\alpha| + 1)/(|BM(a, b)^\alpha| + 1) \ni \alpha \in \text{merge functions}$$

and the combined edge confidence for the interface is

$$N(1, a)_S = \max\{N(a, 1)_S^2, N(1, a)_S^b\}/W_S' \ni S \in \text{Sensor Suite}$$

3. "Rel_OR"—assumes the pixels belong to the same facet if *any* of the sensors confirm it using normalized edge confidences. This function, like "OR," is more appropriate if the sensors being merged have dissimilar modalities. The merged edge confidence is equal to the minimum edge confidence across the sensor suite for that pixel interface, that is,

$$RM(1, a)^{\text{Rel\_OR}} = \min\{|BM(1, a)^\alpha|, \min\{N(1, a)_S\}\} \; \forall S \in \text{Sensor Suite}$$
such that α is the merge function used in the determination of N

4. "Rel_AND"—assumes that the pixels belong to the same facet only if *all* of the sensors confirm it using normalized edge confidences. This function, like "AND," is more appropriate if the sensors being merged have common, or similar, modalities. The merged edge confidence is equal to the maximum edge confidence across the sensor suite for that pixel interface, that is,

$$RM(1, a)^{\text{Rel\_AND}} = \min\{|BM(1, a)^\alpha|, \max\{N(1, a)_S\}\} \; \forall S \in \text{Sensor Suite}$$
such that α is the merge function used in the determination of N

By minimizing with respect to the original merge function results, $|BM(1, a)^\alpha|$, one

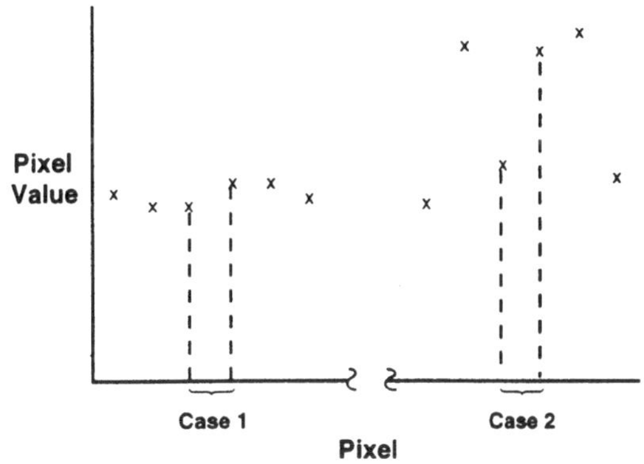

FIGURE 11.6
Hypothetical plot of two pixel value neighborhood cases. Case 1: consistent data yields: low absolute edge confidence, high relative ratio. Case 2: erratic data yields: high absolute edge confidence, low relative ratio.

> reduces the heightened sensitivity in otherwise smooth regions (see Figure 11.6). Recall that this "relative" approach is intended to handle properly both smooth regions (wherein the original merge results would be highly reliable, while the normalization ratio would be inappropriately high) and highly erratic regions (wherein the original merge results are very high in an absolute sense, while the normalized ratio would correctly indicate that the discontinuity is not inconsistent with the neighboring data). Adding one to both the numerator and denominator in the normalized edge confidence equation similarly reduces heightened sensitivity when the denominator is at or near zero.

There are various different flavors of "Rel_OR" and "Rel_AND," based on which merge function ("OR" or "AND") and which affinity function are used to form the initial merged confidence map used to determine the normalized edge confidences. As there are four affinity functions and two nonrelative merge functions, there are eight nonrelative MSI algorithms to compare. Each of these algorithms could be coupled with either of the two relative merge functions, providing an additional 16 MSI algorithms. The empirical evaluation of these 24 approaches will be presented in Section 6.

Region Growing Algorithm

The choice of region growing algorithm is an ongoing research area, and although the latest results are very promising, much remains to be done. Hence, the system described in this chapter is limited to a single region growing algorithm. This algorithm uses a simple, parameter specified threshold to determine whether the merged edge confidence of adjacent pixels warrants merger.

From these set definitions, we derive our search space. The search space consists of a dimension for each sensor in the suite that defines the affinity function to be used, a dimension defining the merge function(s), and dimensions for each parameter used by those functions. Hence, this search space encompasses all MSI algorithms that can be formed by combining those affinity and merge functions and all combinations of corresponding parameter values (within user-specified bounding and precision conditions).

3.2 Implementation Code

Application of a given MSI solution is performed in four separate stages; see Figure 11.4:

1. *Collect statistics*—Each sensor image to be merged is processed separately and meaningful low-level statistics (e.g., change in pixel value, change in slope of adjacent pixel values) are collected for each pixel interface in accordance with required affinity functions. In our implementation, statistics were calculated only once and saved for later, direct access—since during the learning stage the images remain constant; the statistics are repeatedly required; and, in our system, memory space was readily available. This was only to save runtime during the learning stage and in no way affects the system's MSI performance.
2. *Process each sensor*—For each sensor image, its "affinity function" is applied to each pixel interface (e.g., for a two-dimensional image grid, both "horizontal" and "vertical" interfaces would be processed). The resulting numerical edge confidence for each pixel interface is stored for the subsequent stage. This value indicates the "degree of confidence" that the two pixels sharing the interface belong to the same facet according to this sensor image.
3. *Merge sensors*—A single edge confidence value is determined for each pixel interface by applying a merge function to the corresponding affinity values across the entire sensor suite. The process is repeated for each pixel interface and results in a single (merged) edge confidence map.
4. *Form clustered image*—A region growing algorithm is employed to combine pixels whose interface "edge confidence value" is below some parameterized threshold.

3.3 Learning Strategy

Genetic algorithms, GAs, are extremely powerful adaptive global search techniques derived from natural population genetics [20–23]. GAs have been shown to perform well for many types of functions, including those exhibiting very difficult characteristics (e.g., discontinuities, nondifferentiability, multimodality, high dimensionality, huge search spaces, and noise). GAs require no specific, *a priori* function information; only the form of a candidate solution, that is, the number of parameters to be optimized and the desired level of precision (number of bits), and a comparative performance measure for candidate solutions is needed. These capabilities and requirements make GAs well suited to this global optimization task [24, 25].

GAs are a simplified simulation of the natural genetic model. As such, GAs simulate a population of individuals evolving over multiple generations: individuals are specified by a series of genes (bits) that can be independently inherited, reproduction is accomplished by a crossover operation that forms offspring from the genetic material of their parents, and a individual's reproduction frequency is based on his performance in the environment (evaluation function). Thus, trait encoding, sexual reproduction, and "survival of the fittest" propagation are all simulated, and the average performance of the population tends to improve over successive generations. When some user-specified stopping criterion is met, the best individual produced is taken as the GA's solution for function optimization.

Clearly, GAs cannot guarantee discovery of the optimal solution, but they have proved to be powerful global search techniques capable of simultaneously searching extensive regions of the parameter space. The basic GA framework used in this research was provided by the GENESIS [26, 27] GA package with modifications as suggested by Baker [20, 28].

3.4 Evaluation Function

The evaluation function guides and directs the GA's search by providing performance feedback for candidate MSI algorithms. To guide efficiently, the evaluation function should have a high resolution and monotonically encourage clustering quality [20], so that even small improvements will be reflected in the evaluation measure and have a positive impact on the search.

Because the goal is to form pixel clusters that most nearly approximate some user-defined level of detail, the desired MSI results for a given sensor suite's data set are provided as input during the learning stage. This "truth image" has its most meaningful pixels correctly grouped into their separate facets. Erroneous pixels or those corresponding to objects too small to be of interest are associated with a single group understood to be "don't care" pixels; the cluster association of those pixels will have no affect on the evaluation.

In this AMSG system, the GA is executed as a minimization technique. This requires the evaluation function to be written as a "penalty function"; the function increases its evaluation measure with each type of undesirable behavior that it detects. The evaluation function measures how closely the results of a given candidate solution (MSI algorithm) match the desired results (truth image). To do this, the candidate solution is applied to the sensor data; see Section 3.2. Next, each pixel is assigned a "cluster number," which is constant for all pixels within a single cluster and distinct from all other pixels. The actual cluster number associated with any given cluster is arbitrary. Hence, the evaluation function cannot simply compare the number of pixels in each cluster with the correspondingly numbered facet of the truth image. Rather it must determine the best association of candidate clusters to truth facets. These candidate clusters must be evaluated according to how fully they cover their respective truth facets *and* how little they exceed the true facet's boundaries. Thus, the evaluation function must penalize both undesirable fragmentation and undesirable amalgamation.

Specifically, evaluation is performed by considering each true facet in order of decreasing size and associating with each the candidate cluster that has not already been associated and has the greatest representation within that true facet, that is,

$$\begin{aligned} \text{let } & A_j \in \{\text{Candidate Clusters}\} \\ & T \in \{\text{Truth Facets}\} \\ & D = \{\text{Previously Associated Candidate Clusters}\} \end{aligned}$$

then A_K is associated with T if and only if

$$\forall J \ni J \neq K \qquad \text{and} \qquad A_J, A_K \notin D$$

$$|A_K \cap T| \geqslant |A_J \cap T| \qquad \text{and} \qquad K < J \quad \forall J \ni |A_K \cap T| = |A_J \cap T|$$

For each associated cluster, the evaluation function increments the penalty based on the amount by which that cluster failed to cover its true facet and by the amount by which it lies outside its true facet; see Figure 11.7. For example,

$$\begin{aligned} \text{let } & A, B \in \{\text{Candiate Clusters}\} \\ & B \notin \{\text{Previously Associated Candidate Clusters}\} \\ & A = A' \cup A'' \qquad \text{and} \qquad B = B' \cup B'' \\ & T \in \{\text{Truth Facets}\} \\ & T = A' \cup B' \qquad \text{and} \qquad |A'| < |B'| \end{aligned}$$

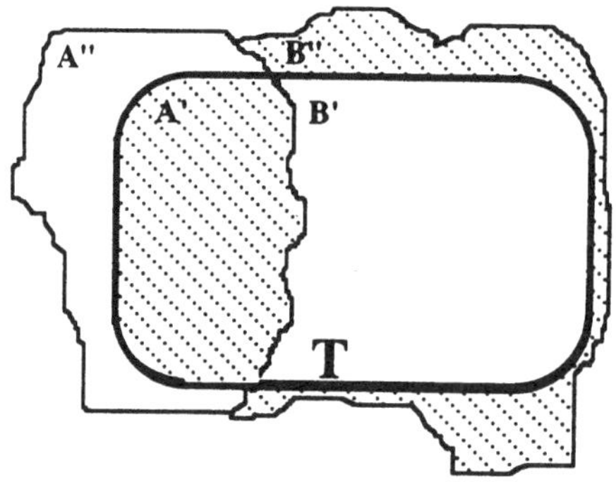

FIGURE 11.7
Cluster matching against truth facet, *T*. Cluster *B* associated with facet *T*, penalty $f(A', B'')$.

then *B* will be associated with *T* and the penalty resulting from *T* is $f(|A'|, |B''|)$. This penalty combination encourages clusters to "grow" within the true facets' boundaries and "shrink" outside those boundaries. Note also that this function makes no distinction between pixels of different facets; every valid, misassociated pixel causes the same amount of penalty. Hence, an "*N*" pixel improvement in a small facet has the same impact as an "*N*" pixel improvement in a large facet, although the two resulting images may appear to have very different levels of "clustering quality," depending on the application. A truly "optimal" evaluation function is highly application specific and cannot be considered within the scope of this chapter. Even so, this general evaluation function should be sufficiently sensitive to direct the search strategy effectively for a wide variety of applications.

4 TARGET SENSOR DOMAINS

Two imaging domains were used to test and validate the AMSG. The first was a laser range camera targeting a "typical indoor warehouse" environment, specifically our autonomous robotic navigation domain. This domain was used for training the AMSG and validating its application robustness. The second imaging domain was a ground conductivity sensor targeting a buried waste field. This domain, while representing an equally important characterization task, provided a highly dissimilar modality and data format for integration. That is, while the laser range camera provided two 2-dimensional images of the surface characteristics targeted, the ground conductivity sensor provided two 0-dimensional data points of the integrated character of the targeted region.

This section presents a description of the physical nature of each of these sensor systems and their targeting environment.

4.1 Ground Conductivity Sensor

The available subsurface data consist of a set of ground conductivity (GC) readings [1] of a cold test pit at the Idaho National Engineering and Environmental Laboratory (INEEL) waste storage site [13]. This set was originally to include ground penetrating radar data, but the ground water and clay soil of that region rendered that sensor modality impotent. Despite that loss, MSI can still be usefully applied by merging the multiple GC data sets and the two measured signals inherent in GC data: quadrature and in-phase strength.

The GC sensor used consists of a magnetic transmitting coil and a receiving coil placed 3.66 m apart [1]. The transmitting coil sets up a magnetic field in the ground that induces eddy currents and in turn a secondary magnetic field, 90° out of phase. The receiving coil measures both the primary magnetic field (in-phase component) and the secondary field (quadrature component). Thus, a GC sensor measures the integrated dielectric constant of the ground in the three-dimensional proximity of the two coils. By its very nature, data from a GC sensor is very ambiguous and unfocused, since a buried conductive object will disturb all readings taken anywhere within its vicinity and the amount of disturbance is a function of the object's size, dielectric constant, orientation with respect to the coils, distance from each magnetic pole, uniformity and distribution of ambient material, the relative position of other magnetic conductive objects, and so on. Hence, from a set of GC data, it is impossible to determine which of the infinite number of perfectly data-consistent possibilities actually corresponds to "reality."

Given GC data's ambiguous and unfocused nature, it would appear a perfect candidate for inverse mapping using MSI "detail enhancement" techniques. However, the inverse mapping function is unresolvable from the GC data alone, because of its inherent ambiguities. Ground penetrating radar was an excellent complementary sensor and, if available, may have permitted the inverse mapping function to be roughly approximated.

The GC data's representation may be improved by using MSI "clustering" techniques. However, since precise truth for the GC data is not available, these data sets cannot be used for designing or training MSI clustering algorithms. This necessitated the use of a second sensor domain, the laser range camera, for the development of general-purpose, MSI clustering algorithms that could eventually be applied to the GC sensor domain. Even so, the GC data is still well suited for the validation of the AMSG's domain robustness and sensor type independence.

4.2 Laser Range Camera

The Odetics laser range camera (LRC) used for this research [29] produces images of 128 × 128 pixels, where each pixel's value is determined by the reflection properties of a directed laser. This LRC measures two values: the reflected light's phase shift, indicating the distance to the target; and its intensity, indicating the target surface's sheen or degree of reflectivity at the camera's operating wavelength of 820 nm. This LRC provides a 60° × 60° field of view with an unambiguous measurement range for targets lying between 3 and 10 m distant. The target chosen for analysis was a portion of our laboratory [30], with distances ranging from 3 to 15 m and scattered, miscellaneous objects (e.g., furniture, boxes, 55-gallon drums). For this domain, MSI could be performed using either multiple LRC images or the multimodal distance and reflectance images from a single view.

The LRC was chosen for MSI algorithm development for four reasons: (1) it is consistent with the requirements of many autonomous and teleoperated robots' environments, including many automated navigation tasks; (2) since its two multimodal images (distance and reflectance) are obtained from the same reflected laser light, they are perfectly registered images, thus eliminating the dependence on separate image registration algorithms; (3) a calibrated LRC was readily available in our laboratory and fully integrated with our computer network; and (4) precise truth measurements could be taken and used for AMSG design, development, training, and evaluation.

5 SENSOR ANOMALY CORRECTION

The LRC used in this research bases its distance values on time-of-flight estimates by comparing the relative phase shift between the original light transmitted and that of its detected reflection. Unfortunately, this method is ambiguous, as it is impossible to determine the integer number of phase lengths traversed (i.e., objects at 0.1, 1.1, and 2.1 phase length distances will all be detected as 0.1 out of phase). This problem produces what is known as "wraparound error" and can be avoided only by limiting the LRC to viewing targets residing within a single phase length of the transmitter. Figure 11.8 displays the raw LRC data when imaging one corner of our laboratory. Figure 11.8(a) presents the distance image shown as a two-dimensional grid with darker color indicating pixels with smaller phase shift fractions. Figure 11.8(b) presents the reflectance image with darker color indicating greater reflectivity (sheen). In Figure 11.8(a), the dark region in the upper left corner actually corresponds to the most distant targets and is an example of a wraparound error.

Figure 11.9 presents the reflectance data of Figure 11.8(b) plotted in three dimensions by using the distance data of Figure 11.8(a). (Note that Figure 11.9's point of view is offset from the original LRC's viewpoint to amplify the three-dimensional effect.) Most areas apparently missing data are the result of object occlusion or wraparound errors. The wraparound error results in a phase length discontinuity and a smearing of data points toward the LRC's view position.

To resolve a sensor anomaly, one must (1) recognize its presence and (2) know how to correct its effect. For the wraparound error, both conditions can be met. First, although erroneous pixels cannot be recognized in isolation, most can be recognized by their distinct

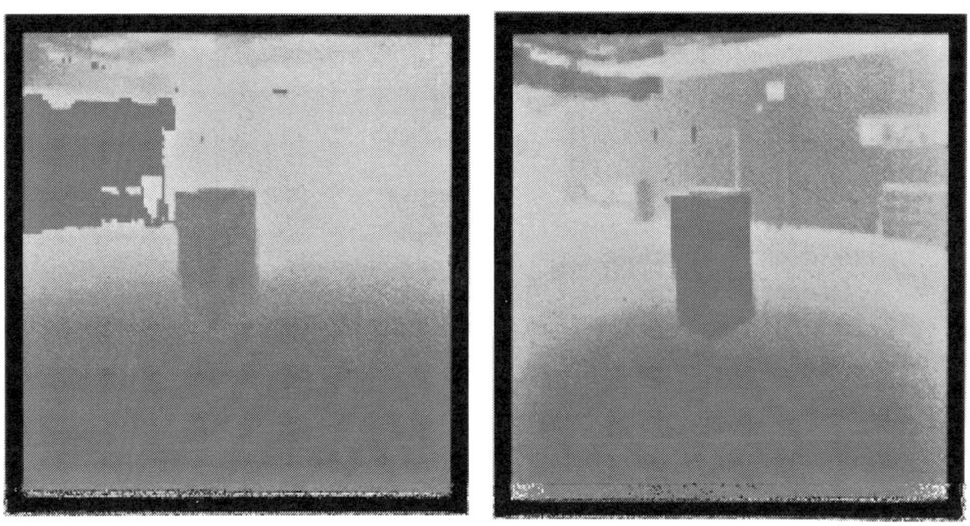

(a) (b)

FIGURE 11.8 (a and b)

Raw laser range camera (LRC) data of a corner of our laboratory, with various objects interspersed, 2-D plots: (a) colored by distance and (b) colored by reflectance. [Note that the dark region in the upper left corner of in (a) is due to a "wraparound" anomaly; i.e., since this LRC determines distance by differential phase shift and since only fractional phase shifts can be measured, targets or slightly greater than one phase length appear to be very close.]

FIGURE 11.9
LRC data of Figure 11.8, 3-D plot colored by reflectance.

interface with nonerroneous pixels (i.e., the presence of drastic discontinuities in distance coupled with insignificant changes in reflectance). Second, our physical understanding of this sensor anomly provides us with a clear knowledge of how to correct it in most cases: add one phase length to the erroneous pixel's distance.

This data enhancement approach is implemented by scanning for pixel interfaces that have very large distance discontinuities (approaching one phase length) and very small reflectance discontinuities (indicating little change in intensity). Whenever such an interface is found, the closer pixel's distance is incriminated by one phase length. The results of this MSI technique are shown in Figure 11.10. By repeating this approach, multiple phase length losses can be resolved, as long as the image contains a somewhat continuous gradient. Note that some situations cannot be correctly resolved with this method. They include adjacent pixels

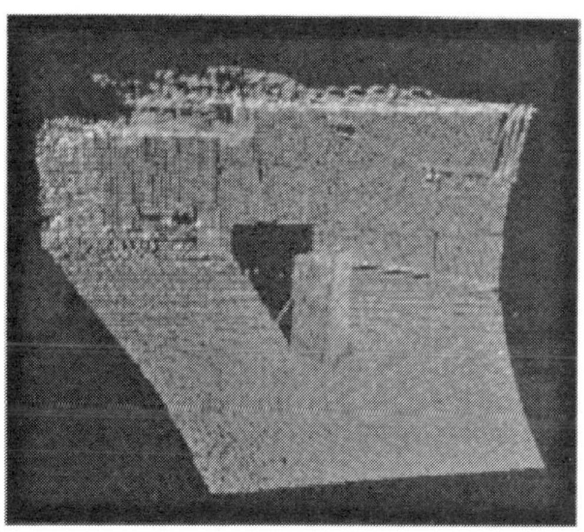

FIGURE 11.10
Wraparound anomaly corrected LRC data of Figure 11.9, 3-D plot colored by reflectance.

corresponding to targets that (1) actually are more than one phase length disjoint, yet have similar reflectance values; (2) are of similar range, yet have a large discontinuity in reflectance values, and the distance value(s) suffer from the wraparound anomaly; or (3) are of multiple phase length discontinuity. There is simply insufficient data to guarantee correct resolution for all possible cases. However, for the indoor domain these situations are rare; the results shown in Figure 11.10 and the following empirical analysis demonstrate the effectiveness of this approach.

6 EMPIRICAL EVALUATION

The evaluation function described in Section 3.4 provides a numerical measure of the quality of a conceptually enhanced image, that is, how well it compares with the desired result. This measure is useful not only for feedback to guide the GA's global search but also for the objective comparison of competing MSI techniques. However, it should be noted that although the GA is a powerful search technique, it does not guarantee any particular proficiency relative to that technique's global optimum, such as a guarantee of achieving **X**% of the technique's optimum performance. For example, in the empirical evaluation presented in this chapter, each search was permitted the same amount of computing resources yet the search space sizes ranged from 2^{10} to 2^{47}. Hence, the competing techniques had significantly different proportions of their search spaces investigated. In each experiment, the GA was permitted to examine 5000 MSI algorithms. For the smaller search spaces, this was sufficient to cover the entire space, but for the largest spaces, the GA could examine less than 2.3 candidate solutions for every trillion defined in the search space. Although the evaluation measure can be used to compare the quality of the resulting images, it can only imply the corresponding technique's relative potential.

For each experiment, the GA was given an initially random population of 50 candidate solutions (MSI algorithms). The GA was permitted to search until it had evaluated 5000 MSI algorithms or its population had stagnated (failed to produce a distinct individual for two successive generations), whichever came first. As described in Section 3, the MSI algorithm to be evaluated was applied to the sensor data and the resulting segmented image compared with the user-defined desired results. To evaluate the relative quality and effectiveness of the various affinity and merging models presented in Section 3.1, 16 separate experiments were performed. Each of the four affinity functions (2-pt, 4-pt General, 4-pt Predictive, and 6-pt) was applied with the two basic merge functions (OR and AND) using the two sensor images (Distance and Reflectance) and they were applied to the two sensor images independently. The single sensor result provides a baseline to verify the advantage and effectiveness of multisensor integration itself. The performances of the overall best solutions are compared in Figure 11.11. (Note that the overall degree of match is a penalty function, and therefore lower values are better.)

Several interesting tendencies are worthy of note. First, and most important, the basic goal of MSI was achieved; MSI led to better representations of the real environment (i.e., more closely in line with the desired results) than the single-sensor systems. By combining the sensors, superior performance was achieved in all four affinity modes (see Figure 11.11). Second, for all modes, the distance image led to better clustering than the reflectance image (see Figure 11.11). This may be a result of primarily planar objects and linear-based affinity functions; the distance image embodies the linearity of the surfaces, while the reflectance image is subject to nonlinear reduction of the intensity signal with distance. Third, there is an interplay between the strictness of the affinity mode (in its size requirements on the interfaced facets, see Table 11.1) and the strictness of the MSI merge function (see Figure

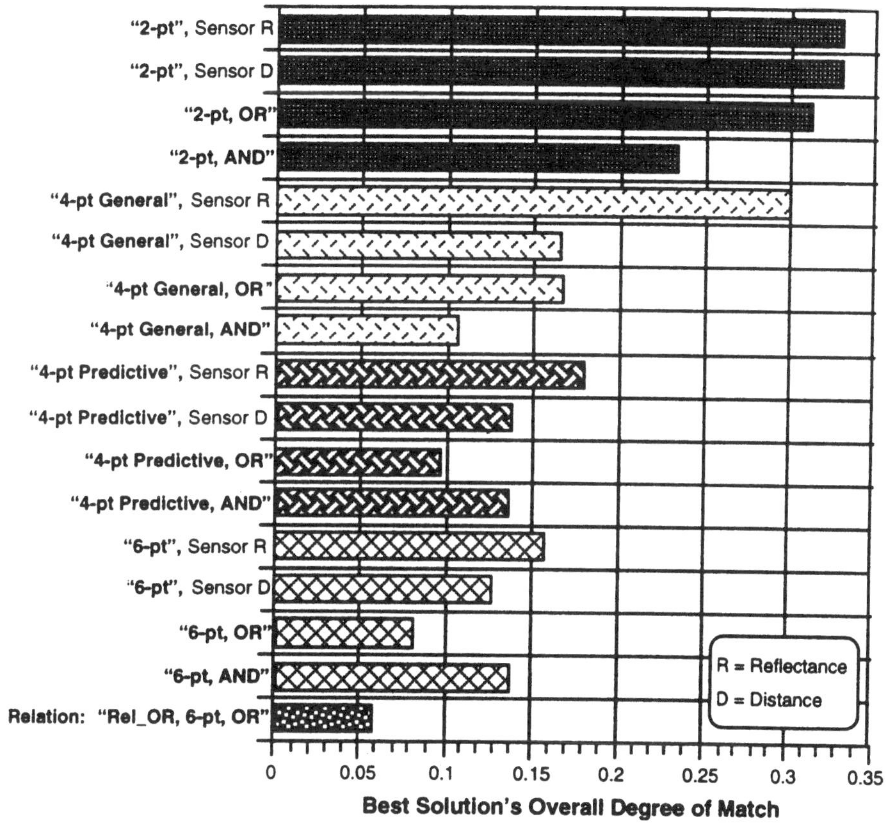

FIGURE 11.11
Comparison of evaluation function results for the best solutions found for each MSI algorithm mode, both for individual images (range and distance) and with merge functions applied to both images.

11.12). For less strict affinity modes (e.g., 2-pt and 4-pt General) the stricter AND merge function leads to better clustering by preventing overamalgamation. For more restrictive affinity modes (e.g., 4-pt Predictive and 6-pt) the less restrictive OR merge function leads to better clustering by preventing overfragmentation. This result implies that the methods investigated span the proper range of facet size requirements for this LRC domain.

The relative merge functions (Rel_OR and Rel_AND) are based on the local uniformity of the affinity mode's results. Hence, there are also 32 different relative methods, corresponding to the 16 nonrelative methods listed in Figure 11.11. Rather than execute the GA an additional 32 times, the choice of method was defined as an additional dimension in the search space to be simultaneously optimized by the GA. The GA determined that the best method for relative operation is "6-pt, R or D," which is also the method found to perform best in the nonrelative mode. Figure 11.11 shows the superiority of relative operation, as it outperforms the best of the nonrelative methods by a significant margin (>27%).

The truth image is shown in Figure 11.13. This image is based on the data shown in Figure 11.8. Its features include (1) floor; (2) a box with two facets facing the camera; (3) a movable wall partition; (4) two facets of the back wall, divided by a supporting pillar; (5) a tool chest; (6) a segment of the left wall; (7) a small crate; (8) a supporting pillar in the middle of the back wall; and (9) a triple-faceted corner support structure.

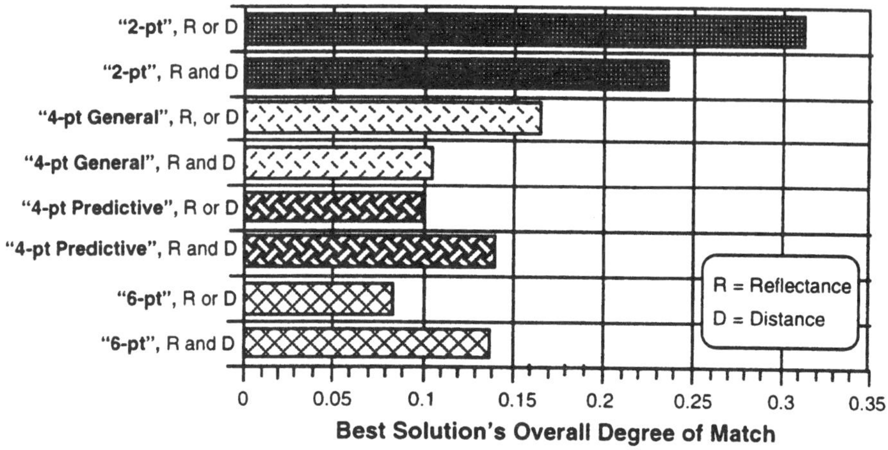

FIGURE 11.12
Comparison of eight basic MSI algorithm modes, showing merge function trade-off with respect to affinity mode strictness.

For human display, adjacent pixels corresponding to the same facet are graphically connected. This display mode makes it easier to discern the facets and their edges. The color of the facets is based on their relative size in pixels. The black space between facets is the result of (1) missing data due to occlusion (e.g., behind the box in the middle of the room); (2) the display mode not graphically connecting adjacent pixels that correspond to different

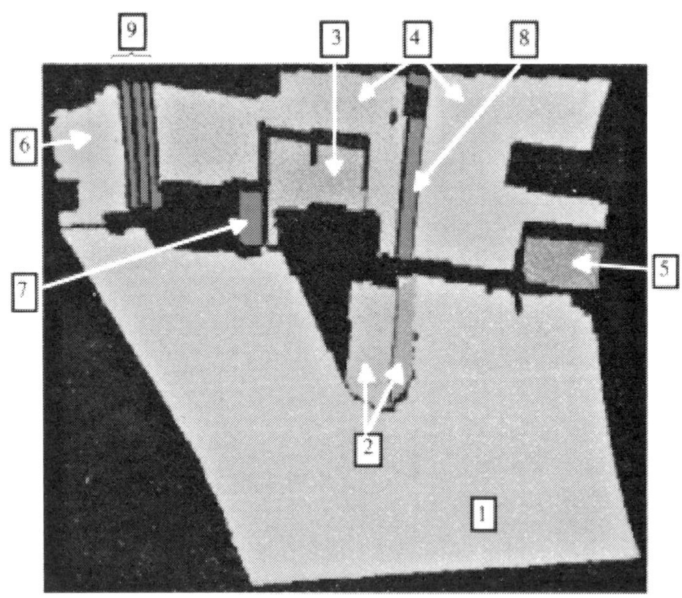

FIGURE 11.13
Truth image (user-defined, desired MSI results) of Figure 11.10, 3-D plot colored by size of facet and numbered for reference.

facets (e.g., separating the two faces of the box and separating the wall supports); or (3) erroneous data labeled as "don't care pixels" (e.g., the random clutter next to facet #7). (Note that the "don't care pixels" were not excluded from processing; the MSI techniques had to deal with them. But the correct clustering of these pixels was not evaluated because we wished to optimize the MSI techniques for their ability to ignore erroneous data, not for their ability to correctly cluster both valid and erroneous data.) In the following figures showing the MSI results, erroneous data is shown with its corresponding clusters; however, to "declutter" the display, only the larger clusters are shown.

Figure 11.14 shows the MSI results of the relative "Rel_OR, 6-pt, OR" method using the GA's optimized parameter settings. This image reveals the major facets (#1–#6) and some of the minor ones, including two of the corner supports (#9). The top surface of the tool chest and a region of clutter in the upper right corner of the image were also correctly differentiated. These facets were considered too small and their data too erroneous to be included in the truth image. Furthermore, despite wraparound affecting facets #1, #4, #6, and #8, this MSI technique was able to form clusters smoothly across the wraparound boundaries and lying within this erroneous region. These results indicate the accuracy and continuity of the MSI wraparound technique described in Section 5.

Although "Rel_OR, 6-pt, OR" failed to differentiate all of the known detail in the image, it was very successful. To get an impression of the quality of these results, one need only review the original images that were merged; see Figures 11.8–11.10. In these images, most of the facets (e.g., #3+) are difficult to discern visually despite our detailed knowledge of them.

Figure 11.15 shows the MSI results of the "6-pt OR" MSI technique using the GA's optimized parameter settings. This technique also differentiated most of the major facets, failing only to amalgamate facet #6. (Because of facet #6's greater distance, its spatial

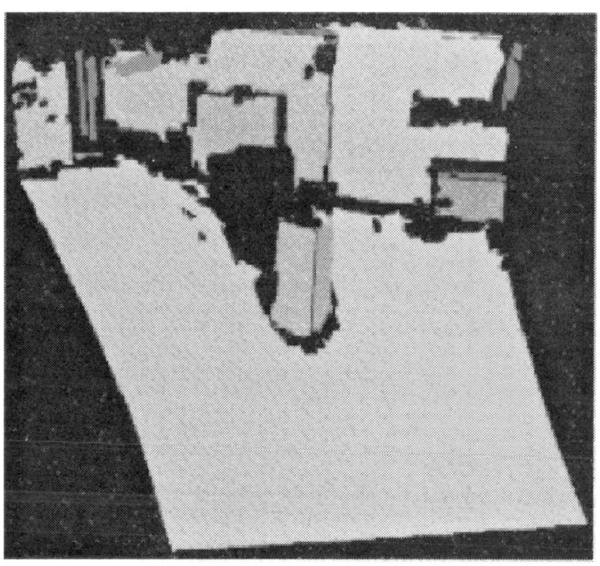

FIGURE 11.14
Clustering results of "Rel_OR, 6-pt, OR" MSI algorithm applied to data of Figure 11.10 and trained by corresponding "Truth Image" (Figure 11.13), 3-D plot colored by size of cluster.

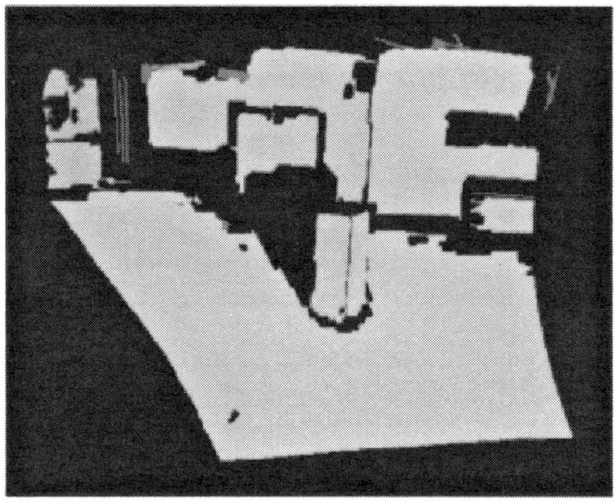

FIGURE 11.15
"6-pt, OR" MSI algorithm results applied to data of Figure 11.10, 3-D plot colored by size of cluster.

sampling resolution is lower. This leads to greater variation in the data, and a non-relative technique that amalgamated those pixels correctly would probably fail to isolate other, more consistent facets. To handle this problem correctly, one must base matching on the local dynamics of the data, i.e., the relative MSI techniques.) All three of the corner supports, facet #9, the top of the tool box, facet #5, and a facet in the upper right corner were found.

It should be noted that the uniform emphasis on pixels rather than any special emphasis on finding each of the primary 14 truth facets explains this technique's "failure" to isolate the smaller facets. However, to do otherwise would require global and/or *a priori* knowledge that would demean the general applicability and robustness of the AMSG. The promise of the AMSG is the quality of its results despite its total lack of global, conceptual shape, or *a priori* knowledge—a common requirement in many robotic applications. These MSI results are of sufficient quality to enhance an autonomous system's world map, a teleoperated display, or a higher level object recognition system.

7 VALIDATION

The AMSG is highly domain and application robust. Domain robustness refers to its suitability for a wide variety of environments and sensor types, once it is tuned to the level of image detail desired by the user. Application robustness refers to the suitability of a given optimized MSI algorithm to a wide variety of situations within a tuned or learned domain. This section validates the AMSG's application robustness by evaluating the performance of its optimized MSI algorithm in the learned domain (i.e., LRCs operating in an indoor environment) and validates its domain robustness by demonstrating its suitability to another, highly dissimilar domain (i.e., GC images of a waste test site).

7.1 Learned Domain, LRC

An MSI algorithm and its control parameters were chosen from an extensive search space (see Section 3) based on a single LRC image set (Figure 11.8) and user-defined truth image (Figure 11.13). The following validation test is to demonstrate the suitability of that MSI algorithm and its parameters to LRC images of other scenes taken within the same domain.

Figure 11.16 shows the LRC's distance and reflectance images of another corner of our laboratory. For clarity of description, the primary facets of interest are numbered: (1) floor; (2) a cylindrical barrel; (3) a movable wall partition; (4) two facets of the back wall, separated by a supporting pillar; (5) a 55-gallon drum; (6) a door set in the right wall; (7) a box with two facets in view; (8) a supporting pillar in the middle of the back wall; (9 and 10) two halves of a windowed door; (11) a suspended hoist and its draped extension cords; (12) three more facets of the back wall separated by #11; (13) a box with one facet in view; (14) a fire extinguisher; and (15) an electrical cable lying on the floor. This image partially overlaps the previous image; facets numbered 1, 4, and 8 match the correspondingly numbered facets of Figure 11.13.

The image in Figure 11.16 is more difficult than the learned image of Figure 11.8. This image contains cylindrical objects (facets #2 and #5), very narrow objects (facet #15 and parts of facet #11), and significantly more detail. The best performing MSI technique from Section 6 ("Rel_OR, 6-pt, OR") and its tuned parameter set were applied to the distance and reflectance images of Figure 11.16.

Figure 11.17 shows the MSI results of the "Rel_OR, 6-pt, OR" MSI technique using the GA's optimized parameter settings. For image clarity, only the larger clusters are displayed. This technique resolved most of the facets enumerated in Figure 11.16, including those corresponding to both the cylindrical and narrow objects. (Note that cylinders are difficult because the various affinity functions defined in Section 3 are based on *linear* extrapolation and the parameters were tuned on an image consisting entirely of planes.) Furthermore, three

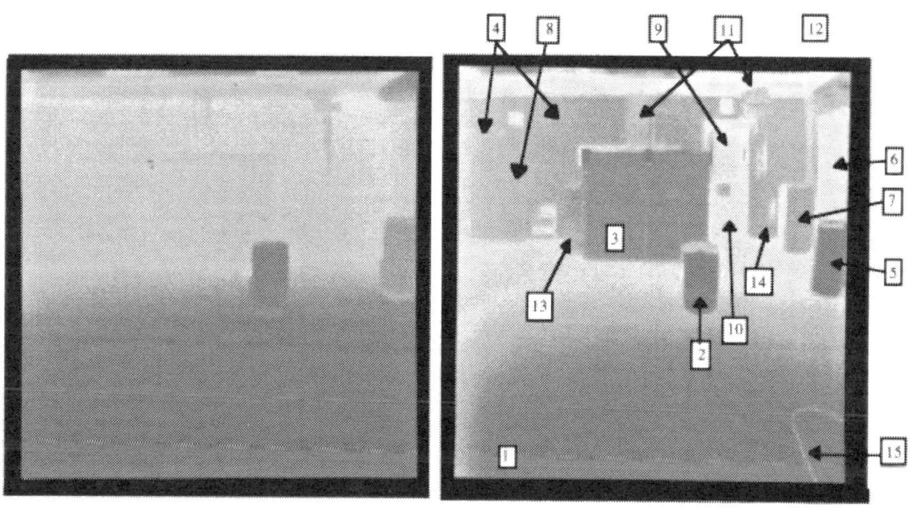

FIGURE 11.16

New scene from the domain of Figure 11.8: raw LRC data of another corner of our laboratory, with new objects interspersed, 2-D plot colored by (a) distance and (b) reflectance.

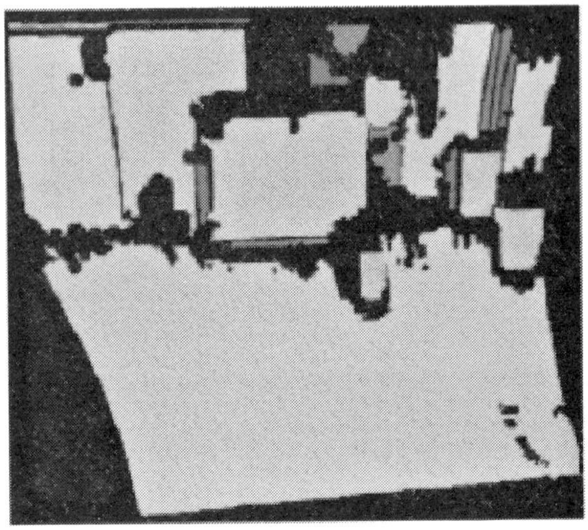

FIGURE 11.17
Clustering results of "Rel_OR, 6-pt, OR" MSI algorithm applied to data of Figure 11.16, using parameters tuned to Figure 11.10, 3-D plot colored by size of cluster and numbered for reference. This example demonstrates AMSF's task robustness by applying a solution from one image to an unknown image from the same domain. The scene in Figure 11.16 is more difficult than the training scene of Figure 11.8. This scene contains cylindrical objects, very narrow objects, and significantly more detail. [Note that cylinders are difficult only because (1) the various affinity functions currently used are based on linear extrapolation and (2) the parameters were tuned on an image consisting entirely of planes, i.e., Figure 11.8.]

facets in the far right corner were unexpectedly differentiated: the door facing, parallel to facet #6; the doorway offset, parallel to facet #12; and a narrow strip of the right wall, again parallel to facet #6. One can also discern the sign and doorknob of facet #9, the wire connecting the two primary components of facet #11, and the speaker and electrical outlet near the top of facet #8. Unfortunately, facet #8 itself was not distinguished from facet #4 and the extreme detail of the hoist and its cables (facet #11) prevented it from being amalgamated into a single, cohesive object, although this technique did correctly isolate it from the other objects. (Note that the support pillar, facet #8, was not distinguished in the training image either, see Figures 11.14 and 11.15, and hence was not expected to be resolved here.)

Figure 11.17 clearly demonstrates the application robustness of the tuned "Rel_OR, 6-pt, OR" MSI algorithm. After the technique was trained to just one application image, it was able to isolate almost all of the significant facets in an *a priori* unknown image. These facets included surface types hitherto never seen: curved surfaces. The success of this MSI technique on curved surfaces, despite its purely linear extrapolation approach and complete lack of experience with this surface type, strongly supports the utility of using local pixel properties to form clusters.

7.2 Unlearned Domain, GC

The MSI algorithms described in Section 3 are designed to be independent of the sensor modalities being merged and the application being addressed. These are general-purpose MSI techniques for determining edge confidences from multiple sensor data sets and merging the results into a single, conceptually enhanced representation. The desired level of image detail, and hence the intrinsic definition of facets, must be defined by the user according to the application and purpose. This definition must be communicated to the system in order for it to tune the parameters for a given application domain. For the LRC, the truth image implicitly embodied the user's definition of facets and of extraneous detail. From this image, the system could infer the relative information content of the sensor modalities being merged and hence optimize the various tuning parameters. If a "truth" definition is not supplied, the user must either rely on general-purpose parameter settings (tuned to a diverse suite of domain types) or "manually" adjust the parameters to achieve the desired level of detail.

The unlearned domain chosen to demonstrate the domain robustness of the AMSG is ground conductivity, GC, images (see Section 4.1) of a buried waste test site. This sensor is an excellent test case because of its extreme dissimilarity with the LRC: it measures internal rather than surface properties; it obtains an integration measure rather than a point measure; it has a wide signal dispersion, leading to sampling overlap; its truth is almost indeterminable, due in part to the effects of varying environmental conditions (e.g., ground water, soil type) across the region being mapped; it is a single rather than an array sensor, and hence its sample positioning must be determined externally; and so on. Since accurate truth for these GC images is unknown, the system cannot be automatically trained as it can for the LRC images. Furthermore, developing general-purpose parameter settings will require images and truth from many diverse sensor modalities and was not the intention of this research. Hence, we must demonstrate the robustness of this MSI approach applied to GC data by examining the results of manually selected MSI algorithms and parameter settings. The purpose here is not to provide a definitive MSI solution for the GC domain but rather to demonstrate that the AMSG is applicable and that, given a "truth image," the system could be trained on this domain as it was on the LRC domain in Section 6.

The raw GC data consists of hundreds of samples taken over a 3-day period. These samples consist of an X-Y position (based on dead reckoning from a given starting point) and two sensor readings: the quadrature component and the in-phase component (see Section 4.1). The data is plotted in Figure 11.18. Because of the unequal spacing of the data points, the MSI techniques described in Section 3 cannot be *directly* applied. The data must first be transformed into a grid representation. A grid resolution and registration are chosen so that all of the internal grid elements are represented in the original data set. The value of each grid element is then defined as the average value of the raw sample points lying within its corresponding area; see Figure 11.19. (Note that the white elements on the lower edges have no corresponding data samples and that the extreme edges of the original data set were discarded; see Figure 11.18.)

Figure 11.20(a) presents the MSI results of the "4-pt OR" technique with a manually selected parameter set. With this MSI algorithm, the main cluster (near the middle of the image) is isolated, as are smaller clusters just below it and in the upper right corner of the image. This result is intuitively consistent with the data in Figure 11.19, based on casual, visual inspection. However, without defined truth (e.g., the relative importance of the two sensor images, an indication of the desired level of detail), other solutions are equally viable. Figure 11.20(b) presents the MSI results of the "6-pt, AND" technique with manually selected parameter set. In this image, the smaller cluster in the upper right corner of the image is considered insignificant, while the main cluster is examined more critically. Again, neither image of Figure 11.20 is offered as a *solution*. Rather, without "ground truth" they represent different, yet equally valid interpretations of the data.

Despite the lack of a truth image to which the approach could be trained and its results evaluated, the two results presented in Figure 11.20 clearly indicate the ability to isolate

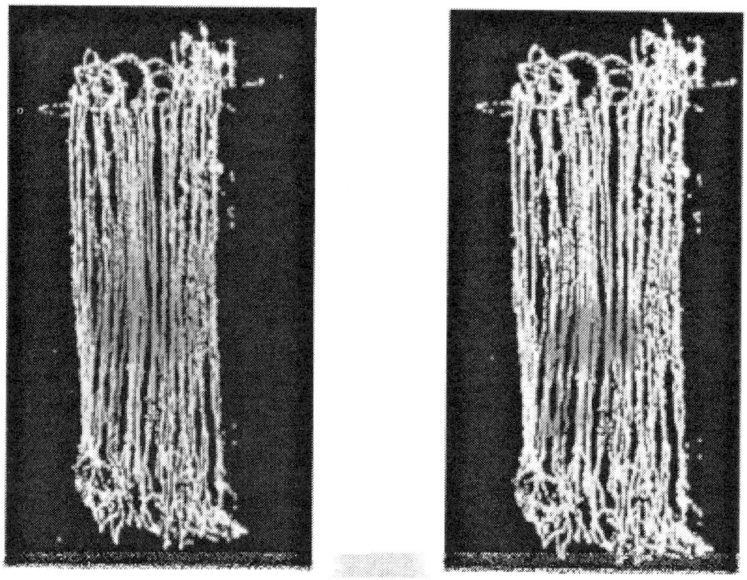

(a) (b)

FIGURE 11.18
Raw GC data of INEEL waste storage site, superimposition of three data sets in a 2-D plot colored by (a) quadrature and (b) in-phase measure.

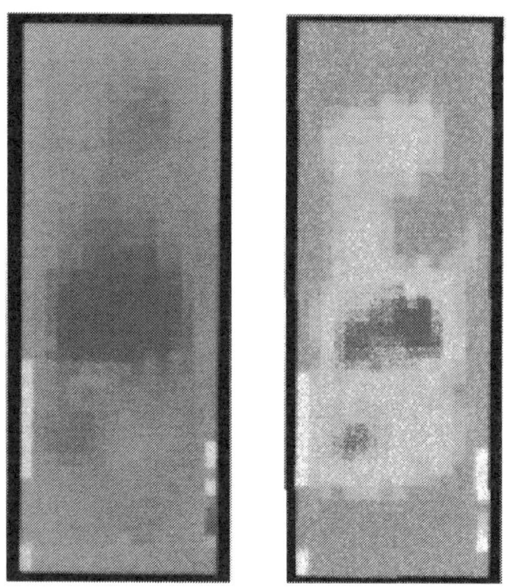

FIGURE 11.19
Raw GC data of Figure 11.18, represented in a 2-D uniform grid colored by (a) quadrature and (b) in-phase measure.

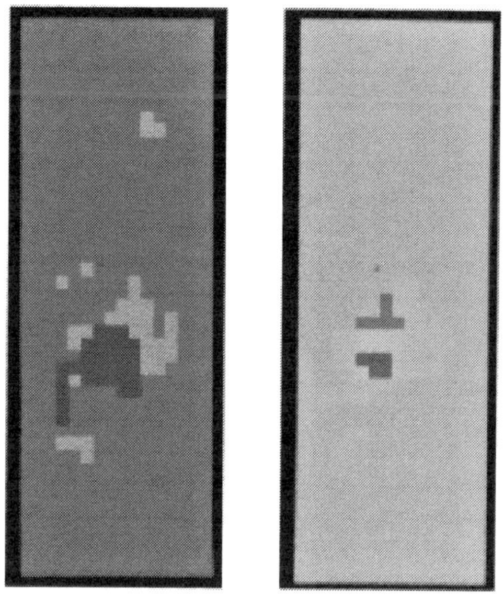

FIGURE 11.20
MSI clustering results applied to data of Figure 11.19 using manually selected parameter values, 2-D plot colored by size of cluster (a) "4-pt, OR" MSI algorithm; (b) "6-pt, AND."

significant facets and yield significantly different conceptual interpretations. As such, these figures demonstrate the suitability of the AMSG to the GC domain and provide an indication of its overall domain robustness.

8 SUMMARY

The Automated MSI Solution Generator (AMSG), presented in this chapter, is designed to take actual sensor data and a user's definition of the desired merged results as input during a training phase and independently find an efficacious method of merging the sensor data. While the basic AMSG approach could be applied to a variety of MSI tasks, we chose to restrict our consideration to the conceptual enhancement problem. Conceptual enhancement is an MSI technique used when the original image contains more detail than is desired, making it difficult to recognize objects of interest. This problem can be resolved by segmenting the image into meaningful clusters of pixels that correspond to conceptually homogeneous regions. The AMSG's underlying MSI approach is based on the premise that given appropriate sensors, clusters of pixels can be quickly and accurately formed by evaluating the continuity between each pair of adjacent pixels based on the local pixel neighborhoods. Hence, the basic direction of this approach is the potentially faster and more robust formation of clusters from pixels rather than segmenting images into clusters. This approach is based on low-level functions that capture fundamental characteristics of continuity and, as such, are applicable to cluster formation regardless of the sensor or application domain. This approach constitutes a robust, automated, MSI methodology.

The AMSG employs a hierarchy of potentially useful definitions for solving conceptual enhancement tasks, from the bottom up. These definitions form a pool of low-level MSI tools that can be automatically selected and optimized for a given sensor suite and application domain. In our implementation, this adaptive search for superior MSI solutions is performed by GA global optimization. Thus, the overall system is domain and application robust and provides a universal interface for diverse MSI tools and techniques to be merged, through standard confidence combination mechanisms.

The AMSG is validated on real distance and reflectance images from a laser range camera (LRC) by its ability to develop a high-quality MSI solution for the merger of this sensor's images in the defined application domain. This MSI solution was able to isolate nearly all of the significant facets of the training image. The domain robustness of this optimized solution was validated by applying it to an unknown LRC image set within the same application domain, containing new types of objects. Nearly all of the significant facets of the new image were correctly clustered. The application robustness of the AMSG was validated by applying the techniques to an extremely dissimilar sensor and application domain—ground conductivity data of buried waste [6].

Future research includes time optimizing the MSI solution on a state-of-the-art digitizer to produce a virtual "multimodal sensor"; extending the sets of definitions being investigated; permitting heterogeneous resolution sensors to be easily merged; improving the conceptual quality of the evaluation function by the addition of facet-based terms; and extending the GA's search to include the actual formation of affinity functions for consideration.

Acknowledgment

Research sponsored by the U.S. Department of Energy under contract DE-AC05-96OR22464 with Lockheed Martin Energy Research Corp.

REFERENCES

[1] J. E. Baker. Adaptive learning of multi-sensor integration techniques with genetic algorithms. *Evolutionary Algorithms in Robotics II, Proceedings of the 5th International Symposium on Robotics and Manufacturing*, Aug. 1994, pp. 383–390.

[2] M. Jamshidi et al., eds. *Proceedings of ISRAM '96, the Sixth International Symposium on Robotics and Manufacturing*, Montpellier, France, TSI Press, Alburquerque, NM, May 1996.

[3] M. Jamshidi et al., eds. *Proceedings of WAC '94, the First World Automation Congress*, Maui, Hawaii, TSI Press, Albuquerque, NM, Aug. 1994.

[4] M. A. Abidi and R. C. Gonzalez, eds. *Data Fusion in Robotics and Machine Intelligence*. Academic Press, New York, 1992.

[5] J. M. Brady Foreword, Special Issue on Sensor Data Fusion, *Int. J. Robot. Res.* 7(5):2–4, 1988.

[6] I. Kadar and V. Libby, eds. *Proceedings of Signal Processing Sensor Fusion, and Target Recognition V*, SPIE Proc. No. 2755, June 1996.

[7] A. Kak and S. Chen, eds. *Spatial Reasoning and Multi-Sensor Fusion, Proceedings of the 1987 Workshop*, Morgan Kaufmann, San Mateo, CA, 1987.

[8] E. R. Dougherty. *An Introduction to Morphological Image Processing*, Vol. TT9. SPIE Optical Engineering Press, Bellingham, WA, 1992.

[9] C. R. Giardina and E. R. Dougherty. *Morphological Methods in Image and Signal Processing*. Prentice-Hall, Englewood Cliffs, NJ, 1988.

[10] R. C. Gonzaliz and P. Wintz. *Digital Image Processing*. Addison-Wesley, Reading, MA, 1983.

[11] R. Kasturi and R. C. Jain. *Computer Vision: Principles*, IEEE Computer Society Press, Los Alamitos, CA, 1991.

[12] M. Beckerman, D. L. Barnett, and S. M. Killough. Ultrasound and visual sensor feedback and fusion, in *Proceedings of the 1990 Japan–U.S.A. Symposium on Flexible Automation*, Kyoto, Japan, July 9–13, 1990, pp. 1315–1319.

[13] J. E. Baker. Image accuracy and representational enhancement through low-level, multi-sensor integration techniques, *Proceedings of Sensor Fusion and Aerospace Applications*, SPIE Proc. 1956, pp. 170–186, April 1993.

[14] J. P. Jones. Real-time construction of three-dimensional occupancy maps. In *Proceedings of the IEEE International Conference on Robotics and Automation*, Atlanta, May 1993, pp. 52–57.

[15] Datacube, Inc., 300 Rosewood Drive, Danvers, MA.

[16] Flock of Birds(TM): Six Degrees-of-Freedom Measurement Device, Ascension Technology Corporation, P.O. Box 527, Burlington, VT, 1996.

[17] *Proceedings of the 1996 Conference on Computer Vision and Pattern Recognition (CVPR '96)*, IEEE Cat. No. PRO7258, San Francisco, June 1996.

[18] *Proceedings of the International Conference on Image Processing (ICIP '96)*, IEEE Cat. No. 96CH35919, Lausanne, Switzerland, Sept. 1996.

[19] *Proceedings of the 13th International Conference on Pattern Recognition (ICPR '96)*, Vienna, Aug. 1996.

[20] J. E. Baker. An analysis of the effects of selection in genetic algorithms, Ph.D. dissertation, Vanderbilt University, Nashville, TN, May 1989.

[21] K. A. De Jong. Analysis of the behavior of a class of genetic adaptive systems, Ph.D. dissertation, University of Michigan, Ann Arbor, 1970.

[22] D. E. Goldberg. *Genetic Algorithms in Search, Optimization, and Machine Learning*. Addison-Wesley, Reading, MA, 1989.

[23] J. H. Holland. *Adaptation in Natural and Artificial Systems*. University of Michigan Press, Ann Arbor, 1975.

[24] L. J. Eshelman, ed. *Proceedings of the Sixth International Conference on Genetic Algorithms*. Morgan Kaufmann, San Francisco, 1995.

[25] S. Forrest, ed. *Proceedings of the Fifth International Conference on Genetic Algorithms*. Morgan Kaufmann, San Mateo, CA, 1993.

[26] J. J. Grefenstette. A User's Guide to GENESIS, Version 5.0. Direct distribution: gref@aic.nrl.navy.mil, Oct. 1990.

[27] J. J. Grefenstette. Optimization of control parameters for genetic algorithms. *IEEE Trans. Syst, Man Cybernet.* SMC-16(1):122–128, 1985.

[28] J. E. Baker. Reducing bias and inefficiency in the selection algorithm. *Proceedings 2nd International Conference on Genetic Algorithms*, Lawrence Erlbaum Associates, Hillsdale, NJ, July 1987, pp. 14–21.

[29] *3-D Laser Imaging System User's Guide*, Odetics Publ No. 8533031-PR, Odetics Inc., 1515 S. Manchester Avenue, Anaheim, CA, March 29, 1990.

[30] Center for Engineering Systems Advanced Research, Oak Ridge National Laboratory, Lockheed Martin Energy Research Corp., Oak Ridge, TN.

CHAPTER 12

Robotics with Perception and Action Nets

SUKHAN LEE

Jet Propulsion Laboratory, California Institute of Technology, Pasadena, California
and
Department of Computer Science, University of Southern California, Los Angeles, California

SOOKWANG RO

Department of Computer Science, University of Southern California, Los, Angeles, California

ABSTRACT

In this section, the problem of intelligent robotic system architecture, referred to here as perception-action Net (PAN), is presented. Connecting sensing and action in real time. PAN automatically synthesizes goal-oriented behaviors under uncertainties, errors, and faults, through task monitoring and replanning.

PAN is composed of the perception and action nets interconnected in closed loops. The perception net connects features of various levels of abstraction or logical sensors in hierarchy. The net is capable of self-calibrating itself by maintaining the consistency of logical sensors based on the forward propagation of sensor outputs and uncertainties as well as based on the backward propagation of errors from constraints. The action net consists of a hierarchy of state transition networks of multiresolution time scales. The net embeds all the feasible system behaviors in various levels of abstraction, such that the system can replan and control its behaviors toward the set goals under errors and faults.

A novel geometric method is presented as a unified framework for computing forward and backward propagations through which the net achieves the self-reduction of uncertainties and self-calibration of biases. The proposed method is applied to the self-calibration of the eye–hand system equipped for a JPL–NASA planetary rover. Simulations and experimental results are shown.

1 INTRODUCTION

Robotic systems aim at achieving intelligence in behavior and dexterity in motion through a real-time connection between sensing and action. Achieving such intelligence and dexterity often requires an integration of distributed sensors and actuators to provide a rich source of sensory and movement patterns that can be clustered into higher levels of concepts and actions. The key to successful integration may be a system architecture that supports computational requirements unique to robotics, including uncertainty management and adaptive error recovery through the interaction among such processes as feature transformation and abstraction, data and concept fusion [1–14], consistency maintenance among data and knowledge, and monitoring and replanning.

In spite of the fact that a decade of research and development in robotics has produced numerous theoretical and experimental results, robots are yet to acquire the level of intelligence and dexterity required for autonomous task execution in unstructured environments. Conventional approaches to building robotic systems without underlying computational principles of integrating sensing, knowledge, and action in real time seem to suffer from limitations in the task complexity they can handle. If robot intelligtence is measured in terms of a power-to-weight ratio, where the power is defined by the product of the complexity and execution speed of tasks and the weight is defined by the product of volume and cost associated with the required hardware and software, an order of magnitude improvement in the power-to-weight ratio seems necessary for the new generation of robotics. A robot's intelligence may be manifested by its extended autonomy. However, the extension should not simply be the result of aggregating additional functional units, which may cause a reduction of the power or power-to-weight ratio by increasing space and time complexity. It is necessary to develop a system architecture that supports extended autonomy without a decrease in the power or power-to-weight ratio. An architecture that embeds system knowledge as well as a general problem-solving paradigm in itself may be desirable.

Planetary science sampling robots should possess extended autonomy with the capabilities of uncertainty management, adaptation to new situations, and fault tolerance. To provide the robot with extended autonomy requires the integration of a high level of discrete event planning and low level of continuous time control in a hierarchy of multiresolution time scales. However, such integration should be done under the limitation of computational power and the requirement of real-time operation. Conventional architectures for intelligent robotic systems, such as the subsumption architecture [15] and Nasrem architecture [16], do not directly address the problem of reducing uncertainties as well as dealing with unexpected events and system faults. Furthermore, the efficacy and efficiency of integrating planning and control in multiresolution time scales are yet to be consolidated.

An architecture of intelligent robotic systems, referred to here as a perception-action net (PAN), is presented for planetary robotic sampling. While connecting sensing and action in real time. PAN automatically synthesizes goal-oriented behaviors or sequences of actions toward the set goals under uncertainties, errors, and faults, through task monitoring and replanning.

In this chapter, we present a method of system uncertainty management based on representing the overall system sensing capabilities by a perception net and propagating uncertainties and errors forward and backward through the net for consistency by a geometric algorithm. A geometric fusion method with a statistical basis can be found in the literature [17]. However, there are number of problems that statistical methods cannot handle, such as uncertainty propagation in feature transformation when nonlinearity is involved and treating system constraints for consistency. The perception net is capable of self-calibrating biases existing in the parameters of transformation modules while carrying out uncertainty minimization. This is important because the modeling of sensors as well as feature extraction algorithms are often subject to biases. These biases are not only from the errors at the time of initialization but also from the physical variations of a system and an environment during operation. The perception net automatically corrects such biases on the basis of the errors at the constraint modules that are exposed further and further as the uncertainties of the net states become smaller and smaller with iterations.

2 PAN ARCHITECTURE

PAN is composed of two major building blocks, the perception and action nets, interconnected in closed loops, as shown in Figure 12.1.

The perception net connects logical sensors or features of various levels of abstraction that can be identified by the given sensor system. In Figure 12.2, the logical sensors or features that can be extracted from the physical sensors, such as camera, proximity sensor, and tactile sensor, are organized in a hierarchy, where the logical and physical sensors are depicted, respectively, as rectangular and elliptical boxes. However, in the perception net, the connections between logical sensors are further elaborated with their relationships in terms of feature transformation, data fusion, and constraint to be satisfied. For instance, Figure 12.3 illustrates the perception net constructed from the logical sensor system of Figure 12.2 as follows: The surface orientation feature may be determined by the distance-to-surface logical sensor based on feature transformation. The same surface-orientation feature may be measured directly by the tactile sensor, so that the feature can be finalized by fusing the two sources of data, one from the distance-to-surface logical sensor and the other from the tactile sensor. By the same token, the hole-3D-position feature can be determined by fusing the tactile sensor output and the result of feature transformation from the hole-2D-position, surface-orientation, and distance-to-surface logical sensors. Furthermore, assuming two holes of the known relative distance, the two hole-3D-position features should be constrained by the known relative distance.

In general, the perception net is formed by the interconnection of logical and physical sensors with three types of modules: feature transformation module (FTM), data fusion

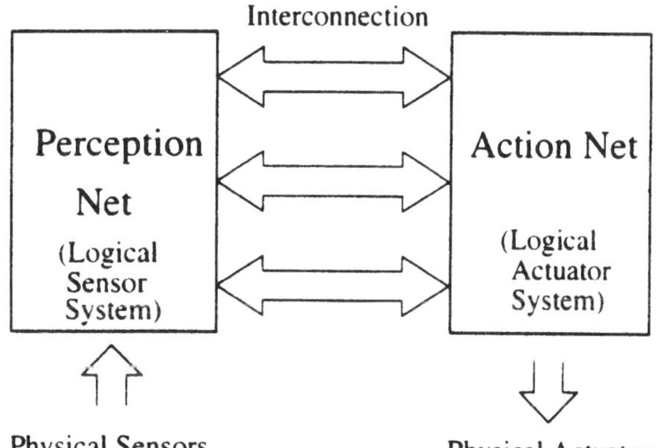

FIGURE 12.1
Two major building blocks of PAN: perception and action net.

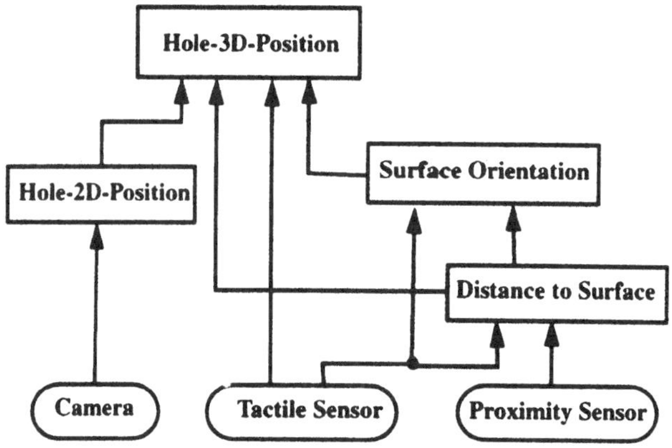

FIGURE 12.2
Schematic illustration of a logical sensor system.

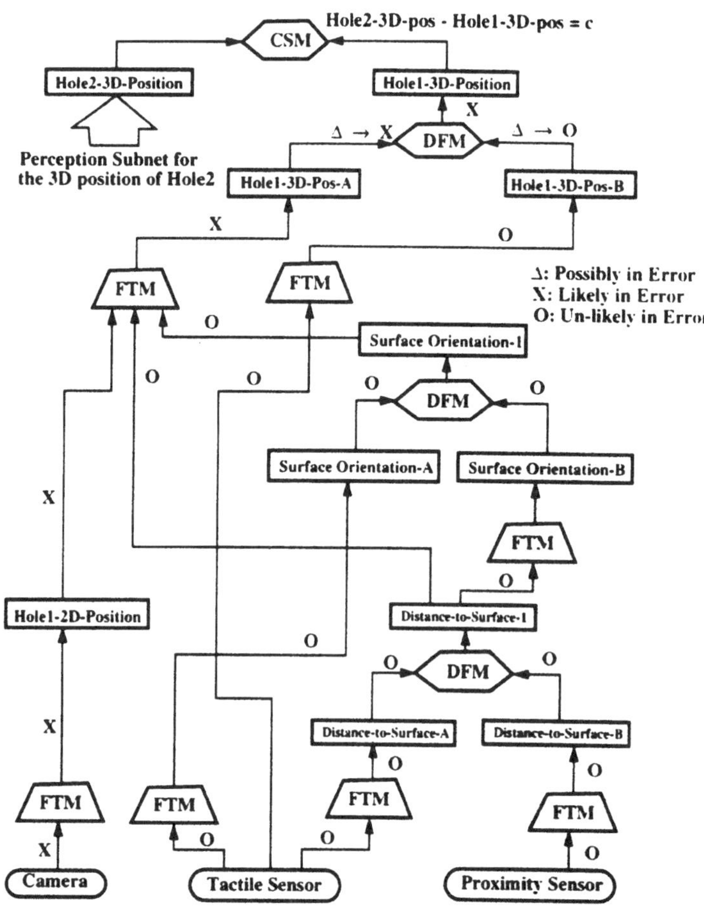

FIGURE 12.3
A perception net representation of a logical sensor system of Figure 12.2. FTM, feature transformation, module; DFM, data fusion module; CSM, constraint satisfaction module. **O**, **X**, and **Δ** indicate the identification of possible error sources by the net.

module (DFM), and constraint satisfaction module (CSM), as shown schematically in Figure 12.4. An FTM transforms a set of primitive features into a more abstract and higher level of feature. A DFM takes multiple data of a feature to generate an optimal estimate of the feature. A DFM may represent either spatial or temporal data fusion: for the spatial data fusion (s-DFM), data are from the single readings of multiple sensors, while for the temporal data fusion (t-DFM), data are from the multiple readings of a single sensor. Each DFM module is responsible for determining which input data are valid for fusion at the current sensor configuration. A CSM represents system knowledge that imposes a constraint upon a set of feature values.

The output of each logical sensor is a tuple representing the current estimates of corresponding feature value and its uncertainty measure and is regarded as the current state of the sensor. Then, the net state is defined as the collection of the states of individual logical sensors. The net is operated in such a way that a state change at a logical sensor propagates to adjacent logical sensors, triggering a chain of state changes throughout the net. For example, the state of a logical sensor can be updated by fusing its current state with a new reading from FTM through t-DFM, as depicted schematically in Figure 12.4.

Note that the propagation of state change is bidirectional, forward and backward, such that the net automatically updates, and maintains the consistency of, its state not only through the forward propagation of state change but also through the backward propagation of state errors to satisfy constraints. In Figure 12.4, the backward signal propagation is explicitly represented by feedback connections from CSMs to the corresponding modules.

Through the bidirectional state-updating process, the net provides not only reduction of uncertainties but also monitoring of errors and faults, based on which decision making and replanning take place in the action net. The perception net presents a formal yet general

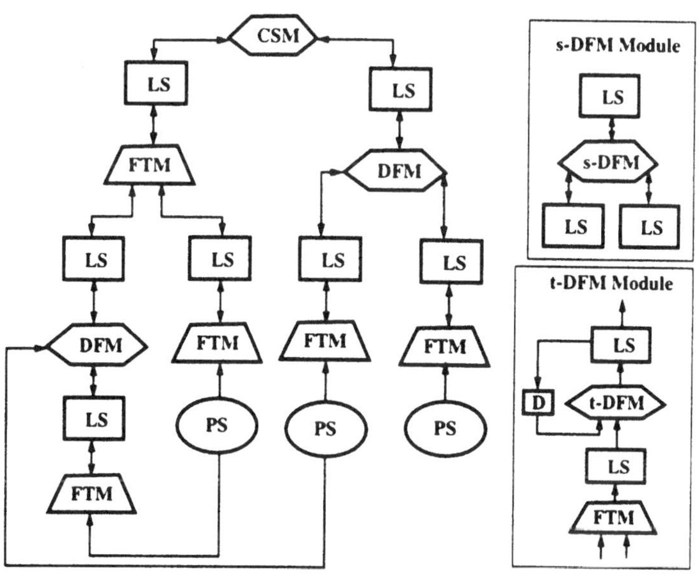

FIGURE 12.4
The architecture of perception net composed of three types of module: FTM, DFM, and CSM, connecting logical sensors (LS) as well as physical sensors (PS). Two types of DFM, t-DFM and s-DFM, are shown, where D implies delay.

architecture for sensor fusion and planning. That is, the net can also be used for curbing uncertainties based on active modification of sensing parameters through sensor planning.

The action net consists of a hierarchy of state transition networks of multiresolution time scales, as shown in Figure 12.8. More precisely, the net represents system dynamics in multiresolution time scales ranging from continuous time to discrete event dynamics, where an action of a higher level of hierarchy is represented by a state transition network of a lower level. The net embeds all the feasible system behaviors in various levels of abstraction. This allows the system to replan and control its behaviors efficiently toward the set goals through errors and faults with feedback to the various levels of action hierarchy.

The action net can be interpreted by analogy to linguistics. The system behaviors that can be generated by the action net are equivalent to the sentences that can be generated by the vocabularies and grammar of a language. Applying planning and control to the action net to generate a goal-oriented behavior for the given task is equivalent to searching for a sequence of grammatical rules to generate a sentence of particular semantics. In this sense, the action net is designed to embed all the feasible behaviors of the system from which a particular goal-oriented behavior can be searched for through planning and control.

In summary, PAN can be considered as a computational knowledge base in which concepts are understood by the system through their interconnections and computational dependences.

3 UNCERTAINTY MANAGEMENT

The management of uncertainties by the perception net consists of the self-reduction of uncertainties and the self-identification of possible biases. The uncertainties propagate in the perception net through the input–output relationships of FTM and DFM modules, as well as through the constraints defined by CSM modules.

3.1 Uncertainty Representation

The uncertainties of logical sensor outputs are due to the random noise and biases involved in measurement data as well as to the biases involved in modeling feature transformations.

Although Gaussian randomness of noise and independence of data measurements are assumed in sensing, the noise involved in a logical sensor output may not be Gaussian because of possible nonlinearity in feature transformation. For convenience, we assume that noise is bounded by an uncertainty hypervolume or hyperellipsoid and that the size of the uncertainty ellipsoid is small enough for a good linear approximation around the nominal point in feature transformation. Formally, we represent the uncertainty, $d\mathbf{x}$, of a logical sensor value, $\mathbf{x}$, as an ellipsoid of the following form:

$$d\mathbf{x}^t W_{\mathbf{x}}\, d\mathbf{x} \leqslant 1 \tag{12.1}$$

where $W_{\mathbf{x}}$ represents a symmetric weight matrix determining the size and shape of the ellipsoid.

3.2 Forward Propagation

Feature Transformation Module

The forward propagation of input data, $(\mathbf{x}_i, W_{\mathbf{x}_i})$, $i = 1, \ldots, m$, through FTM is straightforward once the input–output relationship of FTM is given: the output of FTM, $(\mathbf{y}, W_{\mathbf{y}})$, can be obtained directly from (12.2) and (12.6).

Let us first define the mapping relationship between the input vector, $\mathbf{x}$, and the output vector, $\mathbf{y}$ of an FTM by

$$\mathbf{y} = f(\mathbf{x}, \mathbf{p}) \tag{12.2}$$

where $\mathbf{p}$ represents a parameter vector associated with the module. Then the uncertainty propagation through (12.2) can be approximated as the first-order Jacobian relationship with the assumption that f is smooth and $d\mathbf{x}$ is small, as follows:

$$\mathbf{y} + d\mathbf{y} = f(\mathbf{x} + d\mathbf{x}, \mathbf{p}) \approx f(\mathbf{x}, \mathbf{p}) + \frac{\partial f}{\partial \mathbf{x}} \Delta \mathbf{x} \tag{12.3}$$

Therefore,

$$d\mathbf{y} \approx \frac{\partial f}{\partial \mathbf{x}} d\mathbf{x} = J(\mathbf{x}, \mathbf{p})\, d\mathbf{x} \tag{12.4}$$

where $J(\mathbf{x}, \mathbf{p})$ represents the Jacobian relationship between $d\mathbf{y}$ and $\mathbf{d}x$. The uncertainty of $\mathbf{x}$, represented as an ellipsoid of (12.1), can now be propagated to the uncertainty of $\mathbf{y}$, represented as an ellipsoid in terms of $d\mathbf{y}$, through (12.4). By substituting $d\mathbf{x} = J^{+}(\mathbf{x}, \mathbf{p})\, d\mathbf{y}$, obtained from (12.4), to (12.1), we have

$$d\mathbf{y}^t (J^{+})^t W_{\mathbf{x}} J^{+}\, d\mathbf{y} \leqslant 1 \tag{12.5}$$

where J^{+} represents the pseudo-inverse of J. Equation (12.5) can be rewritten as

$$d\mathbf{y}^t W_{\mathbf{y}}\, d\mathbf{y} \leqslant 1 \tag{12.6}$$

where the symmetric weight matrix, $W_{\mathbf{y}}$, is defined as $W_{\mathbf{y}} \equiv (J^{+})^t W_{\mathbf{x}} J^{+}$. Equation (12.6) is of the same form as (12.1). The forward propagation of uncertainties toward the modules of a higher level of hierarchy can be done with the properly defined weight matrices of their input vectors. In the case in which the input, $\mathbf{x}$, of a module is composed of two or more vectors, $\mathbf{x}_1$ and $\mathbf{x}_2$, with their respective weight matrices defined as $W_{\mathbf{x}_1}$ and $W_{\mathbf{x}_2}$, the weight matrix, $W_{\mathbf{x}}$, of $\mathbf{x}$ can be specified by combining the individual weight matrices $W_{\mathbf{x}_1}$ and $W_{\mathbf{x}_2}$ as

$$W_{\mathbf{x}} = \text{Diag}[W_{\mathbf{x}_1}, W_{\mathbf{x}_2}] \tag{12.7}$$

Note that $W_{\mathbf{y}}$ is a function of $\mathbf{x}$ and $\mathbf{p}$, since J is a function of $\mathbf{x}$ and $\mathbf{p}$.

Data Fusion Module

Because DFM can be represented by an input–output relationship, the forward propagation through DFM can also be done with (12.2) and (12.6). The input–output relationship of DFM can be derived from one of the existing data fusion methods [6]. However, we present here a new geometric method of data fusion to derive the input–output relationship and to propagate the uncertainty ellipsoid in a geometrical basis.

For simplicity, consider the two measurements, $\mathbf{x}_{1m}$ and $\mathbf{x}_{2m}$, defined respectively in the two measurement spaces, $\mathbf{x}_1$ and $\mathbf{x}_2$, where their uncertainty bounds are defined by the weight matrices, $W_{\mathbf{x}_{1m}}$ and $W_{\mathbf{x}_{2m}}$, respectively. The proposed geometric data fusion method starts with defining the augmented space, $\mathbf{z}$, $z = (\mathbf{x}_1, \mathbf{x}_2)^T$, such that the measurement data, $(\mathbf{x}_{1m}, W_{\mathbf{x}_{1m}})$ and $(\mathbf{x}_{2m}, W_{\mathbf{x}_{2m}})$, are represented in an augmented space as $(\mathbf{z}_m, W_{\mathbf{z}_m})$, where $\mathbf{z}_m = (\mathbf{x}_{1m}, \mathbf{x}_{2m})^T$ and $W_{\mathbf{z}_m} = \text{Diag}[W_{\mathbf{x}_{1m}}, W_{\mathbf{x}_{2m}}]$. Then the problem of fusing $(\mathbf{x}_{1m}, W_{\mathbf{x}_{1m}})$ and $(x_{2m}, W_{\mathbf{x}_{2m}})$ is equivalent to finding a point, $\mathbf{y}$, on the constraint manifold, $\mathbf{x}_1 - \mathbf{x}_2 = 0$, defined in the $\mathbf{z}$ space in such a way that the weighted distance between $\mathbf{y}$ and $\mathbf{z}_m$, or $\frac{1}{2}\|\mathbf{y} - \mathbf{z}_m\|^2_{W_{\mathbf{z}_m}}$ is minimum, as shown in Figure 12.5. The uncertainty propagation is a crucial factor for the system convergence and stability property. Overestimating or underestimating an uncertainty bound will cause instability of the system. The Jacobian relationship, as described in the preceding section, is used for FTM due to the nonlinearity of transformation. However, unlike the other modules, DFM has a special constraint, $\mathbf{x}_1 = \mathbf{x}_2$, which is a linear constraint and an exact projection of uncertainty bound onto the constraint, $\mathbf{x}_1 = \mathbf{x}_2$, can be obtained. Once we obtain $\mathbf{y}$ as a function of $(\mathbf{x}_{1m}, W_{\mathbf{x}_{1m}})$ and $(\mathbf{x}_{2m}, W_{\mathbf{x}_{2m}})$, $y = f(\mathbf{x}_{1m}, \mathbf{x}_{2m})$. $W_{\mathbf{y}}$ can be derived based on the geometric uncertainty propagation method as shown in Figure 12.5.

More specifically, the output, $\mathbf{y}$, of DFM with $\mathbf{x}_{1m}$ and $\mathbf{x}_{2m}$ as its inputs can be determined as the vector that minimizes $\Sigma \frac{1}{2}\|\mathbf{y} - \mathbf{x}_{im}\|^2_{W_{\mathbf{x}_{im}}}$:

$$\mathbf{y} = (W_{\mathbf{x}_{1m}} + W_{\mathbf{x}_{2m}})^{-1}(W_{\mathbf{x}_{1m}}\mathbf{x}_{1m} + W_{\mathbf{x}_{2m}}\mathbf{x}_{2m}) \tag{12.8}$$

In the case of multiple inputs, it can be expressed as

$$\mathbf{y} = \left(\sum_i W_{\mathbf{x}_{im}}\right)^{-1}\left(\sum_i W_{\mathbf{x}_{im}}\mathbf{x}_{im}\right) \tag{12.9}$$

Then, as shown in Figure 12.5, the uncertainty bound, $W_{\mathbf{y}}$, associated with $\mathbf{y}$ can be obtained by projecting the uncertainty ellipsoid of $\mathbf{z}_m$ onto subspace $\mathbf{x}_1 = \mathbf{x}_2$.

$$W_{\mathbf{y}} = W_{\mathbf{x}_{1m}} + W_{\mathbf{x}_{2m}} \tag{12.10}$$

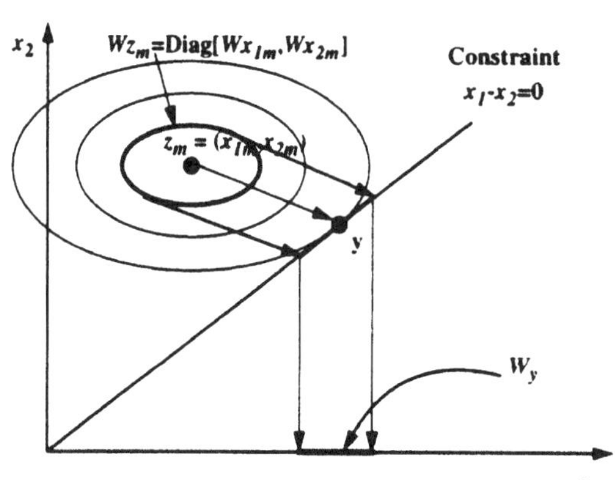

FIGURE 12.5
Proposed geometric data fusion method. The $\mathbf{y}$ is obtained as a function of $(\mathbf{x}_{1m}, W_{\mathbf{x}_{1m}})$ and $(\mathbf{x}_{2m}, W_{\mathbf{x}_{2m}})$, such that the weighted distance between $\mathbf{y}$ and $\mathbf{z}_m$ is minimum. The propagated uncertainty ellipsoid is obtained by the projection of the uncertainty ellipsoid of $\mathbf{z}_m$ onto the subspace, $\mathbf{x}_1 = \mathbf{x}_2$, along the direction of $\mathbf{y} - \mathbf{z}_m$.

In the case of multiple inputs, it can be expressed as

$$W_{\mathbf{y}} = \sum_i W_{\mathbf{x}_{im}} \tag{12.11}$$

Refer to the Appendix for the proof of (12.10) and (12.11). Note that (12.8) is the same result as the maximum Bayesian posteriori probability with the Gaussian assumption of noise.

3.3 Backward Propagation

Constraint Satisfaction Module

The backward propagation process starts with a CSM. For instance, consider that the two logical sensor outputs $\mathbf{x}$ and $\mathbf{y}$ are constrained by $f(\mathbf{x}, \mathbf{y}) = c$. CSM evaluates whether the current estimates, x^f and $\mathbf{y}^f$, of $\mathbf{x}$ and $\mathbf{y}$ from the forward process satisfy the given constraint. If not, CSM updates $(\mathbf{x}^f, W_{\mathbf{x}^f})$ and $(\mathbf{y}^f, W_{\mathbf{y}^f})$, where $W_{\mathbf{x}^f}$ and $W_{\mathbf{y}^f}$ are the weight matrices associated with $\mathbf{x}^f$ and $\mathbf{y}^f$, respectively, into $(\mathbf{x}^b, W_{\mathbf{x}^b})$ and $(\mathbf{y}^b, W_{\mathbf{y}^b})$ in such a way that $\mathbf{x}^b$ and $\mathbf{y}^b$ satisfy the constraint.

More specifically, let us first define the augmented space $\mathbf{z}$ with $\mathbf{x}$ and $\mathbf{y}$, $\mathbf{z} = (\mathbf{x}, \mathbf{y})^T$. Then the constraint of CSM can be represented as a manifold in the augmented space. Furthermore, $(\mathbf{x}^f, W_{\mathbf{x}^f})$ and $(\mathbf{y}^f, W_{\mathbf{y}^f})$ can be represented in the augmented space as $(\mathbf{z}^f, W_{\mathbf{z}^f})$, with $\mathbf{z}^f = (\mathbf{x}^f, \mathbf{y}^f)^T$ and $W_{\mathbf{z}^f} = \mathrm{Diag}[W_{\mathbf{x}^f}, W_{\mathbf{y}^f}]$. Finally, by selecting a vector, $\mathbf{z}^b, \mathbf{z}^b = (\mathbf{x}^b, \mathbf{y}^b)^T$, on the constraint manifold, $f(\mathbf{x}, \mathbf{y}) = c$, in such a way that the weighted distance from $\mathbf{z}^b$ to $\mathbf{z}^f$, $\frac{1}{2}\|\mathbf{z}^b - \mathbf{z}^f\|^2_{W_{\mathbf{z}^f}}$, is minimum. Then $\mathbf{x}^b$ and $\mathbf{y}^b$ can be obtained by solving the corresponding minimization problems in $\mathbf{x}^b$ and $\mathbf{y}^b$ space by applying Lagrange multipliers. By using the Lagrange multiplier, λ, transform the problem to

$$E = \tfrac{1}{2}[\|\mathbf{x}^f - \mathbf{x}^b\|^2_{W_{\mathbf{x}^f}} + \|\mathbf{y}^f - \mathbf{y}^b\|^2_{W_{\mathbf{y}^f}}] + \lambda^T(f(\mathbf{x}^b, \mathbf{y}^b) - c) \tag{12.12}$$

Then we have the following conditions for $\mathbf{x}^b$ and $\mathbf{y}^b$ to be satisfied:

$$\begin{aligned} \frac{\partial E}{\partial \mathbf{x}^b} &= 0: W_{\mathbf{x}^f}(\mathbf{x}^f - \mathbf{x}^b) - \left(\frac{\partial f}{\partial \mathbf{x}^b}\right)^T \lambda = 0 \\ \frac{\partial E}{\partial \mathbf{y}^b} &= 0: W_{\mathbf{y}^f}(\mathbf{y}^f - \mathbf{y}^b) - \left(\frac{\partial f}{\partial \mathbf{y}^b}\right)^T \lambda = 0 \\ \frac{\partial E}{\partial \lambda} &= 0: f(\mathbf{x}^b, \mathbf{y}^b) - c = 0 \end{aligned} \tag{12.13}$$

From Eq. (12.13), $\mathbf{x}^b$ and $\mathbf{y}^b$ can be obtained by applying various search methods, such as the Newton–Raphson method. Lyapunov's method, and the recursive least-squares method, depending on the form of Eq. (12.13). Then the uncertainty bound, $W_{\mathbf{x}^b}$ or $W_{\mathbf{y}^b}$, associated with $\mathbf{x}^b$ or $\mathbf{y}^b$, respectively, can be obtained by projecting $W_{\mathbf{z}^f}$ onto the constraint manifold. To project the uncertainty ellipsoid onto the constraint manifold, linearization around $\mathbf{z}^b$ is needed and the resulting uncertainty bound $W_{\mathbf{z}^b}$ is approximated by linearization with assumption that the uncertainty ellipsoid is small enough for a good linear approximation around the point, $\mathbf{z}^b$, on the constraint manifold as shown in Figure 12.6.

$$W_{\mathbf{x}^b} = W_{\mathbf{x}^f} + (W_{\mathbf{x}^f}(\mathbf{y}^b - \mathbf{y}^f)^{T+}(\mathbf{x}^b - \mathbf{x}^f)^T W_{\mathbf{y}^f})W_{\mathbf{y}^f} \cdot (W_{\mathbf{x}^f}(\mathbf{y}^b - \mathbf{y}^f)^{T+}(\mathbf{x}^b - \mathbf{x}^f)^T W_{\mathbf{y}^f})^T \tag{12.14}$$

$$W_{\mathbf{y}^b} = W_{\mathbf{y}^f} + (W_{\mathbf{y}^f}(\mathbf{x}^b - \mathbf{x}^f)^{T+}(\mathbf{y}^b - \mathbf{y}^f)^T W_{\mathbf{x}^f})W_{\mathbf{x}^f} \cdot (W_{\mathbf{y}^f}(\mathbf{x}^b - \mathbf{x}^f)^{T+}(\mathbf{y}^b - \mathbf{y}^f)^T W_{\mathbf{x}^f})^T \tag{12.15}$$

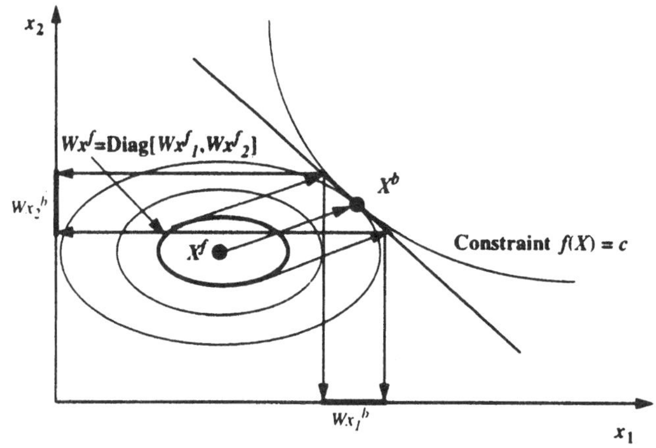

FIGURE 12.6
Proposed geometric method to determine $\mathbf{x}^b$ as a point on the constraint manifold $f(\mathbf{x}) = c$ that minimizes the weighted distance from $\mathbf{x}^f$ to the constraint manifold. The uncertainty ellipsoid is projected onto the constraint manifold along the direction of $\mathbf{x}^b - \mathbf{x}^f$.

where $*^{T+}$ represents the pseudoinverse and transpose of $*$. The proof of (12.14) and (12.15) is very complicated, but it is very similar to the DFM case and not shown here. Refer to the Appendix for the proof of the DFM case.

In general, the constraint of the CSM can be expressed as $f(\mathbf{x}) = c$ with $\mathbf{x} = (\mathbf{x}_1, \mathbf{x}_2, \ldots, \mathbf{x}_m)^T$ and $W_\mathbf{x} = \text{Diag}[W_{\mathbf{x}_1}, W_{\mathbf{x}_2}, \ldots, W_{\mathbf{x}_m}]$. Let the input of a CSM from the forward propagation be $(\mathbf{x}_1^f, W_{\mathbf{x}_1^f}), (\mathbf{x}_2^f, W_{\mathbf{x}_2^f}), \ldots, (\mathbf{x}_m^f, W_{\mathbf{x}_m^f})$, such that $\mathbf{x}^f \triangleq (\mathbf{x}_1, \ldots, \mathbf{x}_m)^f$ and $W_{\mathbf{x}^f} \triangleq \text{Diag}[W_{\mathbf{x}_1^f}, \ldots, W_{\mathbf{x}_m^f}]$.

Then backward propagation at CSM modifies $\mathbf{x}^f$ and $W_{\mathbf{x}^f}$ into $\mathbf{x}^b$ and $W_{\mathbf{x}^b}$ in such a way that $\mathbf{x}^b$ and $W_{\mathbf{x}^b}$ are consistent with the constraint. The proposed geometric method determines $\mathbf{x}^b$ as the point on the constraint manifold, $f(\mathbf{x}) = c$, that minimizes the weighted distance from $\mathbf{x}^f$ to the manifold,

$$\mathbf{x}^b\colon \min_{x^b} \frac{1}{2} \|\mathbf{x}^f - \mathbf{x}^b\|^2_{W_{\mathbf{x}^f}} \text{ subject to } f(\mathbf{x}^b) = c \tag{12.16}$$

Once $\mathbf{x}^b$ is obtained from (12.16) as a function of $\mathbf{x}^f$, $\mathbf{x}^b = g(\mathbf{x}^f)$, then $W_{\mathbf{x}^b}$ can be obtained by projecting $W_{\mathbf{x}^f}$ onto the constraint manifold.

Data Fusion Module

Backward propagation in DFM is straightforward. Unlike other modules, such as CSM and FTM, the constraint of DFM is $\mathbf{x}_1 = \mathbf{x}_2$. Therefore inputs of DFM, $\mathbf{x}_1^b$, and $\mathbf{x}_2^b$, must be same as the output of DFM, $\mathbf{y}^b$, to keep consistency of the network. Also, inputs of DFM must carry the same uncertainty ellipsoid as the output of DFM in backward propagation.

$$\mathbf{y}^b = \mathbf{x}_1^b = \mathbf{x}_2^b \tag{12.17}$$

$$W_{\mathbf{y}^b} = W_{\mathbf{x}_1^b} = W_{\mathbf{x}_2^b} \tag{12.18}$$

Feature Transformation Module

In the backward propagation, the updated logical sensor output plays the role of a constraint for the subsequent module. For instance, assume that the output, $\mathbf{y}^f$, of an FTM, $f(\mathbf{x}^f) = \mathbf{y}^f$, is updated by the backward propagation as $(\mathbf{y}^b, W_{\mathbf{x}^b})$. Then $\mathbf{x}^f$ needs to be updated to $\mathbf{x}^b$ such that $f(\mathbf{x}^b) = \mathbf{y}^b$. Therefore, it is equivalent to say that $\mathbf{x}^f$ is input to a CSM, $f(\mathbf{x}) = c$, with $c = \mathbf{y}^b$.

However, unlike the previous case of CSM where c has no uncertainty, $\mathbf{y}^b$ has uncertainty represented by $W_{\mathbf{y}^b}$. In this case, $\mathbf{x}^b$ is computed in the same way as (12.16) but with $c = \mathbf{y}^b$. But the computation of $W_{\mathbf{x}^b}$ needs to consider the uncertainty of $\mathbf{y}^b$, $W_{\mathbf{y}^b}$, as follows: $\mathbf{x}^b$ obtained by (12.16) can be represented as $\mathbf{x}^b = g(\mathbf{x}^f, \mathbf{y}^b)$. Then, by the same procedure as for CSM, $W'_{\mathbf{x}^b}$ can be obtained as an initial uncertainty bound. According to the uncertainty of $\mathbf{y}^b$, $W'_{\mathbf{x}^b}$ is swept along the $\mathbf{y}^b$ variation according to its uncertainty boundary and then the uncertainty boundary, $W_{\mathbf{x}^b}$, can be generated. $W_{\mathbf{x}^b}$ is now approximated with an ellipsoid that circumscribes the swept region.

3.4 Sensor Planning

It has been shown that the uncertainties of logical sensor outputs can be reduced by the forward and backward propagation of errors through DFM and CSM. However, the rate of the uncertainty reduction based on DFM and CSM tends to reach saturation rather quickly as fusion cycles increase. This may cause, in some cases, an excessive number of fusion cycles in order to reduce the uncertainty bound of a logical sensor to the level required by the action net for decision making. One way to solve this problem is to change the controllable sensing parameters (including sensor positions) during fusion cycles in such a way that the uncertainty in sensing is maximally reduced.

In general, the objective of sensor planning is to change sensor configurations or parameters dynamically in such a way that the system achieves sensing goals, measuring necessary features with desired accuracy, at the minimum sensing cost. Sensor planning has been a subject of interest in robotics and vision [18–24]. Sensor planning offers extended sensing capability with enhanced accuracy and efficiency. However, most work on sensor planning has been based on *ad hoc* methods customized to particular problems. For sensor planning with an integrated sensor system, more formality and generality in problem representation and solution methodology are required. Sensor planning here focuses on reducing the uncertainties of logical sensor outputs to the desired level by controlling system parameters subject to system and parameter constraints at the minimum sensing cost.

Representation of Sensing Goals

The desired accuracy of a logical sensor output can be represented by the desired ellipsoidal bound of uncertainty associated with the output. For example, the desired uncertainty ellipsoid, $E_d(\mathbf{y})$, of an output vector, $\mathbf{y}$, can be represented as $d\mathbf{y}^t W_{\mathbf{y}d} d\mathbf{y} \leqslant 1$, with $W_{\mathbf{y}d}$ being the desired weight matrix. Then a closeness measure between the desired and actual uncertainty ellipsoids of the output vector can be defined as follows: Assume that the desired uncertainty ellipsoid has principal radius vectors of $(1/\sqrt{\lambda_1})u_1$ and $(1/\sqrt{\lambda_2})u_2$, as shown in Figure 12.7.

u_1 and u_2 are the orthonormal singular vectors of W_d, whereas λ_1 and λ_2 are the corresponding eigenvalues. Then we define $\alpha_1 u_1$ and $\alpha_2 u_2$ to represent the vectors from the origin of the actual error ellipsoid to its boundary in the direction defined by u_1 and u_2,

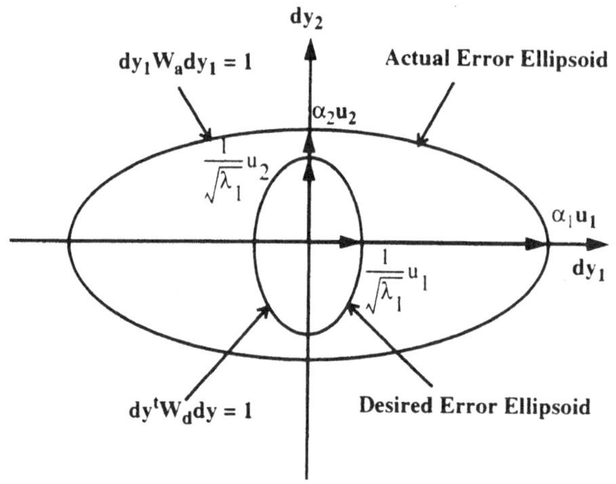

FIGURE 12.7
An example of desired and actual error ellipsoids for the representation of network performance.

respectively:

$$\alpha_i^2 u_i^t W_a u_i = 1, \quad \text{for } i = 1, 2 \tag{12.19}$$

where W_a represents the weight matrix of the actual uncertainty ellipsoid, which is a function of controllable parameter vector **p**. Note that α_i, $i = 1, 2$, can be computed directly from (12.19). Then the closeness of two ellipsoids can be defined by comparing between α_i and $(1/\sqrt{\lambda_i})$ for $i = 1, 2$, such that the sensing performance, defined as the closeness between the desired and actual ellipsoids, can be expressed as

$$\frac{1}{2}\sum_k \sum_i (\alpha_{ik} - 1/\sqrt{\lambda_{ik}})^2 \tag{12.20}$$

where k represents the kth output.

Although the example shown in Figure 12.7 is for a two-dimensional (2-D) case, the method described here is general and can be applied to any dimensional space.

Parametric Sensor Planning

Parametric sensor planning implies the determination of controllable parameters, **p**, that maximize sensing performance, (12.20), under system and parameter constraints at the minimum sensing cost.

We propose an iterative parameter update method for the parametric sensor planning. In the iterative parameter update method, the parameter vector, **p**, is updated iteratively by

$$p(t) = p(0) + \int_0^t \dot{\mathbf{p}}\, d\tau \tag{12.21}$$

for a continuous-time system, or

$$p(k+1) = p(k) + dp(k) \tag{12.22}$$

for a discrete-time system, until an equilibrium is reached when $\dot{\mathbf{p}}$ or $d\mathbf{p}$ is 0.

The parameter update, $\dot{\mathbf{p}}$ or $d\mathbf{p}$, can be generated in the parameter space as the direction that maximally satisfies the following criteria ordered in terms of their priorities:

1. The satisfaction of system and parameter constraints
2. The minimization of network performance measure
3. The minimal sensing cost
4. The avoidance of local minima

To determine the parameter update based on these criteria, let us first derive the three directional vectors, $\dot{\mathbf{p}}_1$, $\dot{\mathbf{p}}_2$, and $\dot{\mathbf{p}}_3$, from each of the top three criteria that can be combined into the final parameter update $\dot{\mathbf{p}}$. The directional vector, $\dot{p}_1$, changes p in a direction that leads the network to satisfy existing system and parameter constraints. To determine $\dot{\mathbf{p}}_1$, we define the following Lyapunov function candidate, V_1:

$$V_1 = \sum_{k=1}^{2} V_{1k} \tag{12.23}$$

where

$$V_{11} = \frac{1}{2} \sum_{i=1}^{l_1} h_{1i}^2(\mathbf{x}, \mathbf{p}) \tag{12.24}$$

$$V_{12} = \frac{1}{2} \sum_{i=1}^{l_2} S^2[h_{2i}(\mathbf{x}, \mathbf{p})] \tag{12.25}$$

V_{11} is defined from the equality constraints of the system and parameter constraints such that $h_{1i}(\mathbf{x}, \mathbf{p})$ represents the ith equality constraint, $h_{1i}(\mathbf{x}, p) = 0$. On the other hand, V_{12} is defined from the inequality constraints of the system and parameter constraints such that $h_{2i}(\mathbf{x}, \mathbf{p})$ represents the ith inequality constraint, $h_{2i}(\mathbf{x}, \mathbf{p}) > 0$, where

$$S[h_{2i}(\mathbf{x}, \mathbf{p})] = \begin{cases} h_{2i}(\mathbf{x}, \mathbf{p}), & \text{if } h_{2i}(\mathbf{x}, \mathbf{p}) < 0 \\ 0, & \text{if } h_{2i}(\mathbf{x}, \mathbf{p}) \geqslant 0 \end{cases} \tag{12.26}$$

From (12.23), (12.24), and (12.25), we cn see that $V_1 = 0$ only when both the equality and inequality constraints are satisfied, that is, only when the system and parameter constraints are completely satisfied. Based on (12.23), (12.24), and (12.25), we can derive the parameter update, $\dot{\mathbf{p}}_1$, that makes the time derivative of V_1 negative, $V_1 < 0$, when $V_1 \neq 0$, as follows:

$$\dot{\mathbf{p}}_1 = -\sum_{k=1}^{2} \frac{dV_{1k}}{d\mathbf{p}} \Bigg/ \left\| \frac{dV_{1k}}{d\mathbf{p}} \right\| \tag{12.27}$$

where

$$\frac{dV_{11}}{d\mathbf{p}} = \sum_{i=1}^{l_1} h_{1i}(\mathbf{x}, \mathbf{p}) \left(\frac{\partial}{\partial \mathbf{x}} h_{1i}(\mathbf{x}, \mathbf{p}) \frac{\partial \mathbf{x}}{\partial \mathbf{p}} + \frac{\partial}{\partial \mathbf{p}} h_{1i}(\mathbf{x}, \mathbf{p}) \right) \tag{12.28}$$

and

$$\frac{dV_{12}}{d\mathbf{p}} = \sum_{i=1}^{l_2} S[h_{2i}(\mathbf{x}, \mathbf{p})] \left(\frac{\partial}{\partial \mathbf{x}} S[h_{1i}(\mathbf{x}, \mathbf{p})] \frac{\partial \mathbf{x}}{\partial \mathbf{p}} + \frac{\partial}{\partial p} S[h_{1i}(\mathbf{x}, \mathbf{p})] \right) \tag{12.29}$$

Equation (12.27) guarantees that only when $V_1 = 0$, that is, when both the equality and inequality constraints are satisfied. The term $\left\|\frac{dV_{1k}}{d\mathbf{p}}\right\|$ used in the denominator of (12.27) allows the value of $\dot{p}_1$ to be exploded in the vicinity of local minima, which helps to avoid local minima.

In the same way as we derived $\dot{\mathbf{p}}_1$, we can define a directional vector, $\dot{\mathbf{p}}_2$, that changes $\mathbf{p}$ in the direction to minimize the closeness measure, V_2, defined by

$$V_2 = \frac{1}{2}\sum_k \sum_i (\alpha_{ik} - 1/\sqrt{\lambda_{ik}})^2 \tag{12.30}$$

where α_{ik} is a function of $\mathbf{x}$ and $\mathbf{p}$. Then, from the condition that keeps the time derivative of V_2 negative when $V_2 \neq 0$, we can determine $\dot{\mathbf{p}}_2$ as follows:

$$\dot{\mathbf{p}}_2 = -\sum_k \sum_i (\alpha_{ik} - 1/\sqrt{\lambda_{ik}}\frac{d\alpha_{ik}}{d\mathbf{p}} \Big/ \left\|\frac{d\alpha_i}{d\mathbf{p}}\right\| \tag{12.31}$$

where

$$\frac{d\alpha_{ik}}{d\mathbf{p}} = \frac{\partial \alpha_{ik}}{\partial \mathbf{x}} \cdot \frac{d\mathbf{x}}{d\mathbf{p}} + \frac{\partial \alpha_{ik}}{\partial \mathbf{p}} \tag{12.32}$$

The directional vector, $\dot{\mathbf{p}}_3$, that incurs the minimum sensing cost can be searched by

$$\dot{\mathbf{p}}_3 = \min_{\dot{\mathbf{p}}} C(\dot{\mathbf{p}}) \tag{12.33}$$

where $C(\dot{\mathbf{p}})$ represents sensing cost in terms of the parameter change, $\dot{\mathbf{p}}$, at the current sensor configuration.

With $\dot{\mathbf{p}}_1$, $\dot{\mathbf{p}}_2$, and $\dot{\mathbf{p}}_3$, defined, respectively, for the constraint satisfaction, the performance optimization, and the minimum sensing cost, the final parameter update, $\dot{p}$, can be determined by combining $\dot{\mathbf{p}}_1$, $\dot{\mathbf{p}}_2$, and $\dot{\mathbf{p}}_3$, as follows:

$$\dot{\mathbf{p}} = \dot{\mathbf{p}}_1 + N(\dot{\mathbf{p}}_1)\dot{\mathbf{p}}_2 + N(\dot{\mathbf{p}}_1 + N(\dot{\mathbf{p}}_1)\dot{\mathbf{p}}_2)\dot{\mathbf{p}}_3 \tag{12.34}$$

where $N(\dot{\mathbf{p}}_1)\dot{\mathbf{p}}_2$ implies the projection of $\dot{\mathbf{p}}_2$ onto the null space of $\dot{\mathbf{p}}_1$. Note that, when the system and parameter constraints are satisfied, then $\dot{\mathbf{p}}_1 = 0$, consequently, $N(\dot{\mathbf{p}}_1)\dot{\mathbf{p}}_2 = \dot{\mathbf{p}}_2$. Equation (12.34) is referred to here as the iterative parametric sensor planning equation. Note that the priority given to $\dot{\mathbf{p}}_1$, $\dot{\mathbf{p}}_2$, and $\dot{\mathbf{p}}_3$ for the computation of $\dot{\mathbf{p}}$ can be changed, that is; a different order of $\dot{\mathbf{p}}_1$, $\dot{\mathbf{p}}_2$, and $\dot{\mathbf{p}}_3$, can be used in (12.34) for computing $\dot{\mathbf{p}}$.

4 ERROR MONITORING AND RECOVERY

DFM and CSM of the perception net make it possible to monitor errors, that is, biases and faults in sensing and action. Upon the identification of biases and faults in sensing and action, the action net invokes error recovery and repairment actions. In PAN, error monitoring and recovery consist of (1) error detection, (2) error identification, and (3) sensor calibration and action replanning for error recovery and repairment.

4.1 Error Detection

Errors may be detected from the following information:

1. The discrepancy between the planned and measured states
2. The inconsistency among the input data of DFM
3. The violation of constraints at CSM

To clarify the meaning of discrepancy, inconsistency, and violation, we need to define quantitatively the thresholds that separate the effect of biases and faults from that of uncertainties.

The inconsistency among the input data of DFM can be evaluated on the basis of the ellipsoidal representation of uncertainty bounds: The input data of DFM are said to be inconsistent if the ellipsoidal uncertainty bounds of input data have no common intersection. The existence of a common intersection among the input data, $(\mathbf{x}_1, W_1)$, $(\mathbf{x}_2, W_2), \ldots, (\mathbf{x}_n, W_n)$, where W_i are the weight matrices associated with $\mathbf{x}_i$ for $i = 1, \ldots, n$, can be evaluated by the following rule:

> If $\|\mathbf{x} - \mathbf{x}_i\|_{W_i}\| \leqslant 1$, for $i = 1, \ldots, n$ with $\mathbf{x} = (\Sigma W_i)^{-1} \Sigma W_i \mathbf{x}_i$, then there exists a common intersection among the ellipsoidal uncertainty bounds of $\mathbf{x}_i$, $i = 1, \ldots, n$.

Similarly, the violation of constraint at CSM can be detected by checking whether the vector, $\mathbf{z}_b$, $\mathbf{z}_b = (\mathbf{x}_b, \mathbf{y}_b)^T$, on the constraint manifold that minimizes the weighted distance from $\mathbf{z}_b$ to $\mathbf{z}_f$. $\|\mathbf{z}_b - \mathbf{z}_f\|_{W_{\mathbf{z}_f}}$, is inside the uncertainty ellipsoid of $\mathbf{z}_f$:

> If $\|\mathbf{z}_b - \mathbf{z}_f\|_{W_{\mathbf{z}_f}} \leqslant 1$, then $(\mathbf{x}_f, \mathbf{y}_f)$ does not violate the constraint where $\mathbf{z}_f = (\mathbf{x}_f, \mathbf{y}_f)$

The discrepancy between the planned and measured states can also be detected by checking whether there is a common intersection between the uncertainty ellipsoids of planned and measure states, where the uncertainty ellipsoid of the planned state can be determined by the expected uncertainty involved in plan execution.

4.2 Error Identification

Upon the detection of errors, it is necessary to identify the source of errors. When more than two input data are involved in DFM, we can check which input data are isolated from the rest in terms of sharing a common intersection. In general, for a DFM with multiple input data, it is possible to identify groups of input data that share a common intersection (based on the method presented before this). The input data that belong to the group of single or small number of members may be considered as a likely source of error.

When error is detected in the input data, $(\mathbf{x}_f, W_{\mathbf{x}_f})$ and $(\mathbf{y}_f, W_{\mathbf{y}_f})$, of CSM, we can check whether $\mathbf{x}_b$ and $\mathbf{y}_b$ are inside the uncertainty ellipsoids of $(\mathbf{x}_f, W_{\mathbf{x}_f})$ and $(\mathbf{y}_f, W_{\mathbf{y}_f})$, respectively. That is, if $\|\mathbf{x}_b - \mathbf{x}_f\|_{W_{\mathbf{x}_f}} > 1$ or if $\|\mathbf{y}_b - \mathbf{y}_f\|_{W_{\mathbf{y}_f}} > 1$, then $\mathbf{x}_f$ or $\mathbf{y}_f$ may be a likely source of error.

Further isolation of error sources can be done through the net hierarchy. By applying the preceding error detection method to DFMs and CSMs distributed in the net, the logical sensors associated with DFMs and CSMs can be classified as either likely-in-error, unlikely-in-error, or possibly-in-error. Then these classifications are propagated through the net to extend the classifications to other logical sensors connected through the hierarchy. The cross-checking of these classifications propagated through the net hierarchy provides further isolation of errors, as shown in Figure 12.3. For instance, assume that the DFM for Hole-3D-Pos detects error due to the inconsistency of its inputs, such that Hole-3D-Pos-A and Hole-3D-Pos-B are marked as δ to indicate that they are in possibly in error. Assume also that the DFMs for Surface-Orientation-1 and Distance-to-Surface-1 detect no inconsistency, such that their inputs, Surface-Orientation-A and Surface-Orientation-B, as well as Distance-to-Surface-A and Distance-to-Surface-B, are marked as **O** indicating that they are in unlikely-in-error. The propagation of **O** markings backward results in the tactile and proximity sensors being in unlikely-in-error (assuming that FTMs are not in error here). The fact that the tactile sensor is not likely in error propagates forward to Hole-3D-Pos-B that Hole-3D-Pos-B is in unlikely-in-error, which forces Hole-3D-Pos-A to be reclassified as likely-in-error. The likely-in-error status of Hole-3D-Pos-A propagates backward to indicate that the camera is in likely-in-error. We can extend the propagation and cross-checking of classifications to the action net, because the discrepancy between the planned and measured states provides additional error detection. If the preceding DFM- and CSM-based error identification method fails to isolate error sources, sensor planning or error-isolation action should take place in the action net for complete isolation of errors.

4.3 Error Recovery

Once error sources are isolated, the system must take actions to repair the errors and to recover from the errors. Two types of actions can take place: (1) calibration of sensors to eliminate biases and (2) replanning the actions to reach the desired goal state under errors. For the first, a predefined sensor calibration routine for the sensor in error will be invoked by the action net. For the second, the action net replans the task based on the PAN modified according to the isolated errors.

5 PLANETARY ROBOTIC SCIENCE SAMPLING

We consider autonomous soil and rock science for planetary robotic science sampling.

5.1 Autonomous Soil Science

Autonomous soil science includes the trenching of planetary soil by a robot to collect and analyze subsurface soil samples. Autonomous soil science is composed of the following activities: trenching site designation by scientists, visual and tactile verification of trenching site by a robot, planning of optimal trenching trajectories with measured soil property, adaptive trenching, planning of optimal scooping trajectory, and adaptive scooping. Undoubtedly, uncertainty management and error monitoring and recovery play an important role.

The PAN architecture for soil science is illustrated in Figures 12.8–12.13.

5.2 Autonomous Rock Science

Autonomous rock science includes

- Rock designation by scientists interactively with the system
- Rock identification, localization

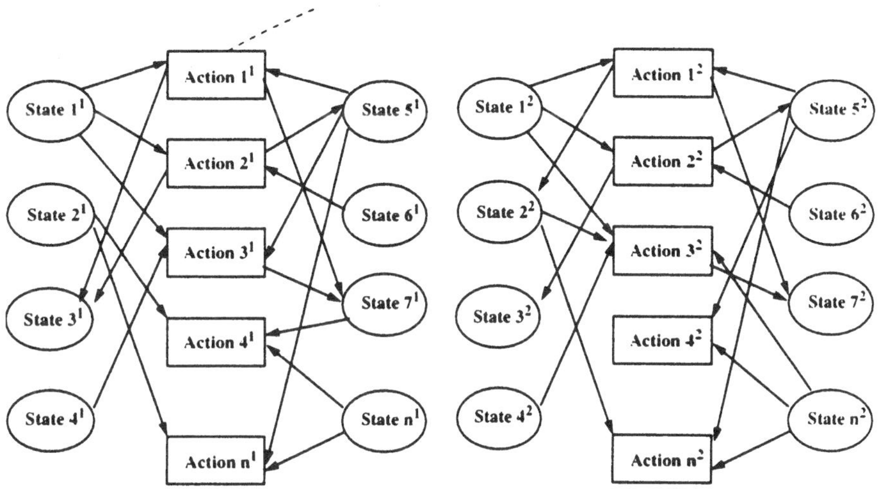

FIGURE 12.8
Action net of PAN.

- Description with visual and tactile exploration
- Classification of rocks as
  1. Graspable and collectable directly
  2. Graspable and collectable indirectly
  3. Graspable and movable but not collectable
  4. Not graspable but movable
  5. Not graspable and not movable
- Grasp and pick-up the rock; push and reorient the rock for grasping; grasp or push to move the rock; and abrade the rock

PAN architecture for rock science based on the preceding steps is illustrated in Figures 12.12 and 12.13.

5.3 PAN Operations

Through the perception net, shown in Figure 12.3, PAN explicitly manages uncertainties. Uncertainties associated with individual logical sensors (depicted as circles) are propagated through such functional modules as DFM, FTM (trapezoidal shape), and CSM (double hexagon). The values associated with logical sensors are updated through the forward and backward process of reaching an equilibrium point.

In the perception net, the reduction of uncertainty in locating the scoop at the designated trenching site is highlighted by the data fusion with the joint encoders, the stereopsis with a marker, and the tactile exploration, as well as the constraint from the trenching plane. In the action net, the top level of the action net of PAN for soil trenching and scooping is shown in Figure 12.10, where actions are depicted by boxes and states are depicted by (double) circles. The lower level of the action net includes the details of actions defined at the higher level, for example, the adaptive trenching state transition network in Figure 12.11.

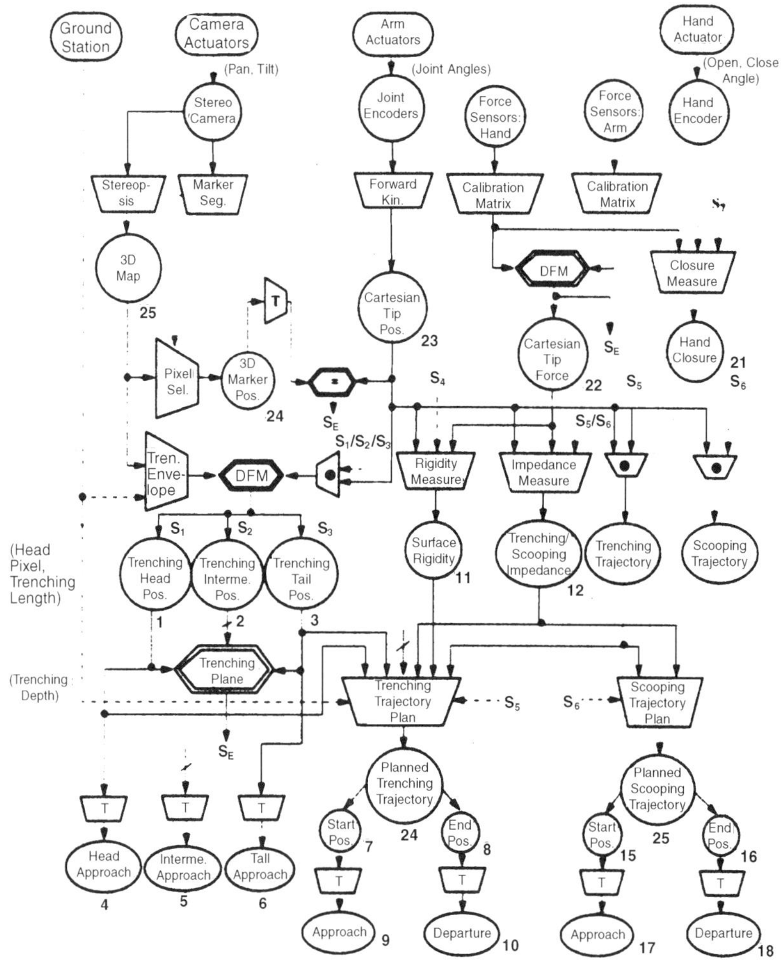

FIGURE 12.9
Perception net of PAN for soil science.

Example: Error Monitoring and Recovery-1

The arm is commanded to move to the preplanned approach point. The following two scenarios can occur:

1. The task is successfully completed; that is, the sensor reading coincides with the planned position (within some allowed error range).
2. The task is not successfully completed, within the expected time); that is, the system fails to reach the planned position.

Now we can check the actual end-effector position by using a 3-D marker.

In the first case, if the 3D-marker reading coincides with the encoder reading, then the system is in a normal operational mode, and data fusion can occur. However, if the 3D-marker reading does not coincide with the encoder reading, we can say that either the

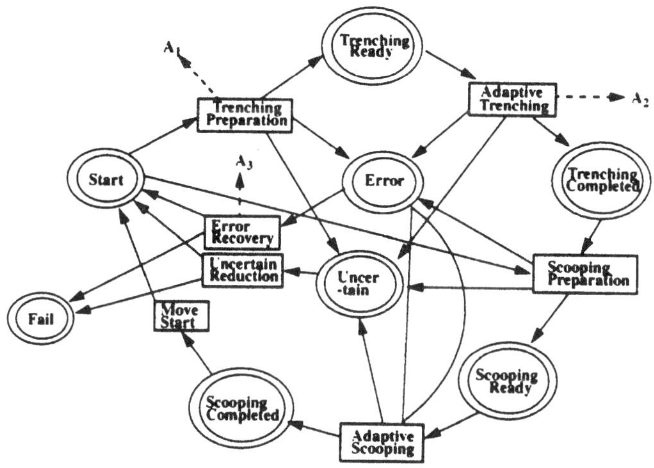

FIGURE 12.10
Action net of PAN for soil science.

encoder is biased or the marker reading (vision) is biased. The encoder bias can be calibrated through the arm calibration procedure (using a zero position or a reference position and potentiometer). Once the encoder bias is calibrated, we know whether the inconsistency is due to the bias in marker reading. If so, we need to follow the vision calibration procedure.

In the second case, if the 3-D marker reading coincides with the encoder reading, it is likely that the actuator or controller is in fault. In this case, the system can make a sense of that motion (for individual joints) to isolate further which actuator is in error. If the 3-D marker reading does not coincide with the encoder reading, the error may be in the encoder and/or in the actuator and/or in the marker reading. Encoder calibration and the actuator fault isolation routines are necessary for further identification of the problem.

Example: Error Monitoring and Recovery-2

During trenching, the abnormal encoder reading or impedance measure indicates that the trenching is not progressing well. This may be a case in which the scoop is stuck in the rigid soil or underground rock site or a failure of adaptive impedance control for trenching. The failure of adaptive impedance control may come from actuators, encoders, power supply, controllers, or force sensor. Abnormal impedance readings mean violation of the preset force and position error relationship. In the case in which the impedance measure is normal but encoder readings indicate a jamming situation, we apply the discrete event control to modify the trenching trajectory. In the case in which the impedance measure is not normal, we perform a series of actions to identify the source exactly, for instance, routines for the calibration of encoders and force sensors and for checking actuator performance. Based on these steps, the problem can be pinpointed.

6 SIMULATION

Computer simulation is performed to test the validity of the proposed method of state estimation and self-calibration. In this simulation, we consider the fixed stereo camera

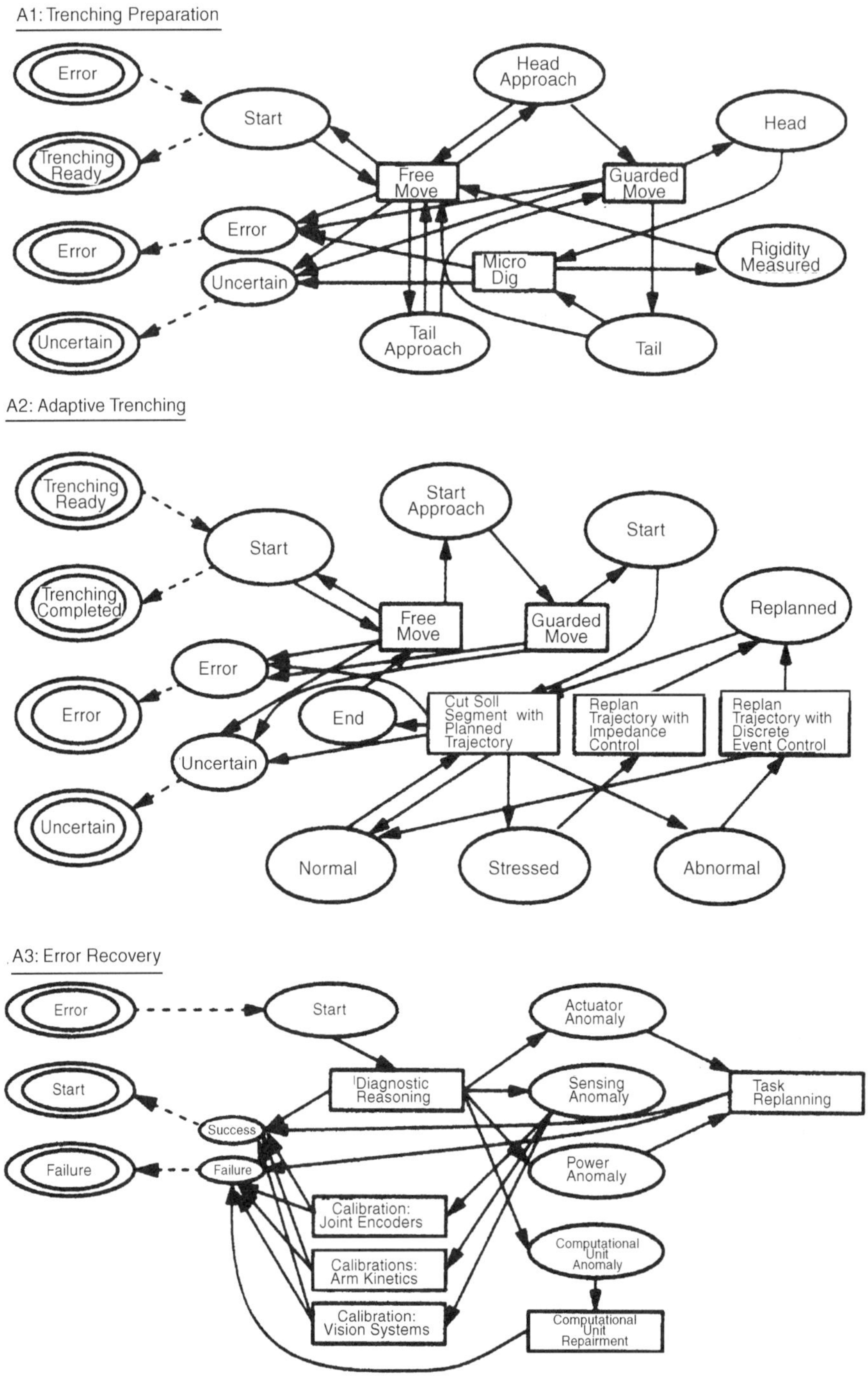

FIGURE 12.11
Higher level: A_1, A_2, and A_3 of Figure 12.10.

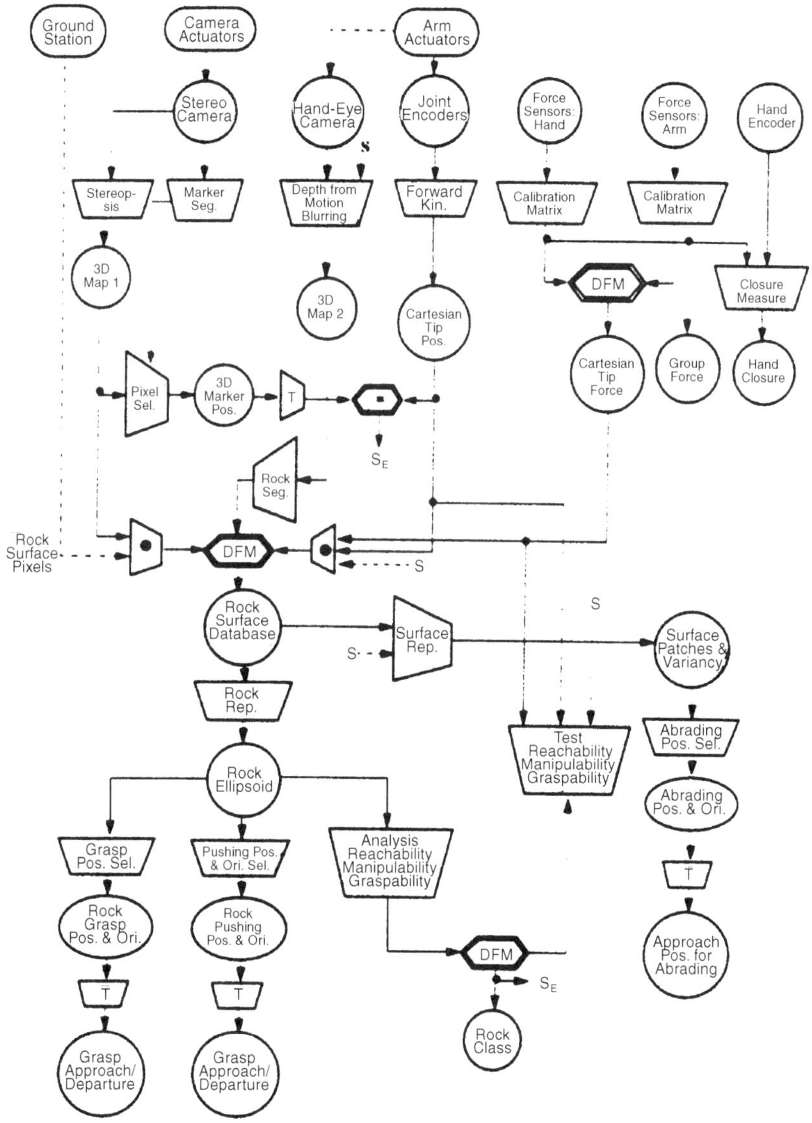

FIGURE 12.12
Perception net of PAN for rock science.

measuring the 3-D coordinates of manipulator as shown in Figure 12.14. The stereo camera is assumed to have a baseline length of 150 mm and a focal length of 35 mm.

A marker on the manipulator end effector is detected by the stereo camera while the manipulator moves. The proposed method is applied to reduce the uncertainty of the manipulator position in terms of feature points as well as to identify the possible biases of the parameters.

In simulation, the true 3-D feature points generated by the computer are projected onto the image planes based on the pinhole camera model. Then white Gaussian noise is added to the projected 2-D image points to simulate uncertainty associated with the measured

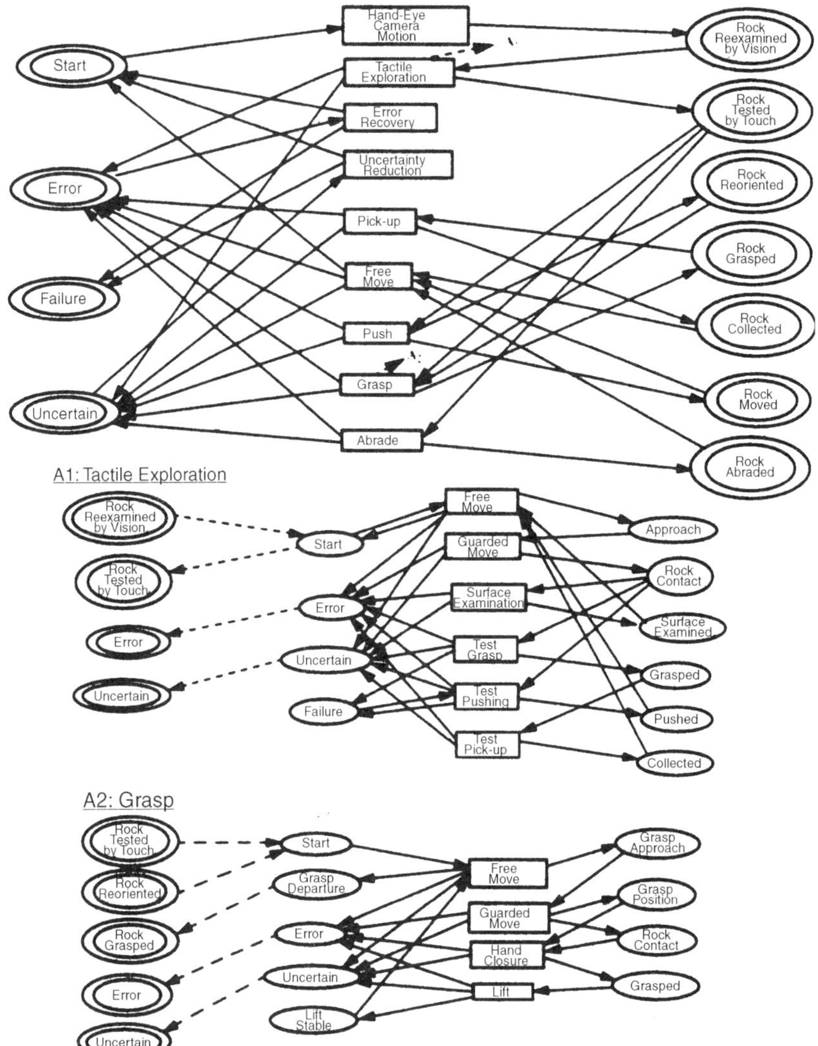

FIGURE 12.13
Action net of PAN for rock science.

feature point. For realistic simulation, the standard deviation of the Gaussian noise on the image planes is selected in such a way that the final error in the detected marker point coordinate is ± 0.01 m. By simple triangulation of the 2-D image points, the measured 3-D coordinates of feature points are reconstructed. Details of the system parameters are shown in Tables 12.1 and 12.2. Then the reconstructed 3-D coordinates are transformed with respect to the base frame, B, for comparison with the 3-D coordinates generated by encoder readings of the manipulator. Bias is added to extrinsic parameters of the stereo camera rig [the displacement from the world coordinate system to the camera coordinate system is represented with rotation and translation and this transform matrix depends on six parameters, here referral as extrinsic: three rotation (α, β, γ) and three for translation (x, y, z)].

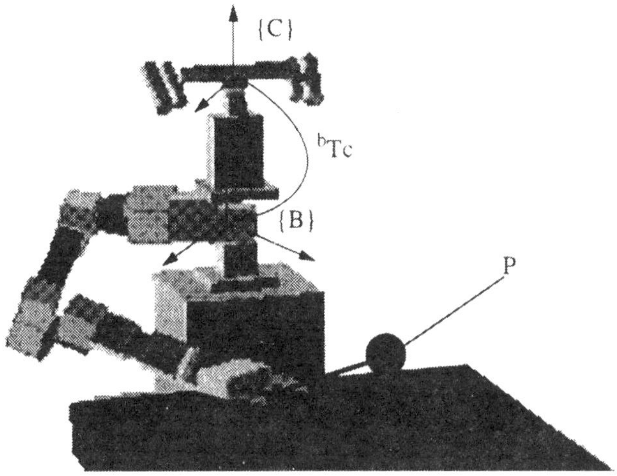

FIGURE 12.14
Simulation setup with stereo camera and manipulator.

Table 12.1. Camera Parameters and Values Used in the Simulation

	Camera focal length	35 mm
	Base line length	150 mm
Cameras	Image size	0.5 inch × 0.5 inch
	Pixel resolution	640 × 480
	Noise standard deviation	10^{-2} m

Simulation data are obtained in the following way. The coordinates of the marker on the manipulator end effector are obtained from encoder readings at time k. The noise added to the encoder reading is assumed to be very accurate and is selected in such a way that it causes ± 0.001 m error on the end-effector position. The coordinates of the marker relative to the camera coordinate frame are determined the basis of on the camera image points at time k updated through forward and backward propagation. The manipulator is assumed to move smoothly and continuously and not too rapidly to capture the camera image at each time instant. Simulation is carried out in two modes: (1) no bias is assumed for extrinsic parameters or camera pose parameters; (2) a bias is added to rotation, R, of the stereo camera rig to verify the capability of the system to locate the source of the bias and to calibrate the stereo camera rig.

Table 12.2. Manipulator Link Parameters Used in the Simulations

Joint	θ_i [rad]	d_i [inch]	a_i [inch]	α_i [rad]
θ_1	0	0	0	0
θ_2	0	3.0	4.0	$-\pi/2$
θ_3	0	3.0	44.0	0
θ_4	0	2.0	43.895	0
θ_5	0	4.38	4.72	0

The perception net representation of the simulation is shown in Figure 12.15. Through FTMs, raw sensor inputs transformed to logical sensor output and through DFM, position informations are fused with the proposed geometric data fusion algorithm. The fact that coordinates obtained from stereo camera and from encoder readings of manipulator must be the same is used as a constraint. In FTM2, output of this module is $P_{k-1} + P'_k - P'_{k-1}$, and in FTM3, output is equal to $P_k - P_{k-1} + P'_k$. These two modules are used to expose the bias effect as well as to enhance the tracking capability of the perception net. The equations used in this simulation are all based on the equations derived in Section 3.

Simulation results are shown in Figure 12.16 through Figure 12.20. As shown in Figure 12.16, data can be processed accurately in few iterations. The error is reduced to less than 1% of noise levels as a result of updating through the net.

Figure 12.17 shows the error reduction with bias added on stereo camera rig parameters. In this case, bias is added on orientation of camera frame. As shown in figure, error increases due to the bias effect. Figure 12.18 shows the values of $x_1^T W_1 x_1 + x_2^T W_2 x_2$. This value is used as a measure of the distance between two ellipsoids in this simulation. As shown in Figure 18, due to the bias the value of $x_1^T W_1 x_1 + x_2^T W_2 x_2$ is huge and never reduced. Note that x_1 is equal to $x'_1 - x_f$ and x_2 is equal to $x'_2 - x_f$, where x'_1 and x'_2 are the estimated values and x_f is the nominal value or fused value.

To calibrate the camera parameter, the parametric update method is applied based on Lyapunov's method. Whenever separation of ellipsoids occurs, parameters are updated such

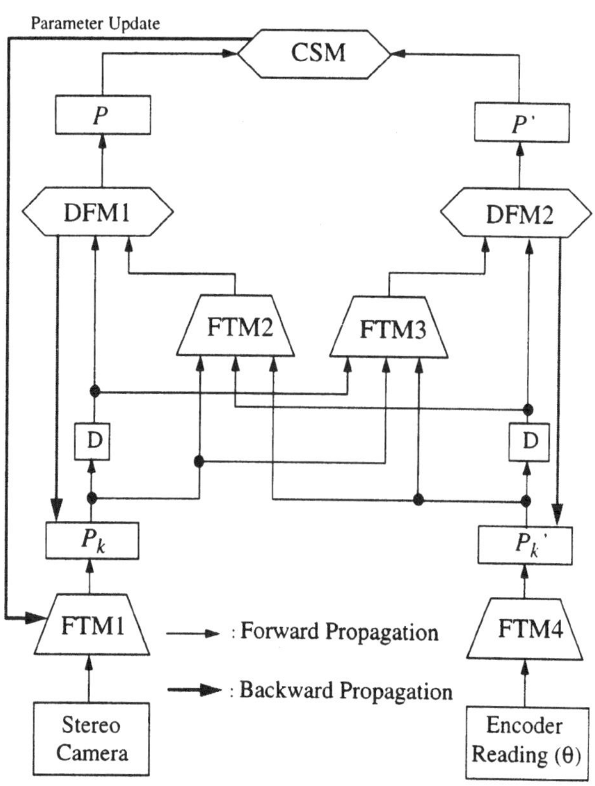

FIGURE 12.15
Perception net representation of a logical sensor system for simulation.

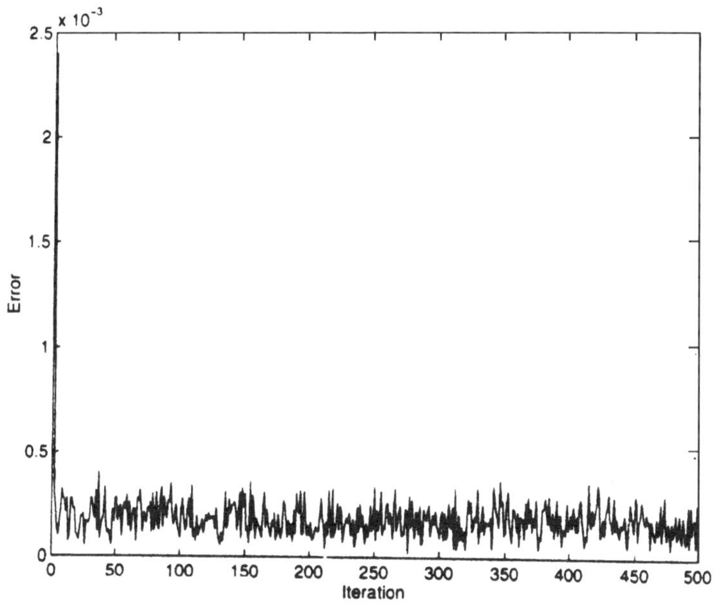

FIGURE 12.16
Error without bias on stereo camera rig parameters, $error = (err_x^2 + err_g^2 + err_z^2)^{\frac{1}{2}}$ at CSM.

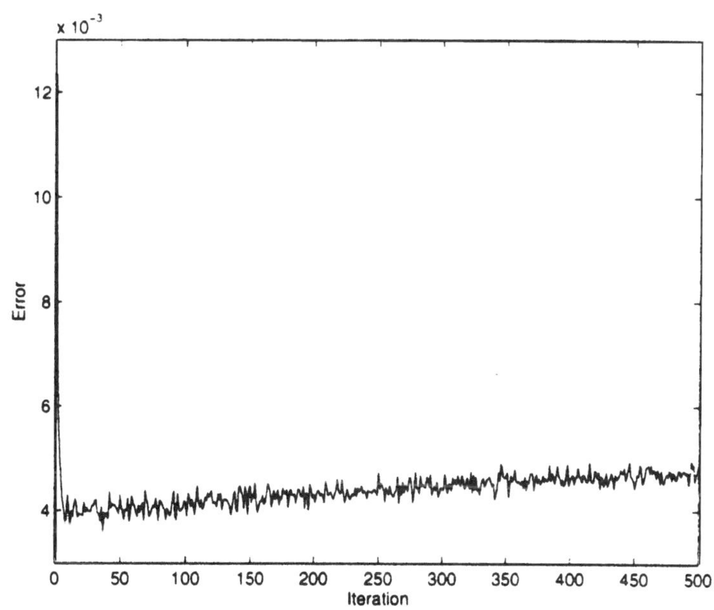

FIGURE 12.17
Error, $error = (err_x^2 + err_y^2 + err_z^2)^{\frac{1}{2}}$, at CSM with bias on stereo camera rig parameters. No parametric update algorithm is applied.

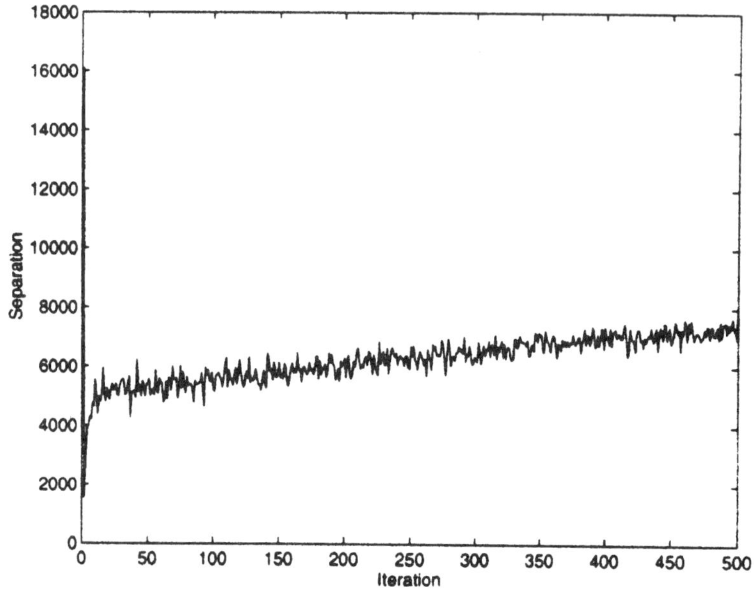

FIGURE 12.18
$x_1^T W_1 x_1 + x_2^T W_2 x_2$: reduction of this value represents the reduction of bias. No parametric update algorithm is applied.

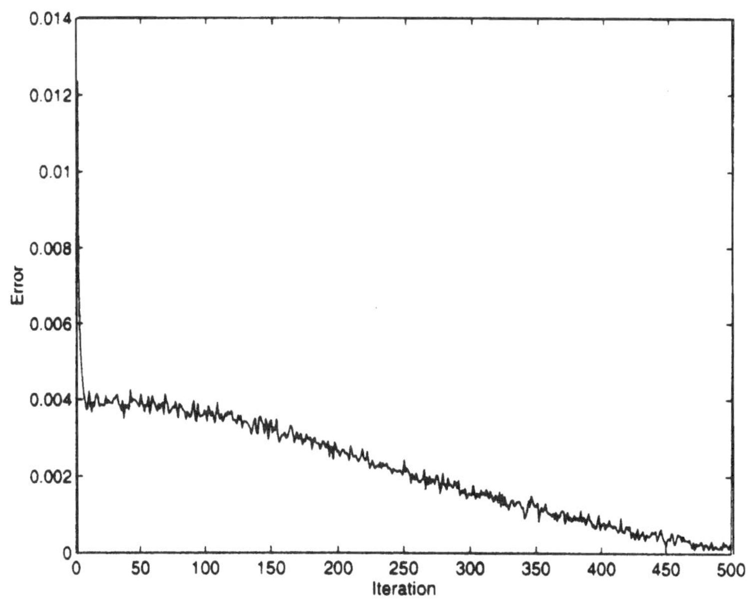

FIGURE 12.19
Error, $error = (err_x^2 + err_y^2 + err_z^2)^{\frac{1}{2}}$, at CSM with bias on stereo camera rig parameters. Parametric update algorithm is applied.

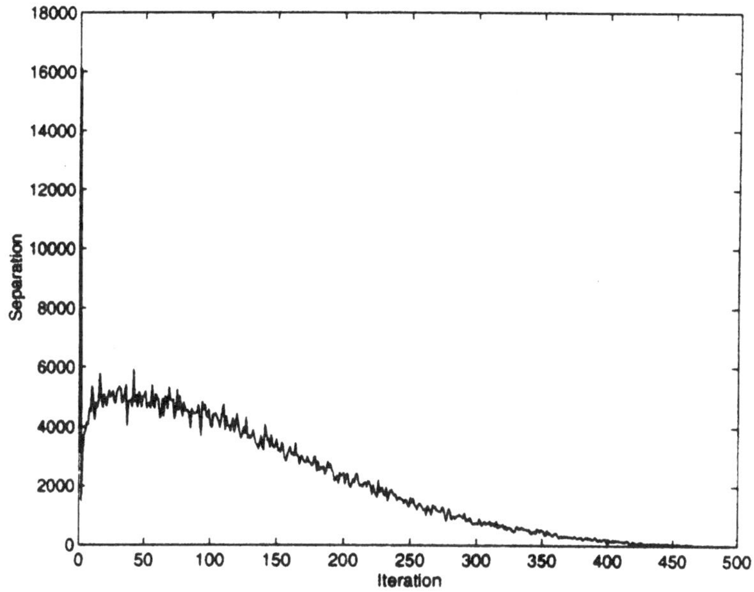

FIGURE 12.20
$x_1^T W_1 x_1 + x_2^T W_2 x_2$: reduction of this value represents the reduction of bias. Parametric update algorithm is applied.

a way that two ellipsoids intersect. The followings are governing equations to update parameters.

$$P = \frac{1}{2} \| x_1^T W_1 x_1 + x_2^T W_2 x_2 \|^2 \tag{12.35}$$

$$J = \frac{\partial P}{\partial \phi} \tag{12.36}$$

$$\theta(k) = \theta'(k) - J^T(P - P') \tag{12.37}$$

If the value of $x_1^T W_1 x_1 + x_2^T W_2 x_2$ is less than or equal to 2, then we assume that there is no bias effect and no update method is applied. Otherwise, an update method is applied because we can assume that there are possible error sources other than noise.

The parametric update method is applied and results are shown in Figures 12.19 and 12.20. As shown in Figure 12.19, the error is reduced and biased parameters are converged to the true value by the parametric update method.

7 EXPERIMENTATION

We applied the perception net–based self-calibration method to the automatic calibration of the stereo camera mounted on the base, which provides 3-D data for the Mars sampling manipulator. More specifically, we intend to remove the biases involved, in particular, in the orientation of the stereo camera with reference to the base frame. It is known that the 3-D

FIGURE 12.21
Picture of the Lightweight Survivable Rover (LSR-1), rover-mounted manipulator, and MicroArm-1. A camera is shown in the bottom of the right corner.

data from the stereo camera are very sensitive to the precise setting of camera orientation in terms of the base frame. The capability of a system to self-calibrate such biases should allow the system performance to be very robust. Figure 12.15 illustrates the perception net configured for the self-calibration of stereo camera pose in terms of the base frame. A sequence of 3-D positions of a single feature point on the manipulator is measured by the stereo camera as well as by the encoders while the manipulator is in motion.

Our experimental platform consisted of the National Aeronautics and Space Administration (NASA)–Jet Propulsion Laboratory (JPL) Lightweight Survivable Rover (LSR-1) and rover-mounted manipulator MicroArm-1 Link parameters are shown in Table 12.4. LSR-1 is a six-wheeled, skid-steered, rocker-bogie design, having approximately half the mass (7 kg) and twice the deployed volume of the Sojourner rover used in the Mars Pathfinder mission. MicroArm-1 is a 1.5 kg, all-composite five-degrees-of-freedom manipulator arm, 0.7 m at full extent, driven by piezoelectric ultrasonic motors, possessing a multifunction powered end effector [25]. The picture is shown in Figure 12.21.

A black-and-white stereo CCD camera pair, with 512×486 resolution, as shown in Figure 12.21, 10-cm baseline, and 130° field-of-view lenses (camera parameters shown in Table 12.3), was mounted on LSR-1 directly (4 cm) beneath MicroArm-1. This camera pair was calibrated using a least-squares calibration technique, producing a camera model that attempts to compensate for radial lens distortion. A black calibration grid having a 16×13 array of 5-mm-diameter white calibration circles with 1-cm center spacing was presented to the cameras in both the horizontal and vertical configurations to provide calibration points for the least-squares camera model estimation [26].

A VME chassis containing one 68040 processor running VxWorks and two Galil motion control boards were used to control the LSR-1 rover and MicroArm-1 robotic manipulator. A Sun Sparc SLC was used as a terminal to connect to the VME chassis. The Sparc SLC

Table 12.3. Camera Parameters and Values Used in the Experimentation

Cameras	Camera focal length	35 mm
	Base line length	100 mm
	Image size	0.5 inch × 0.5 inch
	Pixel resolution	512 × 486
	Noise standard deviation	10^{-2} m

was also connected to a video switcher, which allowed multiplexing of the video feeds into our frame-grabbing hardware, a Silicon Graphics Indy workstation. A command parser process ran on the 68040, accepting network socket connections, over which commands were passed from a remote Java-based graphical user interface running on a Sun Sparc 20.

To acquire a data set for the perception net–based self-calibration algorithm, we placed a black ring-shaped marker with 3.75 cm outer diameter and 1.25 cm inner diameter and white background on the end effector of MicroArm-I. The end effector was then moved in a 5 × 5 × 5 cube pattern, by specifying the desired end-effector position in 3-D space and computing the joint rotations required to achieve the desired end-effector position using the inverse kinematics.

At each location in the 5 × 5 × 5 cube, an image of the end effector was taken from both the left and right stereo cameras. When the motion through the cube pattern was complete, these images were run through our stereo localization code to determine the 3-D location of the end-effector marker. These 3-D marker locations were then translated by a constant vector in the plane of the manipulator so that they could be compared with the end-effector position as known from the manipulator kinematics.

Figure 12.22 shows the error reduction on stereo camera rig parameters. Bias may be present in the orientation or translation of the camera frame to the base frame. As shown in Figure 12.22, the error level increases because of the possible bias effect. Figure 12.23 shows the value of $\mathbf{x}_1^T W_1 \mathbf{x}_1 + \mathbf{x}_2^T W_2 \mathbf{x}_2$. This value represents a measure of the distance between two ellipsoids in this experiment [27]. As shown in Figure 12.23, because of the bias effect, the value of $\mathbf{x}_1^T W_1 \mathbf{x}_1 + \mathbf{x}_2^T W_2 \mathbf{x}_2$ fluctuates and gives peaks that are greater than 2. Note that $\mathbf{x}_1$ is equal to $\mathbf{x}_1' - \mathbf{x}_f$ and $\mathbf{x}_2$ is equal to $\mathbf{x}_2' - \mathbf{x}_f$ in DFM, where $\mathbf{x}_1'$ and $\mathbf{x}_2'$ are the estimated values and $\mathbf{x}_f$ is the nominal value or fused value. This ellipsoid separation checking procedure is performed on each module.

To calibrate the camera parameter, the parametric update algorithm based on Lyapunov's method is applied. In particular, the same update algorithm as for simulation is used for experimentation (Eqs. 12.35–12.37).

Table 12.4. Manipulator Link Parameters Used in the Experimentation

Joint	θ_i [rad]	d_i [inch]	a_i [inch]	α_i [rad]
θ_1	0	0	0	0
θ_2	0	0	2.96	$-\pi/2$
θ_3	0	-2.0	11.21	0
θ_4	0	-2.0	10.9425	0
θ_5	0	-1.82	1.43	0

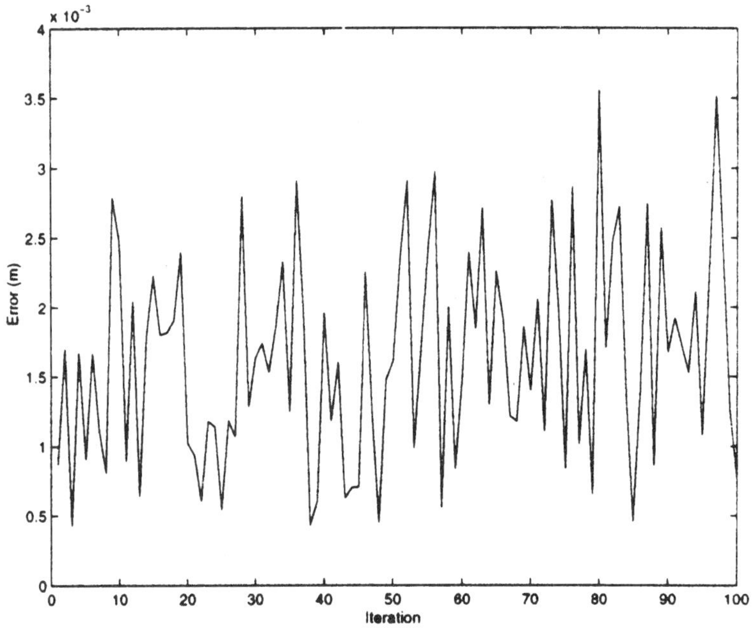

FIGURE 12.22
Error, $error = (err_x^2 + err_y^2 + err_z^2)^{\frac{1}{2}}$, at CSM with bias on stereo camera rig parameters. No parametric update algorithm is applied.

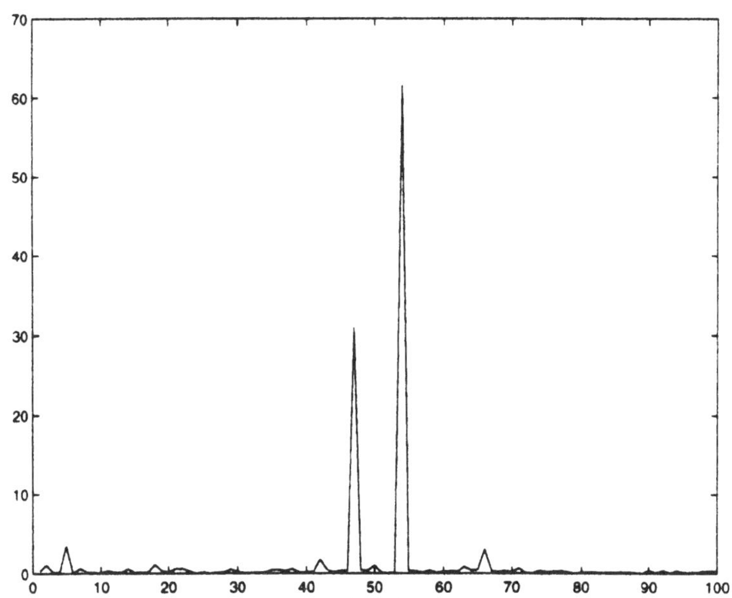

FIGURE 12.23
$\mathbf{x}_1^T W_1 \mathbf{x}_1 + \mathbf{x}_2^T W_2 \mathbf{x}_2$: reduction of this value represents the reduction of bias. No parametric update algorithm is applied.

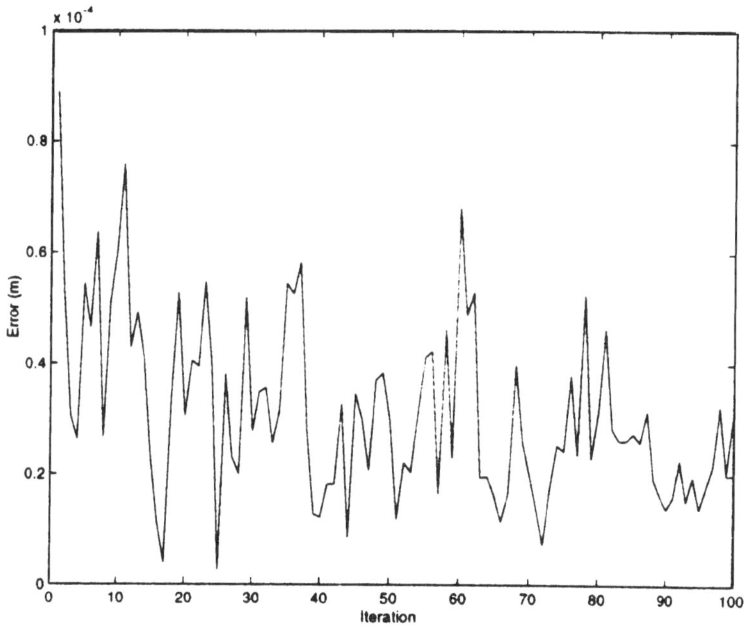

FIGURE 12.24
Error, $error = (err_x^2 + err_y^2 + err_z^2)^{\frac{1}{2}}$, at CSM with bias on stereo camera rig parameters. Parametric update algorithm is applied.

If the value of $\mathbf{x}_1^T W_1 \mathbf{x}_1 + \mathbf{x}_2^T W_2 \mathbf{x}_2$ is less than or equal to 2, then we assume that there is no bias effect and no update algorithm is applied. Otherwise an update algorithm is applied to correct erroneous parameters, which may include systematic errors such as biases and modeling errors.

The parametric update algorithm is applied and the results are shown in Figures 12.24 and 12.25. As shown in Figure 12.24, error is reduced and biased parameters are converged to the true value by the parametric update algorithm. Figure 12.25 shows no ellipsoid separation, and that means the biases are removed.

8 CONCLUSION

The proposed PAN architecture provides a formal mechanism for intergrating sensing, knowledge, and action in real time for intelligent robots. The architecture emphasizes uncertainty management as well as error monitoring and recovery so that the system can provide robots with the capability of generating goal-oriented, yet robust and fault-tolerant, behaviors.

We construct a perception net that connects features of various levels of abstraction, referred to here as logical sensors, with their functional relationships in terms of feature transformations, data fusions, and constraints to be satisfied.

Then we develop a geometric data fusion algorithm on the basis of which the net can maintain the consistency of logical sensors through the forward propagation of uncertainties

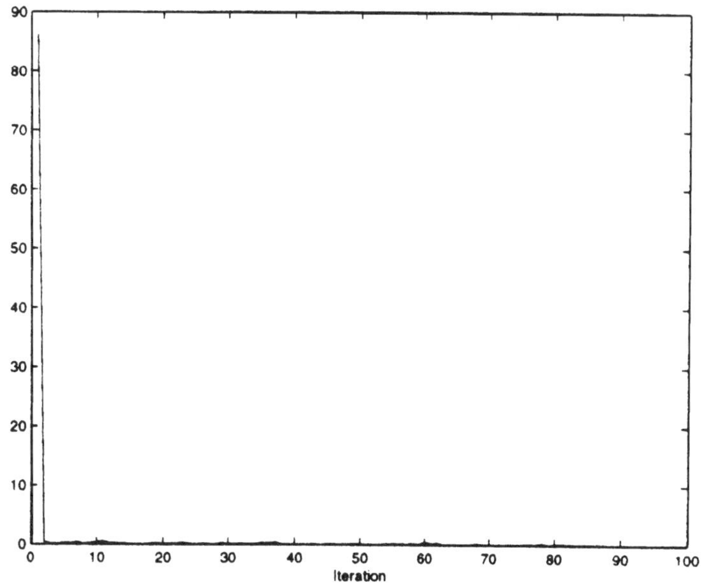

FIGURE 12.25
$\mathbf{x}_1^T W_1 \mathbf{x}_1 + \mathbf{x}_2^T W_2 \mathbf{x}_2$: reduction of this value represents the reduction of bias. Parametric update algorithm is applied.

as well as the backward propagation of constraint errors. The forward and backward process of updating logical sensors toward consistency provides the system with the capability of automatically reducing uncertainties and identifying possible biases. This process may be regarded as a general form of Kalman and extended Kalman estimation.

The proposed geometric method for uncertainty management and error monitoring through the perception net is novel and powerful because of its systematic approach. One might find it interesting to compare it with the existing probability network. The perception net provides a more general but formal way to accomplish sensor fusion and planning.

Simulations and experiments have been conducted by applying the self-calibration of the eye–hand system equipped for a JPL–NASA planetary rover. As shown in Section 7, biases are removed by the updating algorithm with back-propagated errors from the CSM, and the uncertainties are reduced through the net.

The proposed approach is new and promising, but there is a definite need for further development in both theories and implementations. This includes the analysis of performance, stability, and robustness against various types of noises, biases, and faults; the enhancement of systems in terms of theories and implementations; and analytical as well as experimental evaluation of the method. Applications of the proposed method to such complex systems as spacecraft, aircrafts, and power plants are recommended.

REFERENCES

[1] R. C. Luo, M.-H. Lin, and R. S. Scherp. Dynamic multi-sensor data fusion system for intelligent robots. *IEEE J. Robot. Automat.* 4(4):386–385, Aug. 1988.

[2] H. P. Moravec. Sensor fusion in certainty grids for mobile robots. *AI Mag.*, pp. 61–74, Summer 1988.

[3] S. W. Shaw, R. J. P. deFigueiredo, and K. Krishen. Fusion of radar and optical sensors for space robot vision. In *Proceedings of the IEEE International Conference on Robotics and Automation*, 1988, pp. 1842–1846.
[4] T. Henderson and E. Shilcrat. Logical sensor system. *J. Robot. Syst.* 1(2):169–193, 1984.
[5] T. L. Huntsberger and S. N. Jayaramamurthy. A frame work or multi-sensor fusion in the presence of uncertainty. *Proceedings of Spatial Reasoning and Multi-Sensor Fusion Workshop*, 4(4):345–350, 1987.
[6] H. F. Durrant-Whyte. Sensor model and multi-sensor integration. *IEEE J. Robot. Automat.* 1987.
[7] W. Baek and S. Bommareddy. Optimal m-ary data fusion with distributed sensors. *IEEE Trans. Aerospace Electron. Syst.* 31:1150–1152, July 1995.
[8] Z.-Q. Luo and J. N. Tsitsiklis. Data fusion with minimal communication. *IEEE Trans. Inform. Theory* 40:1551–1563, Sept. 1994.
[9] L. Banta and K. D. Rawson. Sensor fusion for mining robots. *IEEE Trans. Ind. Appl.* 30:1321–1325, 1994.
[10] R. Lobbia and M. Kent.l Data fusion of decentralized local tracker outputs. *IEEE Trans. Aerospace Electron. Syst.* 30:787–799, 1994.
[11] L. Chin. Application of neural networks in target tracking data fusion. *IEEE Trans. Aerospace Electron. Syst.* 30:281–286, Jan. 1994.
[12] T. Li and I. K. Sethi. Optimal multiple level decision fusion with distributed sensors. *IEEE Trans. Aerospace Electron. Syst.* 29:1252–1259, Oct. 1993.
[13] M. Kam, Q. Zhu, and W. S. Gray. Optimal data fusion of correlated local decisions in multiple sensor detection systems. *IEEE Trans. Aeospace Electron. Syst*, 28:916–920, July 1992.
[14] G. Hager and M. Mintz. Computational methods for task-directed sensor data fusion and sensor planning. *Int. J. Robot. Res.* 10:285–313, Aug. 1991.
[15] R. Brooks. A robust layered control system for a mobile robot. *IEEE J. Robot. Automat.* RA-2(1): pp. 14–23, March 1986.
[16] J. S. Albus. *Brains, Behavior, and Robotics.* BYTE Books, Peterbourough, NH, 1981.
[17] Y. Nakamura and Y. Xu. Geometric fusion method for multi-sensor robotic systems. In *Proceedings of the IEEE International Conference on Robotics and Automation*, 1989, pp. 668–673.
[18] H. F. Durrant-White. A bayesian approach to optimal placement. *Int. J. Robot. Res.* 9(5):70–88, 1990.
[19] S. A. Hutchinson, R. L. Cormwell, and A. C. Kak. Planning sensing strategies. In *Proceedings of the IEEE International Conference on Robotics and Automation*, Philadelphia, April 1988, pp. 1068–1075.
[20] R. Bajcsy, Active perception vs. passive personal perception. In *IEEE Workshop of Computer Vision: Representation and Control*, Bellaire, MI, 1985, pp. 55–62.
[21] S. Lacroix, P. Grandjean, and M. Ghllab. Perception planning for a multi-sensor interpretation machine. In *Proceedings of the IEEE International Conference on Robotics and Automation*, Nice, France, 1992.
[22] G. Hager and M. Mintz. Task-directed multisensor fusion. In *Proceedings of the IEEE International Conference on Robotics and Automation*, Scottsdale, AZ, 1989, pp. 662–667.
[23] C. K. Crown and A. Bergman. Determining the camera and light source location for a visual task. In *Proceedings of the IEEE International Conference on Robotics and Automation*, Scottsdale, AZ, April 1989, pp. 509–514.
[24] K. Tarabanis, R. Y. Tsai, and P. K. Allen. Automated sensor planning for robotic vision tasks. In *Proceedings of the IEEE International Conference on Robotics and Automation*, Sacramento, 1991, pp. 76–82.
[25] P. S. Schenker *et al.* Dexterous robotic sampling for mars in-situ science. In *Intelligent Robotics and Computer Vision, SPIE*, vol. XVI, Pittsburgh, Oct. 1997, p. 3208.
[26] D. B. Gennery. Least-squares camera calibration including lens distortion and automatic editing of calibration points. In *Calibration and Orientation of Cameras in Computer Vision*, A. Grun and T. Huang, eds. Springer-Verlag, 1997.
[27] H. F. Durrant-Whyte. Consistent integration and propagation of disparate sensor observations. *Int. J. Robot. Res.* 6(3):3–24, 1987.

APPENDIX

Projection onto the constraint begins with putting the uncertainty ellipsoid at the origin for simplicity. The ellipsoid equation and the direction of projection are given by

$$\mathbf{x}_1^T W_1 \mathbf{x}_1 + \mathbf{x}_2^T W_2 \mathbf{x}_2 = 1 \tag{12.38}$$

$$\mathbf{y} = (W_1 + W_2)^{-1}(W_1 \mathbf{x}_1 + W_2 \mathbf{x}_2) \tag{12.39}$$

For conceptual simplicity, the uncertainty ellipsoid is transformed to a hypersphere. Because the weight matrices are symmetric and semipositive definite, singular value decomposition for the uncertainty ellipsoids is possible and is as follows:

$$W_1 = \Sigma_1^T \Lambda_1 \Sigma_1 = (\Lambda_1^{\frac{1}{2}} \Sigma_1)^T (\Lambda_1^{\frac{1}{2}} \Sigma_1) \tag{12.40}$$

$$W_2 = \Sigma_2^T \Lambda_2 \Sigma_2 = (\Lambda_2^{\frac{1}{2}} \Sigma_2)^T (\Lambda_2^{\frac{1}{2}} \Sigma_2) \tag{12.41}$$

Then using (12.40) and (12.41), $(\mathbf{x}_1, \mathbf{x}_2)$ is transformed to $(\mathbf{x}_1', \mathbf{x}_2')$ as

$$\mathbf{x}_1' = (\Lambda_1^{\frac{1}{2}} \Sigma_1) \mathbf{x}_1 \tag{12.42}$$

$$\mathbf{x}_2' = (\Lambda_2^{\frac{1}{2}} \Sigma_2) \mathbf{x}_2 \tag{12.43}$$

and plugging (12.42) and (12.43) into (12.38) and (12.39) gives

$$\mathbf{x}_1^{T\prime} \mathbf{x}_1' + \mathbf{x}_2^{T\prime} \mathbf{x}_2' = 1 \tag{12.41}$$

$$\mathbf{y} = (W_1 + W_2)^{-1} ((\Lambda_1^{\frac{1}{2}} \Sigma_1)^T \mathbf{x}_1' + (\Lambda_2^{\frac{1}{2}} \Sigma_2)^T \mathbf{x}_2') \tag{12.45}$$

By differentiating (12.44) and (12.45), we have

$$\mathbf{x}_1^{T\prime} \Delta \mathbf{x}_1' + \mathbf{x}_2^{T\prime} \Delta \mathbf{x}_2' = 0 \tag{12.46}$$

$$(\Lambda_1^{\frac{1}{2}} \Sigma_1)^T \Delta \mathbf{x}_1' + (\Lambda_2^{\frac{1}{2}} \Sigma_2)^T \Delta \mathbf{x}_2' = 0 \tag{12.47}$$

Projecting onto the constraint space is equivalent to finding the intersection between a tangential line and the constraint space. Therefore, from the condition that (12.46) is the same as (12.47), to find out the tangential line, we have the following relationship:

$$\mathbf{x}_2' = (\Lambda_2^{\frac{1}{2}} \Sigma_2)(\Lambda_1^{\frac{1}{2}} \Sigma_1)^{-1} \mathbf{x}_1' \tag{12.48}$$

Then plugging (12.48) into (12.44), we obtain

$$\mathbf{x}_1^{T\prime} \mathbf{x}_1' + \mathbf{x}_1^{T\prime} (\Lambda_1^{\frac{1}{2}} \Sigma_1)^{-T} (\Lambda_2^{\frac{1}{2}} \Sigma_2)^T (\Lambda_2^{\frac{1}{2}} \Sigma_2)(\Lambda_1^{\frac{1}{2}} \Sigma_1)^{-1} \mathbf{x}_1' = 1 \tag{12.49}$$

Transforming (12.49) to the original coordinates, we have

$$\mathbf{x}_1^T W_1 \mathbf{x} + \mathbf{x}_1^T W_2 \mathbf{x}_1 = 1 \tag{12.50}$$

Finally, rewriting (12.50) in terms of $\mathbf{x}_1$, we have

$$\mathbf{x}_1^T (W_1 + W_2) \mathbf{x}_1 = 1 \tag{12.51}$$

$$W_y = W_1 + W_2 \tag{12.52}$$

Q.E.D.

CHAPTER 13

A Fuzzy Behaviorist Approach to Sensor-Based Reasoning and Robot Navigation

FRANÇOIS G. PIN
Oak Ridge National Laboratory, Oak Ridge, Tennessee

1 INTRODUCTION

Sensor-based operation of autonomous robots in unstructured and/or outdoor environments has proved to be an extremely challenging problem, mainly because of the many uncertainties that are always present in the real world. In robot control, these uncertainties are primarily due to sensor imprecisions and unpredictability of the environment, that is, lack of full knowledge of the environment characteristics and dynamics. In this chapter, we present an approach that attempts to remedy some of these difficulties and derives from two basic principles, or philosophies. The first principle is based on the concept of "minimal models" for accomplishing given tasks and proposes to utilize only the minimum level of information and precision necessary to accomplish elemental functions of complex tasks. This approach diverges completely from the direction taken by most artificial perception studies, which conventionally call for crisp and detailed analysis of every available component in the perception data. The second principle that is proposed is based on the representation of the system's uncertainties using Fuzzy Set Theory–based approximations and on the representation of reasoning and control schemes as sets of elemental behaviors. The next subsection provides an overview discussion of these principles and some motivating examples for introducing them into a generic approach for sensor-based robot reasoning and control. For this and the remaining discussions of this chapter, the context of mobile robot navigation has been chosen to help in providing specific application examples of the developments. Extensions of the developments into other application domains of robotics are, of course, feasible but are not specifically addressed in this chapter. The reader can find aspects of such possible extensions in several of the other chapters of this book.

1.1 Background

The past two decades have seen significant research activity take place in the area of sensor-based robot reasoning and control, in particular for application to the problem of autonomous mobile robot navigation. Several successful applications have been reported in which robots showed robust sensor-based navigation features in nonlaboratory, although fairly well-structured, corridor-type environments such as hospitals, energy producing and manufacturing plants, and/or warehouses [1–4]. One of the greatest challenges facing the research community is to extend the domains of application of autonomous robot navigation to the general class of unstructured environments, that is, environments that are generally dynamic, not fully known *a priori*, and typically unpredictable. The paramount complexity of the sensor-based navigation problem in unstructured environments arises mainly from the uncertainties that exist and become pervasive in the overall system (which includes the robot and its surrounding domain). In addition to the typical sensor inaccuracies (there are no such things as "perfect" sensors), the dynamics and unpredictability of the environment generate very large uncertainties in the perception and reasoning systems; it becomes impossible to generate complete or exact models of the system (robot and environment) and/or of its behavior. These uncertainties, in turn, typically propagate through the control systems and lead to further inaccuracies or errors (e.g., in the robot position, in the sensor orientation) that compound the problem by increasing the uncertainties in the perception. In these conditions, the overall cost (time, computational needs, computing resources, etc.) of achieving the type of precision that was common in structured and static environments jumps by orders of magnitude, either from a requirement for much more refined sensors and perception data or from the need for very time- and computation-expensive methods such as uncertainty analysis and propagation techniques. The impact of this cost increase is particularly important for real-time systems (where real-time is defined as the guarantee of producing a response within a prescribed amount of time), which become much more difficult to design if not impractical to implement in realistic situations.

Humans, on the other hand, seem to cope very well with uncertain and unpredictable environments, often relying on approximate or qualitative data and reasoning to make decisions and to accomplish their objectives. Furthermore, humans seem to gather information in what we will refer to as the "approximate first" fashion: they first look for and perceive some "general-type" information, of a symbolic, iconic, approximate, or even "blurry" nature, and then progressively focus their perception on details, particular regions, or further precision as they judge necessary to supplement the "general" information. This is quite opposed to the approach that conventional artificial perception techniques seem to pursue; that is, precision-aimed analysis of every piece of acquired data is performed first (e.g., every pixel in every region of every image produced by a CCD camera), followed by progressive extraction of more and more "general" information through successive applications of "filters" or mathematical models (e.g., edge finders, region extraction, analyses of geometric properties). In this progressive processing, each extractive pass generates more "general" but also more "approximate" information because of the cumulative propagation of the basic uncertainties existing in the original data. In other words, humans seem to perceive and reason using information cycles progressing, on an "as needed" basis, from "general and blurry" to "precise and detailed," whereas conventional artificial perception systems seem to operate using the opposite direction, from "detailed analysis of every data" to "general" information extraction. Quite a difference, indeed.

1.2 Need for Minimal Models

One of the significant features that humans exhibit when they reason is their capability to make rapid *trade-offs* among various aspects of their inferencing. In particular, humans are very apt at trading the quantity and precision of the information they could acquire or derive (but chose not to) for some other characteristics of the reasoning process that they deem much more essential in accomplishing their objectives. Speed of achieving a decision, for example, is one of the most intuitively known factors for which humans will trade off precision or quantity of information. There are, of course, many other factors that humans will trade off in their reasoning, but for the sake of analogy with real-time sensor-based robotic systems, speed versus quantity and precision will be the focus of our discussions here.

It is interesting to note that people who are consistently successful in resolving previously unencountered challenges in changing environments are those who deliberately change or adapt their reasoning to correspond to the situation at hand; that is, they consistently look for a reasoning strategy that has a chance of solving the problem, find what the "right amount" of information to implement that reasoning is, and then adapt their information-gathering process to focus (and generally in an exclusive fashion) on acquiring that particular information. The key, therefore, to successfully and consistently accomplishing objectives over a variety of tasks in real-time conditions seems to consist of finding what the "*minimal* amount" of information needed to reach the desired goals is and what other nonessential process can be traded off so that focus or priority can be given to acquiring the essential information first. For robotic systems, this translates into capabilities to (1) select, on a task-dependent basis, the minimal information, be it qualitative or quantitative, that can be assembled (with available resources or from memory) and is needed to reach *any* decision that would satisfy the goal and (2) process that information, with a reasoning scheme *commensurate with the acquired qualitative or quantitative information*, to produce a decision that meets the objective. It is clear that the selected "reasoning scheme" is also a part of the needed information and therefore becomes an integral part of the "minimal model" necessary to reach the objective successfully.

To provide one illustrative example of what a minimal model may be for a specific task, consider the problem of a camera-equipped mobile robot having to follow a moving object such as an automobile, while maintaining a safe separation distance. The conventional artificial vision approach would consist of analyzing every pixel of the image and every detail of the background environment to identify the car and try to determine its motion parameters (e.g., through a very computation-intensive optic flow method). This seems quite an inefficient perception procedure, since the only desired information for the task is the approximate size in the image plane of the "form" that is ahead and has "approximately this color." On the other hand, if the system could first acquire an indication of the approximate size in the image plane of the "form" that has this approximate color, it could quickly determine its approximate location and recent size variation (e.g., the size in the image plane increases slowly, decreases rapidly) and take appropriate control actions relative to the task (e.g., speed up, slow down) using a few simple and computationally inexpensive rules, with the consequence that the system could react and operate much faster than in the previous case. Thus the capability to rapidly extract data that is approximate but has high information content, directly from perception systems, is one of the items that we argue would put "intelligent machines" on a better footing (with respect to speed, survivability, efficiency/cost, etc.) for operation in changing and unstructured environments by allowing the use of the necessary, albeit *minimal*, models to accomplish their goals.

1.3 Need for Approximate Reasoning

From the discussion of the previous section, it is clear that the information to be processed during the reasoning phase may appear in either quantitative or qualitative form. Furthermore, significant uncertainty ranges may be associated with the data, be it of a quantitative or qualitative nature. Methods for representing, encoding, and processing quantitative data are quite advanced today with the computer technology explosion. This is not the case, however, for qualitative data or information, for which few methods of representation, encoding, or processing are formally or firmly established.

Qualitative reasoning (also termed approximate reasoning) refers to a set of methodologies that have been developed to provide decision-making capabilities in systems where not all uncertainties can be fully engineered away (e.g., there are limits on maximum sensor precision, on the predictability of the environment). These methodologies attempt to capture some aspects of the reasoning methods typically exhibited by humans when controlling systems; that is, they aim at implicitly incorporating uncertainties in the information gathering and reasoning processes, rather than explicitly determining and propagating them through numerical calculations or representations. Several approximate reasoning theories and associated mathematical algebra have been developed over the past two decades (e.g., see methods and references in [5]), the one most commonly used today for applications to control systems being Zadeh's Theory of Fuzzy Sets [6–8]. This theory is at the basis of very successful implementations varying from control of subway cars, elevators, cement kilns, washing machines, cameras and camcorders, and inverted pendulums, to painting processes and color image reconstruction, and even to Ping-Pong–playing robots [9–15].

One of the important factors that have prevented the widespread utilization of approximate reasoning methodologies in real-time systems has been the lack of computer hardware allowing processing and inferencing directly in terms of approximate or linguistic or "fuzzy" variables (e.g., far, fast, slow, left, faster) and approximate rules (e.g., if obstacle is close, then go slower; if temperature is high and pressure is increasing, then decrease power a lot). Prospective implementations thus had to rely on simulations of the approximate reasoning schemes on conventional computers based on "crisp" processing. The result was a significant penalty in speed of operation, typically prohibiting applications in most "hard real-time" systems. Over the last decade, however, several innovations have allowed some bridging of this gap; in particular, unique computer boards have been developed that use custom-designed VLSI fuzzy inferencing chips [16–18] on VME bus–compatible boards. These systems can be directly programmed in terms of qualitative variables and rules and, when incorporated in a control system, can directly communicate and interface with robotic hardware (e.g., with motors, actuators). Such computer hardware developments have proved extremely useful in supporting the developments needed in the area of approximate reasoning for real-time "intelligent" machines and have been a strong basis [18, 19] for the activities reported here.

2 FUZZY BEHAVIORIST APPROACH AND RULE GENERATION FOR VEHICLE NAVIGATION

In everything that follows, we have selected the problem of autonomous mobile robots navigating in *a priori* unknown and unpredictable environments as an illustrative context to exemplify the methodologies for development of qualitative reasoning systems. This choice was motivated by the fact that the characteristics of the navigation problem rank very high on the list of criteria that typically indicate a strong need for resolution through qualitative reasoning and decisional trade-offs:

- The input to the control system, particularly when provided by sonar range finders and odometric wheel encoders, is inaccurate, sparse, uncertain, and/or unreliable.
- No complete mathematical representation exists of the process termed "navigation," although, as demonstrated by humans, a logic for this process exists that can typically be represented and successfully processed in terms of linguistic variables.
- The approximations involved in the numerical representation of the system and its environment (e.g., geometric representations, map discretization in grid) are significant.
- By its nature, the behavior of an outdoor environment is unpredictable, leading to large uncertainties in its representation and frequent need for trade-off of speed versus precision.

Several research groups have studied approximate reasoning techniques, in particular fuzzy logic, to mimic human reasoning capabilities in navigation tasks [18–22]. In all these applications, the sensor-based decision-making process has been implemented as a set of fuzzy rules that, together, express the desired navigation decisions of the robot for various combinations of the input data. Very successful results have been achieved when the number and complexity of the rules were small. When these increased, however, and/or the perception system grew more sophisticated (i.e., more sensory input data was provided), the typical difficulties encountered with large rule base systems emerged: the lack of established formalism for the development of rule bases—in particular with respect to completeness, interaction, and redundancy of the rules—made the actual coding of the fuzzy rules an iterative empirical process, requiring lengthy trial-and-error experiments.

In an attempt to alleviate this general shortcoming of rule-base system development, we proposed the "Fuzzy Behaviorist Approach" (FBA), which provides a formalism for the development of fuzzy rule systems for control of autonomous robots [19, 23–26]. The basic premises underlying the FBA are as follows:

1. Each action of a robot results from the concatenation of elemental behaviors.
2. Each elemental behavior is a direct mapping from a single stimulus mode to a single output control.
3. Each behavior is represented by one or a set of fuzzy rules that are defined by the membership functions of the rule's antecedent (stimuli) and consequence (output controls).
4. Each mode of stimulus corresponds to a single dimension of the input space and is independent, in a *possibilistic* framework, of other stimulus modes.
5. Each type of input data provided by the sensors is fuzzified with a membership function expressing, as a *possibility* distribution, the uncertainty associated with the specific measurement or calculation (see Section 2.3).
6. Triggering of any behavior takes place when the current input data and the antecedents of the behavior's fuzzy rules have nonempty fuzzy intersection. The triggering strength transferred to the output control is dictated by the fuzzy intersections of the input data and the rule antecedents (see Section 2.1).
7. In each of the output dimensions, the merging of all triggered behaviors is accomplished using the Union Operation of Fuzzy Sets. In the system discussed here, a typical "center of area" defuzzification scheme is then used to generate "crisp" control set points for those output variables that represent direct actuator commands.

8. For behaviors affecting *the same output control dimension*, the possible conflicts between behaviors with stimuli that overlap in the multidimensional input space must be resolved through the expression of the respective dominance between the various behaviors (see Section 2.4).

In the following subsections, the implications of these basic axioms for the development of FBA-consistent rule bases are discussed in detail.

2.1 Basic Fuzzy Inferencing

The inferencing methodology utilized in the FBA is inspired by the Theory of Fuzzy Sets (TFS). In what follows, the reader is assumed to have only a very basic knowledge of this theory. If necessary, Refs. [6] to [9], as well as many textbooks, can also provide further understanding of the foundation of this theory. In the TFS, the function $\mu_X(x)$ defining the membership of an element x in a subset X of a universe of discourse U can take any value in the interval [0, 1], rather than only the discrete {0, 1} values (0 for does not belong, 1 for belongs) used in conventional (crisp) set theory. The function $\mu_X(x)$ thus defines the *degree* of membership of the element x in X. Such a subset X of U is termed a fuzzy (or approximate, or linguistic, or qualitative) variable for reasoning on the universe of discourse U.

In the FBA, reasoning is embodied in programmable "production rules" operating on qualitative input and output variables, as in

$$\text{IF } (A \text{ is } A_1 \text{ and } B \text{ is } B_1 \text{ and } C \text{ is } C_1 \text{ and } D \text{ is } D_1 \text{ and} \ldots), \quad \text{THEN } (E \text{ is } E_1 \text{ and } F \text{ is } F_1 \text{ and} \ldots) \tag{13.1}$$

where $A_1, B_1, \ldots, F_1$ are qualitative variables whose representative membership functions define the rule, and $A, B, C, \ldots, F$ are the time-varying qualitative input data and output variables analogous to memory elements in conventional production systems.

The Fuzzy Set theoretic operations can be directly applied to qualitative variables and their membership functions on the same universe of discourse: given two subsets A and B of U,

$$\mu_{A \cap B}(x) = \min(\mu_A(x), \mu_B(x)) \tag{13.2}$$

$$\mu_{A \cup B}(x) = \max(\mu_A(x), \mu_B(x)) \tag{13.3}$$

The laws of logical inferences including modus ponens, Cartesian product, projection, and compositional inferences (e.g., see Refs.[6], [7], and [8] for detailed description of these laws of inferencing) can also be applied to multivariable systems. In particular, the extension principle is used in the mapping between a set A of the input universe of discourse U and its extension through F to the output universe of discourse V, as:

$$\mu_{F(A)}(v) = \operatorname*{Sup}_{u} (\mu_A(u)) \tag{13.4}$$

where $v = F(u)$, $u \in U$, $v \in V$.

For VLSI implementations such as those described in Refs. [16] and [17], each fuzzy variable is represented by its membership function discretized over a (64×16) array of $(x, \mu(x))$ values. Equations (13.1)–(13.4) can thus be easily implemented using series of min. and max. gates, with all rules operating in parallel. Figure 13.1 schematically represents an

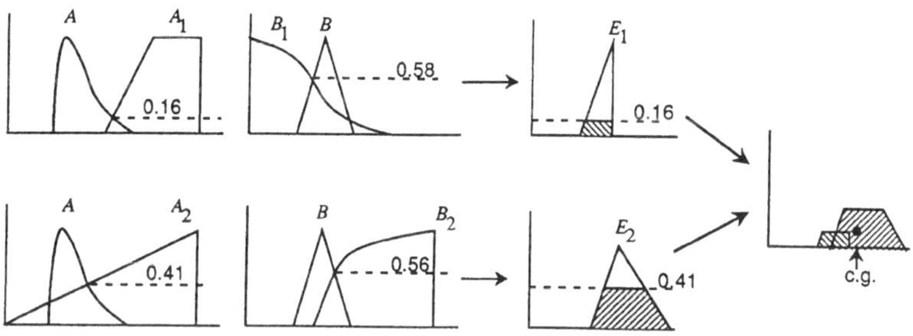

FIGURE 13.1
Schematic of a fuzzy inference with two rules operating on two input and one output channels.

inference with two rules: IF (A is A_1 and B is B_1) THEN (E is E_1), and IF (A is A_2 and B is B_2) THEN (E is E_2), operating on two fuzzy input A and B and producing a composite membership function for E.

The data from the sensors, A and B, are first matched within their corresponding IF clause in each rule to determine the degree to which they verify each rule (e.g., $\mu_{A \text{ is } A_1}(x) = \mu_{A \cap A_1}(x)$). The extension principle [Eq. (13.4) and Eq. (13.2)] to implement the logical AND are then used to produce the truncated membership function (e.g., E_1) for each rule. Merging of all the rules' output membership functions into a single membership function for each output dimension is done using the Union operation [Eq. (13.3)]. If needed, for an actuator command for example, a "defuzzification" scheme, such as the "center of area" illustrated in Figure 13.1, can be used to generate single control set points from the output membership functions.

2.2 Elemental Behaviors

The overall inferencing, I, of the (robot) reasoning system represents a relationship between the input space (or stimulus space) U and the output space (or response space) V. The input space U is multidimensional, with each dimension representing a type of input data (i.e., a universe of discourse) on which the inferencing can act (e.g., distance to obstacle to the right of route, direction to the goal, distance to the goal). In other words, each input dimension is a mode of stimulus, s_i, that can excite the inferencing. Similarly, the output space, V, is multidimensional, with each dimension representing a type of output data, that is, a type of control, c_j, (e.g., turn control, motor #1 speed) that can be implemented. Thus we have

$$I\colon U(s_1, s_2, \ldots, s_i, \ldots, s_n) \rightarrow V(c_1, c_2, \ldots, c_j, \ldots, c_m) \tag{13.5}$$

The total numbers of possible stimulus modes, n, and of control modes, m, are of course dependent on the sensory and actuation capabilities implemented on the robot. Each dimension s_i or c_j is a one-dimensional space on which fuzzy sets can be defined using membership functions in the conventional manner [6–9]. An elemental behavior, B_{ij}, is thus defined as a direct mapping from s_i to c_j:

$$B_{ij}\colon s_i \rightarrow c_j \tag{13.6}$$

which is represented by one or several fuzzy rules relating fuzzy subsets of s_i to fuzzy subsets of c_j. As an example, assume s_3 represents the "direction to the goal" input dimension and

c_2 represents the "turn control" output dimension; then the behavior B_{32}, "turn control as a function of the goal direction," would include fuzzy rules of the type: IF (direction to the goal is *left*) THEN (turn value is *positive*), where *left* and *positive* are fuzzy subsets defined by their membership functions on s_3 and c_2, respectively.

Note that the fourth requirement of the approach specifies that the input space be designed such that the stimulus modes (input dimensions), s_i, are independent of each other, that is, such that the possibility for the ith input to be any fuzzy subset in s_i is completely independent of the possibility for any other input to be any fuzzy subset on their stimulus mode. In other words, the possibilities for any and all stimuli to occur are unrelated and independent of each other. This allows a behavior B_{ij} to be extended to a mapping B^*_{ij} from U to V as

$$B^*_{ij}: U(\#, \#, \#, \ldots, s_i, \#, \ldots, \#) \rightarrow V(\phi, \phi, \ldots, c_j, \phi, \ldots, \phi) \tag{13.7}$$

where the # and ϕ signs represent "nonsignificant" input and output dimensions, respectively. By definition, s_i and c_j are the "significant" input and output dimensions of behavior B^*_{ij}. For example, the behavior B_{32} has 3 (for input) and 2 (for output) as significant dimensions and would be extended to a behavior B^*_{32} on the multidimensional input and output spaces with fuzzy rules now expressed as:

> IF (input 1 is *anything* and input 2 is *anything* and direction to the goal is *left* and input 4 is *anything* and ...) THEN (output 1 is *do nothing* and turn value is *positive* and output 3 is *do nothing* and ...) (13.8)

and the fuzzy subsets *anything* and *do nothing* have membership functions uniformly equal to 1 and to 0 over the entire range of their respective input and output dimensions.

2.3 Membership Functions

One of the major challenges when implementing a Fuzzy Set–based approach is the generation of membership functions. In our system, there are two types that require attention: the membership functions representing the various qualitative variables used in the rules, such as $A_1, B_1, \ldots, E_1$, in Eq. (13.1) or Figure 13.1, and the membership functions of the input data, such as A or B in Eq. (13.1) or Figure 13.1.

For the first type, conventional approaches to fuzzy rule development usually define a set of basic qualitative variables and their membership functions first and then use conjunction or disjunction of these variables to express the rule condition variables (such as A_1, B_1 in Eq. (13.1)). Here, we diverge from this approach and allow each $A_1, B_1 \ldots$ variables in each of the rules to be specifically designed on a behavior-by-behavior basis. This allows direct tailoring of the qualitative variables in each rule with respect to their specific behavior, with the added benefit of a dramatic saving in the number of rules needed to describe a behavior. Examples of behavior-specific membership functions will be found in the section on experimental investigations.

For the second type, input data membership functions, it should be noted that conventional sensors typically provide data in "crisp" form, that is, they provide a single number that does not reflect the uncertainty that is always involved in any measuring process. In order to reflect the imprecisions or uncertainties of the measuring process that are neglected in the "crisp" numbers, it is desirable to add this uncertainty on the data, effectively mapping it to a fuzzy variable, prior to processing through the fuzzy reasoning scheme. This step

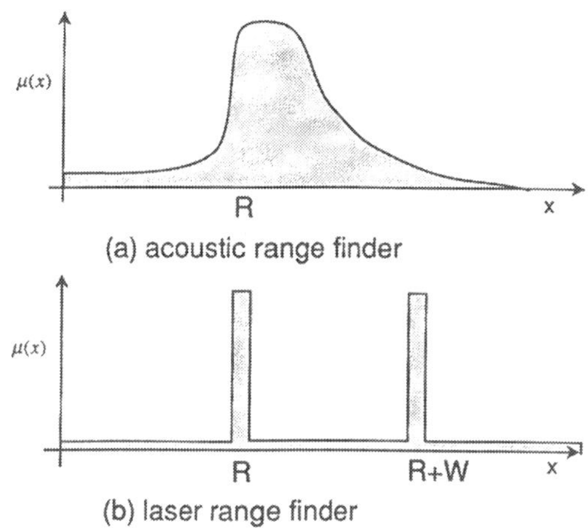

FIGURE 13.2
Example of fuzzification membership function for typical "distance to obstacle" data provided by (a) an acoustic and (b) a laser range finder.

(which has been termed fuzzification) is of course not necessary if the data is already in the form of a fuzzy set.

However, if the data provided by the sensors is "crisp," then the intent of the fuzzification process is to add on the input data some level of information expressing the imprecision or inaccuracies involved in the measurement. This is of course sensor dependent and generally also environment dependent and, according to the Fuzzy Set framework [6–9], should be expressed in terms of *possibilities* rather than probability distributions. As an example, Figure 13.2 shows two membership functions that could be associated with a distance measurement, R, provided by a typical acoustic range finder and by a laser range finder working with phase shift. For the acoustic range finder, the example membership function expresses the fact that the possibility of the object being closer than R is small (e.g., due to specular reflections) and that it is very high at R and greater distances, decreasing with increasing distances.

For the laser range finder with an ambiguity (wraparound) interval, W, of about 10 m and 9 bits data (512 levels), the measurement, R, could correspond to any location within the resolution value of about 2 cm (1000 cm/512) at any of the wraparound distances. This is expressed by the successive square functions of width 2 cm at the distances $R + nW$, $n = 0, 1, 2, \ldots$ In addition, if there are possibilities for the laser light to be fully reflected by the object surface, the measurement could correspond to absolutely any location. Depending on the environment characteristics, this possibility may be small, as that shown in Figure 13.2, very significant (e.g., in mainly metallic environments), or nonexistent, and the respective membership functions should reflect these differences. One key point, however, is that "there are no such things as uncertainties on the uncertainties," and our approach has been to use single, most conservative membership functions for each sensor and the environment considered.

2.4 Resolving Potential Interrule Conflict with Dominance Concepts

A very important aspect of the FBA formalism that needs to be emphasized here is the requirement for independence and nonconflict of the stimuli of the behaviors affecting the *same output controls*. This requirement simply expresses that only one action command can be sent to a single output control for any given stimulus (a single point in the input space). This leads to the concept that certain behaviors must "dominate," or "be dominated by," some of the other behaviors in one or more regions of the input space. This concept of dominance between behaviors, which exhibits itself in almost every action of our everyday life, is illustrated here using a simple example: consider two behaviors acting on the *same* output control, say the two object-holding fingers of a child or, for analogy, the gripper of a robot. Initially, the child or the robot has only one behavior, which is a "don't get burned" behavior and which could consist of one fuzzy rule expressing that IF (the object being held is hot) THEN (release the object). The input dimension for this behavior is "temperature of object being held," and the fuzzy set "hot" can be represented with a typical membership function as on the left-hand side of Figure 13.3. Suppose that the child is now being taught the value of things, or assume that the robot is being given a new perception device so that it can recognize the value of objects. A new behavior could be given to the robot or taught to the child, stating that IF (the object being held is expensive) THEN (don't release the object). The input dimension for this behavior is "value of the object being held," and the fuzzy set "expensive" can be represented with a typical membership function as that shown at the bottom of Figure 13.3. Taken separately, the two behaviors are fully independent and each has a different one-dimensional input space. They could therefore be developed and exhibited independently of each other. When these two behaviors are merged to create a more complex reasoning system, then the new input space for the system becomes two-

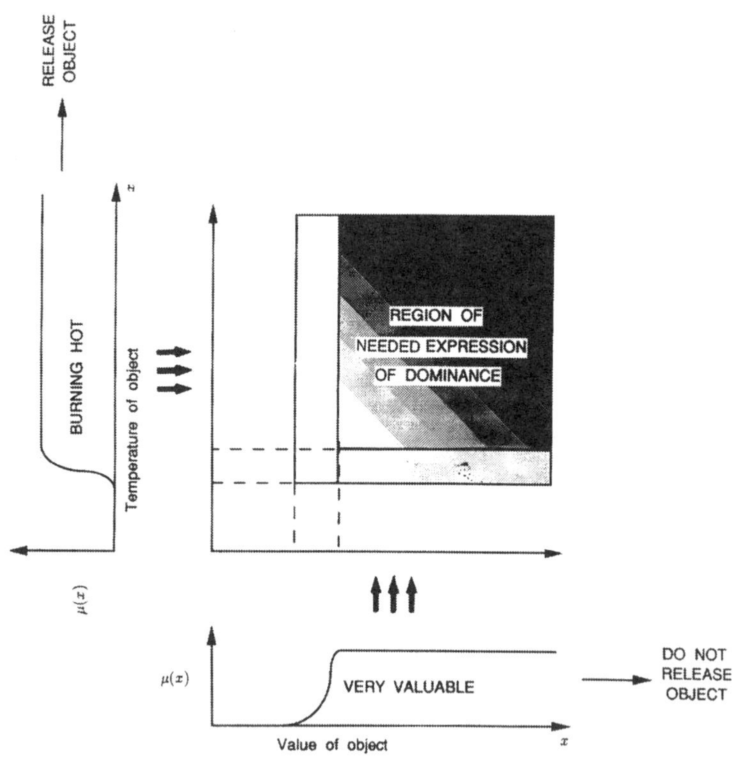

FIGURE 13.3
Example of possibly conflicting behaviors operating on different input dimensions and the same output dimension.

dimensional. Within that new system, the two behaviors are still independent because they trigger from stimuli on different input dimensions that do not affect each other. However, every time the child or the robot will handle one of great grandmother's priceless china cups filled with very hot tea, the implicit dominance of one behavior over the other in their overlapping region of the input space will be exhibited: the cup will, or will not, be dropped, signifying respectively that the "don't get burned" behavior dominates, or is dominated by, the "don't drop valuable objects" behavior at that particular point in the input space.

It is clear that which behavior dominates the other may vary over the region of overlap (see Figure 13.3) and that (for both the child and the robot) some instruction, reinforcement learning, experience, and so forth may cause the dominance process and its resulting outcome to vary over time. However, it is also clear that an indication of what behavior dominates the other in what region of their overlapping areas in the input space *does exist at all times within the child or the robot*. For the robot, therefore, this dominance information must be included *at the time the second behavior is added and merged with the other into the system's reasoning module*. In the FBA formalism and in the automated system discussed in the next section, this is accomplished either through the "suppression" mechanism, in which the output membership function of the dominant behavior is modified so that its weight will appropriately overpower that of the other behavior in the center of area (c.a.) calculation, or using the concept of "inhibition," in which the input membership function of the "weaker" behavior is partially truncated so that the dominant behavior always triggers with greater strength. Although both suppression and inhibition mechanisms result in expressing the desired dominance, their concepts are quite different: one basically operates on the relative weight of the behaviors in the output space, whereas the other modifies the triggering conditions of the behaviors in the input space. This difference, which is transparent in the automated system, becomes significant when one seeks to couple the system with learning modules for refinement of the behaviors and/or of the rule membership functions through reinforcement learning.

Figure 13.4 schematically shows these very intuitive suppression and inhibition mechanisms on a simple example in which two rules, each with two inputs and one output, act on

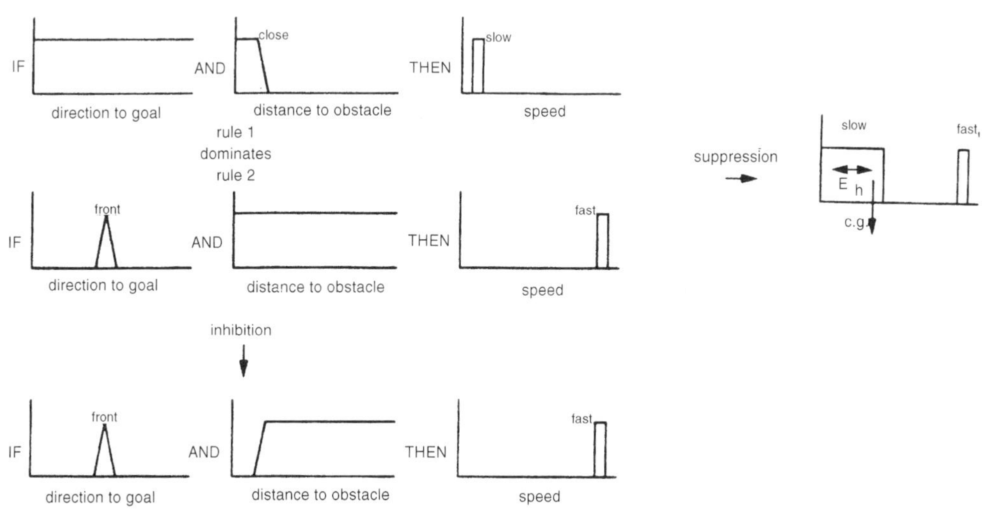

FIGURE 13.4
Schematic of the suppression and inhibition mechanisms.

the speed control of a robot. The first rule, which may be part of a behavior for speed control as a function of frontal obstacle proximity, states that if the obstacle is *close* (and the goal direction is *anything*) then the speed should be *slow*. The second rule states that if the goal is *straight ahead* (and the obstacle distance is *anything*) then the speed is *fast*. The membership functions uniformly equal to 1 over their entire range represent the nonsignificant input dimensions of each behavior. In the example of the figure, the output membership functions of both rules initially are similar in weight at both ends of the speed spectrum. In situations in which both behaviors will trigger, that is, when the goal is straight ahead and an obstacle is very close in the frontal direction, then the merging of the two behaviors will result in speed that will be in the center of the range, say medium. However, it is clear that a medium speed may be too large to maintain the safety from obstacles embodied in rule 1 and that rule 1 should dominate rule 2 in that situation. If using the suppression mechanism to express this dominance, the output membership function of the dominant rule, rule 1, is modified so that it overpowers the output membership function of rule 2 when the merging and defuzzification is performed; that is, it is made sufficiently large to "attract" the resulting center of area within an E_h value of the initial output center of rule 1, as shown on the right-hand side of Figure 13.4. This suppression mechanism can be thought of as analogous to "conditioning" in which, for example, parents would repeatedly reinforce the outcome of one behavior of a child over another of his or her behaviors in specific ambiguous (or conflicting) situations.

If the inhibition mechanism is used to express the dominance, then the nonsignificant input dimension of the dominated rule is truncated to prevent the dominated rule from triggering in the particular conflicting situation. This is essentially equivalent to removing some type of stimulus from a particular behavior and can also be seen as analogous to some inhibition conditioning in learning processes. A mathematical representation and implementation of these two dominance mechanisms is presented in the following subsections.

Suppression Mechanism

Given a rule i and its output function $\mu_i(y)$ on the y output dimension, the center of mass y_i and weight m_i of its output are given by

$$y_i = \frac{1}{m_i} \int y\mu_i(y)\, dy \tag{13.9}$$

$$m_i = \int \mu_i(y)\, dy \tag{13.10}$$

and, by definition, $m_i > 0$ and $y_{\min} \leqslant y_i \leqslant y_{\max}$, where $y_{\min}$ and $y_{\max}$ are the extremum values of the y output dimension. Assume that two rules, which control the same output y, have respective output membership functions $\mu_i(y)$ and $\mu_j(y)$, with corresponding centers of mass and weights given by Eqs. (13.9) and (13.10). When both rules trigger under the same input conditions, the resulting output membership function is the union (or max operation) of $\mu_i(y)$ and $\mu_j(y)$ and, assuming a "center of area" defuzzification scheme is used (other schemes and their respective representation could of course also be used as alternatives), the overall output y_o is

$$y_o = \frac{m_i y_i + m_j y_j}{m_i + m_j} \tag{13.11}$$

From Eq. (13.11), it is clear that the *relative* value of the output weight of the rules can change the output y_o. In particular, if the weight of one rule is strong enough, the output y_o can be "attracted" *within* the output membership function of this rule, *even though the other rule still "works" and contributes to the overall output*. This constitutes the basis for implementing the suppression mechanism.

In the general case, a rule, labeled rule h, may be required to suppress a set S of other rules $i, i \in S$. The suppression mechanism can be expressed as the requirement that, when all the rules h and $i, i \in S$, trigger, their combined output y_o must fall within an "allowable error," E_h, of the center of mass y_h of rule h. Typically, a small value E_h indicates a "strong" dominance of rule h over the rules in S, while a large E_h represents a "weak" dominance or what we will refer to as a "preference." The suppression condition can thus be written as

$$|y_o - y_h| \leqslant E_h \tag{13.12}$$

Since

$$y_o = \frac{m_h y_h + \sum_{i \in S} m_i y_i}{m_h + \sum_{i \in S} m_i} \tag{13.13}$$

we have

$$\frac{\sum_{i \in S} |y_h - y_i| m_i}{m_h + \sum_{i \in S} m_i} \leqslant E_h \tag{13.14}$$

from which a minimum value for m_h can be calculated. Since the distances between the centers of mass $|y_i - y_h|$ are always less than or equal to $Y_r = y_{\max} - y_{\min}$, the selection of m_h as

$$\left(\frac{Y_r}{E_h} - 1\right) \sum_{i \in S} m_i \leqslant m_h \tag{13.15}$$

guarantees that rule h suppresses the other rules $i, i \in S$. From m_h, and the selected shape (e.g., rectangular, triangular) of the output membership function, the width of, and/or the full $\mu_h(y)$, can easily be determined.

Inhibition Mechanism

In some cases, the suppression mechanism may not be implementable because the output membership function of the dominant rule cannot be made sufficiently overpowering to obtain the desired control result. This will typically occur when E_h is specified to be very small [see Eq. (13.15)] or several dominance mechanisms need to be implemented within the rule base, resulting in progressively large weights calculated from Eq. (13.15). In those cases, the inhibition mechanism can be used instead of the suppression mechanism, and the dominance of rule h over rules $i, i \in S$, is forced by appropriately truncating some of the *input* membership functions of rules i so that these rules do not trigger when rule h does. The only membership functions that are truncated are those in the dimension that is significant for rule h and nonsignificant for rules i. For example, if $(A_1, B_1\, C_1, D_1, E_1, F_1)$ defines a rule [see Eq. (13.1)] that has B as a significant input dimension and dominates rule $(A_2, B_2, C_2, D_2, E_2, F_2)$, then B_2 is the overlapping input to be truncated. The truncated condition B_2^* is expressed as $B_2^* = B_2 \cap \bar{B}_1$, where $\bar{B}_1$ represents the complement of condition B_1. Note that this effectively removes the overlap of the rules' input in the multidimensional input space (and therefore the conflict between the rules) and that no truncation is necessary in the other input dimensions (that are nonsignificant for the dominant rule).

2.5 FBA Control Architecture

With the FBA features described in the previous subsections, rule bases can be generated, as will be discussed in the following section. At this point, it is important to note that all behaviors are acting simultaneously and in parallel as illustrated in Figure 13.5. The user should keep this in mind when designing rules and/or creating strategies for rule base assembly, as this concurrency of operation is what allows complex reasoning to be performed through simultaneous execution of multiple rules, but it also may lead to some of the conflicting situations resolved using the dominance concept described in Section 2.4.

As illustrated in Figure 13.5, it may sometimes be desirable to group sets of rules in individual rule bases, each rule base corresponding to a particular task that the robot could perform. Aside from providing a clean organization of the overall rule base for robots that can perform many and varied tasks (e.g., navigation, manipulation, vision, and assembly), this grouping also illustrates how a complete rule base corresponding to an entirely new task can be added to the system through simple superposition. A modular hierarchy thus becomes apparent in a reasoning rule base, where rules can be grouped into behaviors, behaviors can be grouped into tasks, tasks can be grouped into activities, and so on.

From Figure 13.5 and Eqs. (13.7) and (13.8), it is apparent that all input data is available to every rule in the system. When a new sensory mode is added to the robot, for example, when a new task (and its corresponding rule base) using a new sensor or a different input calculated from existing sensor data is added to the system, then the dimension of the input space is increased. The resulting modifications needed to the previously existing rule base are

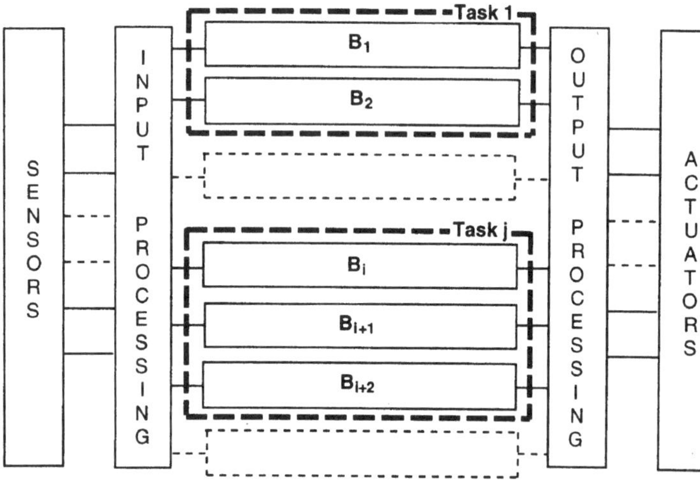

FIGURE 13.5
Schematic of the FBA basic control architecture.

very simple, because the new input dimension is "nonsignificant" for every one of the rules in the existing rule base. Therefore, according to Eq. (13.7), membership functions uniformly equal to 1 over their entire range need to be added to each existing rule in the new input dimensions. Similarly, if new actuation modes and corresponding action commands are added to the robot, then the dimension of the output space of the system is correspondingly increased, and membership functions uniformly equal to 0 over their entire range simply need to be added to each existing rule in the appropriate output dimension, according to Eq. (13.7). Growth of an FBA-type system, be it by addition of sensory modalities, output capabilities, or new behaviors, tasks, or activities to be performed, is thus straightforward, as it requires only a simple superposition of rules or entire rule bases on the already existing rule base and/or simple additions of "standard" membership functions to all the existing rules.

3 RULE BASE GENERATION METHOD AND AUTOMATED SYSTEM

Just as different people may use different strategies, different rules, and different qualitative variables to express their navigation process and still navigate efficiently "in their own way," several strategies may be used to embody a particular process in a rule base; that is, there is not a single or unique rule base representation of a given process. For example, a rule base for obstacle avoidance may be built on the basis of a distance-to-obstacle strategy, as was done in Refs. [18] and [19], with behaviors organized and developed for input conditions in which obstacles are very near, near, far, very far, and so forth. It could also be built on the basis of the direction to obstacles, with behaviors organized and developed for obstacles located on the right, center, or left of the travel direction. Because of the requirement of our FBA for each behavior to trigger from a single input dimension, the expression in rules of various possible strategies may appear quite different, even though the overall process and resulting actions of the robot may be similar. Thus, one of the very first activities that a user should perform when developing an FBA-based rule base is to develop a "strategy" for representing the reasoning scheme embodied in the rule base. This will include the defining of what variables will constitute the input and output of the system, depending on the type of data and control modes available in the hardware system. From this strategy, an initial expression of the behaviors and rules, possibly written in "qualitative" terms, can be produced, for example, IF goal is to the *left* (alternatively, *straight ahead*, *right*), THEN turn *left* (alternatively *nothing*, *right*), or IF obstacle is *far* THEN speed is *fast*. At this point, the user should also define the qualitative values, or fuzzy sets, such as left, right, far, that are used in the qualitative expression of the behaviors or rules, by defining their membership functions (e.g., see Figures 13.1–13.4 and Section 2.3). When generating the list of rules in "qualitative" form, the user needs to verify the proper implementation of the FBA principles listed in Section 2. This is generally easily done and the method can be automated. In fact, to ease the development and modification of rule bases for our experiments, we have developed a computer system to automate this process. In what follows we will present the methods and processes for the development of FBA based rule bases by describing our automated system, so that the reader can better perceive both the FBA formalism and its methodical implementation.

In the current version of our automated system, each rule is assumed to be of the form

$$\text{IF } (A \text{ is } A_1 \text{ and } B \text{ is } B_1 \text{ and } C \text{ is } C_1 \text{ and } D \text{ is } D_1) \quad \text{THEN } (E \text{ is } E_1 \text{ and } F \text{ is } F_1) \quad (13.16)$$

therefore operating on four input and two output channels. This configuration was chosen in our version of the automated system because it corresponds to what is available on the custom-designed VLSI fuzzy inferencing chips and boards [16–18] that we utilize in our experimental work. However, extension to any number of input and output channels is possible, and the reader can very easily tailor an automated system for his or her particular conditions and applications if desired.

In our automated system, the user inputs the strategy for the rules in a "qualitative" form using the format shown in Figure 13.6. The five-line format on the left and the four-line format on the right in Figure 13.6 both describe one rule. In either case, the first line gives the "reference name" of the rule, and the last two lines specify what the desired input and output of the rule on the "significant" input and output dimensions of the behavior are. For example, the five-line-format rule on the left of Figure 13.6 specifies that IF the input data in the second input dimension is *near*, THEN the output membership function in the first output dimension is centered at 20% of the x-axis range, while the four-line-format rule on the right specifies that IF the input data in the third input dimension is *far*, THEN the output membership function in the second output dimension is centered at 10% of the x-axis range. In the notation of Eqs. (13.5) to (13.7), $n = 4$, $m = 2$, and the rule on the left of Figure 13.6 corresponds to a behavior B_{21} while that on the right corresponds to a behavior B_{32}.

If the subject rule must dominate other rules, then the five-line format on the left of Figure 13.6 is used, with the second line listing the names of the rules or behaviors that are suppressed (or inhibited) by this rule and the third line giving the corresponding suppression parameter E (see Section 2.4). Use of the character "?" in the suppression list of line 2, such as "G?" in the example of Figure 13.6, indicates that the rule suppresses all other rules whose name includes the other character, in this case all rules whose name begins with the letter "G." If the subject rule does not suppress other rules, then the four-line format on the right of Figure 13.6 is used, where the first character of line 2 is the letter M followed by a number expressing the desired initial weight m of the output membership function (discussed in the following paragraphs), and the line giving the suppression parameter E is omitted.

When the user has listed all the rules of the desired behaviors in the format of Figure 13.6, the automated system can generate a "skeleton" of the rule base and check whether it verifies the input-related requirements of the approach. In particular, the system constructs the four-dimensional input spaces for each of the two output dimensions, so that it can evaluate completeness of, and redundancy in, the rule base and report all instances to the user. For any region of incompleteness, that is, regions of the input space not covered by any of the behavior stimuli, the user decides on either the addition of a behavior to cover these possible stimuli, extension of the current behaviors (through extension of their input membership function) to include these input regions, or no modification if input data within these uncovered regions or "blind spots" are never expected to occur (for example, if these regions correspond to values outside the operating range of the sensors). For the regions of redundancy, that is, areas where stimuli from two or more behaviors are overlapping, the

```
name : RN                                        name : LF
suppressing list : G? LF                         M       : 0.3
E                : 50                    or      Outputs : - 10
Outputs          : 20 -                          Inputs  : - - Far -
Inputs           : - Near - -
```

FIGURE 13.6
Input format for the automated rule generation system.

system reports every rule for which a dominance has not been specified but may be required because of the input overlapping. The user can then interactively add or modify the dominance specifications in lines 2 and 3 of each rule, until all requirements of the approach are verified and all desired dominances are expressed in the rule base. The process explained in this paragraph is iterated until no further modification by the user is needed in the list of rules in "qualitative" format.

The actual generation of the rule base, including the suppression and/or inhibition mechanism, can then proceed as follows: initially, all rules are given a "standard" output membership function with a weight of 1 or, if specified, the weight following the letter M in line 2. The output membership function is centered at the percent value of the x-axis range specified in the one-before-last line of the "qualitative" expression of the rule (see Figure 13.6). The system then checks the sets of rules that are affecting the *same* output dimension. If no suppression mechanism has been expressed between the rules because dominance is not necessary, then the output membership functions are unconstrained and they remain at their "standard" value. If a dominance has been expressed between two or more rules, then the dominant rule is the one that is modified if suppression is possible; otherwise the dominated rules are modified using inhibition, as explained in Section 2.4. The output membership functions are first finalized from the application of Eqs. (13.9) to (13.15) for the suppression mechanisms. As mentioned in a previous section, the membership functions defining the significant input fuzzy sets are defined by the user on a behavior-by-behavior basis (and can possibly be stored according to their "name" in a "membership function library"). Using these, the input membership functions in the nonsignificant dimensions that are affected by inhibition mechanisms are then appropriately modified. The full set of membership functions for the entire rule base can thus be generated by the automated system. Examples of such automatically generated rule bases can be found in the next section on experimental results.

4 SAMPLE EXPERIMENTAL RESULTS

The FBA and associated automated system were utilized to generate rule bases for the sensor-based navigation of autonomous robots and the resulting rule bases were tested on navigation problems in a variety of *a priori* unknown environments. In this section, sample results from some of these experiments are presented to illustrate the approach and the rule base generation process including the suppression and inhibition mechanisms. Actual paths taken by the robot in test environments are displayed to show the sensor-based navigation behaviors resulting from various rule bases. Note that, except for specifying the goal, no information on the environment is given *a priori* to the robot, nor is any map generated during the motion. The navigation, therefore, is purely reactive, that is, it involves no memory or real-time information storage of any type. Purely reactive systems are known to have limitations, and Section 5 will discuss an approach to remedy some of them.

The first series of experiments took place in laboratory-type environments using our recently designed omnidirectional platform [27], pictured in Figure 13.7. The photograph in the figure shows the two batteries (rear right and rear left) and the seven-slot VME bus (center front), which hosts the fuzzy inferencing board. On top of the six threaded poles visible in the figure fits a ring of 24 acoustic range sensors mounted at the edge of the platform. The control system of the platform (detailed in Ref. [27]) includes a velocity loop servoing at 100 Hz on the commanded motor velocities. The motor velocities are very simply calculated from the desired translational and rotational velocities of the platform [27]. These latter velocities will be referred to hereafter as the platform speed control and turn control, respectively.

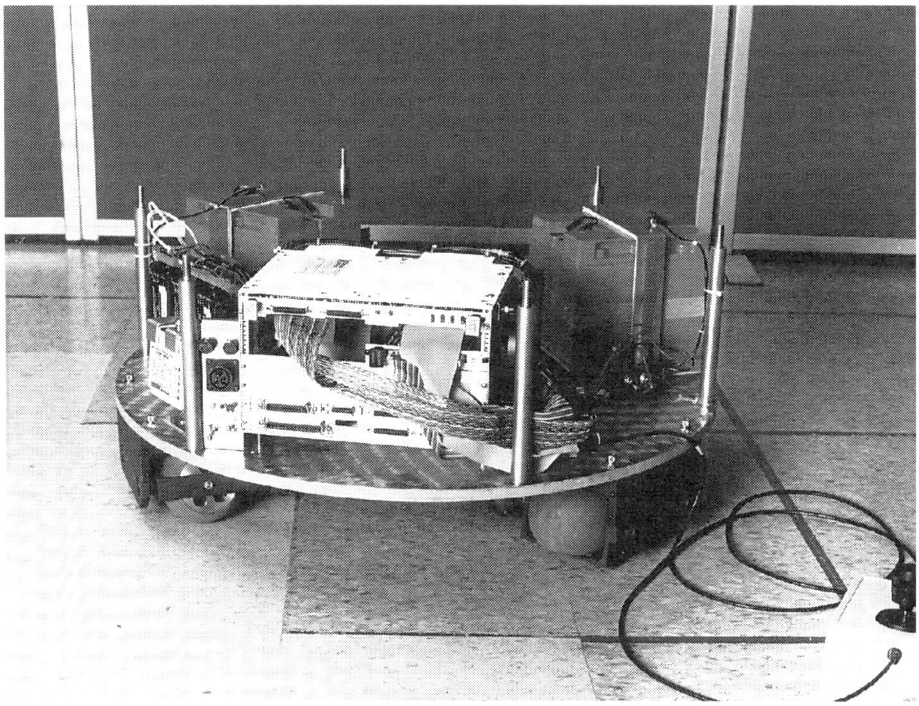

FIGURE 13.7
The omnidirectional holonomic robotic platform prototype.

The fuzzy inferencing chips and boards described in Refs. [16], [17], and [18] were used in this experimental work. Because these boards allow only inferencing on four input variables to produce two output variables, the sensory inputs to the fuzzy inferencing were selected as the goal direction and obstacle proximity in three sectors located at the left, center, and right of the travel direction. As shown in Figure 13.8, each sector was selected to encompass five sonars with an overlapping of one sonar between the sectors. In each sector the distance returns from each of the five sonars are weighed by a factor proportional to their firing direction and the smallest value is utilized to indicate obstacle proximity within the sector. Effectively, this corresponds to giving the platform the equivalent of three "very wide and blurry" eyes, each 75° wide. For added flexibility in the series of experiments, the navigation goal specification was made user selectable with capability to specify the goal as a point or as a heading to be maintained (see the right-hand side of Figure 13.8) at the beginning of the experiment. When the goal is a point, the odometry system updates the position of the robot at each loop rate and calculates the relative direction to the goal point as input to the inferencing system. When the goal is a heading, a compass is used to provide directly the relative goal direction as the difference between the platform current heading and the goal heading. The two output channels of the inferencing were selected as the speed and turn control of the platform. Thus, with these input and output channels, behaviors corresponding to speed control (SC) and turn control (TC) as functions of goal orientation (GO) and obstacle proximity (OP) could be developed.

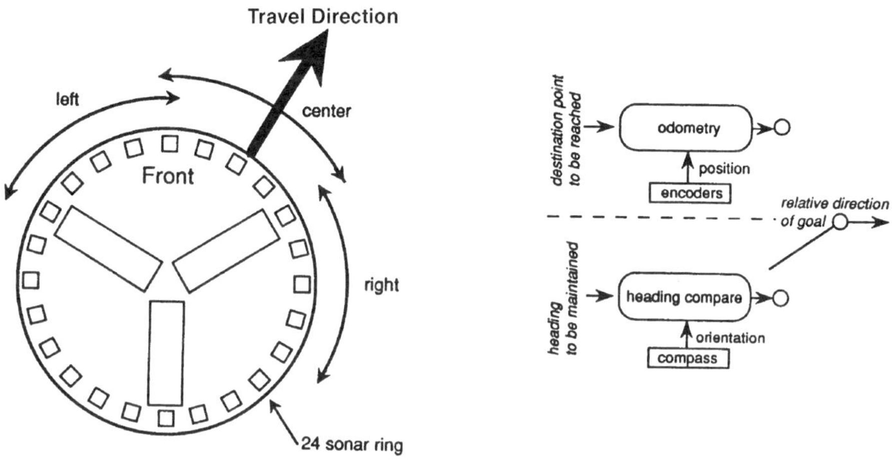

FIGURE 13.8
Schematic of the three 5-sonar sectors providing obstacle proximity input data and of the two methods for calculating the goal direction depending on the mode of goal specification.

4.1 Example of Basic Navigation Behaviors

As mentioned previously, several rule bases, representing various "strategies," may be developed to solve a complex problem or to embody a complex behavioral process. Figure 13.9 shows the rules for one of the rule bases we investigated for sensor-based navigation, in which front obstacle proximity is used only for speed control, while side obstacle proximity is used only for turn control. The behaviors shown in Figure 13.9 are thus organized as follows:

GO → TC (3 rules)

GO → SC (1 rule)

"front" OP → SC (4 rules)

"left" OP → TC (4 rules)

"right" OP → TC (4 rules)

The input file that was used to generate the rules through the automated system is shown on the left of the figure. Each group of four "qualitative" rule descriptions (in the format of Figure 13.6) of the input file is presented next to the automatically produced rule membership functions graphically displayed on the right side of the figure. Each of the 16 rules in the figure is displayed as a vertical arrangement of six graphs of membership functions. The top four graphs in each rule show the membership functions corresponding to the input variables A_1, B_1, C_1, and D_1 in Eq. (13.16), that is, the direction angle to the goal and the distances returned by the left, center, and right "wide blurry eyes," while the bottom two graphs correspond to the output variables E_1 and F_1, that is, the turn command and the speed command. The vertical axis of each membership function is labeled in bits (for the

```
Behave : GF
M     : 1.0
Outputs : 50 -
Inputs  : Front - - -

Behave : GL
M     : 1.0
Outputs : 10 -
Inputs  : Left - - -

Behave : GR
M     : 1.0
Outputs : 90 -
Inputs  : Right - - -

Behave : GB
M     : 0.5
Outputs : - 0
Inputs  : Back - - -
```

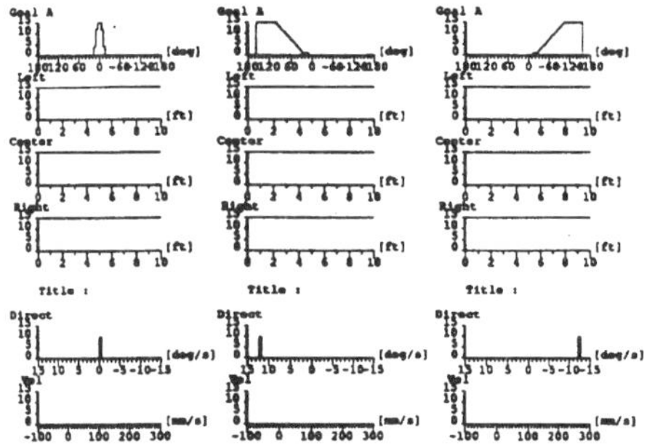

GO ⟶ TC

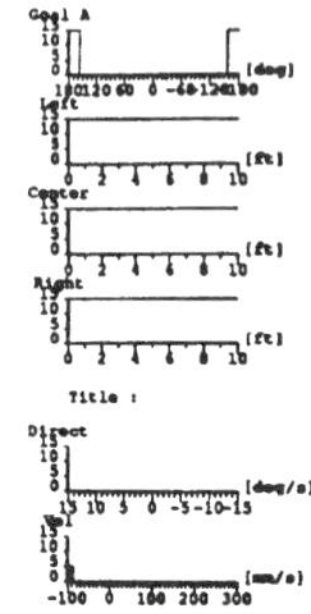

GO ⟶ SC

```
/***   Front    ***/
Behave : FC
Sup   : G?
E     : 10
Outputs : - 25
Inputs  : - - Close -

Behave : FN
Sup   : G?
E     : 50
Outputs : - 50
Inputs  : - - Near -

Behave : FF
M     : 0.5
Outputs : - 80
Inputs  : - - Far -

Behave : FNo
M     : 0.3
Outputs : - 100
Inputs  : - - No -
```

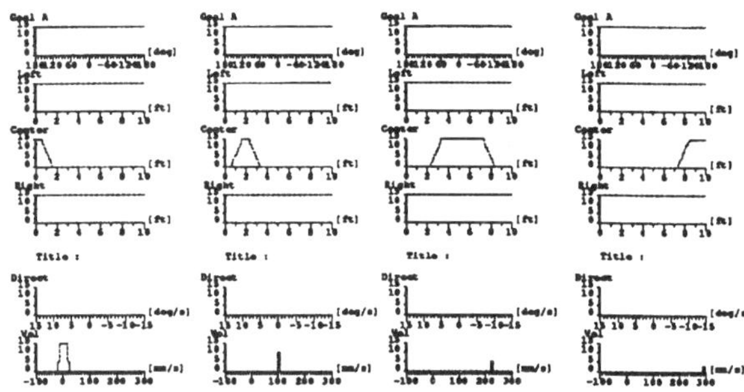

front OP ⟶ SC

```
/***   Left    ***/
Behave : LC
Sup   : RN
E     : 20
Outputs : 90 -
Inputs  : - Close - -

Behave : LN
Sup   : G?
E     : 20
Outputs : 80 -
Inputs  : - Near - -

Behave : LF
Sup   : G?
E     : 100
Outputs : 70 -
Inputs  : - Far - -

Behave : LNo
M     : 0.3
Outputs : 50 -
Inputs  : - No - -

/***   Right    ***/
Behave : RC
Sup   : LN
E     : 20
Outputs : 10 -
Inputs  : - - - Close

Behave : RN
Sup   : G?
E     : 20
Outputs : 20 -
Inputs  : - - - Near

Behave : RF
Sup   : G?
E     : 100
Outputs : 30 -
Inputs  : - - - Far

Behave : RNo
M     : 0.3
Outputs : 50 -
Inputs  : - - - No
```

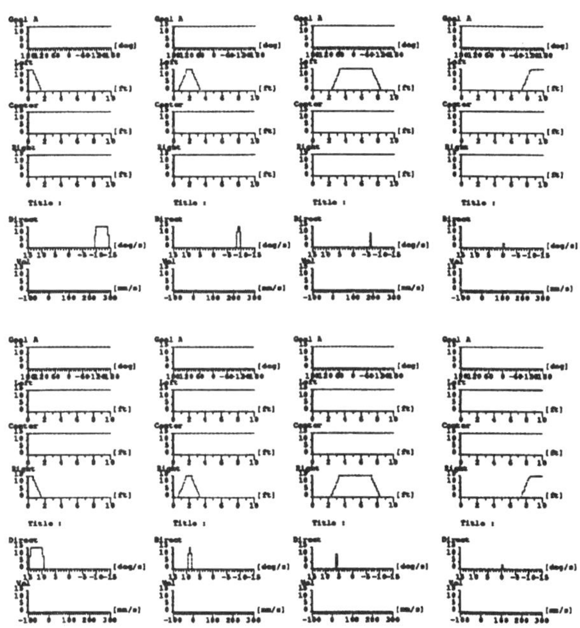

side OP ⟶ TC

FIGURE 13.9

Sample rule base for sensor-based navigation.

implementation on the VLSI chip) where the [1, 0] interval is discretized over the 16 $\{0, \ldots, 15\}$ bit values.

Note that every behavior, and consequently every rule, represents a mapping from only one input dimension to only one output control. In the other input dimensions, membership functions uniformly equal to 1 over their entire range signify that the behavior is not affected by stimuli in these dimensions or, in other words, that the data in these dimensions is "unimportant" or "can be anything." For the output controls that are not affected by a given behavior, a membership function uniformly equal to 0 over the entire range of that output ensures no contribution of the behavior to that particular actuator control.

The three rules of the GO → TC behavior simply express that if the goal is to the center, left, or right, respectively, then the robot should make a zero, positive (i.e., left), or negative (i.e., right) increment of turn, respectively, and there is no command for speed change. The GO → SC rule states that if the goal is toward the back sector of the robot, the robot should slow down (i.e., apply a negative contribution of speed). The four rules of the front OP → SC behavior express that if the obstacles in front are very far, then a speed close to the maximum possible can be applied, and that the closer the frontal obstacles, the slower the robot should go, eventually stopping when the obstacles are dangerously close. Note that the weight of the velocity command (i.e., the membership function) increases with "increasing danger." Similarly, the eight rules of the OP → TC behaviors express that the closer the obstacles on a side, the greater the increment of turn in the opposite direction and the "heavier" the turn command should count in the output control calculation.

Note the large weight of the output membership function of the "very near" OP → TC rules, which results from this behavior having been selected as suppressing the GO → TC behavior, that is, expressing that when obstacles are very close, their avoidance is always of greater importance than tracking the goal. It is clear that without this expression of dominance, the GO → TC and OP → TC behaviors would often result in deadlock or oscillatory situations in which the robot would not turn at all or would oscillate between two orientations. This type of situation constitutes one of the very serious drawbacks of the navigation methods using potential field techniques and has been alleviated here using the suppression mechanism.

Figures 13.10 and 13.11 show plots of actual runs made with the robot to illustrate the overall reactive navigation obtained with the automatic generation of fuzzy rules. These plots are also given here to provide an example of the effect on the navigation behaviors that a dominance mechanism (suppression or inhibition) can produce. In the figures, the shaded areas represent the obstacles that were placed in the room, while the path of the robot is illustrated using the succession of circles showing the position of the robot every 20 loop rates. In Figure 13.10, the rules shown in Figure 13.9 were used, which embody a very strong dominance of the obstacle avoidance (OP → TC) rules over the goal tracking (GO → TC) rules. Consequently, due to the almost constant proximity of the corridor walls, the suppression mechanism is quite effective in the early part of the run and the robot wanders around for quite a long time, guided principally by obstacle avoidance. It eventually gets positioned ideally to enter the corridor and then turns right, in a direction closest to the goal. Clearly, the dominance of the obstacle avoidance rules over the "move to the goal" behavior may be too strong in this environment.

For the sample run shown in Figure 13.11, this dominance has been decreased; that is, the suppression parameter [E_h explained in Eqs. (13.12) to (13.15)] has been doubled from 20 to 40 in the obstacle avoidance rules named LC, LN, RC, and RN (see input file on the left-hand side of Figure 13.9), and a corresponding rule base has been generated using the automated system. For comparison with the original rule base of Figure 13.9, Figure 13.12 shows the

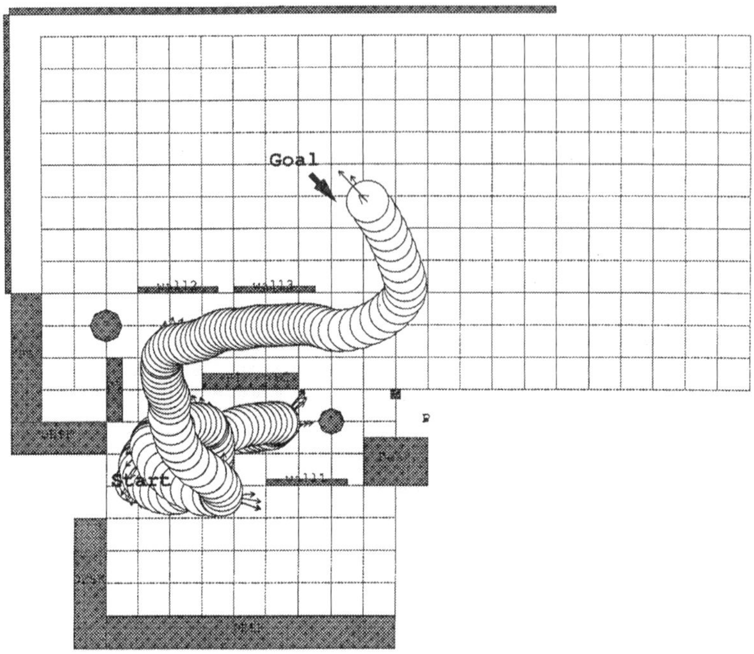

FIGURE 13.10
Actual run of the robot with strong behavioral dominance of obstacle avoidance over goal tracking.

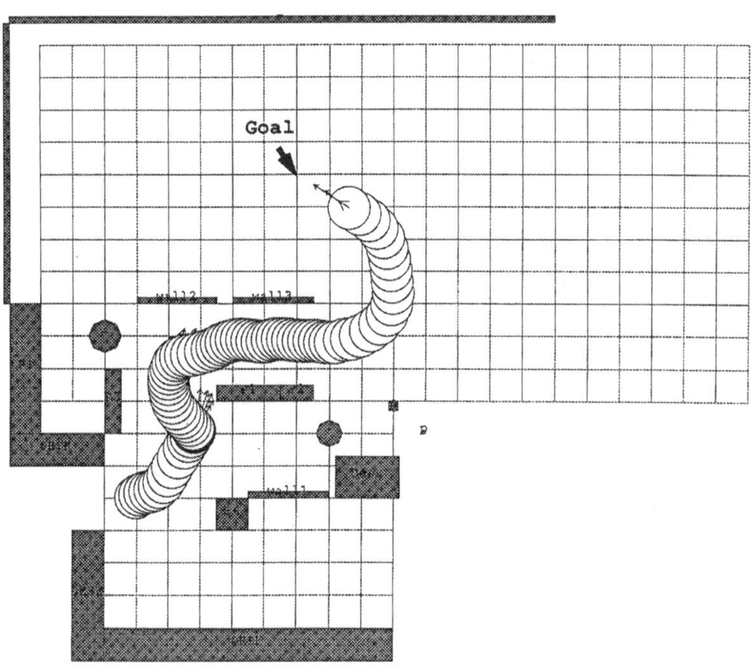

FIGURE 13.11
Same as Figure 13.10 with lesser behavioral dominance of obstacle avoidance over goal tracking.

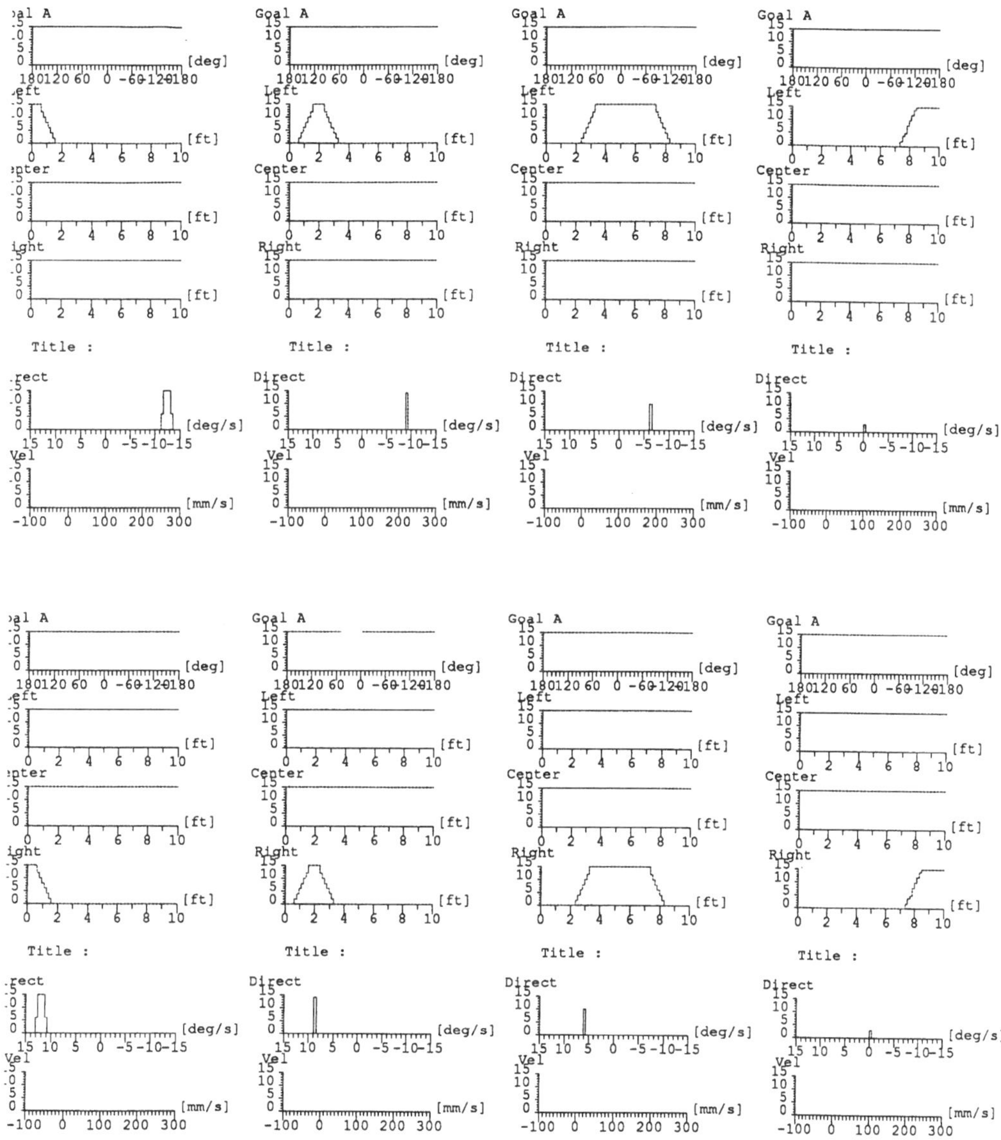

FIGURE 13.12

Modified OP → TC rules due to a change of suppression parameter from 20 to 40 in the RC, RN, LC, LN behaviors of the rule base of Figure 13.9.

OP → TC behaviors after they have been modified due to the change of the suppression parameter. Only the turn control output membership functions of the LC, LN (the two rules on the left of the top row in Figure 13.12) and RC, RN (the two rules on the left of the bottom row) rules have been modified, now exhibiting a lesser weight on the turn control. In Figure 13.11, the robot is seen to negotiate the entrance of the corridor much more rapidly because of the greater effect of the goal-tracking behavior, resulting in a much shorter run to the goal.

From an overall behavioral point of view, the simple change of the dominance of the "obstacle avoidance" over "move to the goal" behaviors in the rule base has transformed the "shy" robot driven mainly by obstacle avoidance in Figure 13.10 into a "much braver" robot proceeding more rapidly toward its goal in Figure 13.11. Note that this "transformation" has been accomplished without modification of the *qualitative logic* embodied in the elemental behaviors. Also note that the change in speed when the robot exits the corridor and faces the open space in the upper right section of the environment is very similar in both sample runs, illustrating that, as expected, the effect of the speed control behaviors on the robot motion has not been affected by the change of relative dominance in the turn control behaviors.

4.2 Extension to Robots with Kinematic Constraints

One of the expected strengths of the FBA using elemental "human-like" behaviors is that the *linguistic logic* embodied in the behaviors should be invariant among systems of similar characteristics. In other words, for robots with similar perceptive abilities and motion capabilities, the linguistic expression of given behaviors, and therefore their representation in the fuzzy framework, should be the same for compatible input and output. For example, a "goal-tracking" behavior connecting the perceived goal direction to a rate of turn [e.g., IF (goal is to the right) THEN (apply increment of turn to the right)] should be invariant for any robot that has a means of perceiving the goal direction and performing the required turn. Using this property (and realizing that the rate of turn of a car is proportional to the steering angle of the wheels [2]), all navigation behaviors developed for the laboratory omnidirectional platform appear directly applicable to the driving of a car of similar size, except for behaviors that require a rate of turn too large for the car to perform because of its limited steering angle. The "very near" OP → TC rules (see Figure 13.9), which require the platform to perform high rates of turn (using its omnidirectional capability) when obstacles are detected at dangerously close ("very near") distances on the left or right of the travel direction, are the only ones that therefore would require attention when changing from the omnidirectional platform to a car with a small maximum steering angle.

As a demonstration of this transportability of invariant behaviors from one system to another, the same behaviors and the same fuzzy rules that were utilized for the omnidirectional platform were used to implement the autonomous control of a car with a large limit on the steering angle, on the basis of the same three "wide and blurry" eyes and goal direction input. Figure 13.13 shows a simulation example of such a navigation in which the car has to reach a goal (in the upper right section) and then return to its start position (in the lower left section). Note that the out and return paths are different. Also note that the large maximum steering angle that has been selected for the car in this simulation allows a small radius of turn (e.g., see the sharp turn in the upper right section) and therefore prevents situations with very near obstacles.

To take into account the car's limited radius of turn and to complete the navigation rule base for the driving of the car, a behavior had to be included to handle the situations in which very near obstacles are detected. Another strength of the FBA is its capability for superposition of elemental behaviors allowing progressive addition of behaviors to the system to resolve situations of increasing complexity. Because the other basic behaviors assume collision-free navigation amid far and near obstacles, the situations involving very near obstacles would occur only when the car does not have enough space to complete a turn away from obstacles because of its limited steering angle and radius of turn and thus would require some maneuvers using reverse gear. By observing human reactions to such stimuli, a human-like response was created that can be expressed as follows: IF [obstacle is very near on right (alternatively, left)] THEN [steer right (alternatively, left) AND (back up)]. This response was further divided into a steer control behavior, "very near" OP → TC, and a speed

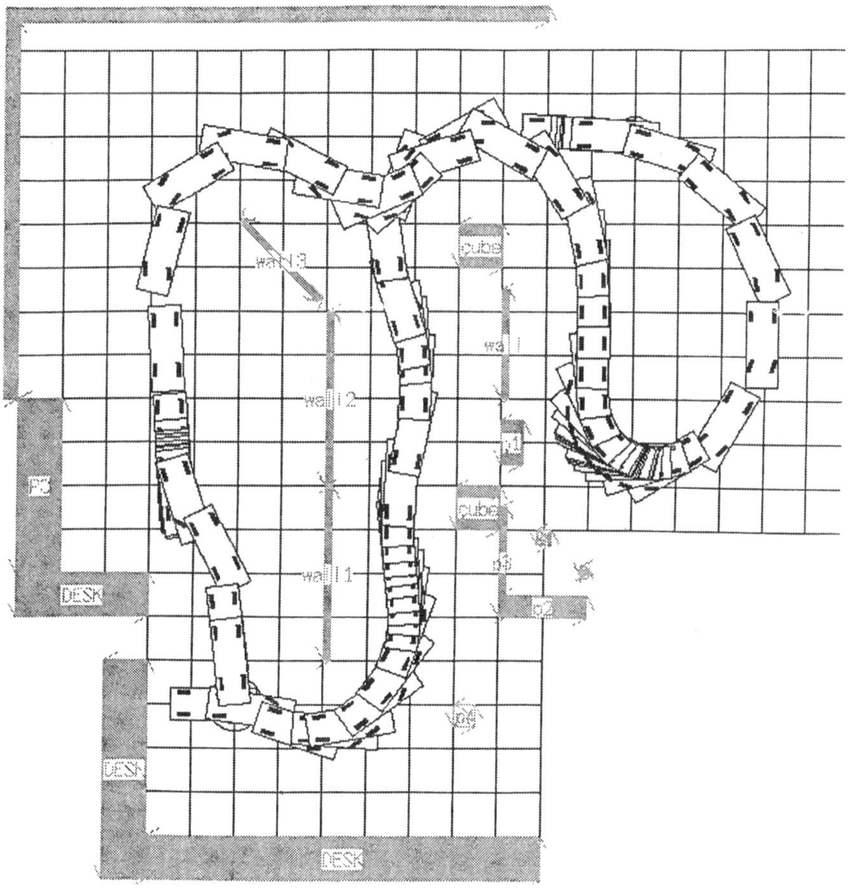

FIGURE 13.13
Simulation example of the autonomous navigation of a car using three "wide" sonars and the same invariant navigation behaviors as for the omnidirectional platform.

control (back up) behavior, "very near" OP → SC, to respect our approach's requirement for a single input dimension of behaviors.

Both simulation and outdoor experiments were performed with the maneuvering behavior–augmented rule base. These experiments and results are presented in detail in Ref. [23] and are only briefly reviewed here to illustrate the augmentability and transportability of FBA-compatible rule bases. Once the basic fuzzy rule base (e.g., see Figure 13.9) augmented with the maneuvering behaviors had been tested for autonomous navigation in simulated environments [23], the system was implemented on a real car for outdoors experiments. Because of the unavailability of a car with automated actuation, no autonomous navigation tests were performed outdoors. However, the system was investigated for use as a "driver's aid" using one of the company cars. In this driver's aid mode, the same sensors and inferencing system as in the previous experiments are used; however, the commands produced by the rule base are displayed to the driver to guide him or her in driving the car. The driver conventionally uses the gas and brake pedals and the steering wheel to implement the commands that are displayed on a portable computer screen located next to him or her in the cabin (see Figure 13.14). During these tests, the driver is prevented from seeing the

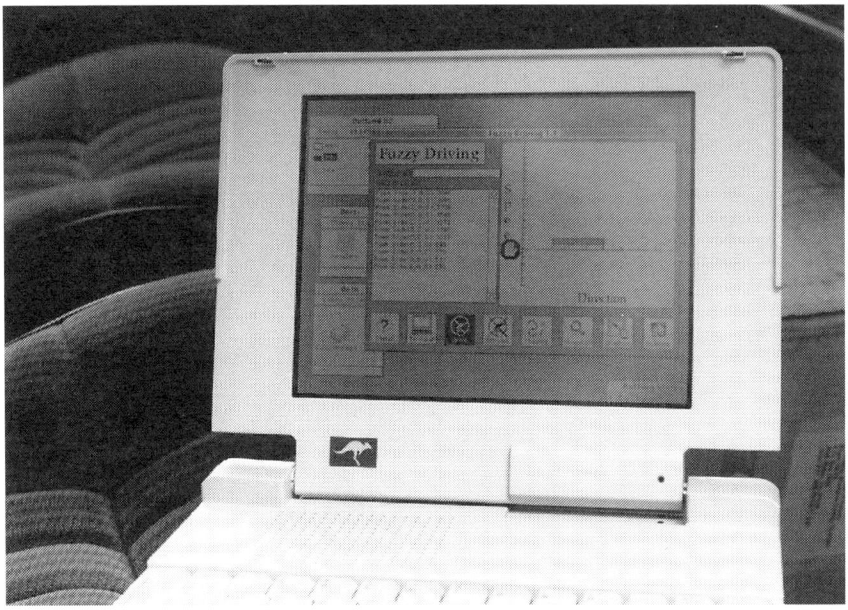

FIGURE 13.14
Photograph of the inferencing results displayed as "qualitative" commands on the computer screen located inside the cabin.

environment while driving by having to drive in the "backward" direction while looking at the screen in the cabin. Note that the commands are not displayed to the operator as crisp control values but as bars of variable lengths over the generic speed and steering scales, effectively providing only the direction of the command (left or right, forward, or back) and the *relative strength* (i.e., more steering, faster, slower, etc.) which the driver should apply on the controls. Had a speech synthesizer been available on the portable computer, these commands could have been given in linguistic form using a few simple words.

The types of environments in which the tests were performed were the diversely occupied parking lots of Oak Ridge National Laboratory (ORNL), as can be seen in the background of the photograph in Figure 13.15. In these types of nonengineered environments, the car was successfully driven in the "blind driver" aiding mode. With the nonnegligible reaction time of humans and the short clearances, sometimes less than 3 m, provided by the parking lot corridors, dynamics and safety considerations called for speeds of up to but not exceeding 10 km/h to be used in the experiments. It was interesting to observe each operator develop his or her own interpretation of, and response to, the relative and qualitative commands displayed on the computer screen, leading to different routes and maneuvering situations for the same start and goal positions. From the system's development point of view, this inclusion of the human in the control chain effectively consists of including a source of unpredictable noise and delays in the actuation system of the autonomous operation mode. The successful operation of the rule base in this mode of driving thus provided a stringent robustness test of the qualitative inferencing scheme and navigation system.

FIGURE 13.15
Photograph taken during one of the outdoor sensor-based navigation experiments with driver's aid mode in one of the ORNL parking lots.

5 AUGMENTING THE SYSTEM WITH MEMORY AND MEMORY-PROCESSING BEHAVIORS

In the previous sections, we discussed the FBA and its use for the development of approximate sensory data–based reasoning systems for robot control. The resulting fuzzy rule bases consist of a superposition of elemental fuzzy behaviors that represent direct mappings from the perception systems to the motion controllers. This essentially produces a purely reactive system that has no possibility for temporal reasoning because it does not involve any on-line memorizing or storing of information. Due to this totally reactive nature, such reasoning schemes can encounter problems such as infinite loops and limit cycles. Adding a memory and memory-processing capabilities to an existing reactive system is, of course, one method for remedying such undesirable phenomena.

In this section, we discuss how to perform the addition of memory-related behaviors to an FBA-based robot control system. It is important that these new behaviors be developed in such a way that they conform to the formalism of the FBA in order to preserve the parallelism of the existing control system's architecture and other desirable features (e.g., augmentability, transportability, dominance concepts) of the FBA. As in the previous sections, we will present the approach through an illustrative example within the context of an autonomous robot navigating in *a priori* unknown environments, dealing in particular with the recognition and avoidance of limit cycles. In this example, the robot identifies a limit cycle when it recognizes that its current position and motion direction approximately correspond to some that it previously encountered. The robot then uses "virtual obstacles" to avoid these regions of local minima. Other ways of detecting and avoiding navigation limit cycles are certainly feasible; however, this method is selected here mostly for the purpose of providing a simple illustrative implementation of the concepts.

5.1 Architecture

Figure 13.16 shows the conceptual architecture and corresponding typical flow of control for enhancing the basic reactive system with memory and memory-processing capabilities. The lower part of the figure shows the architecture of a basic reactive system similar to that which was described in Figure 13.5 of Section 2.5. In Figure 13.16, memory (or information storage capability) is added to the basic reactive system, and memory feeding behaviors are responsible for processing and sending sensor data to the memory. This data is stored and maintained in memory in an orderly, useful manner by memory management behaviors. When necessary, data from the sensors and/or the memory are used by memory utilization behaviors to perform various memory-dependent functions, for example, recognize limit cycles or infinite loops so they may be avoided, as will be discussed here in the illustrative example. When these memory utilization behaviors are activated, they can act in three ways: (1) send commands to the actuators, (2) modify the data stored in memory, or (3) modify the processed sensor data that serves as input to the basic reactive system. Note that, in all cases, the reactive behaviors themselves are not modified; only the input, output, or memory data is affected.

A very important aspect of this architecture is that it preserves the basic parallelism of execution of all the behaviors in the system, thereby maintaining the FBA principles and associated formalism and ensuring suitability for implementation on very fast, parallel processing–based VLSI fuzzy inferencing chips and boards (e.g., see Ref. [16] or [17]). Although not conspicuous in the diagram of Figure 13.16, this parallelism is quite obvious in the system data flowchart of Figure 13.17. In this chart, the memory is clearly seen as a "fact-providing" device, analogous to the robot sensors. Together, these two devices constitute the robot perception system at any given instant in time (i.e., a given loop rate). The raw perception data (from memory or sensor) is processed to produce the input data to the behavior-based system. In a similar fashion, the output of the behaviors, that is, the action commands, are processed to generate either an actuator move or a modification of stored information. Together, the motion generation and information-changing devices constitute

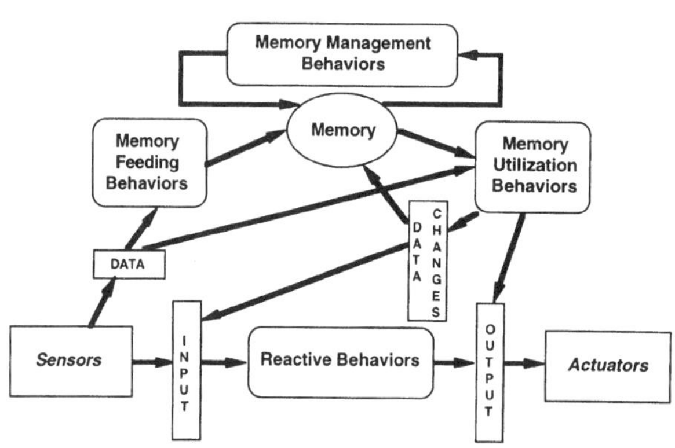

FIGURE 13.16
Overall architecture of memory-enhanced FBA-based robot control systems.

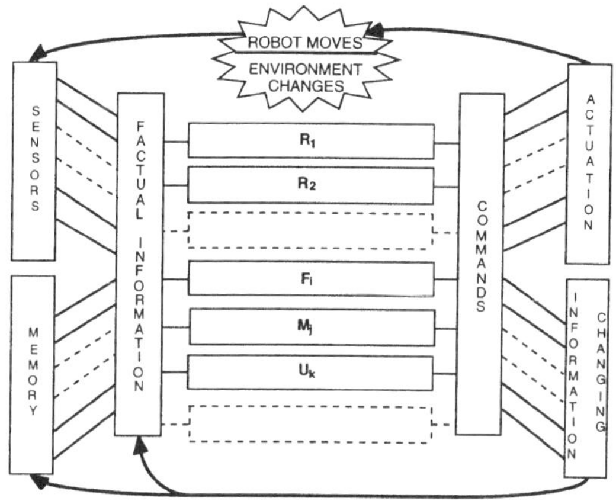

FIGURE 13.17
Flow of control and data in a memory-enhanced FBA system.

the robot response system. In Figure 13.17, example behaviors that are called reactive, memory feeding, memory management, and memory utilization in Figure 13.16 are labeled R, F, M, and U, respectively. Table 13.1 shows typical input–output data for each of them, illustrating their roles and functioning. However, it should be clear that for the reasoning system, all behaviors are similar in nature and operate in the same fashion. They process, concurrently and in parallel, data from the perception system to generate data (action commands) for the response system. In the following section, a sample implementation of this approach for limit cycle detection and avoidance is presented to illustrate the functioning of the proposed architecture in a very simple case. Of course, other methods or behaviors could be implemented for the same purpose using the overall memory-enhanced FBA approach because, as discussed in previous sections, several reasoning strategies are usually available to resolve a particular problem successfully. However, the efficiency of this simple illustrative implementation should provide a clear example of the role of the memory-enhanced FBA in increasing the reasoning capabilities of "intelligent" robots.

5.2 Sample Implementation for Detection and Avoidance of Navigation Limit Cycles

For the illustrative implementation described in the following subsections, we use the same computational and experimental framework as was described in the previous sections, in

Table 13.1. Typical Input–Output Data for Various Types of Behaviors

Type of Behavior	Type of Perception Data	Type of Response Data
Reactive	Sensor	Acuator Command
Memory Feeding	Sensor	Modify Information
Memory Management	Memory	Modify Information
Memory Utilization	Sensor and/or Memory	Actuator Command and/or Modify Information

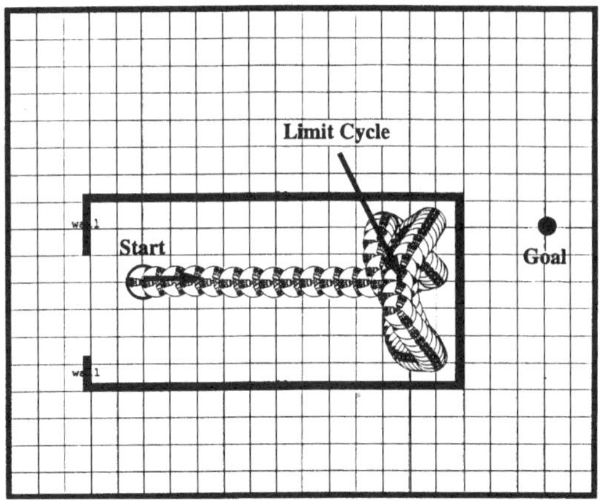

FIGURE 13.18
Sample run without the memory behaviors. The robot rapidly enters a limit cycle oscillating against the far wall of a local minimum nearest the goal.

particular, the autonomous omnidirectional platform equipped with a set of 24 ultrasonic sensors. The shortcomings of a purely reactive navigation system with respect to the potential occurrence of limit cycles can be very easily illustrated: Figure 13.18 shows a navigation run with a purely reactive scheme, that is, without the use of memory or temporal information. The robot is placed in a room that constitutes a very strong local minimum. After initially moving straight toward the goal, the robot quickly finds itself oscillating back and forth in a limit cycle against the far wall of the room. The robot will remain in a limit cycle in this strong local minimum area due to its lack of temporal reasoning; without memory, it cannot "remember" and detect that it accomplishes the same path repeatedly. Indeed, a system with no memory or temporal information storage capability of any type cannot detect limit cycles and therefore cannot avoid them.

The basic principle of limit cycle detection and avoidance that is used in the illustrative implementation here consists of giving the robot the capability to memorize the current status of absolute position and speed so that, at a later date, it can detect limit cycles as "having passed here already" and consequently perform appropriate reasoning to avoid further looping. The robot identifies a limit cycle when it recognizes that its current position and motion direction approximately correspond (in the sense of Fuzzy Sets) to some previously encountered. Thus, memory-feeding behaviors can be developed that are responsible for generating the robot's position and speed data at each sampling period and for sending them to the memory for storage. Memory management behaviors can also be developed that are responsible for storing and maintaining this data in an orderly, easily retrievable manner. Then memory utilization behaviors can use both stored and current sampling data to identify previously visited positions, leading to the recognition of limit cycles. The following paragraphs give examples of how these behaviors can be very simply implemented.

Memory-Feeding Behaviors

For its navigation, the robot already uses its sonar and wheel encoder readings to calculate its x, y position coordinates, speed, and orientation with respect to the absolute coordinate frame (taken as the initial position and orientation of the robot) using simple odometry calculations. The function of the memory-feeding behaviors is thus simply to obtain this data, pass it through an intermediate buffer, and eventually transfer it to the memory management behaviors for storage. The use of such a buffer for temporary storage of new position information prevents the control system from mistaking two consecutive sets of position data as being approximately the same position with the same orientation. In our example implementation, this buffer queue holds 10 sets of position data, essentially producing the necessary "delayed remembrance of positions" while the robot navigates.

Memory Management Behaviors

To manage and ease the retrieval of previous robot positions in memory, a conventional hashing technique is used here as an example: a memory management behavior stores each new set of data into memory according to the following formula:

$$hashvalue = \tfrac{1}{2}X^2 + XY + \tfrac{1}{2}Y^2 + \tfrac{1}{2}X + \tfrac{3}{2}Y \tag{13.17}$$

where $X = x/\langle\text{size of grid}\rangle$ and $Y = y/\langle\text{size of grid}\rangle$. Position memory is split into four sections representing the four quadrants in a two-dimensional Cartesian plane. Each quadrant is broken into grids in accordance with formula (13.17) as shown in Figure 13.19. A linked list holds all positions within a particular grid. These linked lists are stored in an array indexed by the hash values. These indices represent the different grid elements sketched in Figure 13.19.

As described in the following subsection, the system uses the concept of "virtual obstacles" to avoid areas detected as dead ends or limit cycles. Once a virtual obstacle is created, all stored positions falling within the bounds of that virtual obstacle are no longer needed and should be forgotten by the control system. Another memory management behavior frees the precious memory space containing all of these positions. Because of the large number of positions typically stored during a limit cycle recognition, large portions of memory are freed up upon creation of a virtual obstacle. This behavior may be thought of as an example of

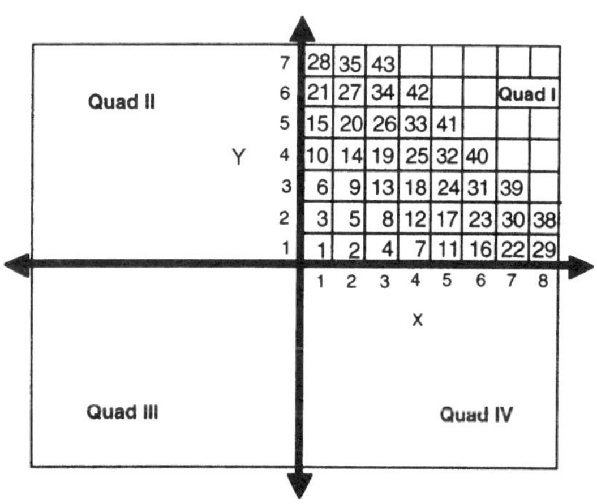

FIGURE 13.19

The robot's surrounding environment is represented as a two-dimensional Cartesian plane broken into grids in accordance with a hash formula.

"information sorting" behaviors that essentially "clean" the memory of no-longer-needed data.

Separately from the position grid, the main navigation goal, specified as a relative x, y position and orientation of the robot with respect to its position and orientation at the start of the navigation experiment, is stored in the bottom of a stack. When the stack is empty, the robot has achieved its main goal and the navigation stops. A memory management behavior is also responsible for adding and removing from the stack the subgoals that are generated with the virtual obstacles, as explained in the following paragraphs.

Memory Utilization Behaviors

Memory utilization behaviors are developed to determine whether or not the robot has visited a particular position before, with or without the same orientation. In this illustrative example, these memory utilization behaviors consist of two fuzzy rules (the rules in the other memory utilization behaviors are simple "crisp" production rules triggering a conventional numerical procedure). These fuzzy rules compare the current position and motion direction of the robot with those stored in memory. One rule establishes recognition of the x, y position of the robot while the other determines whether or not the same motion direction exists. In these fuzzy behaviors, as well as all other behaviors involving checking for the relative distance or orientation between two robots' configurations (e.g., the "goal reached" or "goal proximity" behaviors discussed in later sections), the rules take the form

$$\text{IF } [\textbf{FUNCTION}\,(P_1, P_2) \text{ is } \textit{zero}] \quad \text{THEN} \quad [\textbf{INFORM-MOD}] \tag{13.18}$$

where FUNCTION represents the appropriate norm difference operator between the components of the two configurations P_1 and P_2, INFORM-MOD represents a modification of memory information, and *zero* is one of the fuzzy sets representing the thresholded and approximate proximity. The interesting aspect of using fuzzy rules here, just as in the basic reactive system, is that the full strength of approximate reasoning using membership functions can be utilized. For example, the threshold of recognition of "having been here before" (in the fuzzy set *zero*) can be adjusted based on the precision (uncertainty) of the odometry sensors and the length of the already executed journey (during which position uncertainty increases).

In this illustrative implementation, the robot detects that it is in a limit cycle or infinite loop through two consecutive recognitions of a previously visited position with the same orientation; that is, a point in the memory is revisited twice with the same motion direction. For this, the fuzzy rules just described increment a counter in each set of position and orientation data hashed in memory. When the counter reaches 2, a behavior activates a buffer in which the successive robot x,y position data are stored. When the counter reaches 3, a limit cycle has been recognized, and the avoidance scheme behaviors to exit and avoid reoccurrence of the limit cycle are triggered. These behaviors trigger the creation of a virtual obstacle and of a temporary subgoal to be placed on top of the goal stack. The robot then continues its navigation process, but now trying to reach this subgoal and taking into account this virtual obstacle. The virtual obstacle is formed by using the x,y coordinate values of all positions collected in the buffer between the two consecutive encounters, that is, while in the limit cycle. The minimum and maximum of these values form the points representing the four corners of the virtual obstacle.

The subgoal is established on the basis of the position of the virtual obstacle just created. The space surrounding a virtual obstacle is divided into eight sectors and the sector opposite the current goal is determined [see Figure 13.20(a)]. If that sector is one of the four corner sectors, the subgoal is placed at a 45° angle from the corner in the opposite direction of the current goal, 1500 mm from the virtual obstacle [see Figure 13.20(b)]. Otherwise the subgoal is placed 1500 mm from the middle of the virtual obstacle wall facing the chosen sector as seen in Figure 13.20(c). As mentioned in the previous paragraphs, subgoals are stored in memory in the goal stack and are removed in accordance with their accomplishments. The subgoal at the top of the stack is the active navigation subgoal.

This method for computing a subgoal will sometimes create one in an unreachable position, that is, too close to an obstacle. To remedy this, a "goal reached?" behavior constantly checks for reachability of all existing subgoals (e.g., is a subgoal within a fuzzy distance of one robot diameter?). If a goal or subgoal that supersedes the current goal (i.e., is lower on the goal stack) is reached, then the current and all superseded goals are removed from the goal stack, and navigation continues.

Following their creation, the virtual obstacles' data are kept in memory. At each sensor sampling period, a behavior checks each of the existing virtual obstacles, determines the sonar distances to them, compares these distances with the real sonar distances, and outputs the shorter of the two as the sonar readings to be used for the current time step reactive inferencing. In calculating distances to virtual obstacles, it is important to realize that because of their rectangular shape, at most two walls of each virtual obstacle will be used at any given time. These are determined by the distances from the current robot's position to the four corners of the virtual obstacle being considered. Two possible cases exist: only one side of the virtual obstacle, that which faces the robot, is needed [i.e., when the robot falls in one of the four side sectors shown in Figure 13.20(a)], or two walls, both facing the robot, are needed [i.e., when the robot falls in one of the four corner sectors of Figure 13.20(a)].

In the case in which only one wall is needed, the shortest distance between the current robot's position and the virtual wall is calculated. Then, based on the current orientation of the robot with respect to the initial orientation, the orientation in space of each sonar facing toward the virtual obstacle is determined. Assuming a 15° cone angle for each sonar sound wave, the shortest sonar returns from the virtual obstacle can be easily computed. In the second case in which the robot faces two walls (i.e., a corner) of the virtual obstacle, the calculation method is the same as described for a single wall, but with the determination of which particular sonar intersects the corner being done first, to simplify the treatment of the other surrounding sonars on the two walls.

There is a particular case that requires attention when creating virtual obstacles in the

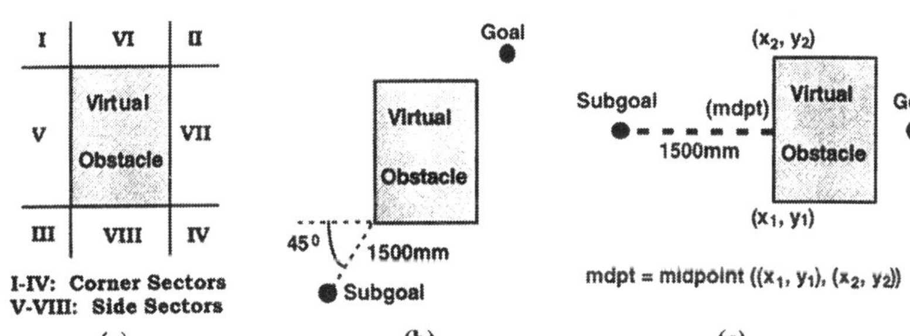

FIGURE 13.20
(a) Environment surrounding a virtual obstacle broken into eight sectors; (b) and (c) examples of the location of a subgoal based on the position of the navigation goal and a virtual obstacle just created.

way described here: the robot may end up "surrounded" by virtual obstacles. In this case, the "discrete minima" behavior (which is included in the totally reactive system) is invoked. When the robot gets "stuck" in a discrete minimum, that is, all of its wheel velocities are zero, this behavior essentially "looks around," checking all of the sonar distances (including toward the back), finds the direction of the largest free distance, and sets a subgoal in that direction, three-fourths of the free distance away. Until it reaches this subgoal, the robot ignores all virtual obstacles. Once this subgoal is achieved, the robot continues its navigation using all virtual obstacles.

5.3. Sample Experimental Results

Experiments were performed in which the new memory-related behaviors were added to the existing totally reactive navigation system discussed in the previous sections. Figure 13.21 shows a navigation run in the same environment conditions as those in Figure 13.18, but with the memory-related scheme and behaviors added to the control system. After two instances of noting an already visited position and orientation, the robot recognizes it is in a limit cycle. Virtual Obstacle 1 is created on the basis of the minimum and maximum x,y coordinate values collected. Subgoal 1 is created according to the position of the new virtual obstacle, and the robot navigates toward that subgoal. The robot reaches the subgoal and navigation then continues with the real sonar readings being overridden by the distances to the virtual obstacle when necessary. A very similar limit cycle occurs again, this time involving the creation of Virtual Obstacle 2 and Subgoal 2. With the control system repeating the process while keeping all virtual obstacle data in memory, the robot is progressively "forced" out of the "dead-end" chamber. This "dead-end" chamber is eventually "filled" with virtual obstacles, preventing reentrance by the robot, which can then successfully continue toward its original goal.

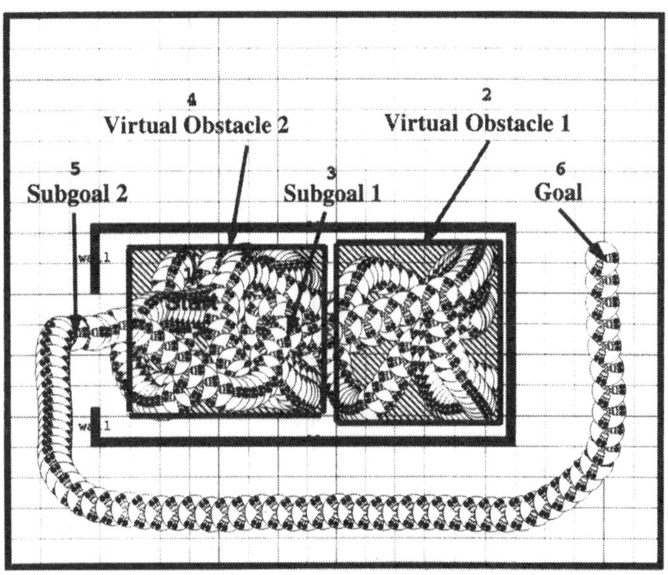

FIGURE 13.21
Sample run with the memory-related behaviors, illustrating the use of virtual obstacles and subgoals to force the robot out of a "dead-end" chamber.

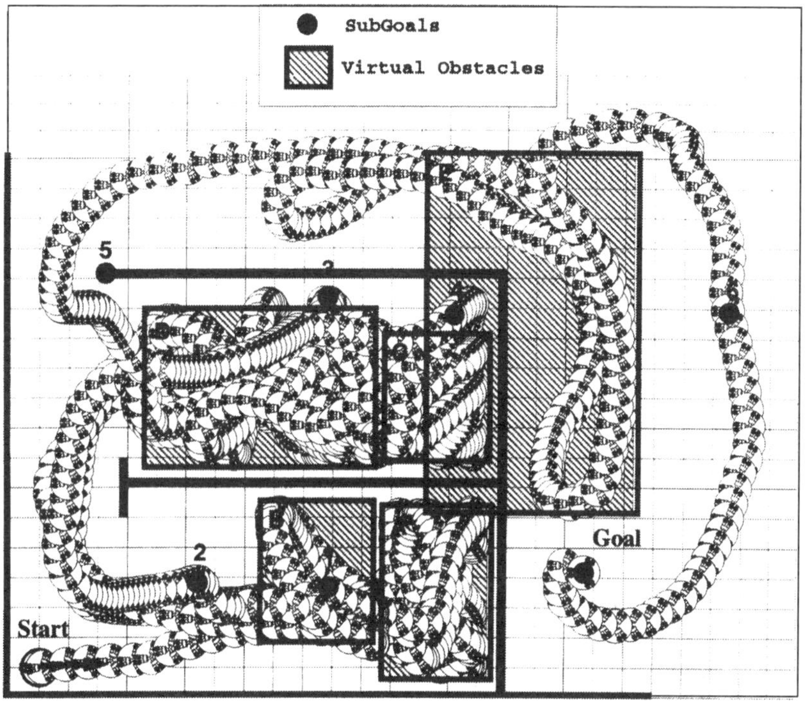

FIGURE 13.22
Sample run with the memory-related behaviors illustrating the use of virtual obstacles and subgoals, the "discrete minima" behavior, and the relative precedence of goals and subgoals.

A slightly more complicated test environment can be viewed in Figure 13.22. This navigation test run exemplifies many different capabilities of the new memory-using navigational system. Navigation begins with the point labeled start. The initial orientation is shown by the arrow pointing right. The robot identifies a first limit cycle and creates Virtual Obstacle A and Subgoal 1. Upon reaching Subgoal 1, navigation continues with the use of Virtual Obstacle A. A new limit cycle is recognized and Virtual Obstacle B is created, along with Subgoal 2. Subgoal 2 is reached and navigation once again continues with the robot exiting the first "dead-end" chamber. A third Virtual Obstacle C and Subgoal 3 are created in the second chamber, but while attempting to reach Subgoal 3, the robot gets "stuck" in the upper right-hand corner of that chamber. The "discrete minima" (or "look around") behavior triggers, Subgoal 4 is created, and all virtual obstacles are ignored for the time being. Once Subgoal 4 is reached, the robot again looks for Subgoal 3. After reaching Subgoal 3, again using all virtual obstacles, the robot finds itself in another limit cycle and creates Virtual Obstacle D and Subgoal 5. However, Subgoal 5 is positioned in an undesirable location within an existing wall. In an attempt to find Subgoal 5, the robot generates Virtual Obstacle E and Subgoal 6. Subgoal 6 is reached and the robot continues to look for Subgoal 5. In doing so, it identifies reachability of the main goal and reaches it. Because the goal holds precedence over Subgoal 5, that is, it is lower on the goal stack, Subgoal 5 is pushed off the stack and forgotten. Navigation ends since the main goal is reached and the goal stack is empty.

6 CONCLUDING REMARKS

Autonomous robot control in *a priori* unknown, unpredictable, and dynamic environments requires many calculational and reasoning schemes to operate on the basis of very imprecise, incomplete, sparse, or unreliable data, knowledge, or information. In such systems, for which engineering all the uncertainties away from the hardware is not currently fully feasible, approximate reasoning may provide an alternative to the complexity and computer requirements of conventional uncertainty analysis and propagation techniques.

The concepts of Minimal Models and the Fuzzy Behaviorist Approach have been presented for the development of fuzzy rule bases embodying "human-like" behaviors in sensor-based decision-making systems. The concepts of suppression and inhibition of behaviors to resolve possibly conflicting behaviors have been described. An automated system has been developed that embodies all the FBA formalism and principles. This automated system can generate fuzzy rules from the user-provided qualitative description of a reasoning process. Examples of the use of the automated system to generate fuzzy rule bases for the sensor-based navigation of autonomous robots have been discussed. Sample runs of the robots have been presented to illustrate the overall navigation behaviors as well as the effect of a change in the interbehavior dominance expressed through the suppression and/or inhibition mechanism. Experiments with a real car have also been discussed to illustrate the capability of readily adding behaviors to the fuzzy rule base to resolve situations of increasing complexity and, as shown in the driver's aid feasibility study, the straightforward "linguistic" interfacing capability of the fuzzy behavior–based system.

An approach to remedy some of the shortcomings of purely reactive systems has also been presented. This proposed approach calls for the addition of memory and memory-processing behaviors to the system, but, respecting the principles of the Fuzzy Behaviorist Approach in order to utilize the strengths of both Approximate Reasoning and the Behaviorist Theory in uncertainty-prone decision-making conditions. The proposed memory enhancement method also provides for preserving the existing system's architecture and its parallelism. Three forms of memory-related behaviors, memory feeding, memory management, and memory utilization, have been discussed for addition to existing, totally reactive, FBA-based robot control systems. The overall memory-enhanced FBA has been illustrated through a sample implementation for the detection and avoidance of limit cycles in the sensor-based navigation of an autonomous robot in *a priori* unknown environments.

A variety of lessons and observations can be drawn from these experiments. Several of these, relevant to the topics dealt with here, are listed and/or discussed in the following.

- Entire FBA-consistent navigation codes or schemes can be developed that consist of about 20 to 30 fuzzy rules. Compared with the 50,000 or more lines of "crisp" coding that were previously utilized to accomplish the same task, the efficiency and gain in code development time, code and data storage space, and so on, of the approach can be great indeed. However, it should be noted that not all robotic processes may be as well suited for resolution through approximate reasoning as is the navigation task. The criteria listed in Section 2 can provide a good evaluation in that respect.
- Our observations showed the approximate reasoning scheme to be much more robust and reliable than the previously used "crisp" codes when faced with sensor inaccuracies and environmental uncertainties. This seems to support the claim that, for situations in which precision is not the primary goal, robotic tasks can be very efficiently accomplished using an approximate reasoning scheme but can also be made more robust with

respect to uncertainties through an implicit "folding" of these uncertainties within the approximate variables of the reasoning scheme.

- As illustrated through the various experiments involving small omnidirectional and indoor robots and/or an outdoor car, approximate reasoning schemes can be built to embody very generic functions; that is, neither the sensory data nor the code itself need to be specifically developed for a particular robotic platform. This seems to indicate that, contrary to what is typically obtained with "crisp" logic codes, approximate reasoning schemes need not be system specific but rather can be *function specific*: they can embody a reasoning *strategy* rather than a specific instantiation of the strategy.
- An implication of this genericness of the approximate reasoning schemes is the resulting straightforward transportability of codes among various systems with *similar* perception and motion *means*.
- Another significant consequence of this genericness of approximate reasoning schemes is, of course, their scale-up capability. As shown in the experiments, basic navigation schemes could be augmented and/or enhanced with additional behaviors *without* rewriting the previously tested behaviors. The gain in development time resulting from this property of approximate reasoning schemes is expected to be substantial.
- With respect to the minimal model concept, it should be noted that the sparsity of the data typically utilized with approximate reasoning schemes may be somewhat misleading. This sparsity exists only on the time scale of one sampling period. If accumulated over time, however (e.g., using memory-related behaviors), a very large part of the environment is sampled because the robot and consequently the sensors are translating and rotating. This obviously brings the question of trade-off between the type of information which is minimally needed at the sampling rate versus that which is minimally needed at the overall task rate. It is clear that in making such trade-offs, consideration must be given to the amount of information that will result from accumulation over time of the data, rather than simply snapshot-type information.

REFERENCES

[1] T. Skewis, J. Evans, V. Lumelsky, B. Krishnamurthy, and B. Barrows. Motion planning for a hospital transport robot. In *Proceedings of the 1991 IEEE International Conference on Robotics and Automation*, Sacramento, CA, April 1991, pp. 58–63.

[2] H. A. Vasseur, F. G. Pin, and J. R. Taylor. Navigation of car-like mobile robots in obstructed environments using convex polygonal cells. *Robot. Auton. Syst.* 10(2–3):133–146, 1992.

[3] R. A. Brooks. Elephants don't play chess. *Robot. Auton. Syst.* 6(1–2):3–15, 1990.

[4] C. Andersen, *et al.* Navigation using range images on a mobile robot. *Robot. Auton. Syst.* 10(2–3):147–160, 1993.

[5] L. N. Kanal and J. F. Lemmer, eds. *Uncertainty in Artificial Intelligence*. North-Holland, Amsterdam, 1988.

[6] L. A. Zadeh. Fuzzy set. *Inform. Control* 8:338–353, 1965.

[7] L. A. Zadeh. Outline of a new approach to the analysis of complex systems and decision processes. *IEEE Trans. Syst. Man Cybernet.* SMC-3(1):28–45, 1973.

[8] L. A. Zadeh. Fuzzy logic. *IEEE Comput.* 21(4):83–93, 1988.

[9] L. A. Zadeh, K. S. Fu, K. Tanaka, and M. Shinamura, eds. *Fuzzy Sets and Their Applications to Cognitive and Decision Processes*. Academic Press, New York, 1975.

[10] S. Yasunobu, S. Miyamoto, T. Takaoka, and H. Ohsihima. Application of predictive fuzzy control to automatic train operation controller. In *Proceedings of the IECON '84*, 1984, pp. 657–662.

[11] L. P. Holmblad and J. J. Ostergaard. Control of a cement kiln by fuzzy logic. In *Fuzzy Information and Decision Processes*, eds. M. M. Gupta and E. Sanchez. Elsevier Publishing, Amsterdam/New York, 1982, pp. 389–399.

[12] L. I. Larkin. A fuzzy logic controller for aircraft flight control. In *Industrial Applications of Fuzzy Control*, ed. M. Sugeno. North-Holland, Amsterdam, 1985, 87–103.

[13] H. Ono, T. Ohnishi, and Y. Terada. Combustion control of refuse incineration plant by fuzzy logic. In *Proceedings of the 2nd International Fuzzy Systems Association Congress*, July 1987, pp. 345–348.
[14] T. Yamakawa. Stabilization of an inverted pendulum by a high-speed fuzzy logic controller hardware system. *Fuzzy Sets Syst.* 32:161–180, 1989.
[15] K. A. Hirota, Y. Arari, and S. Hachisu. Fuzzy controlled robot arm playing two dimensional Ping-Pong game. *Fuzzy Sets Syst.* 32:149–159, 1989.
[16] H. Watanabe, W. Dettloff, and E. Yount. A VLSI fuzzy logic inference engine for real-time process control. *IEEE J. Solid State Circuits* 25(2):376–382, 1990.
[17] J. R. Symon and H. Watanabe. Single board system for fuzzy inference. In *Proceedings of the Workshop on Software Tools for Distributed Intelligent Control Systems*, Sept. 1990, pp. 253–261.
[18] F. G. Pin, H. Watanabe, J. R. Symon, and R. S. Pattay. Autonomous navigation of a mobile robot using custom-designed qualitative reasoning VLSI chips and boards. In *Proceedings of the 1992 IEEE International Conference on Robotics and Automation*, Nice, France, May 10–15, 1992, pp. 123–128.
[19] F. G. Pin and Y. Watanabe. Navigation of mobile robots using a fuzzy behaviorist approach and custom-designed fuzzy inferencing boards. *Robotica* 12(6):491–503, 1994.
[20] J. Yen and N. Pfluger. A fuzzy logic based robot navigation system. In *Proceedings of the AAAI Symposium on Applications of Artificial Intelligence to Real-World Autonomous Mobile Robots*, Cambridge, MA, Oct. 23–25, 1992, pp. 195–199.
[21] T. Takeuchi, *et al.*, Fuzzy control of a mobile robot for obstacle avoidance. *Inform. Sci.* 45:231–239, 1988.
[22] M. Sugeno *et al.* Fuzzy algorithmic control of model car by oral instructions. *Fuzzy Sets Syst.* 32:207–219, 1989.
[23] F. G. Pin and Y. Watanabe. Steps toward sensor-based vehicle navigation in outdoor environments using a fuzzy behaviorist approach. *Int. J. Intell. Fuzzy Syst.* 1(2):95–107, 1993.
[24] F. G. Pin and Y. Watanabe. Automatic generation of fuzzy rules using the fuzzy behaviorist approach: The case of sensor-based robot navigation. *Intell. Automat. Soft Comput.* 1(2):161–178, 1995.
[25] F. G. Pin and S. R. Bender. Adding memory processing behaviors to the fuzzy behaviorist-based navigation of mobile robots. In *Proceedings of the World Automation Congress (WAC '96)*, Montpellier, France, May 28–30, 1996, Vol. 3, pp. 647–652.
[26] F. G. Pin and S. R. Bender. Adding memory processing behaviors to the fuzzy behaviorist approach: Resolving limit cycle problems in autonomous mobile robots navigation. *Intell. Automat. Soft Comput.* 5(1): in press.
[27] F. G. Pin and S. M. Killough. A new family of omnidirectional and holonomic wheeled platforms for mobile robots. *IEEE Trans. Robot. Automat.* 10(4):480–489, 1994.

Index

Bold page numbers indicate main entries.